D1217889

COLONIAL ARCHITECTURE OF ANTIGUA GUATEMALA

Memoirs of the
AMERICAN PHILOSOPHICAL SOCIETY
Held at Philadelphia
For Promoting Useful Knowledge
VOLUME 64

SIDNEY DAVID MARKMAN

The American Philosophical Society
INDEPENDENCE SQUARE · PHILADELPHIA · MCMLXVI

Library of Congress Catalog Card Number 66–13634

For
Malvina
Sarah Dinah
Alexander Jacob
Charles William

FOREWORD

IT IS indeed fitting that these first lines should be devoted to expressing my thanks for the support given me over the years without which this book could not have been written. From 1950 on, the Duke University Research Council subsidized my work and many trips to Guatemala and Central America to study the monuments at first hand, and finally provided a generous subvention toward defraying part of the costs of publication. The Council on Hispanic Research of Duke University enabled me to return to Guatemala in the summer of 1963 to take the photographs which now illustrate the book. In 1959 I was the recipient of a travel grant from the American Philosophical Society for the purpose of pursuing my research in Spain. A Fulbright Research Grant for the academic year 1961–1962 made it possible for me to continue my investigations in Spain and to bring the manuscript to a conclusion there.

The sympathetic interest taken in my work by John Tate Lanning and Robert Sidney Smith, colleagues at Duke University, proved a constant source of encouragement and moral support. I am especially beholden to my wife who shared my interests and who took on the tedious task of typing the manuscript over and over again. And finally, I wish to express my deep feeling of gratitude to the American Philosophical Society for making it possible for my work to come to light. I wish to thank Dr. George W. Corner, Executive Officer and Editor of the Society, who more than extended himself on my behalf; Miss Marie A. Richards, Assistant to the Editor of the Society, who edited the manuscript and who oversaw the production of the book; and Miss Julia A. Noonan, also of the Society, who, though she may consider it of little moment, opened a door for me.

S. D. M.

Duke University
Durham, North Carolina
April, 1965

CONTENTS

Part II. The Antiguan Style: Analysis of Elements

ILLUSTRATIONS

TABLES

COLONIAL ARCHITECTURE OF ANTIGUA GUATEMALA

INTRODUCTION

BY the eighteenth century the Spanish dominions in the New World were divided into four viceroyalties and four captaincies general. Antigua was the principal city and capital of the Captaincy General of Guatemala, sometimes also referred to as the Reino de Guatemala, which comprised most of present-day Central America including the Mexican state of Chiapas and the republics of Guatemala, Honduras, El Salvador, Nicaragua, Costa Rica, and a small area beyond the border in Panama. (FIGS. 12–14.) Though considered part of Central America today, Panama was in the viceroyalty of New Granada. The region is a sinuously narrowing ribbon of land stretching between two continents with the Atlantic and Pacific framing the broken, mountainous terrain so that from some peaks both oceans can be seen. Guatemala was brought under Spanish control by 1524, but the first permanent European settlement was not founded until 1527.

Antigua, as the capital of the Capitanía General de Guatemala and the seat of the Audiencia de Guatemala, nurtured a building tradition which, though rooted in the Iberian peninsula, resulted in a distinctive style reflecting the peculiar social, economic, and political forces which operated there for some 250 years of colonial history. The principal houses of the monastic orders, which were the chief instruments for the conversion of the native population, were all located in Antigua. The regular clergy, monks, were the administrators of the parish churches in the country at large until the middle of the eighteenth century when this work was taken over by the secular clergy. Monks were often the designers of the churches of the Indian towns under their ecclesiastical administration, or most certainly the patrons for whom the architects or practical builders worked.

When compared to the great viceroyalties of Mexico and Peru, the Captaincy General of Guatemala was poor and of minor importance indeed, a condition reflected not only in the social and economic spheres, but in the building arts as well. The wealth of Guatemala lay not in its gold and silver mines, the product of which was insignificant, to say the least. The amassing of fortunes in portable wealth was hardly possible. The first settlers and *conquistadores* soon learned that in Guatemala one did not "get rich quick" and depart for the homeland Spain. More than a century passed before it was realized that the wealth of Guatemala lay in her topsoil. The economic base of Guatemala was its agriculture, as it continues to be even today. Wealth was measured only in terms of land and what it produced with the labor of the native population. It was not until the third or fourth generation after the conquest in the seventeenth century that the native-born Guatemalan Spaniard began to think of the New World as his true homeland rather than far-off Spain which he had never seen. And it is only in the seventeenth century that an architecture of permanent character begins to appear. But it is an architecture commensurate with the restricted means of an agricultural economy in contrast to the sumptuous buildings undertaken in Mexico and Peru, rich in gold and silver mines.

The contemporary literature is replete with complaints of the general poverty of the region throughout the colonial period, the principal cause of which is invariably given as the restriction of commerce arbitrarily imposed by the powers in the Iberian Peninsula. The disastrous effects on the prosperity of Central America resulting from the policy of favoring the trade of one region over another need not be examined here. Still, the restricted economic means of the governmental and ecclesiastical authorities and the generally impoverished condition of the lower classes had a strong limiting effect on the building tradition.[1] Not by the

[1] For archival references to the various edicts, laws, and regulations which restricted overseas commerce between Guatemala and the rest of Spanish America and Spain itself see *Efem*, pp. 85, 90, 96, 98, 100, 113, 119, 134, 136, 143, 166, 167; also Larreinaga, *Prontuario*, pp. 6 ff., who cites a series of *cédulas* under the heading of prohibited commerce dated in 1672, 1675, 1679, 1710, 1711, 1712, 1715, 1717, 1725, 1750. For some nineteenth-century comments on the subject see García Peláez, **2**: pp. 35 ff., "Restricciones del Comercio," and Larrazabal, *Discurso*. On the problem of the drainage of money and the imbalance of imports over exports see Díaz Durán, *Historia de la Casa de Moneda*, also *Efem*, p. 16. For the deleterious effects of piracy on overseas commerce see García Peláez, **3**: p. 13; Gage, ch. XXI, p. 421. For overland trade and the impossible road conditions see Gage,

farthest reach of the imagination can the history of the architecture of Reino de Guatemala be considered in the same terms as that of Mexico and Peru where, early indeed and as a matter of course, the construction of buildings was commonly undertaken on a scale commensurate with those of Spain during the most prosperous moments of her history. During the greater part of the sixteenth century even the principal churches and governmental buildings of the capital city of Guatemala were constructed of the humblest materials and were for the most part of a nondescript and temporary nature designed to serve the grossest utilitarian ends of the moment. It was not before the seventeenth century that a more permanent and monumental architecture began to develop. But, because of the restricted economic means and the sparse population, a popular or vernacular character ever typifies the building tradition throughout the rest of the colonial period. The architecture of colonial Central America is based on brick and stucco of which not only the structure itself is built, but even decorative architectonic details and ornamental sculpture are carried out as well.

Another important factor in shaping the Guatemalan style throughout its history was the scant population, both Indian and European, resulting in the construction of a small number of buildings in contrast to the thousands of churches and civil buildings representing the architectural styles of the populous viceroyalties of Mexico and Peru. At the end of the eighteenth century it is estimated that the total population of the Reino de Guatemala was about three-quarters of a million. The total number of churches consisted at the most of some 750, of which 450 were located in the present-day republic of Guatemala. Approximately 45, or 6 per cent of the total in all Central America, were standing in Antigua in 1773 when the city was destroyed by the most devastating earthquakes of its entire history so that it was abandoned soon after.[2] These 45 religious and civil buildings comprise the prototypes for the architecture in the rest of Reino de Guatemala, for it was in Antigua where the style originated and whence it spread to the rest of the country.

The small figure of about 6 per cent representing but 45 churches out of a total of 750 is not commensurate with the importance of those buildings for the development of the overall Central American colonial architectural style. This is equally the case when one considers the ratio of about one church per thousand people for the country as a whole. The population of colonial Central America, culturally speaking, was not homogeneous in the sense that it was in the Iberian Peninsula. The 750 churches or so were located in the urban centers, the *loci* of European culture. One must not lose sight of the fact that the far greater proportion of the population, the indigenes representing the agricultural labor force, was never completely assimilated to the culture and religion of the relatively scant few European settlers. Prior to the conquest the Indians lived scattered and dispersed over vast extensions of broken, accidented, ever-erupting volcanic land, the work of the devil according to the early missionaries who found it so difficult to gather the natives into towns in order to convert them to Christianity and a European way of life. The Indian preferred to live in his own cornfield. "Cada indio en su milpa . . . ," Ximénez noted more than 250 years ago.[3] This preference still persists. The Indian way of life has not completely disappeared in remote regions of Central America where the majority of the population continues to remain Indian in culture but, as is to be expected, with European accretions. This is especially true in Chiapas and highland Guatemala where the native languages and customs persist among the majority of the population, Spanish being still a foreign tongue. De Córdova, writing at the end of the eighteenth century concerning the low state of economic conditions, points out that so few Indians, mestizos, Negroes, and mulattoes are integrated into the Western or European way of life that there is really no opportunity for the artisan to prosper. If only 5 per cent of the population, that is, the Spaniards, need houses and clothes, where are the carpenters and tailors to make a living?[4]

Nor was the number of urban centers of colonial Guatemala as great as that in the viceroyalties of Mexico or Peru. Neither was Antigua, capital city though it was, ever more

---

ibid.; de Córdova, *Utilidad*; Villacorta, *Historia*, p. 183, quoting Irisarri, also in *ASGH* **10** (1934/35): p. 386; Fuentes y Guzmán, **2**: pp. 60, 72, 180, and **3**: pp. 88 ff.; Ximénez, **2**: p. 120; Remesal, **2**: p. 429; Larrazabal, *Apuntamientos*, p. 101; Vásquez, **4**: p. 40; Ponce, *Relación*, pp. 421 ff.; Osborne, *Arterias comerciales*, pp. 320–325; *Efem*, pp. 24, 37, 40, 172, and the following documents: *AGG*, A 1.23 (1574) 1513–523; A 1.2–4 (1541) 15752–52; A 1.23 (1546) 1511–33; A 1.23 (1549) 1511–125; A 1.23 (1549) 1511–142; for the same documents see Pardo, *Prontuario*, pp. 26 and 28; also A 1.17.3 (1756) 38.302–4501; A 1.17.3 (1793) A 1.25–21389–2603 fol. 4; A 1.1 (1797) 24094–2817.

[2] See ch. VII, "Building Activity" below.

[3] Ximénez, **1**: p. 48.

[4] The problem of integrating the native Indian population as well as that of mixed ancestry to western ways is one that still remains to be solved in many of the more remote regions of Central America. See de Córdova, *Utilidad*; also *AGG*, A 1.16 (1798) 2905–149, "Fr. Antonio de San José Muñoz solicita la impresión de una memoria acerca de las ventajas que resultan a los indios y ladinos de que se calcen."

than a small provincial town in comparison with Lima, Quito, and Mexico City, for example. At the end of the eighteenth century, according to Juarros, there were twelve cities and twenty-one *villas*, making a total of about thirty-three urban centers in all the Reino de Guatemala where Western culture predominated.[5] The importance, then, of Antigua with its forty-five buildings as the source of the Central American architectural style cannot be underestimated, for from this center the style perforce spread to the remaining thirty-two Europeanized cities of the region.

Nor is the architecture of Antigua and the rest of Central America as sumptuous or on as grand a scale as that seen in contemporary Mexico and Peru. When compared to the principal cities of the Spanish New World, Antigua, even at its most prosperous moment in the eighteenth century, was a small, impoverished town plagued by constant destruction from earthquakes which put a financial strain on the already meager resources of private citizens, civil authorities, and religious communities who were forever required to repair the repeatedly damaged buildings. Construction on a magnificent scale was beyond their means because of the constant outlay required for repairs. According to Arana, writing of the earthquake of 1717, in the 170 years or so since its founding, the city had been ruined on nine different occasions. An outlay of at least 100,000 *pesos* had been needed annually to maintain the buildings in repair which, even when not added to the original cost of the construction, represented a total of about 18,000,000 *pesos* spent from about 1541 to 1717.[6]

II

THE colonial architecture of Guatemala and Central America is as yet not as well known as is that of other Hispano-American styles. No monographic literature exists and, in general, so little has been published that the architectural historian is perforce obliged to make preliminary archaeological, that is, physical, investigations of the actual monuments prior to undertaking any critical study of either the internal development of the Guatemalan style as a whole or its relation to the architecture of the rest of Hispano-America and Spain. Besides gathering and reporting all the strictly factual historical and physical data relative to the buildings themselves, the architectural historian has still another task, that of discovering the ambient or nonarchitectural determinants of the Guatemalan style. Since these are also not as well known as those of Europe in the first place he is, therefore, obliged to take close account of socio-historical phenomena not always directly related to the architectural monuments in question, but germane to an understanding of the style as a whole. It is not surprising then, but rather a normal circumstance, that the sources of information resorted to by the architectural historian are the very same as those sought by the historian; for, like the latter, he too must take into account the social, political, religious, and economic institutions of Guatemala and Central America of which the architecture is a direct reflection.

Important sources of information are to be found in the contemporary literature dealing both directly and indirectly with Guatemala and Central America. It should be borne in mind, however, that in general the authors of the colonial period were not professional writers, but rather *conquistadores*, travelers, ecclesiastics, settlers, and government officials whose literary activity was peripheral to their main occupation. Furthermore, none treat of the architecture *per se*, that is, as a phenomenon directly within the realm of their main purpose. As a result, information relative to the architecture in general and to individual buildings in particular is often found in the least expected places and most frequently as a purely marginal concern of the author.

From the sixteenth century the writings of Bernal Díaz del Castillo, one of the soldiers who accompanied Pedro de Alvarado in the conquest of Guatemala and one of the first permanent settlers of Ciudad Vieja and Antigua, provide one of the first objective descriptions of the conquest, the country, and its native inhabitants based on personal experience. Other sixteenth-century authors such as the Licenciado Palacios, Montero de Miranda, Pineda, and Ponce visited the region for shorter periods of time and wrote accounts from which information relative to the architecture is gained in passing. Something is learned also from sixteenth-century professional authors such as Acosta, Oviedo, López de Gómara, López de Velasco, none of whom visited Guatemala and all of whom except Oviedo based their scant descriptions on secondary sources. Besides, they were writing general histories of the New World.[7]

Sources from the following century include the English friar Thomas Gage who lived in Guatemala during the second two decades of the seventeenth century and who wrote a very entertaining description of the country, its inhabitants and customs including brief accounts of some of the buildings of Antigua. But unfortunately, like Herodotus of old, he believed everything he heard and in some instances

[5] Juarros, I: pp. 66 ff., for a chart dealing with the figures for 1776.
[6] Arana, *ASGH* 17 (1941/42): p. 240.

[7] See the bibliography for exact titles and dates of the works of the writers mentioned above as well as those below.

even invented stories which are a delight to read but irritatingly inaccurate. The Dominican monk Antonio de Remesal is the first truly objective historian of Central America. His whole point of view may be summed up in a statement he makes in the prologue to his history, completed in 1615 and first published in 1617, "... el fin de la historia, no es escribir las cosas para que no se olviden, sino para que enseñan a vivir con la experiencia, maestra muda, que es la utilidad y bien público."[8] His book is of prime importance to the architectural historian because it is replete with information of sixteenth-century Guatemala and Chiapas.

From the end of the seventeenth century comes the first history and geographical description of Guatemala from the pen of a layman, Fuentes y Guzmán, who writes in the true spirit of his time and his class; for he was a Spaniard born in Antigua whose ties to Spain were of secondary consideration since he thought of Guatemala and not Spain as his native home. However, his descriptions of the churches very frequently are vague and follow a standard formula since he was not directly concerned with the architecture as such. Yet his data on population, languages, customs, and physical character of the country are reasonably accurate. In many instances the documentary and literary sources from which it would seem he drew his information are still available and corroborate his statements. Other seventeenth-century sources pertinent for the architectural history of Guatemala, but not of the intimate sort as those mentioned above, may be found in the writings of Vásquez de Espinosa, who wrote a short description of the Audiencia de Guatemala, and Herrera, González Dávila, Torquemada, and Díez de la Calle, who deal briefly with Guatemala in more general works basing their information on secondary sources.

The outstanding historians from the eighteenth century are two monks, Ximénez, a Dominican, and Vásquez, a Franciscan, who wrote historical accounts of the two most important monastic orders in Central America. In addition to these monkish chronicles, there are also some secular sources for the history of Guatemala, as for example the report of the military engineer Diez de Navarro in 1744 containing a geographical description of the whole of Central America, published a century later.[9]

In 1740 the crown actually ordered that geographical surveys be made of the whole of the Reino de Guatemala.[10] In due course this order was complied with, to judge by the number of documents existing in the Archivo General del Gobierno in Guatemala City under the general classification number A 1.17 and covering all parts of Central America.

Another first-hand report of inestimable value was written by the archbishop Cortés y Larraz in the latter part of the eighteenth century after visiting the greater part of present-day Guatemala and the bordering area of El Salvador. His observations of the people and his descriptions of the country as well as his statistics on population are reliable. From about the same time there is the work of Alcedo but which is based on secondary sources rather than personal experience. By far the most important eighteenth-century source for the architecture of Guatemala is the history of Domingo Juarros who was born in Antigua and who was on the scene during its destruction and abandonment in 1773. He actually wrote and published his book early in the nineteenth century. With him the roster of contemporary authors of the colonial period comes to an end. His geographical index is an invaluable source for the location and identification of many towns whose names were altered in postcolonial times. In addition, he gives a rather minute and detailed history of the city of Antigua and its principal architectural monuments.

To judge by the number of narratives of journeys written and published by English and American travelers in Guatemala and Central America during the first half of the nineteenth century, it would seem that the country was being discovered for the first time by the English-speaking world. Hale in 1826, Roberts and Dunn in 1827, Thompson in 1829, Montgomery in 1838, Dunlop in 1847, and Baily in 1850 are but a few. The most informative of all these early nineteenth-century English-speaking travelers, with regard to both pre-Columbian and colonial monuments and especially a description of the city of Antigua, is John Lloyd Stephens. At mid-century there appeared a history by a native Guatemalan author, García Peláez, who in spirit, however, is quite close to his colonial predecessors and draws frequently on their works. From the second half of the century there are general descriptive works by two German-speaking authors, Scherzer who published in 1857 and Stoll whose work appeared in 1886. An important contemporary French author who was more interested, however, in pre-Columbian antiquities, Brasseur de Bourbourg, also gives some information relative to the colonial architectural monuments.[11]

In the last half of the nineteenth century histories and

[8] See his *Historia general de las Indias Occidentales* **1**: p. 15.

[9] See also *AGG*, A 1.17.3 (1744) 17508–2335.

[10] See *AGG*, A 1.17 (1740) 112–6.

[11] See his "Antigüedades guatemaltecas etc.," in *ASGH* **22**, 1 and 2 (1947): pp. 99–104, and *Histoire des nations*, etc.

general descriptions of Central America began to be written in a more detached or objective vein, as for example the works of Squier in 1855, Bancroft in 1875 and 1882, Boddan-Whetham in 1877, Bates in 1878, Maudslay in 1899, and by Central Americans such as Ayón, León Fernández, and Gámez, all in 1889, and Milla in 1879, whose work was continued by Gómez Carrillo until 1905. In 1868 the Guatemalan geographer Gavarete published a modern geography of the republic of Guatemala proper. In 1876 the world-renowned photographer Eadweard Muybridge made a trip through the whole of Central America and Panama bringing back about 144 photographs, many of which include not only items of scenic and general cultural interest but also views of some of the colonial architecture, especially of the city of Antigua.

When compared to the number of titles available from the preceding century, relatively few histories of Guatemala and Central America have been published in the twentieth. That of Batres Jáuregui does not add to what had been written since the history of García Peláez appeared, while that of Villacorta, published in Guatemala City in 1942, is largely derivative in nature. He often follows closely in the tracks of his nineteenth-century predecessors both as regards selection of facts and their interpretation. Many smaller works, mainly of a popular nature, and fewer of more specialized interest have appeared in this century, but none at all on the architecture of the colonial period.

The architectural history of Guatemala and Central America when included in more general works, as is to be expected, is at best treated only peripherally. In those books directly concerned with the architectural history of Latin America, the Guatemalan style is summarized in a single or at most a few chapters, as for example, in such scholarly works as Angulo and Marco Dorta's *Historia del arte hispanoamericano*, Kubler and Soria's *Art and Architecture in Spain and Portugal etc.*, and Kelemen's *Baroque and Rococo in Latin America*, or else in books of more popular interest such as Miguel Sola's *Arte hispano-americano*. The only work which deals exclusively with the art and architecture of Guatemala is that of Victor Manuel Díaz, *Las bellas artes en Guatemala*. Unfortunately, the author does not organize the monuments within any historical scheme or stylistic context. Villacorta who does make such an attempt, devoting an entire chapter to the colonial architecture in his history of Guatemala, employs the terminology which had been established earlier in the century for the classification of other native American styles by Guido, Noel, Toussaint, Dr. Atl, and Buschiazzo. He was unfortunately ignorant of the fact that the sum total of Antiguan architecture dates from 1650 to 1773, a century and one-quarter.[12]

III

PERHAPS the most important sources for the architectural history of Guatemala and Central America are to be found in the colonial archives which exist in all the principal cities of Central America. But the most complete and best-organized collection is that in Guatemala City where a greater number of pertinent documents are, furthermore, more readily available than those in the great archives in Seville. The archives in the individual churches and in the cathedrals of Central America, especially that in Guatemala City, are other important storehouses of documents. But unfortunately, except for the archives of the cathedral in Guatemala City, none have been organized on modern scientific library lines, so that it is extremely difficult and excessively time consuming to find and to make use of pertinent material. The Archivo General del Gobierno in Guatemala City, thanks to its late director, Joaquín Pardo, is endowed with a card catalogue in which the documents are classified in categories as to contents so that one may rapidly and conveniently find great numbers of them referring directly to such matters as building contracts, petitions for help in construction and repairs, as well as sundry matters of importance to the architectural historian. Furthermore, these documents represent an accumulation pertaining to the Reino de Guatemala as a whole, and consequently are rich in materials relative to the colonial history not only of Guatemala, but also of all the present Central American republics and the Mexican state of Chiapas. A considerable number of documents have also been published by Pardo over a period of about ten years in the *Boletín del Archivo General del Gobierno*, and from time to time in *Anales de la Sociedad de Geografía e Historia*. Others have also appeared, but to a far lesser extent, in *Antropologia e Historia de Guatemala*.

Many documents in the archives of both Guatemala and Seville have been published at different times since the latter part of the nineteenth century. In 1856 Rafael Arévalo, secretary of the Ayuntamiento of Guatemala City, published a small book of the minutes (*cabildos*) of the very earliest city council meetings in Ciudad Vieja from 1524 to 1530, which was reissued in 1935. Of the many publications of documents in the Archivo de Indias in Seville the most use-

[12] See his *Historia*, pp. 306 ff.

ful is that edited by Manuel Serrano Sanz as volume VIII in *Colección de Libros y Documentos etc.* which deals with Central America. A small index of documents in the Archivo de Indias related to matters of Guatemalan interest appeared as a short article in 1939 in the *Anales de la Sociedad de Geografía e Historia*, "Indice de documentos, etc." In the mid-nineteenth century, Miguel Larreinaga published his *Prontuario de todas las reales cédulas* which gives the contents in very brief form of some of the most important documents for the history of Guatemala between the years 1600 and 1818. A work of similar nature which has been of first importance and indispensable in tracing the architectural history of Antigua is that by the director of the archives in Guatemala City, Joaquín Pardo's *Efemerides* published in 1944. His index of the documents in the archives of Guatemala published in 1945, *Indice de Documentos etc.*, is unfortunately not complete, covering but a small fraction of the card catalogue. Pardo's smaller work, *Prontuario de reales cédulas*, fills in the gap before 1600 in the work of Larreinaga and covers the years 1529 to 1599. Finally, the monumental series of volumes containing material from the Archivo de Indias, *Colección de documentos inéditos etc.*, which appeared in two series from 1864 to 1884 and from 1885 to 1926, was made more readily useful through the work of Schäffer, whose *Indice de la colección de documentos inéditos*, published in 1946/1947, made the search for material dealing with Guatemala a less time-consuming and tiresome one.[13]

IV

BUT the study of the monuments of Antigua requires special techniques since all the buildings were destroyed or badly damaged in 1773 and later in the numerous earthquakes, which continue to be a common occurrence in Central America. Because the city had been abandoned very soon after 1773, the ruined buildings have retained their original colonial character more or less intact and unaltered by postcolonial repairs or alterations, a unique condition and most fortunate for the architectural historian. The year of 1773 is then, an important *terminus ante quem* for dating the end of the Antiguan style just as the year 1541 is the *terminus post quem* when it begins.

Therefore, it was a matter of prime importance first of all to examine and describe the physical remains of these ruined buildings of Antigua. This activity precluded, for the time being, the introduction of any preconceived ideas with regard to dating the buildings by means of stylistic comparisons with monuments either in Spain or elsewhere in Hispano-America. The sole intention was to report the facts observed by means of the physical examination of the actual architectural remains. Once the architectural data had been gathered, this was correlated with the historical information relative to each building which had been garnered from the contemporary literature and archival documents. Employing these purely objective criteria it was thus found that the architecture of Antigua falls into four historical or stylistic periods between 1541 and 1773. The architectural or stylistic changes which are manifest in the buildings themselves were corroborated by the historical data. The results of the historical investigations and the reports of the actual physical studies of the individual monuments have been presented in Part III. Each building is treated separately as an independent monographic study which is inserted in the text in the proper chronological order as determined by the historical evidence. This part of the book is not designed for continuous reading but is meant, rather, to serve as a ready reference tool for the architectural historian interested in the Antiguan style for its own sake or its relation to the Hispanic building tradition in general.

13 The Archivo de Indias in Seville have been thoroughly searched for documents pertinent to the colonial architecture of Guatemala by Francisco Xavier Mencos Guajardo-Fajardo. His doctoral dissertation, presented to the Faculty of Philosophy and Letters of the University of Madrid about 1952, dealing with his findings is still unpublished.

V

PART I is of more general nature and meant for continuous reading. The first chapter which deals with the history of the city of Antigua was included so that the reader might have some notion of the urban development of the city where the prototypes of Central American colonial architecture originated. The remaining chapters in Part I deal in a general fashion with such matters as materials and methods of construction, labor, architects and practical builders, and building activity. The purpose here has been to relate as much as possible of the peculiar conditions which shaped the Antiguan style, conditions by which it may be judged and evaluated within a context proper to it. This does not mean to imply that the Antiguan style is something apart and different from the mainstream of the Hispanic architectural developments in the New World, but rather to indicate on what grounds it may be considered as one of the integral parts of the Hispanic tradition as a whole.

Though designed for continuous reading, Part II is an analysis part by part of the elements of the style. It deals with such matters as plans, retable-façades, pilasters, door openings, niches, plaster ornament, and other details of façade treatment without regard necessarily to any chronological context. The different elements are examined separately and presented as a basic vocabulary or alphabet, so to speak, by which stylistic comparisons between Antiguan buildings may be made. By reducing the basic architectonic elements of the buildings into stylistic categories or types, an objective tool is provided which, it is hoped, will be of use to those architectural historians interested in the problem of establishing the relation of the Antiguan style to that of Hispano-America and Spain, a task which is outside the immediate scope of this book. The problem of stylistic connections between the architecture of Antigua and the Hispanic world requires a special study. The author trusts that this book, especially the factual material contained in Part III, will be of use to those architectural historians who may undertake this complicated task, especially since the buildings are destined to eventual obliteration.

THE detailed descriptions of the extant remains, the architectural data, given for each monument appearing in Part III, are really in the nature of factual archaeological reports. In carrying out his program of excavations, the field archaeologist often is faced with the unhappy task of destroying the site in order to study it. More frequently than not, the published excavation reports remain as the only evidence of the history of the site investigated. In Antigua, however, though the investigator is not called upon to destroy the buildings in order to study them, still the constant and ever recurring earthquakes will. The reports containing the archaeological data in Part III will in time be all that is left of the architecture of Antigua. And if earthquakes do not throw the remains still standing to the ground, then the well-intentioned, but misguided, restorers will forever obliterate their pristine character, as has already happened to some buildings in the nineteenth century, and also in recent years, but with even more disastrous results.

PART I

History of Antigua, Materials and Methods of Construction

I THE HISTORY OF ANTIGUA GUATEMALA

1. *The Founding of the City*

ANTIGUA was neither the first nor the last city where the capital of the Spanish colony in Central America was located. From 1527 to 1541 the first capital of the Reino de Guatemala was in the valley of Almolonga about three miles from Antigua in a town now known as Ciudad Vieja. Very soon after 1541 when the latter town was destroyed in an earthquake and a flood, the capital was established in a new city, now known as Antigua. There it remained until shortly after 1773 when once again, as a result of a series of devastating earth movements, the seat of government was moved some twenty-five miles distant, continuing there until the independence from Spain in 1821 and on to the present day.

The official name of colonial Antigua was Santiago de los Caballeros de Guatemala. The last colonial capital, present-day Guatemala City, was named "La Nueva Guatemala de la Asunción," while the former capital began to be referred to as Antigua Guatemala. The pre-1541 capital in the valley of Almolonga is known as Ciudad Vieja. In the contemporary literature and documents from before 1773 Antigua is simply referred to as Guatemala or, less frequently, as Santiago.

The city council, *ayuntamiento*, actually began to function officially in the new location only in 1543, that is, not until two years after the destruction of Ciudad Vieja.[1] Though the decision to relocate had been rapidly taken, not all parties concerned responded at once. The ecclesiastical authorities were somewhat hesitant and did not move to the new site until May of 1543. Two official town plans, *trazas*, for the new capital were made the year of the destruction of Ciudad Vieja, one in October and another in November. Sites for the church, the jail, the town hall, and building plots for the citizens were all allocated. The plan drawn in November included a few minor revisions in order to mollify the complaints of some of the residents who were dissatisfied with the parcels of land granted them.[2] The new capital, then, was not occupied immediately after the destruction of the old, about two years intervening before all the civil and ecclesiastical authorities and principal citizens had established themselves permanently in the new site.

Hard on the destruction of Ciudad Vieja in 1541, the architect Juan Bautista Antonelli is said to have arrived in Guatemala.[3] He was commissioned to inspect various sites where a new capital might be located.[4] Among other reasons, the availability of building materials in the Valley of Panchoy was a determining factor in his choice of the new site where the capital city of Antigua was finally established.[5] The town plan of Antigua has been ascribed to him. Though this is quite possible, no actual evidence for this conclusion is available.

2. *The Sixteenth-Century City*

THE appearance of Antigua in the first decade after its establishment must have been more like a frontier town with many of the building sites still vacant and others occupied by structures of less than permanent character. The building lots which had been distributed to the Spanish residents in 1541 had not yet been fenced, as required in token of having taken possession, by May of 1542. The time limit for doing so was thereupon extended to Easter of 1543.[6] But even as late as May 22, 1555, twelve years after the expiration of the period previously extended, many of the sites were still neither fenced nor occupied by buildings.[7] In 1566, when Felipe II, as a mark of his appreciation, granted the new capital the high-sounding title of "la Muy Noble y Muy

[1] Juárez Muñoz, *Peregrinación*, p. 148; see also Chinchilla Aguilar, *El ayuntamiento colonial*, pp. 21 ff. For an account of Ciudad Vieja see Seczy, *Santiago de los Caballeros.*

[2] Juarros, **2**: p. 178. See also his apparent source, Ximénez, **1**: pp. 157 ff.

[3] Fuentes y Guzmán, **1**: pp. 130 ff.; see also Angulo, *Bautista Antonelli*, for a biographical account.

[4] *ASGH* **18** (1942/43): pp. 73 ff.

[5] Juarros, **2**: pp. 178 ff.

[6] *Efem*, p. 7.

[7] *Ibid.*, p. 14.

Leal Ciudad de Santiago de los Caballeros de Guatemala," he could hardly have done so in recognition of Antigua's material worth or the monumental character of its edifices.[8] López de Velasco who wrote his book about this time describes the city saying that it was well built because of the abundance of good materials such as pine and cypress lumber, stone, lime, plaster, tile, and brick.[9] To judge by the list of materials, the houses and public buildings were built of brick, plaster, wood, and tile; in fine, a popular or vernacular type of construction, but hardly of a monumental character commensurate with the new aristocratic and imposing title of the city.

A somewhat more detailed description of the appearance of the city and its buildings is given by Ponce who was in Guatemala in the 1570's.[10] He says that the town was of good size, but smaller than Puebla in Mexico and that residing there were many noble though not very rich people, and that the houses were built of tamped earth with some buttresses of brick and stone laid in lime mortar, and roofed with tile. Among the religious establishments located there he enumerates the Cathedral, the convent of nuns of La Concepción, and three convents of monks; namely, La Merced, Santo Domingo, and San Francisco. The Franciscan establishment, originally solely of tamped earth, was in the process of being rebuilt during Ponce's visit. He describes the new construction as being of tamped earth, but with many buttresses of stone and brick laid in lime mortar, and states that the presbytery of the church was roofed with a dome of brick. It may be safely concluded that the majority of the buildings of sixteenth-century Antigua were constructed of the same humble materials and hardly of formal architectural character.

Sixteenth-century Antigua was probably not populous either. According to López de Velasco, there were five hundred *vecinos*, heads of Spanish households, resident in Antigua, of whom seventy were *encomenderos* and the rest *tratantes*.[11] If his report is correct, it would seem then that the total Spanish population, provided all five hundred *vecinos* were married to Spanish women and had children, which is hardly likely, could not have been more than two or three thousand people. Nor were the ecclesiastics a large factor in the population for, according to him, there were about twenty monks resident in the convent of Santo Domingo, fifteen to twenty in San Francisco, and about twelve to fifteen in La Merced. He is silent on the number of nuns in La Concepción, but does report that there were two hospitals, one for Spaniards and one for Indians.

[8] *Ibid.*, p. 18.
[9] *Geografía*, pp. 286 ff.
[10] Ponce, *Relación*, pp. 410 ff.
[11] *Geografía, loc. cit.*

The number of Indians residing in Antigua must have been no less than that of Spaniards, if not actually greater. The Franciscans maintained a special chapel for Indians adjacent to their convent.[12] Pineda who was in Guatemala in the late sixteenth century informs us that adjacent to each monastery in Antigua, including two others besides the Franciscan, there was a neighborhood of Indians who were former slaves liberated by Lic. Cerrato, the first president of the *audiencia*, and that they were all craftsmen engaged in various[13] trades. Corroborating the number given by López de Velasco, Pineda also gives the total of Spanish householders, *vecinos*, as five hundred, a remarkable increase in the fifty years or so since 1529 when there were but 150 *vecinos* in the first capital of Ciudad Vieja according to a later but reliable source.[14] This can hardly represent a natural increase in the Spanish population and must be accounted as the result of immigration from Spain. Evidence of the increase in the population of Antigua is also indicated by the fact that already as early as 1559 the city limits had to be extended beyond the original boundaries of the official town plan. Royal license to distribute additional building lots and to open new streets to the north of the town limits arrived on the sixteenth of July of that year.[15]

One may assume, therefore, that Antigua in the sixteenth century was a small town of about four to at most six thousand inhabitants, both Spanish and Indian, with little if any monumental architecture at all. Public services also were quite primitive and at a minimum during the first decades after the establishment of the capital in the new location, for an order from the city fathers in November of 1559 made it unlawful to empty sewage from the houses directly into the streets and required the householders to build cesspools in the patios of their houses.[16] Sewage disposal continued to be a problem, it would seem, for twenty years later in 1579 an ordinance was proposed to require that all sewage be carried in pipes below ground.[17] When compared with contemporary Mexico, sixteenth-century Antigua was a small town of rather nondescript appearance.

3. *The Seventeenth-Century City*

DURING the seventeenth century the population increased and, as a result, the physical size of the city was again enlarged. An increase in building construction is also noted in this century. But much of this activity was in effect rebuilding, for already at the beginning of the century the many

[12] Ponce, *Relación, loc. cit.*
[13] *Descripción*, p. 330.
[14] Remesal, 1: p. 14.
[15] *Efem*, p. 15.
[16] *Ibid.*, p. 15.
[17] *Ibid.*, p. 24.

recurring earthquakes frequently resulted in damage to, and destruction of, the nondescript sixteenth-century buildings. These either required extensive repairs or rebuilding from the ground up. An earthquake on October 9, 1607, caused such widespread havoc that a special order was emitted instructing certain Indian villages of the surrounding region to provide laborers to be employed in the repair of the damage, a custom which was to become traditional, lasting to the end of the colonial period.[18] This was neither the first earthquake nor the last to plague Antigua in the seventeenth century; one of them almost completely destroyed the city in 1586.[19] González Bustillo, referring to the earthquake of 1773 which led to the abandonment of Antigua, relates that earthquakes were so frequent in the seventeenth century that for a period of about sixty years from 1590 to 1659 the inhabitants were so terrified that they did not dare erect buildings of any value.[20] He no doubt got this information from Vásquez who makes the same observation using almost the same language.[21]

If population figures for the sixteenth century derived from contemporary sources may be accepted as approximately correct, then the number of inhabitants of Antigua in the middle of that century was about five hundred Spanish *vecinos* plus an indefinite number of Indians, at least as many as, if not more than, the Spaniards. A nineteenth-century writer,[22] probably relying on a tax list compiled in 1604,[23] believes that by the beginning of the seventeenth century the population was about 4,450. He derives this number from the 890 *vecinos* given in the tax list, allowing five members for each family. But he does not take into account the Indians, not considered *vecinos*, who must have been numerous. Gage, known to be given to exaggeration, writes that *ca.* 1627 there were about 5,000 Spanish families in Antigua plus about 200 Indian families living in a neighborhood of their own called Santo Domingo.[24] He may have meant individual Spaniards and not families. It would seem that, since the middle of the previous century, the Spanish population of Antigua had increased by 390 households, and more than quadrupled since the first Spanish establishment in Almolonga where in 1529 there were but 150 *vecinos*.[25] Herrera who gives a very vivid description of the whole region but seems to be relying on the reports of López de Velasco, for he never was there himself, says that in his day (the late sixteenth and early seventeenth centuries) there were 600 *vecinos españoles* living in Antigua and that the town had 25,000 *indios tributarios*.[26] He does not say where these Indians lived, but there can be no doubt that they resided in their own villages in the surrounding region and comprised the economic production force on which the Spaniards depended.

That the growth of the population did not remain static for very long in this century is also known from other sources, perhaps more reliable ones: namely, the records of the town council. The city limits had to be extended in 1641 at which time the *ayuntamiento* also petitioned the *audiencia* requesting an allotment of Indians to be employed for both public and private works then necessary as a result of the increase in population.[27] In 1650 it was also necessary to enlarge the slaughterhouse because of the increased number of inhabitants.[28]

Various epidemics struck the city during the seventeenth century causing a high mortality. The worst occurred in 1686 when about one-tenth of the total population died including poor Spaniards, ordinary people, mestizos, mulattoes, and Indians without number.[29] But the growth in population continued despite this disaster, for by the end of the century, in 1693, the city limits were once again extended and new streets at least five *varas* wide laid out in the neighborhood, *barrio*, of Chipilapa.[30] Four years later, in 1697, the *ayuntamiento* requested permission from the *audiencia* to lay out even more building sites in the Llano de los Remedios and the Prado del Calvario, two areas in the southern section of the town, and to proceed with the sale of these lots.[31]

The century also witnesses a remarkable increase in the number of ecclesiastical establishments. The English friar Thomas Gage who was in Antigua *ca.* 1627 remarks that the churches were not as fair as those of Mexico and that there was but one parish church.[32] Elsewhere he gives its name as San Sebastián.[33] Actually there were two parish churches in his time, the other being Los Remedios. He mentions the Cathedral as standing in the chief market place, that is in the Plaza Mayor, and goes on to say that all the other churches belong to the various religious orders, specifically the Dominican, Mercedarian, Augustine, and Jesuit, and two of nuns, La Concepción and Sta. Catarina. This would make a total of about nine churches including Los Remedios which he does not mention in this connection. Fuentes y Guzmán writing at the end of the seventeenth century, *ca.*

18 *AGG*, 1.1 (1607) 1–1.
19 Vasquez, **1**: p. 265.
20 *Razón particular.*
21 Vásquez, **1**: p. 264.
22 García Peláez, **1**: pp. 203 ff.
23 *AGG*, A 1.2.6 (1604) 11810–1804.
24 Gage, ch. XVIII, p. 280.
25 Remesal, **1**: p. 14.
26 *Historia general* **1**: p. 84.
27 *Efem*, p. 55.
28 *Ibid.*, p. 61.
29 Vásquez, **4**: p. 388.
30 *Efem*, p. 113.
31 *Ibid.*, p. 120.
32 Gage, ch. XVIII, p. 283.
33 *Ibid.*, p. 265.

1690, says there were twenty-four churches in town including the Cathedral, ten of the conventual organizations, three parish churches, five hermitages, and four churches belonging to *beaterios*. The number actually falls one short of the sum total he gives, but still shows an increase of fourteen religious buildings over Gage's estimate in the space of about seventy-five years or less.[34]

Only after the middle of the century did the appearance of the city improve, mainly because of the building activity of the various religious orders and the concern of the *ayuntamiento* with providing more adequate public services such as water supply, sewage disposal, street repair, and the like. The Plaza Mayor where the Cathedral and buildings for the *audiencia* and the *ayuntamiento* stood had already been embellished toward the end of the sixteenth century when the city government agreed to build a fountain there in October of 1580.[35] (FIGS. 19–21.) The Cathedral which still exists today, partly in ruins and partly reconstructed in the early nineteenth century, was inaugurated in 1680. In addition to the Plaza Mayor, the main square, at the end of the seventeenth century there were ten other plazas, most of which were located in front of churches. The total of eleven are as follows: (1) the Plaza Mayor used for pageants and bullfights; (2) the plaza of the Escuela de Cristo; (3) the plaza near San Pedro; (4, 5, and 6) three plazas around the convent and hospital of Belén; (7) the plaza of the church of La Candelaria; (8) the plaza near the convent of Carmelitas Descalzas (Santa Teresa); (9) the plaza near San Sebastián; (10) that near San Jerónimo; and (11) the plaza in front of the church of Espíritu Santo. Most of these plazas still exist even today.[36] (FIG. 15.) By the latter part of the century a university was established, but it is hardly likely that it had a very imposing or monumental building. The university building existing today actually dates from the last half of the eighteenth century.

Sewage disposal and other public services also improved in the seventeenth century.[37] Likewise, streets were maintained, repaired, and even small bridges built over the streams that flank the city.[38] The water supply was a constant concern of the city government, and as early as 1617 water mains of masonry were installed.[39] The colonial archives in Guatemala City are replete with documents concerned with the development of this public service which throw light on the colonial method of water distribution and sale by "*paja*," a custom still prevalent in Guatemala City.[40] In the middle of the century the water system was improved with the construction of masonry distribution tanks with bronze outlets.[41] By the end of the century more than one source of water was required to supply the city's needs. Water was brought in from six different sources, one being as far off as Pastores.[42]

The increase in population of the city in this century brought about the growth of the *barrios* or neighborhoods outside the official *traza* or town plan where members of the non-Spanish races divided into distinctive social castes lived. Some of these peripheral neighborhoods were frequently slumlike, impoverished areas with houses of the very simplest construction of adobe, roofed with thatch. The Spaniards lived in the center of the city in houses built of more formal and permanent materials. The lower castes in the surrounding slums continued to live in thatched houses to the end of the colonial period.[42a] After the earthquake of 1751 it is noteworthy that straw to value of 500 *pesos* for making repairs was distributed to the poor of the *barrios* who had lost their houses.[43]

Fuentes y Guzmán lists ten *barrios* or wards into which the city was divided in 1686 as follows: (1) San Francisco, inhabited by Indians, probably the first to be established after the founding of the city, for it is also mentioned by sixteenth-century travelers; (2) El Tortuguero; (3) San Sebastián, the most populous with water supplied from Pamputic and its own parish church; (4) El Manchén to the east of San Sebastián; (5) San Jerónimo to the west of San Sebastián, not as populous and inhabited by the poorer classes; (6) Santiago, which he labels an *arrabal*, a slum; (7) Espíritu Santo; (8) Santo Domingo; (9) La Chácara; (10) La Candelaria where some Spaniards as well as *ladinos*, *mulatos*, *negros*, and *indios* resided, all of whom were masons, carpenters, and metal workers. On the outskirts of the city he mentions two more, Chipilapa and Santa Cruz, which were not very populous.[44]

The residents of these *barrios* were rarely if ever members of the Spanish caste. The Spaniards presumably lived in the

34 Fuentes y Guzmán, I: pp. 138 ff.

35 *Efem*, p. 25.

36 Fuentes y Guzmán, *loc. cit.*

37 *Efem*, pp. 24 and 70.

38 *AGG*, A 1.2 (1663) 951–39 and A 1.2–1 (1684) 25064–284.

39 *Efem*, p. 44.

40 Chinchilla Aguilar, *El ramo de aguas*, pp. 19–31.

41 *Efem*, p. 57.

42 Fuentes y Guzmán, I: p. 134.

42a The meaning of the word "caste" is the same as the Spanish word *casta* as used during the colonial period in Guatemala. See ch. IV, fns. 26, 32, and 78, below. For an exhaustive treatment of the population of colonial Central America, see Baron Castro, *La Población*.

43 *Efem*, p. 205.

44 Fuentes y Guzmán, I: pp. 136 ff.

center of town. Fuentes y Guzmán specifically mentions some of the *barrios* administered by the Mercedarian order along the western edge of the city as being the worst slums, populated with Indian dyers, saddlers, and shoemakers. In one particularly, Espíritu Santo, the Indians dressed in the Spanish manner but were given to drunkenness. In the *barrio* of Santiago, located on the banks of the Magdalena River on the western limit of the city, lived the poorest classes, mainly *ladinos* with a few Indians.[45]

The problem of the population of Antigua at the end of the century, however, is hardly clarified by Fuentes y Guzmán.[46] Quoting some parish lists made for tax purposes, he says that there were 6,000 households, *vecinos*, in Antigua numbering 60,000 inhabitants and including only those persons between the ages of fourteen and sixty. His figures cannot be accepted unless the tax list he quoted included all the towns of the area around Antigua. Such a large number of inhabitants could hardly have been contained within the city even a century later when it extended considerably further. The approximately 5,000 Spanish inhabitants in 1604 no doubt increased in number as did the Indians and other elements too. But this increase could hardly have reached the figure given by Fuentes y Guzmán.

4. *The Eighteenth-Century City*

THIS century marks the culmination of the building activity begun in the last half of the seventeenth. The great acceleration in construction, however, was not due so much to religious fervor or economic prosperity as it was to sheer necessity. In 1717 and 1751 many seventeenth-century structures were damaged or completely destroyed by earthquakes, making it necessary in many cases to start anew from the foundations.[47] From an official report rendered by the *ayuntamiento*, dated February 3, 1719, it is learned that most of the damage in the center of the town resulting from the earthquake of two years before had been repaired.[48] The destruction had been greatest in the surrounding neighborhoods and these were presumably still in bad shape. The religious communities, owing to lack of funds, had not yet been able either to do much work by way of repairs or to undertake new constructions. For a short period of time after the earthquake of 1717, since the destruction had been so widespread, certain official quarters gave serious consideration to a plan for abandoning the site and moving the capital to a new location. But the opposition was so great that the idea was given up. However, after the earthquake of 1773, the forces for moving the city won out.

The city grew in size and improved in appearance in the eighteenth century. Vásquez describes the town of 1700 with its streets laid out in a grid running north-south and east-west.[49] The distance from east to west was more than one mile and a little more than that from north to south, so that from the church of Los Remedios to that of San Sebastián it was more than half a *legua* and the same from La Candelaria to Santa Lucía.[50] (FIG. 15.) All of the area described was populated and enclosed within a perimeter of two *leguas* not counting some of the *barrios* outside the official city limits. Within the city limits, he continues in another place, that is within a circumference of not more than a half a *legua*, are fifty churches, hermitages, and sanctuaries with twenty permanently established *sagrarios*, sacristies.[51]

Arana in reporting the damages of the 1717 earthquake claimed the town was two *leguas* in circumference. Ximénez, on the other hand, strongly differed with him, insisting that the city was not even a quarter league in circumference and, if the nearby neighborhoods were added, not even half a league.[52] Ximénez was of the party opposed to moving away and always minimized the damages as an argument in favor of his views. Ximénez's smaller dimensions probably refer to the measurement on a straight line, that is, the diameters of the more or less oval shape of the city, or perhaps only to the central part within the official town limits excluding the surrounding neighborhoods. Vásquez actually gives both the perimeter, "... *bojea dos leguas* ...," corroborating the dimension given by Arana, and a straight line or north-to-south diameter measurement from and including the neighborhoods of San Sebastián to Los Remedios. In fine, it would seem that Antigua was about one English mile square at the beginning of the eighteenth century.

By 1721 the city grew to the south of Los Remedios and even beyond El Calvario. The *ayuntamiento* decided to tap the water of the nearby town of San Juan del Obispo and connect this pipe line with the water mains of Santa Ana in order to supply this new neighborhood located at a higher elevation because there was not sufficient pressure in the

[45] *Ibid.*, p. 388.
[46] *Ibid.*, p. 151.
[47] Juarros, **1**: pp. 63 ff.
[48] *Isagoge*, pp. 402 ff.
[49] Vásquez, **4**: p. 352.
[50] The true length of the *milla* (mile) and the *legua* (league) referred to by Vásquez is not known with exactitude. The *legua*, a unit of distance used in the colonial period, is also employed in modern Guatemala where it is thought of as being about three English miles or five kilometers. Elsewhere its length varies from 2.4 to 4.6 miles or 3.9 to 7.4 kilometers. The actual distance from the two points north to south he gives is about 1,700 meters or just over a mile. See fig. 15 for a plan of Antigua.
[51] Vásquez, **4**: p. 385.
[52] Ximénez, **3**: p. 352.

mains from San Juan Gascón and Pamputic to reach there.[53] In the same year the residents of the neighborhood known as El Tortuguero in the southwest part of town asked for certain vacant building sites which had already been assigned others but were still unoccupied. These were needed for house construction in view of the fact that the population of the neighborhood had grown.[54] About fifteen years later two other neighborhoods likewise had increased in population and required more land for house building.[55] The same trend continued right up to the time of the great earthquake of 1773 when the neighborhoods of both Los Remedios and Santa Cruz, also in the southern part of the city, required more water to fill the needs of the increased population.[56]

Some specific figures concerning the number of buildings occupied by the various ecclesiastical organizations, the number of domestic residences, and the population of Antigua and of the surrounding towns under the jurisdiction of the city are given in an interesting document from the middle of the eighteenth century.[57] Thirty-one ecclesiastical establishments are listed, the majority of which are also known from other sources as well. The population resident in the towns near Antigua is given as 3,530 individuals comprising 1,320 Spaniards, 1,420 mulattoes, 690 mestizos, and 100 Negro slaves.

The same source gives the number of houses standing in Antigua as 2,952, of which 1,802 were roofed with tile and 1,150 with thatch. Residing in these houses was a population of 6,610 comprising 2,240 Spaniards including children, 2,570 mulattoes, 1,800 mestizos. This number does not include Negro slaves and other individuals of the "lower classes." The population figure of 6,610 probably does not include children of the non-Spanish castes and, except for the Spaniards, only heads of families are included. If the figure 6,610 is meant to represent individuals, this would make the average of about 2.24 individuals per house, that is, 6,610 divided by 2,952 houses, which can hardly represent the total population of all races and classes resident in Antigua in 1740. A more reliable figure for the total population of Antigua might be obtained by taking the number of houses as a guide. The population may then be supposed to have been about 12,000 allowing approximately four individuals per household, or at most 15,000 if five individuals are considered to have comprised the average household.

The population apparently did not remain static during the rest of the eighteenth century and probably increased, but not nearly as much as some contemporary reports purport. For example, González Bustillo claimed that at the time of the earthquake of 1773 the number of houses in Antigua had grown to between five and six thousand, a figure which would make it appear that the city population in terms of households had doubled in size in the space of thirty-three years since 1740.[58] Though not certain, this remarkable increase in the number of households may be explained by the fact that González Bustillo might have included the *barrios* or neighborhoods contiguous to Antigua, but outside the official city limits, as part of his total. Yet his figure is quite conservative when compared with that reported by the city authorities at the same time. In a letter dated August 31, 1773, from the *ayuntamiento* to the crown relating the destruction resulting from the earthquakes during July of that year the number of private houses given is 8,000. This number is about two and one-half times more than that in 1740 and, therefore, must certainly have included the surrounding neighborhoods.[59] It is not stated whether the houses were roofed with tile or thatch, an important clue in determining their location. Multiplying the number reported by four individuals per house gives a total of 32,000 inhabitants living in Antigua in 1773, a number far too great for the known size of the city at that time and, therefore, unacceptable. It would seem that the municipal authorities wished to exaggerate the damage in order to win the sympathy of the crown in their need for immediate aid. The figures given by González Bustillo more nearly represent the facts though they are not exact either.

Cortés y Larraz gives a more objective estimate of the population some four years earlier in 1769, and also under far less trying circumstances, stating there were 12,354 people resident in Antigua.[60] He does not include the suburban neighborhoods in his count, but only the four parishes within the city proper. A census taken right after the first destructions of 1773 and submitted the twenty-second of October gives the number of inhabitants of these same four parishes as 9,044.[61] A year later one Manuel Galistea certified that 9,144 persons still remained in Antigua. This presumably included the neighborhoods on the periphery of the city, for he also states that 5,917 had already emigrated.[62] On the basis of these latter figures an original total of 15,061 inhabitants is indicated as residing in Antigua and its imme-

53 *Efem*, p. 151.
54 *Ibid.*, p. 165.
55 *Ibid.*, pp. 178 and 180.
56 *Ibid.*, p. 243.
57 *AGG*, A 1.17.1 (1740)5002–210, see also *BAGG* I (1935/36): pp. 7 ff.
58 *Razón particular.*
59 *AGG*, A 1.10 (1773) 18.773–2444.
60 *Descripción*, fol. 8 ff.
61 *AGG*, A 1.10 (1773) 1535–55.
62 *Efem*, p. 257.

diate environs just prior to the earthquake, a number more in keeping with the physical size of the city about that time and one which also corroborates that given by Cortés y Larraz as of 1769. The 8,000 houses alluded to by the city fathers in their letter of August, 1773, must be catalogued as an exaggeration and not borne out by the October census. Nor can the resulting estimated figure of 32,000 inhabitants be accepted even if the population of the surrounding neighborhoods and villages be included, for in 1740 it numbered only 3,630 individuals. Even if this latter figure is doubled and added to the October, 1773, count, the total for both city and environs may be estimated to have been from twenty to about twenty-two thousand people at most, which is not likely at all.

Cadena, describing the devastation of 1773, says that in a scant twenty-two years since the previous earthquake of 1751 the town had been completely rebuilt, so that in truth one could say that all Guatemala was new.[63] No doubt many new private houses were built, but they hardly would have doubled the size of the city since 1740 as the figures given by González Bustillo purport, or tripled it as the *ayuntamiento*'s figures imply. A conservative estimate would place the population of Antigua and its immediate environs in 1773 somewhere between more than fifteen and less than twenty thousand people. The foregoing evidence does point to the fact that the city had grown somewhat between the middle of the century and 1773, especially toward the southern limit. The east-west growth was hindered by the two rivers, the Pensativo and the Magdalena, bordering the town on either side.

This city was both the ecclesiastical and civil capital of the Reino de Guatemala which the *ayuntamiento*, in the letter of August, 1773, referred to above, said consisted of three bishoprics, eleven "*ciudades y muchas villas* . . . 900 *pueblos* . . . 17 *Reales de Minas* . . . 23 *Gobiernos y Alcaldías Mayores* . . ." and that surrounding the capital were seventy-two *pueblos*. And capital though it was of an area covering most of present-day Central America, only toward the end of the eighteenth century were its streets beginning to be paved, and at that only the center of town, except for the Plaza Mayor which had been paved for the first time in 1704.[64] A series of *cabildos* from 1764 to 1770 records the specific streets which were paved with cobblestones, all in the center of town.[65] The streets of the outlying areas were narrow, twisted, muddy in the rainy season, and dusty in the dry, just as they are still today.[66] After the earthquake of 1717 all the alleys and streets of the *barrios*, since they were so narrow, were obstructed with debris from the fallen houses and had to be cleared.[67]

The Plaza Mayor must have presented a rather dismal aspect throughout the eighteenth century. It is true the square was flanked by the Cathedral, the Capitanía, and the Ayuntamiento, the latter two with their arcaded façades. It had been laid out according to instructions promulgated by the crown as far back as 1573.[68] But the side opposite the Cathedral was occupied by one-story business establishments which continued around the north side of the square and shared half a block with the Ayuntamiento directly opposite the Capitanía. These private structures were in complete ruin after 1717. Some of their roofs were supported with timbers that had rotted out and so could not withstand the tremor.[69] The present-day parklike appearance of the Plaza Mayor is the result of work done in the twentieth century. (FIGS. 19–21.) Even in the late nineteenth century the square was rather bleak looking, its one embellishment being the fountain, still standing today, in the center. It had been built anew in 1738 since the former one located off to one side was in a bad state of repair.[70] This new fountain was designed and built by Diego de Porras, *maestro mayor de obras*, who a year later was paid one hundred *pesos* in partial payment and given a special note of thanks for his fine work.[71]

As late as 1760 sewage from the archbishop's palace adjacent to the Cathedral on the northeast corner emptied out into the plaza. The order was then given that a waste sewer line be constructed below ground.[72] The center of the plaza was choked with stalls of the market which was held there. The custom of using the main square as a market went back to the seventeenth century and possibly earlier.[73] Gage refers to the plaza as the main market place.[74] The crowding of the stalls, *cajones*, actually impeded the passage of carriages. In 1718 the *ayuntamiento* firmly ordered the ". . . cajoneros de la plaza mayor . . ." to arrange their stalls in some systematic fashion so that carriages might be able to pass through.[75] Finally in 1719 an annual rental was charged the vendors who maintained stalls in the Plaza Mayor. The money received was designated for cleaning, paving, and maintenance of the plaza.[76]

[63] *Breve descripción.*

[64] *AGG*, A 1.2–2 (1704) 11.780–1786.

[65] *ASGH* **25** (1951): pp. 155, 165, 169, 210, 212, 226, 229, 230, 235, and 239.

[66] Juarros, **1**: pp. 63 ff.

[67] Ximénez, **3**: p. 387.

[68] See Stanislawski, *Early Spanish Town Planning*, pp. 90–105, and *Origin and Spread*, pp. 105–120; also Zucker, *Town and Square*, pp. 132 ff. For some plans of other Spanish colonial cities see Chueca Goitía, *Planos, passim.*

[69] Ximénez, **3**, p. 354.

[70] *Efem*, p. 184.

[71] *Ibid.*, p. 186.

[72] *Ibid.*, p. 217.

[73] *AGG*, A 1.2 (1697) 15793–2211.

[74] Gage, ch. XVIII, p. 283.

[75] *Efem*, p. 149.

[76] *Ibid.*, p. 151.

5. *The Nineteenth-Century City and the Moving of the Capital after 1773*

THE earthquakes of 1773 were not the first of great violence the city had experienced. Others had occurred in 1565, 1577, 1586, 1607, 1651, 1663, 1689, 1717, and 1751.[77] A specialized literary form evolved in Guatemala, "the earthquake account," sometimes written as a factual report but more often embellished with a layer of fanciful prodigies of divine or infernal origin. These were often published as books or pamphlets, though many consisting of eye witness accounts or official reports still in manuscript are to be found in the Archivo General del Gobierno in Guatemala City as well as the Archivo de Indias in Seville. All contemporary authors from Remesal to Juarros give accounts of those earthquakes through which they lived or learned of from others.[78]

The ruined buildings of the abandoned capital were further destroyed by earthquakes in the nineteenth and twentieth centuries. For example, the church of the Cruz del Milagro which was still partly standing in the first quarter of the twentieth century is now completely obliterated. The cupola of Santa Cruz was also still in place until the early part of this century, but a gaping hole now marks its former existence. Many other buildings no doubt suffered further destruction in the last 175 years or so. It is highly probable that the original ruin may not have been as great as appears from the condition of the remains visible today. Such a conclusion is warranted if one takes greater stock of the reports of the ecclesiastical as opposed to those of the civil authorities. This being the case, it may very well be that the damage in 1773 was no greater than that occasioned in the earthquake of 1717 when the alternative of moving the capital was proposed but turned down.

After the earthquake of 1773 the forces in favor of moving the capital won out, though their purpose was not accomplished all at once. The main contestants at first were the *ayuntamiento* and the *audiencia*. The latter wished to move out of the site immediately, but the members of the *ayuntamiento* protested that there were no lodgings for government officials in the new site, let alone any for private citizens. They complained further that the hire of mules for carting belongings and other paraphernalia had gone up from three *reales* to from eighteen to twenty *reales* per day. Martín de Mayorga, president of the *audiencia*, nevertheless, imposed his will on the *ayuntamiento* and moved out the sixth of September, 1773.[79]

A royal order arrived in January of 1774 approving the provisional transfer of the city to La Ermita in the Valle de las Vacas, the present site of Guatemala City. A year and one-half later, in July of 1775, a royal *cédula* arrived making the transfer permanent.[80] But it was not until December of that same year that the provisions of the *cédula* were formally obeyed.[81] Still many private people and the ecclesiastical authorities resisted the order, and understandably, for it meant abandoning all their real property and taking up unimproved sites in the new city. The impasse lasted until June of 1777 when the president of the *audiencia* published an order prohibiting anyone from continuing to live in Antigua and allowing the ecclesiastical communities, the municipal authorities, and the private citizens, *vecinos*, but two months to be out.[82]

The president of the *audiencia* was so determined to move the capital from the fateful site in the valley of Panchoy that as early as 1773 he issued an order to dismantle much of the Capitanía and other public buildings for the double purpose of making these buildings altogether unserviceable and obtaining ready materials for the new government buildings in La Ermita.[83] Juarros relates that the destruction of 1773 was not as bad as it was painted by the engineers, architects, and scribes believing the city could have been rebuilt.[84] The neighborhoods of La Candelaria, Santo Domingo, Chipilapa, and part of San Sebastián had been very badly hurt, but the damage to private buildings in the center of town was not great. It seems that the truth lies somewhere be-

[77] Juarros, **1**: pp. 63 ff.

[78] A partial list of "earthquake accounts" includes: "*1541, Septiembre. Santiago de Guatemala.—Relación anónima de un testigo de vista: Sobre la tempestad de aguas que en 10/9/1541 destruyó la ciudad de Santiago de Guatemala. Con enumeración de las personas que perecieron.*" Schaeffer *Índice* **2**, no. 2413, in *Colección Muñoz* **82**, **1**, 3: pp. 378–386; "*1541. Septiembre. Santiago de Guatemala.—Relación del Obispo de Guatemala, D. Francisco Marroquín: Sobre la tempestad que en 10/9/1541 destruyó la ciudad.*" Schaeffer, *op. cit.* **2**, no. 2414, in *Colección Muñoz* **82**, Extracto de J. B. Muñoz, **1**, 3: pp. 386–388; Rodríguez, Juan, *Relación*, also a description of the 1541 destruction; Hincapie Meléndez, *Relación*, a description of the 1717 earthquake; *AGG*, A 1.18 (1718) 1400–2021, "*Breve y verdadera historia del incendio del volcán . . . y terremotos de la ciudad . . . 27 de Septiembre de 1717*"; Arana, *Relación*, also of the 1717 earthquake; Caxiga y Rada, *Breve relación*, a description of the earthquake of 1751; Cadena, *Breve descripción*, a report of the 1773 earthquake; González Bustillo, *Extracto*, *Razón puntual*, and *Razón particular*, all three dealing with the ruin of 1773; and *La Ciudad mártir* contains transcriptions of Cadena and González Bustillo.

[79] *Efem*, pp. 248 ff. [80] *Ibid.*, p. 257. [81] *Ibid.*, p. 258.

[82] *Ibid.*, p. 261. For an account of the abandonment of Antigua and the founding of Guatemala City, see Pérez Valenzuela, *La Nueva Guatemala*.

[83] *AGG*, A 1.10.2 (1774) 1642–66; see also A 1.10 (1777) 4524–69 and A 1.10.3 (1777) 4571–76. [84] Juarros, **2**: pp. 179 ff.

TABLE I

SUMMARY OF THE POPULATION OF ANTIGUA, GUATEMALA

Ciudad Vieja before 1541	Sixteenth Century	Seventeenth Century	1740	1769	1773	1800	1838
150 *vecinos* (Spaniards) other castes unknown	500 *vecinos* (Spaniards) other castes unknown 2,000–3,000 Spaniards 4,000–6,000 total all castes	890 *vecinos* (Spaniards) other castes unknown 4,000–5,000 Spaniards in 1604 Indians unknown	2,952 houses or 12,000–15,000 all castes	12,354 all castes according to Cortés y Larraz	15,061 all castes	7,000–8,000 all castes	10,000–12,000 inhabitants

tween the extremes of Juarros and González Bustillo, mute evidence for which is still visible in the ruins of Antigua today.

The city was never really totally abandoned despite the royal orders and the willful dismantling of public buildings. The principal Spanish families moved away as did the governmental and ecclesiastical authorities, but the populace at large remained, especially those of the *barrios*, the poor neighborhoods whose mud and thatch huts could be readily rebuilt with little work and less money. After 1779 the city was classified as a *villa*, made the capital of the province of Sacatepéquez, and divided into three parishes which also included some of the villages on the outskirts. About 1800 the population according to Juarros was between seven and eight thousand people, most of whom were mulattoes, but including some few Spaniards too.[85] When the English traveler Montgomery passed through Antigua in 1838 he found the town still populous with from ten to twelve thousand inhabitants.[86] The American traveler and diplomat, the indefatigable John Lloyd Stephens, also passed through the town about this time accompanied by the artist John Catherwood who made a drawing of the Plaza Mayor.[87] (FIG. 19.) Except for some errors in architectural details of the façade of the Capitanía and the placing of the Volcán de Agua to the east, the illustration gives a clear picture of the appearance of the town in the mid-nineteenth century. The Cathedral which had been partially restored earlier in the century and served as the parish church of San José is clearly depicted. The main area of the plaza was bare while the seemingly eternal market vendors with their *cajones* block the view of the Capitanía. The whole east half of the two-story arcade of the building facing the plaza, however, was still in ruins. From the last quarter of the century an interesting photograph by the renowned photographer Eadweard Muybridge taken from the upper story of the Capitanía about 1876 shows the plaza as an open unpaved dirt area with a typical Guatemalan market in progress. (FIG. 21.) The Ayuntamiento stood as it stands today, but adjacent to it was a row of one-story tile-roofed houses, set in front of which were some wooden stalls of more or less permanent but nondescript character with tile roofs.

Various buildings were partially restored and altered in the nineteenth century, particularly the Cathedral, the convent buildings of Santa Catalína, the church of La Merced, and the church and hospital of San Pedro. A women's prison was located in the convent of Santa Teresa until the 1950's. The Ayuntamiento is still used by the municipality as a city hall, and part of the Capitanía houses various government offices of the Department of Sacatepéquez. Part of the Cathedral serves as a church, the convent of Belén as a

85 *Ibid.* I: p. 56.
86 *Narrative*, p. 153.
87 *Central America* I: p. 266.

hotel, the University as a museum, the Seminario Tridentino as a private dwelling, and the Compañía de Jesús as the public market. In recent years some work of restoration has been begun in various other buildings: Santa Cruz, Las Capuchinas, San Francisco, Escuela de Cristo, and others. Some of the former ecclesiastical buildings are privately owned today, but most have been declared national monuments as has the urban complex of Antigua as a whole.[88]

[88] Though the city had been declared a national monument in the early 1940's in order to preserve its colonial character, some of the buildings have, nevertheless, been misguidedly "restored" thus violating the very purpose of the law. Some of the vaults of the church the church of Santa Cruz have been rebuilt with reinforced concrete! The fountain of the Dominicans, though not on its original site, was removed from the place where it had been located for many years. The church of San Francisco has served as the setting for musical concerts and fashion shows of women's clothes. Work to rebuild the vaults with concrete and place the building in use as a church again was well under way in 1963. For some information on Antigua as a national monument, among other matters, see Annis, *El plano de una ciudad colonial.*

II MATERIALS OF CONSTRUCTION

THE materials and the methods in which they were employed in Antigua are not unique there, for the same were used throughout the whole of Central America during the colonial period. In many instances the building methods described as being common in the sixteenth century in such outlying regions as Chiapas, for example, fill the gap in our information for Antigua. One must look beyond the capital city for examples of buildings more or less still in their pristine state, be they from the sixteenth, seventeenth, or eighteenth centuries, for none have been preserved in this condition in Antigua.

1. *Wood*

THE site of the first capital in Ciudad Vieja was selected from among others largely because of the availability of wood. To judge by the discussion *pro* and *con* on the choice of the site at Almolonga over that of Tianguecillo, the deciding factor was that of the abundance of wood in the former. Gonzalo Dovalle testified in a meeting dated November 21, 1527, "... que en los llanos no hay madera para edificar, ni leña sino muy lejos . . . y el asiento del valle es alegre y vistoso . . . montes muy cerca para edificios, y leña en mucha cantidad. . . ." The fact that at the other site of Tianguecillo there were many old Indian structures of stone and the availability of stone in the nearby hills was not considered as strong an argument as that of the abundance of wood in the valley of Almolonga.[1] The same arguments prevailed about fifteen years later when it was decided to remove the capital after the disaster of 1541. At that time the choice was between the present site of Antigua in the valley of Panchoy, the valley of Chimaltenango, and again Tianguecillo, for wood was the principal building material for both public and private constructions during the better part of the sixteenth century.[2]

When Pineda was in Guatemala at the end of that century, he commented on the fact that the recently freed Indian slaves inhabiting the towns surrounding Antigua had found ample employment in cutting and sawing timber. He says, "... son cortadores y aserradores de vigas y tablas y alfaxias y calcontes para las casas de los españoles de dicha ciudad de Guatemala."[3] This observation is also made by López de Velasco who says that the buildings of Antigua were well constructed because of the abundance of good materials, mainly wood, enumerating the types available as: "... pino y encina y ciprés . . . ," as well as plenty of stone, lime plaster, tile, and brick.[4] It is noteworthy that these types, pine, evergreen oak, and highland cypress are still cut and widely used today.

In some parts of Central America, as for example in San Cristóbal de las Casas, Chiapas, and in Viejo, Nicaragua, whole logs were employed.[5] The area around San Cristóbal de las Casas is still an important source of building lumber for the region. In other parts of Guatemala proper, in Verapaz, for example, there was also an abundant supply of pine which served as the chief building material in the sixteenth century.[6]

But it was soon noted that the highland timber was not very durable for it was attacked by termites and dry rot. Oviedo, though speaking specifically of conditions in the isthmus of Panama, describes this process very vividly, calling the termites ants which "... comen la madera, y asimismo las paredes hasta dejarlas tan huecas como un panar."[7] He warns that great care must be exercised to avoid the utter destruction of a house for that animal is the same for the house as the moth is for cloth, "... para la casa es aqueste animal no otra cosa que la polilla para el paño." Tropical hardwoods which abound on the north coast of Guatemala and in the lower elevations above sea level throughout Central America were hardly employed in the

[1] *ASGH* 4 (1927/28): pp. 95 ff., "Fundación de la Ciudad de Guatemala en Almolonga"; see also Remesal, I: pp. 39 ff.

[2] Fuentes y Guzmán, I: pp. 126 ff.

[3] Pineda, *Descripción*, p. 330.

[4] López de Velasco, *Geografía*, pp. 286 ff.

[5] Ponce, *Relación*, pp. 352 and 478 ff.

[6] Montero de Miranda, *Relación*, pp. 342–358; see also Pineda, *Descripción*, *loc. cit.*

[7] *Sumario*, pp. 190 ff.

highlands. This is not strange when one considers the great difficulties which must be overcome even today in the cutting and transporting of the lumber from the inaccessible rain forests of the lowlands and coast. Highland timber was near at hand and most readily sawn into lumber. It took more than a century of experience of trial and error before it was realized that the soft evergreens of the highlands were not adequate for the construction of permanent public buildings or churches and monasteries. Ximénez was only stating the facts when he said that one of the reasons so many of the buildings of Antigua had been damaged in the earthquake of 1717 was that the supporting timbers of the roofs of many had rotted out in the course of time.[8] The conventual church of San Francisco in Antigua was rebuilt in the late seventeenth century when it was inadvertently discovered that the timbers had rotted out and new tile could not be set until the whole framework was replaced.[9]

Wood was used so widely in the sixteenth century in Antigua that by 1579 some conservation remedies were proposed in order not to destroy the surrounding forests.[10] Apparently the Indians had become accustomed to the wasteful practice of dressing logs with axes which meant that only a single board could be rendered from each. It was suggested that the Indians be required to change their methods and use saws and thus produce a number of boards from each log. But the excessive cutting of timber for construction and fuel too did not cease, and about six years later Gaspar Arias de Avila made a plea in the *ayuntamiento* that some means be found to conserve the forests surrounding Antigua.[11] By the seventeenth century it would seem that all the good timber nearby had been cut off and it had to be brought from as far off as Tecpán.[12] This lumber was mainly pine, cedar, and highland cypress. Tecpán was still an important source of the same type of lumber even in the late eighteenth century[13] and continued to be so even after the colonial period during most of the nineteenth century.[14] Another source somewhat nearer than Tecpán, in the Valle de las Vacas where Guatemala City is presently located, began to supply building lumber in the late seventeenth century too.[15] After the earthquake of 1689, in the name of equity and in order to prevent profiteering due to the great demand, the *ayuntamiento* was obliged to regulate the price of wood and lime as well.[16]

Wood remained one of the most widely employed building materials throughout the colonial period, but by the eighteenth century it was rarely used for major structural elements except in small churches of lesser importance which continued to be roofed with timber and tile. By the eighteenth century almost all ecclesiastical and civil constructions in Antigua employed vaulted roofing. Yet, in out-of-the-way places like the north coast it is recorded that the Fort of San Fernando in Costa Rica, built in 1743, was protected by a wooden palisade.[17] Nor was wood scorned in so rich a country as colonial Mexico even as late as the seventeenth century. One hundred years after the conquest Gage, who arrived in Vera Cruz in September of 1625, describes the city saying,

> Of the buildings little we observed, for they are all, both Houses, Churches and Cloisters, built with Boards and Timber, the Walls of the richest man's house being made but of Boards, which with the impetuous Winds from the North, hath been for the most part of it burnt down to the ground.[18]

2. *Thatch and Wattle and Daub*

JUDGING by the contemporary descriptions of the conventual establishments of even such important orders as the Franciscan and Dominican, all of which were built of thatch and wattle and daub at first, it is small wonder that so little of early sixteenth-century architecture has survived, if indeed architecture it can be called when carried out in such ephemeral materials. The majority of the buildings of the first capital in Ciudad Vieja were uniformly roofed with thatch which would flare up in flames, one such general conflagration being recorded in 1536.[19] Discussing the convent of Santo Domingo which had been established about 1538–1539, Ximénez says that the common method of construction in those early days was to place four corner posts in the ground and cover the walls with mud and the roof with straw.[20] Wattle and daub seems to have been most commonly employed in the early days after the conquest. The first Central American cathedral was built of this material. Even as late as 1544 the Dominican convent, by then already located in Antigua, was a collection of huts built of cane covered with mud and the church was no more than four corner posts supporting wattle and daub walls roofed with thatch.[21] The Spaniards did not introduce thatching, for it

[8] Ximénez, **3**: p. 353.
[9] See ch XIII, no. 1, below.
[10] *Efem*, p. 24.
[11] *Ibid.*, p. 28.
[12] Vásquez, **4**: p. 42, the editor quoting *Archivo Arzobispal de Guatemala*, A 4.5–2.
[13] Cortés y Larraz, fol. 218 ff.
[14] Irisarri, *Cristiano errante*, ch. V, pp. 249 ff.
[15] Fuentes y Guzmán, **1**: p. 281.
[16] *AGG*, A 1.2–2 (1689) 11778–1784.

[17] Juarros, **1**: p. 46.
[18] Gage, ch. VIII.
[19] Juarros, **1**: p. 160.
[20] Ximénez, **1**: p. 146, quoting almost verbatim from Remesal, **1**: pp. 437 ff.; see also **2**: pp. 39 and 128.
[21] Remesal, **1**: p. 385. See Mencos, *Arquitectura*, ch. II, for information on the first cathedral in Ciudad Vieja.

was well known to the Indians of all Central America. Thatch was the principal roofing material even for important monumental preconquest structures. Oviedo, describing the types of Indian houses he noted in the New World, says the thatch is "...paja o yerba larga, y muy buena y bien puesta, y dura mucho, y no se llueven las casas, antes es tan buen cubrir para seguridad del agua como teja."[22] He goes on to say that the Spaniards use the same type of thatch too, for it is not unlike that used to roof the houses in the villages and hamlets of Flanders, but that the local straw or grass is much better than the latter.

Another important building material, especially in the hotter climates of Central America, which was also used even at high elevations in very cool climates, is cornstalks. As a matter of fact, the Dominican convent in Antigua is described as a collection of straw-covered huts with walls of cornstalk.[23] Even today many of the more impoverished Indian villages consist of houses built exactly in this manner: (1) four logs or posts are set at the corners; (2) these are connected by purlins halfway up and by a plate on top, the horizontal members thus forming a timber framework; (3) the walls, really fillers only, are formed of cornstalks placed vertically side by side and made fast to the framework by horizontally set stalks to which the upright ones are tied with string made of corn husks; (4) the roof is framed of lighter timber and thatched. It would seem, therefore, that the early settlers of Guatemala and Central America sometimes employed native materials and building methods in the sixteenth century. This was especially true in the construction of private houses and, in some instances, even of public buildings as a temporary measure until more permanent structures could be undertaken. In the mining towns there was a continuing lack of interest in building permanent structures even though they could well afford them. In Gracias a Dios, for example, where the seat of the *audiencia* was located about 1544 and which was a town rich in minerals, its churches were roofed with thatch.[24]

If the architecture of the capital cities, first Ciudad Vieja and later Antigua, was of nondescript character and of the most ephemeral materials for the better part of the sixteenth century, one may safely assume that this condition was even more common in the outlying regions and prevalent there well on into the seventeenth century. Most of the towns in Chiapas, for example, Chiapa, Zinacantán, and Copanaguastla which had been founded with Indians forcibly brought in from the surrounding countryside, had trouble keeping them there, for they would run off and return to their old homes. It is no wonder, then, that the churches where these Indians gathered for worship were of the most rudimentary and temporary nature and described thus: ". . . la Yglesia que era del tamaño de una celdita, de palos y barro, que mas parecía casa de gallinas que Yglesia . . . ."[25] Even the conventual houses of the Dominicans in most of Chiapas during the sixteenth century were built of the same materials, wattle and daub and thatch.[26] Conditions of this sort continued to the end of the seventeenth century even in many parts of Guatemala though it was the most heavily populated, advanced, and developed region of Central America. In the whole curate of Jutiapa there was not a single church built of permanent materials, all having thatch roofs and walls of wattle and daub, known as *bajareque*, a term still employed today.[27] Even in a town quite near Antigua, San Pedro Sacatepéquez, the church was built of the same materials at the end of the seventeenth century.[28]

But in Antigua by the end of the sixteenth century conditions had improved so that more permanent structures began to be built. For example, the thatch parish church of Los Remedios was rebuilt with a wood and tile roof, that is, an *artesón*.[29] In general, however, it was not before the seventeenth century that more permanent materials began to be used as a normal matter of course, not only in Antigua but also increasingly so in the provinces. Even the small church which had burned to the ground in such an out-of-the-way place as Zamayaque, a town no longer existing, was rebuilt with an *artesón* and roofed with tile.[30] The existence or absence of tile and wood roofs was considered an important factor in classifying towns in the various censuses taken; and rightly so, for pertinent conclusions relative to economic conditions and the process of acculturation of Indian towns toward eventual Europeanization could be deduced from such facts. Fuentes y Guzmán almost always indicates whether the houses of a given town are roofed with tile or thatch, as do Cortés y Larraz, Juarros, Ximénez, and others. It was a mark of progress to install tile roofs, for thatch roofs were associated, as they still are, with the "inferior" Indian culture. Some towns, for economic reasons, had thatch-roofed churches even late in the seventeenth century, as for example, San Agustín Acasaguastlán where the local authorities, in 1667, petitioned to have their church rebuilt with an *artesón* to replace the thatch roof.[31] In gen-

22 *Sumario*, p. 134 ff.

23 Ximénez, *loc. cit.* A reed called *caña de Castilla* is also used today.

24 Fuentes y Guzmán, **2**: p. 12.

25 Ximénez, **1**: p. 383, quoting the MS of Fr. Tomás de la Torre.

26 *Ibid.* **1**: pp. 421 ff.

27 Fuentes y Guzmán, **2**: p. 201. The term *quincha* is used in the Andean area, especially in Peru.

28 *Ibid.* **3**: p. 186.

29 *AGG*, A 1.23 (1586) 1513–660; also Pardo, *Prontuario*, p. 21.

30 Vásquez, **4**: p. 275.

31 *AGG*, A 1.10.3 (1667) 31.253–4046.

eral, when Indian towns had public buildings and houses roofed with tile, such as in Patzún, this was considered a mark of progress.[32]

Roof tiles were made very early in Antigua. But when glazed tile for floors and walls were needed, as was the case in the construction of the Capilla de Loreto in the conventual church of San Francisco about 1600, these had to be brought from Mexico 300 *leguas* distant.[33] And for truly monumental constructions where copper sheets were required, as in the church and convent of La Recolección, even as late as 1703 this material had to be imported from Oaxaca.[34] The houses of the *barrios*, the poorer neighborhoods, of Antigua were ever of rather rudimentary character, roofed with thatch and destroyed time and time again in the almost countless earthquakes. The typical lower-class houses of the *barrios* were described after that of 1717 as being no more than "... cuatro tapias de buena suerte ...," four hit-or-miss tamped-earth walls.[35] And even in the middle of the eighteenth century not all the private houses of Antigua were yet roofed with wood and tile, for of the total of 2,952 houses given in a census of 1740 only 1802 were permanently roofed while the remaining 1,150 were covered with thatch.[36]

3. *Stabilized Earth: Adobe and Tamped Earth*

THE use of stabilized earth either in the form of sun-dried mud brick, adobe, or tamped earth has a history going back to remote antiquity in the Mediterranean world. It is more likely than not that the Spaniards introduced the use of both adobe and tamped earth in Central America, the latter called *pisón* and walls of this material *tapias*. It is possible, however, that the use of sun-dried brick may have been known even before the conquest, there being a possibility that this building material was employed in the southwest of the United States.[37] Adobes of pre-Spanish date have also been found in South America where they were used by the Incas.[38] These are described as being about eight inches square and about thirty-two inches long with a high proportion of straw to mud. This is a rather odd-shaped brick and more like a mud log. The large amount of straw in proportion to mud was no doubt necessary to make certain that there would be enough voids to allow for contraction in drying such a cumbersome mass. Drying of such an adobe must have been rather bothersome because of the thickness of eight inches in comparison to the small exterior surfaces. The common adobe brick used in Guatemala since colonial times measures approximately two and one-half inches thick by twelve wide by twenty-four inches long, thus providing an exterior surface more than five times its thickness so that the brick can be thoroughly dried in a relatively shorter time than the pre-Spanish Inca type.

Sun-dried mud brick was used from the first in Ciudad Vieja. The main walls of the Franciscan convent were of adobe, though the partitions separating rooms were of *bajareque*, wattle and daub, and stood intact until about 1565.[39] When the Franciscan convent in San Salvador was constructed sometime around 1580, it too was built of adobe.[40] The first Dominican convent in Antigua, built soon after the catastrophe of 1541 in Ciudad Vieja, was also of adobe. In fact, one of the monks, Fr. Matías de Paz who died in 1579, "... andaba todo el día haciendo adobes, asentándolos ..." and also instructed the Indians how to make and lay adobes.[41] This was a new building material for the Indians of Guatemala, or at least a new use for mud, for they were accustomed to build their huts of cane or cornstalks which they sometimes daubed with mud as described time and time again by the contemporary chroniclers of Central America.

At the time of the conquest one of the gravest problems which confronted the civil and ecclesiastical authorities was that of gathering the Indians into towns for both economic and religious ends. Remesal speaks with great feeling about this saying it was difficult to uproot them, for they loved their homes and thought more of the hovels they lived in than of the richest palaces. Their houses he says, "... son de poca costa y embarazo, cuatro horcones hincados en tierra, el tejado de paja, las paredes de caña cubiertas con lodo, puertas ni ventanas ... en cuatro horas se hacía una casa, y en dos días un pueblo," that is, of little cost or trouble, four posts stuck in the ground, the roof of straw, the walls of cane covered with mud, no doors or windows, in four hours a house was built, in two days an entire town.[42] Ximénez says that only after 1549 when the Indians of Chiapas were gathered into towns did they begin to live in houses of adobe. Even the Dominican convent in San Cristóbal de las Casas was first built of this same material.[43] By the seventeenth century Indian houses were generally built

[32] Fuentes y Guzmán, **1**: p. 383.

[33] Vásquez, **4**: p. 223.

[34] García Peláez, **3**: p. 27.

[35] Ximénez, **3**: p. 353.

[36] *AGG*, A 1.17.1 (1740) 5002–210.

[37] Meighan, Clement W., in a letter dated May 3, 1955.

[38] Rowe, "Inca Culture," *Handbook of the South American Indians*, Bureau of American Ethnology, Bulletin no. 143, **2**: pp. 226–227.

[39] Vásquez, **1**: p. 115.

[40] *Ibid.*, p. 221.

[41] Remesal, **2**: p. 494.

[42] *Ibid.*, p. 244.

[43] Ximénez, **1**: p. 482; see also Markman, *San Cristóbal*, p. 59.

of adobe in the Spanish manner,[44] except in *tierra caliente* where wattle and daub were used,[45] as they frequently are even today.

When the new capital was laid out in the Valley of Panchoy in 1542 after the destruction of Ciudad Vieja, the preparation of materials for construction began almost immediately and these were in large part adobes.[46] One of the considerations in choosing the site of the valley of Panchoy for the new capital city Antigua, was the abundant supply of earth particularly appropriate for the making of adobes.[47] In fact, the building plots which had been laid out in the town plan and which were assigned to the citizenry in 1542 were ordered to be fenced with either adobe or tamped earth walls as a sign that possession had been taken.[48]

There is no doubt that tamped earth, *pisón*, was a well-known building material in the Iberian peninsula as far back as antiquity when it was probably introduced by the Phoenicians from the Near East. Walls of tamped earth were referred to as *tapias* and were known almost immediately after the conquest in other parts of the New World. Oviedo compares the houses of Santo Domingo with those of Barcelona which were of stone but also of "... hermosas tapias y tan fuertes, que es muy singular argamasa . . . ."[49] Throughout the whole of the sixteenth century both tamped earth and adobes as well as wattle and daub were frequently employed for private dwellings of the poorer classes in Central America, depending on local conditions and experience. López de Velasco describes the houses in Sonsonate as being of tile and adobe[50] while Ponce, writing *ca.* 1586, states that in Viejo, Nicaragua, they were of wattle and daub, and thatched.[51] In various places in his report, Ponce gives descriptions of building materials common in his day. For example, in the highlands the houses of the Indians are of adobe and thatched, but in hot country of wattle and daub; the houses of San Salvador are of tamped earth; in San Cristóbal de las Casas, the houses are of logs roofed with tile, while the Franciscan convent in construction is of adobe; in Izalco, Salvador, the large church there is built of tamped earth with a thatch roof and a masonry façade; in Sonsonate, the Franciscan convent is of adobe and tamped earth roofed with tile, houses are of the same materials; in Zamayac (Zamayaque), the Franciscan convent is of adobe and thatch.[52]

These humble building materials never really went completely out of use for public buildings and were frequently employed even as late as the end of the seventeenth century and, in some instances, the mid-eighteenth century. The first building occupied by the Universidad de San Carlos in Antigua was constructed of tamped earth, *pisón*.[53] The raw materials for the construction of the *Salón Mayor* were taken right from the site itself so that the rector of the university was obliged to register a complaint before the *audiencia* in August of 1681 asking that the main patio of the university be filled and leveled "... de que se sacó mucha cantidad de tierra para la fábrica y an quedado muchos oyos y derrumberos de suerte que con el desaliño, parece un corral muy feo . . . ."[54] Even so important a civil monument as the Capitanía, the seat of the *audiencia*, had portions of the walling built of tamped earth, some of which caved in during the earthquake of 1717.[55] The now completely obliterated church of the Beaterio de Indias was built sometime in the mid-eighteenth century of a combination of tamped earth with brick reinforcing.[56]

4. *Architectural Character of Sixteenth-Century Buildings*

ONE can hardly speak of formal architectural character when dealing with such ephemeral materials as wood, wattle and daub, adobe, cornstalks, and thatch. The architectural character of sixteenth-century Central American buildings was, by and large, nondescript. There were possibly a scant few monumental structures in one or two cities which were cities in title only, being in fact no more than villages without any formal architectural character.[56a] Only when brick, stone, and lime began to be produced in sufficient quantities, and at that hardly before the seventeenth century, did a formal architecture evolve in Central America.

It would seem that from the very first the royal authorities in Spain were aware of the nondescript and rough-and-ready frontier character of the architecture in the new settlements in Central America. In 1538 a royal order arrived

[44] Remesal, *loc. cit.*
[45] Ponce, *Relación*, p. 385.
[46] Vásquez, **4**: p. 384; see also Remesal, **2**: p. 46.
[47] Fuentes y Guzmán, **1**: pp. 126 ff.
[48] Remesal, **2**: p. 45.
[49] Oviedo y Valdés, *Sumario*, p. 88.
[50] López de Velasco, *Geografía*, p. 296.
[51] Ponce, *Relación*, p. 352.
[52] *Ibid.*, pp. 385, 398, 402, 403, 434, 478 ff.

[53] *AGG*, A 1.3 (1763) 1157–45; also *ASGH* **17** (1941/42): pp. 376 ff.
[54] *Efem*, p. 95.
[55] Ximénez, **3**: p. 354.
[56] Angulo, *Planos*, pl. 154, **2**: pp. 48 ff., and **4**: pp. 412 ff., *AGI*, Est. 100—Caj. 7—Leg. 22 (6), also Torres Lanzas, *Planos*, no. 187.
[56a] Recinos, *La ciudad*, pp. 57 ff., emphasizes the fact that the first establishments of the Spaniards in Guatemala during the sixteenth century in general were nondescript in the extreme and that "La frase de Alvarado 'hize y edifiqué en nombre de su magestad una ciudad de españoles,' es un tanto exagerada. Los conquistadores ocuparon simplemente las casas de los indios de Iximché y no tuvieron tiempo para edificar cosa alguna . . . ."

in Ciudad Vieja instructing that only stone, brick, and tile be employed in house construction and that rooms be of ample size with patios open so that sunlight might enter.[57] This was followed by a second order of the same tenor giving the town authorities six months in which to carry out the instructions of the first.[58] To judge by the verbal descriptions in the contemporary literature and by the almost complete lack even of vestiges of sixteenth-century buildings in Guatemala and the rest of Central America, which it must be borne in mind was not completely pacified until well into the seventeenth century and in some outlying parts even as late as the eighteenth, these impractical royal orders were honored more in the breach than in the observance. The royal hopes for better housing were unrealistic because it was impossible to carry them out in many regions such as Honduras which had not yet been pacified at the time. The Indians on whom the materialization of the royal wishes depended, were still not as yet gathered into towns and converted to Christianity, let alone taught European crafts. Their labor, therefore, could not be utilized in the construction of houses of stone and wood ". . . conforme a la calidad de las personas de cada uno de los conquistadores y pobladores . . . ."[59] Missionary activity began "to bear fruit" in parts of Honduras and Nicaragua beginning only in the seventeenth century where the Indians lived mainly by hunting and fishing and practiced very little agriculture.[60]

Of the fourteen *pueblos* already established in the province of Verapaz by 1575, according to Montero de Miranda who was there at that time, six of which were in *tierra templada* and the remaining eight in *tierra caliente*, only one had a formal street plan, the others were laid out "sin concierto ni orden de calles anchas sino con unas sendas como de venados . . . ." Some of the churches, however, were built of stone, probably the one in Cobán, and others of adobe, all roofed with timber and tile and some furnished with fine ornaments and sculptures and paintings. The last part of the statement is probably an exaggeration.[61] Even in Chiapas, less remote than Verapaz, during the sixteenth century most churches were hardly more than wattle and daub huts lacking even the most elementary furnishings for the divine cult.[62] As a matter of fact, with regard to Franciscan establishments, it was expressly ordered in 1541 that their convents be of the simplest and humblest kind.[63] The shocking nondescript character of so many of the churches in Guatemala in the mid-sixteenth century led López de Cerrato, who had come as a special emissary of the crown, to encourage the religious orders to expend more time and money on the construction of buildings worthy of the newly introduced religion so that the Indians might be attracted to it and leave behind their heathen idol-worshiping.[64] It was only after López de Cerrato's time in the late sixteenth century that church-building took a turn for the better and the wattle-and-daub meanly furnished churches were slowly replaced by more formal structures, though still hardly of monumental character.

5. *Brick Piers with Tamped Earth Walls*

ONLY after some royal orders had been expedited in 1573 and 1578 did the Franciscans, who had previously been specifically instructed to build convents of the simplest kind, begin to rebuild their establishments on more formal and permanent lines in those towns of the *Real Corona*; namely, Comalapa, Tecpán Guatemala, Sololá, Quetzaltenango, and San Miguel Totonicapán.[65] In 1586 when Ponce visited there, many of the Franciscan convents which were spaced across the length of Central America were still in construction. Many were still roofed with thatch but this was being replaced with tile.[66] In some instances the walling material is described as *tapias con rafas de piedra y ladrillo*,[67] that is, tamped-earth walls with reinforcing piers or buttresses of brick and stone. This is a method described by Fr. Lorenzo de San Nicolás whose works were first published in 1633. Fray Lorenzo was probably describing a method of building already well known, rather than inventing a new one. He gives four methods of laying up walls, one of which is is to use piers of brick with ". . . tapias de tierra . . . que en edificios angostos es buen modo de edificar . . . ."[68]

Ponce was especially impressed with the Franciscan convent of Ciudad Vieja which had been rebuilt after the catastrophe of 1541. It had a two-story cloister, a garden, and a church which ". . . es todo de tapiería de rafas de piedra, cal

[57] *AGG*, A 1.2–4 (1538) 2196–138; also Pardo, *Prontuario*, p. 28.

[58] *AGG*, A 1.2–4 (1538) 15752–44 vuelto; also Pardo, *op. cit.*, p. 41.

[59] Pedraza, *Relación*, p. 138.

[60] Vásquez, **4**: p. 109; see also Markman, *San Cristóbal*, p. 23, fn. 6, for some documentary sources on missionary activity in Honduras, Vera Paz, and Costa Rica in the seventeenth and eighteenth centuries.

[61] Montero de Miranda, *Relación*, p. 355.

[62] Remesal, **2**: p. 39.

[63] Vásquez, **1**: p. 113.

[64] Fuentes y Guzmán, **2**: p. 370.

[65] Vásquez, **1**: p. 246.

[66] Ponce, *Relación*, p. 403, Izalco; p. 434, Zamayaque; p. 439, Quetzaltenango; p. 440, Totonicapán; p. 442, Sololá; p. 449, Tecpán.

[67] Ponce, *op. cit.*, p. 402, Izalco; p 403, Sonsonate; p. 421, Ciudad Vieja; p. 434, Zamayaque; p. 439, Quetzaltenango; p. 440, Totonicapán; p. 450, Comalapa; p. 398, San Salvador; p. 446, Santiago Atitlán had adobe, *tapias*, and stone laid in mud mortar.

[68] *Arte*, pt. 1, ch. xxxv, p. 86.

y ladrillo, hízole el rey, y es el mejor que entonces había en la provincia . . . ."[69] He also describes the Franciscan establishment at Quetzaltenango saying the church ". . . llevaba buen edificio de tapiería con rafas de piedra y ladrillo . . . ."[70] The same description is given for the Franciscan churches at Comalapa,[71] at Granada, Nicaragua,[72] and at San Salvador.[73]

Evidence of the practice of strengthening walls with buttresses or piers of masonry is revealed in the actual architectural remains. In many brick and stone buildings of the seventeenth century it is noteworthy that the buttresses have no direct relationship to the existing interior vaulting which was actually added later, as for example, the churches of Santa Cruz and Los Remedios in Antigua, San Felipe in San Cristóbal de las Casas, and San Bartolomé Milpas Altas in Sacatepéquez not far from Antigua. The buttresses frequently do not rise to the total height of the wall and serve only as a reinforcement of the wall itself rather than as a counterforce to the interior vaulting. The church of Santa Catalina in Antigua was laid out from the very first in this manner with sections of masonry and sections of tamped earth. In fact, the walls of Los Remedios still preserve vestiges of this type of construction.[74]

Toward the end of the sixteenth century some stone and brick began to be used sparingly, there being only about two or three examples mentioned by Ponce.[75] In Verapaz at the end of the sixteenth century only one church, that of Cobán, was built of stone. Many more had been begun but had not advanced beyond foundations, while services were still commonly held in churches of the most rustic sort.[76] The stone Dominican churches and convents in Copanaguastla and Tecpatán, Chiapas, may possibly date from the end of the sixteenth century, but this is not altogether certain.[77]

But no sooner did the building activity begin to take on a character based on formal architectural designs carried out in materials of more permanent nature, than did the earthquakes, an almost daily occurrence in Central America, begin to lay waste what had been erected. The disastrous effects of earthquakes were a matter of grave concern and constant fear. Remedies were sought, but to no avail. Vásquez, writing in the early eighteenth century, pointed out the dire fact that the deeper the foundations the more susceptible were the buildings to be affected by the earth tremors, the uppermost parts suffering the most. The ever present threat of earthquake resulted in a paralyzation of building activity from 1590 to 1650 when ". . . no osaban edificar templos ni casas de suntuosidad, porque cuanto más recias las fábricas, tanto menos seguridad tenían, y en los edificios más fuertes era mayor el estrago."[78] It would seem that those churches resulting from the accelerated building activity noted by Ponce, Pedraza, and Montero de Miranda and others toward the end of the sixteenth century were short-lived indeed, for the new type of walling, tamped-earth walls reinforced with masonry piers, was not adequate to withstand the shock of earthquakes.

6. *Masonry Materials: Brick, Stone, and Lime Mortar*

It was only in the seventeenth century that brick and stone buildings began to be undertaken as a common practice. The use of tamped-earth and brick piers, however, did not altogether disappear. These were employed where economy was an important factor, as for example in the church of Santa Catalina in Antigua. Dressed stone as an exclusive building material for walls and vaults was never used in Antigua. There are about four cases, however, where it was employed as a veneer in the eighteenth century in Antigua. Though ordered to do so by the crown, in the sixteenth century stone was almost never used largely because of the lack of specialized craftsmen who knew how to handle this material. Spanish settlers were scant in numbers and stonecutters among them even fewer, while the Indians had not yet been trained in this trade.[79] The Dominican church and monastery of Cobán may possibly have been built of stone by the middle of the sixteenth century,[80] and, if entirely of stone, which is most unlikely, it is the only example extant from this period.

Even in the late seventeenth century the newly discovered quarries in the hills near San Felipe and San Cristóbal just outside of Antigua could not be worked for lack of skilled mechanics.[81] The stone bridge at Los Esclavos over the river of the same name was built in 1592 and repaired in 1636. In the contemporary literature it is always spoken of

[69] *Relación*, pp. 421 ff.
[70] *Ibid.*, p. 439.
[71] *Ibid.*, pp. 450 ff.
[72] *Ibid.*, p. 325.
[73] *Ibid.*, p. 421.
[74] See ch. XII, nos. 1 and 2, below.
[75] *Relación*, p. 402, Izalco, only the façade of stone; p. 449, Tecpán, brick church in construction; p. 481, Chiapa de Indios (Chiapa de Corzo), the fountain; p. 440, Totonicapán, the brick *capilla mayor* completed, the rest of church of brick and still in construction.
[76] Pineda, *Descripción*, p. 350; also Viana, *Relación*, for an account dated 1574.
[77] Olvera, *Copanaguastla*, pp. 115–136, and *Joyas de la arquitectura*, pp. 14–17, 29; also Berlin, *Convento de Tecpatán*, pp. 5–13.
[78] Vásquez, I: p. 264.
[79] Pedraza, *Relación*, p. 138.
[80] Montero de Miranda, *Relación*, p. 355; also Pineda, *Descripción*, pp. 347 ff.
[81] Fuentes y Guzmán, I: p. 135.

as a great feat of engineering.[82] This opinion was still held far and wide even as late as the middle of the nineteenth century and commented on by a European traveling through the country in 1854 who says in part, "... it is very characteristic of the state of the arts in this country that this very simple and ordinary structure is considered the finest bridge in Central America, and looked on by people here as a most outstanding performance."[83]

Toward the beginning of the seventeenth century brick and terra-cotta roof tile also began to be used more and more as a natural outcome due to the abundant supply of clay and wood for firing. Brick had probably begun to be used to a small extent even in the sixteenth century in Antigua when permission was granted for the establishment of two brickyards in 1556.[84] The famous domical brick fountain in Chiapa de Corzo (Chiapa de Indios) was already standing by the middle of the sixteenth century.[85] And earlier still, the cathedral of San Cristóbal de las Casas was begun in 1533 of brick and tile, the price of which then was four gold *pesos* the thousand for brick and four *pesos* two *tomines* for tile.[86] Even in such a small town as Tequecistlán, north of Antigua toward Vera Paz, a three-nave brick church was in construction at the end of the sixteenth century.[87] But it was only in the seventeenth century that there was an accelerated use of brick which at first was almost wholly employed for the reinforcing piers in tamped-earth walls. After the middle of that century, as Indian labor became more and more experienced and costs reduced, was brick widely used for the first time. By the end of the seventeenth century the Indians in some of the nearby towns were independent entrepreneurs in brick manufacture; for example, the contract to supply the brick for La Recolección in 1703 was undertaken by an Indian from one of these towns.[88]

The inhabitants of Chimaltenango and the nearby villages became specialists in the manufacture of brick and tile which they carried to sell in Antigua. The brick from the towns of San Lorenzo el Tejar, San Sebastián el Tejar, and San Miguel el Tejar, the latter now known as San Miguel Morazán, was much sought after in Antigua and cost about five *pesos* more per thousand than that from the nearby *barrios* or suburban villages of Jocotenango and San Felipe.[89] It is noteworthy that the names of the towns in the valley of Chimaltenango alluded to above still bear the descriptive attribute "El Tejar" and, in fact, are still centers of brick and tile manufacture today. At the time the two brickyards were presumably established in Antigua in 1556, the price was set at four *pesos* the thousand for tile and four and one-half *pesos* for brick.[90] It is not stated whether these were gold or silver *pesos*, but it is more likely than not that gold may have been implied which would have made the price in Antigua about the same as it was a few years earlier in Chiapas.[91] The exact price of brick and tile in the seventeenth century is not known for certain, but it would seem that it had come down since the common currency by then was the silver *peso*. At least by the middle of the eighteenth century, bricks were relatively cheap, for one Eugenio Alberto Bocanegra agreed to supply the brick for the Ayuntamiento in Antigua in 1740 at the rate of twelve and one-half *reales* per thousand, or about one-third the sixteenth-century price if gold *pesos* were meant and considerably less if silver. Brick and tile were relatively cheap throughout the eighteenth century. After the destruction of Antigua and the removal of the capital to Guatemala City when so much brick and other building materials were needed for the new construction work, the brickmakers began to agitate for higher prices for their products.[92]

Brick was employed in a variety of ways: (1) in combination with tamped-earth walls, already described above; (2) as leveling courses in walls of rubble stone; (3) for vaults, arches, and pilasters; and (4) as the core material for architectonic decoration in stucco, including moldings and sculpture. In both the seventeenth and eighteenth centuries, especially in Antigua, it was a common practice to construct walls of a combination of brick and rubble or uncut stone laid in a thick lime mortar. The same combination of materials was used in vaults, especially to fill the haunches. This method is described by Fr. Lorenzo de San Nicolás in connection with the construction of pillars.[93] He advises that the piers should be laid alternating one course of stone with two courses of brick. This same method was already used in 1571 in the construction of the church of Atitlán and described as "... una Iglesia de cantería i de madera labrada i cubierta de teja i solada de ladrillo i cal ...."[94] The method is related to that of the construction of tamped-earth walls, the difference being that rubble is substituted for the earth. Walls of *pisón*, that is, tamped earth, require shutters or wooden forms to hold the material in place as it is thrown

[82] *Ibid.*, p. 135 and **2**: pp. 129 ff.; Gage, ch. xxi, p. 414; Juarros, **2**: p. 74.
[83] Scherzer, *Travels*, **2**: p. 243.
[84] *AGG*, A 1.2–4 (1556) 2196–127; also Pardo, *Prontuario*, p. 21.
[85] Remesal, **2**: p. 422; Pineda, *Descripción*, p. 346.
[86] Remesal, **1**: p. 392; Markman, *San Cristóbal*, p. 47.
[87] Pineda, *op. cit.*, p. 346. No town of this name appears on Juarros' list, **1**: pp. 75 ff., nor does any exist today.
[88] García Peláez, **3**: p. 27.
[89] Fuentes y Guzmán, **1**: pp. 345 ff.
[90] *Efem*, p. 14.
[91] Remesal, *loc. cit.*
[92] *AGG*, A 1.10.3 (1776) 4544–75.
[93] *Arte*, pt. 1, ch. xxxv.
[94] *Relación de los caciques*, p. 437.

in damp and tamped down, rocks frequently being added to give greater bulk. After the mud and aggregate have dried sufficiently, the shutters are removed and set above the completed section and the process repeated. In Antigua, and also in some of the buildings in San Cristóbal de las Casas, large boulders and uncut stones laid in massive quantities of sand and lime mortar alternate with brick leveling courses. The wooden shutters necessary to keep the wall plumb and stable during construction when mud is the mortar material are thus eliminated. Since the masonry material, rubble stone, brick, and mortar, could be stabilized without shutters, it was possible to lay the walls up without interruption. The leveling courses of brick were all that was necessary to keep the wall plumb and strong during construction. Examples of this type of walling, probably of Moorish origin, abound in southern Spain, indicating that this was an old practice there and not necessarily introduced into Central America only after Fr. Lorenzo de San Nicolás had described it in the seventeenth century. At any rate, walls of rubble and brick were economical, saving considerable material and labor. An unskilled *peon* could handle the rubble sections of the wall which would be then plumbed and leveled by the master bricklayer with one or two courses of brick.

Brick vaults did not begin to be built on a widespread scale until the eighteenth century. Prior to that time vaulted structures were rare even in Antigua. By that century, the structural or load-bearing character of the wall was of minor importance in vaulted buildings. The wall was conceived of as a screen between the massive piers supporting the arches and vaults above. The piers were most often treated as buttresses on the exterior and as pilasters on the interior of the building. The vaults were also constructed of brick, frequently only skin deep, for only the soffit of the vault was laid radially, if at all, while the haunches were filled with rubble of broken brick and boulders.

Brick was used most ingeniously, however, for architectural decoration and even for sculpture. The core of the shape to be executed was always formed of brick bonded to the main structure, be the detail a simple molding or a complicated rib pattern on the soffit of a vault or even statues for the niches on the façades. Time and time again Fr. Lorenzo de San Nicolás states his preference for brick work of this kind, especially in the design of façades with applied architectural orders.[95]

As attested to by the many monuments of pre-Columbian date still visible today, mason's lime as a mortar for laying stone was known to the indigenous inhabitants of Central America. The ancient methods employed in burning lime have not been radically changed even today. It was only with the opening up of a portland cement and lime factory in Guatemala City before the Second World War that any real competition to the traditional methods was registered. In the provinces the old methods still prevail. The stone to be broken down into lime is heaped up in an open field and covered with firewood which is then set ablaze and kept going until the stone is converted to lime. The lime from the Valle de las Vacas, where Guatemala City is now located, was by the seventeenth century perhaps the best in all Central America. Enormous quantities were produced without too much difficulty because of the abundance of good stone and sufficient firewood for the kilns.[96] The lime produced remained of excellent quality even during the postcolonial period as was noted by a mid-nineteenth-century traveler who marveled at the public fountains in the villages of western Guatemala which were built of brick and covered with a solid and beautiful coat of lime mortar which was actually waterproof.[97] As a matter of fact, even today in modern reinforced concrete buildings, roofs are commonly treated with a layer of broken brick and a coat of lime mortar troweled down to a hard, smooth finish and laid with the proper pitch as a means of waterproofing.

The preparation of lime as described by Fr. Lorenzo de San Nicolás, the common usage in seventeenth-century Spain, is exactly the same as that used today in Guatemala.[98] He advises that after the lime has come from the kiln it be slaked with water and left in some shady humid spot covered with some sand so that it may keep for a long time. In making mortar he advises that sand "de minas," that is, excavated sand, be employed and not river sand. The method used in Guatemala today also is precisely like that specified in the contract for the building of the Church of Santa Catalina in 1626.[99] The mixture described was to be used for filling foundation trenches and is referred to as the one then common in Antigua; namely, four parts sand, two parts earth, and one part lime. The earth referred to is no doubt the material known today as *terrón*. It comes in large lumps and is a soft porous material of volcanic origin not altogether free of organic matter and is still used for foundations, that is, for filling the trenches up to the level where the brickwork wall begins. The mixture for mortar suggested

[95] *Arte*, pt. I, ch. LVI, p. 148.

[96] Fuentes y Guzmán, I: p. 281.

[97] Irisarri, *Cristiano errante*, ch. v, in *ASGH* 10 (1933): pp. 250 ff.

[98] *Arte*, pt. I, ch. XXV, p. 53.

[99] *AGG*, A 1.20 (1626) 757, also *BAGG* 10 (1945): pp. 21 ff.

by Fray Lorenzo is almost the same proportion specified in the 1626 contract, five of sand to one of lime. The sand commonly employed in Guatemala today for mortar as well as stucco work is known as *arena amarilla* which is inorganic in composition and of volcanic origin. When used for finish work it is screened just as it seems to have been during colonial times. The final coat on all walls was invariably pure lime plaster troweled smooth and then painted.

Lime plaster for sculpture and the finishing coat over masonry walls was also employed in Central America in pre-Columbian times. This does not mean to imply that the Indian plasterer's craft carried over into the colonial period, or that the tradition of plaster decoration on church façades and interiors was but a continuation of a former indigenous practice. What is implied is that native workmen naturally took to this type of craftsmanship under the instruction of Spaniards who also had an ancient tradition in the use of this same humble material. The practice of applying a veneer of ornament in plaster, so common on the churches of Antigua and elsewhere, was very well known in *mudéjar* construction in the Iberian peninsula.[100] The custom of using plaster, *argamasa*, to work geometric ornaments began in Mexico in the seventeenth century and continued to the end of the colonial period, as it did in Central America too.[101] It is a humble and cheap material extremely easy to manipulate and yet has a great durability. It can be modeled, cast, incised, and painted, which indeed it was in the hands of the able workmen of Antigua who thus gave an air of grandeur and monumentality to the humble brick churches of their city.[102]

[100] Bevan, *Spanish Architecture*, pp. 105, 112; see also Contreras, *Historia del arte hispánico* **2**: pp. 444 ff., and Torres Balbás, *Arte almohade*, pp. 53 ff., 172 ff., and 368 ff.

[101] Toussaint, *Arte colonial*, p. 123.

[102] See Bankart, *Art of the Plasterer*, pp. 3 ff.

III METHODS OF CONSTRUCTION

IT was not until the seventeenth century when permanent materials became more readily available that the building construction of Antigua and the other cities of Central America began to take on formal architectural character. By this time too, the numbers of craftsmen increased, both native Indian and mestizo, as well as mulattoes and even some Negro slaves. For the most part, the craftsmen of the sixteenth century mentioned in the contemporary literature and documentary records were Spaniards, both professional and amateur builders. Many monks, perforce, had to engage in building operations since formally trained workmen from the Iberian peninsula were few in numbers. The question of how many craftsmen actually had come from Spain is directly related to the problem of the ethnic character of the total population inhabiting Central America during the time of the colony. Not only Spanish, but also the Indian and Negro elements as well as their social and economic status must be taken into account as pertinent and germane to an understanding of the development of the local architectural style.

Be the answers to the population problems what they may, there is, at any rate, a sharp demarcation between sixteenth- and seventeenth-century building activity with regard to the number of new constructions undertaken and the consequent appearance for the first time of buildings of a formal architectural style. The uppermost question which comes to mind, pertinent to explaining the changes which took place, is the possible role played by architects' and builders' handbooks such as that published by Fr. Lorenzo de San Nicolás. It is an open question whether books of this sort served as guides to the inexperienced *maestros de obras* who by that time were men born in Guatemala but who lacked the European training and experience which the few Spanish craftsmen of the previous century obviously had. Actually, architectural handbooks had not been too common even in sixteenth-century Spain, the earliest one being that of Diego de Sagredo published in 1526.[1] That of Fray Lorenzo, the first edition of which appeared in 1633, was perhaps one of the best of its kind. In it he evaluates a number of previous authors including Sagredo and the well-known Italians such as Palladio and Vignola.[2]

Without doubt Fray Lorenzo was not inventing new building methods altogether but was, rather, describing well-established practices in terms of his personal experience to serve as a guide to those not fully conversant with contemporary building methods. In other words, the building methods he describes were common knowledge and normally employed in Spain during his time. It would be logical to suppose that these building practices were carried to the New World and introduced in Guatemala for the first time during the seventeenth century when buildings of formal architectural character began to appear. The question of the introduction of Iberian building methods bears no positive relation to whether or not the handbook of Fray Lorenzo was known in Guatemala. It probably was, though we have no proof of this. The building methods he describes and which were common knowledge in Spain, were most certainly also known in Guatemala. Proof of this is demonstrable, especially with regard to his instructions for masonry construction and plastering, in the contemporary Antiguan monuments still extant as well as in documents dealing with building contracts where they are specified. In fact, such details as the laying out of buildings, the digging of foundation trenches, the selection and employment of walling materials, the mixing of mortar, and the combining of tamped-earth walls with masonry reinforcing piers can be found in most seventeenth- and eighteenth-century buildings in Antigua. His instructions on how to fill foundation trenches were so ingrained in Guatemala, that they persist to this day and are followed even in connection with the construction of reinforced concrete structures. Trenches are filled with *terrón* (large chunks of soft volcanic stone) or *ripio* (brick and stone from dismantled buildings) mixed

[1] Sagredo, Diego de, *Medidas*; see Sánchez Cantón, "Diego de Sagredo y sus medidas del Romano," *Arquitectura*, p. 120 as quoted by Calzada, *Historia de arquitectura*, p. 219.

[2] See Kubler, *Arquitectura*, p. 80.

with *mezcla*, a mortar of sand with a large proportion of lime. It would seem, then, that by and large the building methods employed in colonial Guatemala were really Spanish and not indigenous.[3]

1. *Foundations*

EVEN in remote Verapaz as early as the sixteenth century foundations were laid in a manner described by Fray Lorenzo and his predecessor Sagredo, a method which persisted in later times. Local Indians were employed to dig and fill enormous foundations. Pineda describes them as follows, "... y los cimyentos muy hondos, que tendrán dos estados, y de ancho diez pies, y estos los hinchen de piedra, cal y tierra que los yndios traen acuestas."[4] That foundations should be so deep and wide is not surprising, for Fray Lorenzo advises that the depth of foundation trenches for churches be one-third the width of the building and one-quarter for private houses.[5] In any case, the trenches must be excavated as deep as it is necessary to get to solid earth, that is, undisturbed ground. If solid earth cannot be reached feasibly, he suggests that one follow the counsel of Vitruvius and throw an arch over the weak portions of the trench and also drive piles of *alamo negro* (black willow), olive, oak, or *sauce* (a species of willow that grows on river banks), all of which are water resistant. These piles are to be charred and rammed in with sledge hammers until solid earth is reached. His instructions on how to fill trenches actually describe the exact method used in Antigua; namely, both large and small stones should be employed, the first course of which is laid dry on the ground and each succeeding course then covered with lime after first wetting the stone well.[6] If sand is to be added, lots of water should be used and the stone well rammed. In Antigua and in the surrounding region, in lieu of quarried stone a soft volcanic pumice-like material that can be dug with pick and shovel and which comes out in large lumps, known as *terrón*, was and is still used today. The specifications for the foundations of the church of Santa Catalina in Antigua dated 1626, and described in the specifications of the contract as a type well known and commonly employed at that time in that city, were wider than deep, "... seys quartas de ancho y cinco de profundidad," or as deep as necessary and to be filled with rocks and lime mortar.[7] By the eighteenth century the extreme width of foundations was abandoned, for it was apparently understood that depth down to solid ground was the more important factor, as for example in the construction of the Casa de Moneda adjacent to and integrated with the Capitanía in Antigua.[8] (FIG. 197.) And finally, at the end of the eighteenth century, after the destruction of Antigua, in an effort to make buildings somewhat more earthquake resistant, building codes dealing specifically with foundations and construction regulations in general were published.[9]

2. *Walls*

WALLS of tamped earth reinforced by masonry piers did not go out of practice altogether in the early seventeenth century, as for example the church of Santa Catalina, referred to above, and the church of Los Remedios. (FIG. 24.) The tamped-earth walls of the latter are still in evidence today. But by the end of this century, the use of tamped-earth for civic and religious buildings was abandoned in Antigua, though stabilized earth, either as *pisón* or adobe, continued in use for private houses and for the less important churches of the small villages in the outlying regions of Central America.

The new material now employed for walls was mainly uncut stone and boulders laid in a manner reminiscent of tamped-earth methods with the addition of brick leveling-courses at regular intervals up the height of the wall. This method was also known in Mexico, and widely used in the city of Puebla in the seventeenth century.[10] As vaults began to be employed more and more, especially in the eighteenth century after the earthquake of 1717, the walls began to lose their structural character serving as screens between the massive piers as seen, for example, in the churches of San Francisco, El Carmen, and San José el Viejo. (FIGS. 51, 116, 167.) The building was conceived of as a skeleton of piers and arches supporting the vaults which only rarely, as in the case of Santa Clara, were set so as to project toward the inside only leaving a more or less unbroken surface on the exterior. (FIGS. 141, 142.)

In general, seventeenth-century walls are frequently very thick, sometimes as much as 2.00 m. and even more in some cases. By the eighteenth century these ponderous dimensions are abandoned and a uniform thickness of 1.50 m. seems to have been most commonly employed. And even these relatively thin walls are more often than not further reduced by niches so that the screen between piers is often hardly more than 0.80 m. thick. The churches of El Car-

[3] See no. 6 of this chapter.

[4] Pineda, *Descripción*, p. 350.

[5] *Arte*, pt. I, ch. XXIV, p. 52.

[6] *Ibid.*, pt. I, ch. XXVI, p. 54.

[7] *AGG*, A 1.20 (1626) 757, also *BAGG* **10** (1945): pp. 221 ff.

[8] Díaz Durán, *Casa Moneda*, pp. 217 ff.

[9] *AGG*, A 1.10 (1776) 31361–4049, "Método regular en formación de los cimientos," "Sobre colocación de horcones," and "Instrucciones para construir."

[10] Atl, *Iglesias de México* **4**: p. 20.

men, Las Capuchinas, San José el Viejo, Santa Rosa, all in Antigua, are but a few examples of this practice.

The principal façades of churches are handled in a special manner in conformity with the requirements of the retable treatment. In general, retable-façade walls are as much as 3.00 m. thick at the top of the foundations since the applied orders and other architectonic decoration are bonded with the wall proper. The first story is normally of greater thickness than the second which is always set back slightly thus reducing the thickness of the wall in the upper parts. In addition, the niches and the spaces between the applied orders are also deeply inset so that, in a sense, the façade wall is honeycombed with voids thus reducing its massive character. The parts exposed to the front are wholly of brick so that the retable itself may be conceived of as a brick veneer bonded to the main bearing wall which is composed of rubble and brick leveling courses. The third story of the retable-façade is usually freestanding almost like a roof comb and is even thinner than the story below. (FIG. 134.) It too is entirely of brick.

The other three elevations of church buildings have little architectural interest. In those buildings roofed with wood and tile, the piers or buttresses, if employed at all, have no direct relation to the interior plan, nor do they necessarily run the total height of the wall. Frequently these piers were added at a later date to strengthen the walls, as was so often necessary in stabilized earth constructions.[11] In some cases courses of brick, *verdugos*, are introduced in tamped-earth walls, as for example late in the eighteenth century in San Jerónimo.[12] (FIG. 180.) In the case of vaulted structures, the massive piers which support the vaulting project only on the exterior wall facing the street or on whichever one no other structures abut. (FIG. 127.) These piers appear as buttresses on the exterior (diminishing in size above the springing of the interior vaulting) and as pilasters on the interior to which the screen walls between are bonded. In the case of the church of Santa Clara, as mentioned above, the exterior wall is completely devoid of buttresses though the interior is vaulted. The piers actually project only into the interior of the building. (FIGS. 141, 142.)

Another interesting practice in wall construction, no doubt inherited from stabilized earth building methods, was to fill the space at ground level between the exterior buttresses with a mass of masonry sometimes as much as 2.00 m. high. The fill usually consists of rough uncut stones without brick leveling-courses. The topmost surface is always pitched away from the wall. In walls of stabilized earth the stone masonry fill serves a double purpose, that of reinforcing the wall at the bottom and also that of keeping it dry.

Where other constructions abutted on church walls, as was common in monastic establishments, it was the customary practice to build what amounted to a double wall up to the height of the springing of the interior vaults. Above this level, buttresses of reduced size emerge corresponding in dimension to the upper parts of those on the opposite exposed exterior side of the building. The convent cloister usually shares this party wall with the church, and in those cases where it is two stories high, the buttresses begin above the roof of the upper corridor. In a sense then, the cloister wall is added to the church wall, as actually happened when the churches of San Francisco and La Merced were rebuilt. (FIGS. 51, 62.) The two types of brick are still visible in the party wall of the former.

In observing the actual remains of the churches of Antigua and then reading the instructions of Fr. Lorenzo de San Nicolás relative to the building of walls, one is amazed to learn that the methods he recommends were those actually employed from the late seventeenth through the eighteenth century.[13] He warns that if walls are too thick they will crack and if too thin will fall down and recommends that they be at least one-third the width of the church. If to do so is inconvenient, as frequently happens when the sides of the church are on a street, then the best solution is to build *estribos*, that is, piers or buttresses, since this added thickness is required to support the vaults of stone. He suggests that then the wall can be one-sixth and the pier the other sixth. In cases where the walls are of brick, since this is a lighter material the walls may, therefore, be thinner, that is, one-seventh. These piers should be set at intervals no more than half the width of the church. An examination of the plans of the majority of the churches of Antigua, especially those built after the earthquake of 1717, reveals a surprisingly close conformity to these recommendations. (FIGS. 116, 145, 167.)

3. *Plastering and Wall Finishing*

THE structural materials of walls were never left exposed but were plastered with a special surface finish in stucco or sometimes treated with a stone veneer, as for example, Santa Clara, Las Capuchinas, and the Ayuntamiento. (FIGS. 141, 146, 156.) In one instance, that of the 1626 church of Santa

[11] *Efem*, pp. 154, 164, 181.

[12] Angulo, *Planos*, **2**: p. 78, pls. 163, 164, no. 2, also Torres Lanzas, *Planos*, no. 192, referring to *AI*, 101–4–12(1).

[13] *Arte*, pt. 1, ch. xx, p. 46.

Catalina, the walls were finished with stucco in imitation of a stone veneer.[14] There is but one example, the fountain in Chiapa de Corzo in Chiapas, where the brickwork, recalling a *mudéjar* trait, forms the decorative surface treatment. Both the interior and exterior of buildings were always plastered, even the architectural details such as moldings, ribbing on the soffits of vaults, archivolts, and architectural orders applied to retable-façades were worked in this material. In addition, the overall surface decoration, known as *ataurique* and consisting of both abstract geometric and floral patterns, was worked in this material. The façades of the churches of La Merced, San Sebastián, La Candelaria, and Santa Cruz, among others, are tastefully decorated with intricate designs in this humble material lending the buildings a monumental and rich architectural quality despite the simplicity of plan and construction. (Figs. 67–69, 82, 123, 124, 135–137.) This particular type of craftsmanship in plaster, doubtlessly of *mudéjar* origin, is well known in the Iberian peninsula and was already flourishing in Mexico by the seventeenth century.[15] However, the Antiguan repertoire of motifs is quite different, nor may it be related with any certainty, except superficially, to the style brought to southern Spain by the Italian stucco workers in the seventeenth century.[16] The earliest extant remains of this type of decoration in Antigua are inside the Cathedral, dated 1680, and on the façade of San Sebastián from the last decade of the seventeenth century. (Figs. 47–50, 82.)

The methods of plastering walls were already standardized in the first quarter of the seventeenth century and are described in the contract of the 1626 church of Santa Catalina, alluded to above, which reads in part ". . . de ser encalada a dos manos, la primera de cal y arena y tierra saharrada, y la segunda de cal blanco, conforme el uso de encalado . . . ." The interior of the church was to receive two coats of plaster: the first, a rough or scratch coat consisting of lime and sand and screened earth; and the second, a finish coat of pure white lime. The earth referred to was probably either clay or a type of sand of volcanic origin known today in Guatemala as *arena amarilla*. The 1626 contract further specified that the exterior was also to receive two coats of plaster and be finished off with bands of white and dark lime interspersed to imitate masonry. Later constructions, those of eighteenth-century date in particular, seem to have been treated with three coats of plaster. Beginning first with a rough scratch coat, a second coat with finer-screened yellow sand was immediately applied before the first had completely set up. The finish coat of pure lime apparently was applied later in a separate operation well after the first two had completely dried and set up hard and were well bonded to the wall, a custom still followed in Guatemala today.

Just how the workman applied the plaster during colonial times can be learned from two separate sources, both of which corroborate each other; from Fr. Lorenzo de San Nicolás' handbook and from the observation of contemporary twentieth-century practices. Both are almost exactly the same.[17] Fray Lorenzo advises that for plastering the mix, *mezcla* (a term still employed today), should have a smaller proportion of sand than that used for mortar and, furthermore, that this mix should be stored for a longer time to cure than that allowed for mortar. He also instructs the workman to keep the wall plumb while plastering by first placing small pieces of wood, *maestras* (a term still employed in Guatemala today), about four feet apart all adjusted and set in the same plane with a plumb bob. The plaster is then laid on with a trowel and smoothed off with a straight edge, *regla* (a term still common in Guatemala), using the *maestras* as guides.

He also recommends three coats of plaster, the first rougher than the second and the second rougher than the last coat, and says that, according to Vitruvius, each coat should be no thicker than a leather hide. If an earth wall is to be plastered, it is well to roughen the surface first and to cover it with a *lechada*, that is, a soupy mixture of the same *mezcla*. Brick and stone walls, he continues, should be dusted off first and then washed well with water. As a matter of fact, in present-day Guatemala, especially during the dry season, the wall is usually drenched with water before the first rough coat is applied to insure slow drying and, therefore, firm bonding. Were this not done, the water content of the plaster would be absorbed by the stone or brick wall making a firm bond impossible. The final coat to be applied, according to Fray Lorenzo, could either be lime or plaster of Paris, and if lime is used then some ground alabaster stone should first be added and this coat should also be rather thin. In Guatemalan colonial buildings the finish coat is almost pure lime, or if anything at all is added, then it is a very finely screened river sand. In any case, the finish coat is very thin indeed and troweled smooth as advised by Fray Lorenzo. In some instances, for example, in the cloister of the Palacio Arzobispal and on the façade of the church of the Compañía de Jesús, the finish coat was incised with a design and paint applied while probably still damp. (Figs. 103, 83, 85.) The

[14] *AGG*, A 1.20 (1626) 757.

[15] Toussaint, *Arte mudéjar*, pp. 11, 42 ff. See ch. II, fn. 100, above.

[16] See Schubert, *El Barroco*, pp. 288 ff., and Kubler, *Arquitectura*, pp. 36 ff.

[17] *Arte*, pt. I, ch. XLVI, pp. 121 ff.

technique does not seem to be fresco, for the wall was troweled again after the design had been painted, or perhaps was burnished with a flat stone, a method advised by Fray Lorenzo to gain a shiny effect. In a Spanish manuscript from the end of the eighteenth century such a method is described in which a thin wash of stucco is applied to the surface and then rubbed with a cloth over and over again until the wash is rubbed away.[18]

4. *Wood and Tile Roofs*

THE earliest religious buildings, even of the most important monastic orders, were commonly roofed with thatch which was replaced in time by roofs of wood framing covered with terra-cotta tile. By the second half of the seventeenth century, thatch had completely gone out of use for churches and was being replaced by wood and tile. Even in such out-of-the-way places as San Agustín Acasaguastlán a portion of the tax contributions of the town was requested to replace the thatch roof with an *artesón*.[19] The colonial archives in Guatemala City are replete with petitions from numerous towns, dated toward the end of the seventeenth century, requesting help to convert from thatch to tile. In Antigua, however, all public buildings of seventeenth-century construction were invariably roofed with permanent materials from the very first, except possibly some of the smaller chapels and hermitages in the outlying, poorer neighborhoods. Wood and tile roofs had already been employed in the sixteenth century for the more important buildings, not only in Antigua but also in Chiapas where the church of the Dominican convent in San Cristóbal de las Casas was so constructed by mid-century.[20] In the more remote parts of Central America, as for example in Comayagua, Honduras, sometimes the monks themselves supervised the actual cutting and preparation of the timber.[21] This form of roofing never really went completely out of use in Central America even in the eighteenth century although vaults began to be employed almost exclusively in new construction in Antigua. Many buildings there were partly roofed with wood and tile as well as with vaults, the vaulted parts representing later additions. Santa Cruz, San Sebastián, and Los Remedios are but three examples of this combination of roofing. The rotted-wood roof over the nave of San Francisco, one of the most important monastic churches in all Central America, was replaced as late as 1675.[22]

[18] Díez, *Arte de hacer estuco*, p. 254.
[19] *AGG*, A 1.10.3 (1667) 31253–4046.
[20] Remesal, **2**: p. 428; Markman, *San Cristóbal*, pp. 60 ff.
[21] Vásquez, **3**: p. 133.
[22] *Ibid.*, **4**: p. 329.

Just what these early wood and tile roofs in Antigua looked like, especially how they were finished on the interior, is difficult to ascertain since all have disappeared. One would hardly expect to find the complicated carpentry treatment normally found in the *artesonados* of contemporary Spain. It is more likely that the rafters and ridge beams were left exposed, that collar beams and ties formed simple trusses resting on the side walls, and that the soffits lacked the paneling and interlacing moldings associated with Spanish examples. Just such a simple system of carpentry is specified in the 1626 contract for Santa Catalina.[23] It is described as "... una armadura de par y nudillo, llana con sus tirantes doblados y su lima por sobre el altar, haziendo la forma de la dicha lima y por testero el coro, de mojinete entablada y tejada conforme se use en esta ciudad," which is to say, a simple armature of rafters and collars and ties set on a plate running from the choir to the altar and the gables sheathed with planks. The contractor adds the parenthetical note that it is to be done in the manner commonly employed in the city. It is more likely than not that the framing methods in vogue at that time in Antigua were not too different from those described by Fray Lorenzo of which some examples of uncertain date still exist in Guatemala and El Salvador.[24]

In the town of Patzún, Department of Chimaltenango, Guatemala, the parish church is roofed with an *artesón* built in 1788 and not unlike the type specified for the 1626 church of Santa Catalina. The framing is carried out by means of rafters which rest on a plate running along the top of the side walls, each pair of rafters joined by horizontal collar beams. At regular intervals, pairs of tie beams connect the rafters to form a simple truss. But unlike Santa Catalina which had wooden gables, the choir end of the ridge abuts on the façade wall and the other end on a masonry wall behind the altar. Apparently the rafters were laid out on the ground first, for those just in front of the choir are numbered from one to six and bear a consecutive inscription in black paint.[25]

[23] *AGG*, A 1.20 (1626) 757, also *BAGG* **10** (1945): pp. 221 ff.
[24] *Arte*, pt. 1, ch. XLIV, pp. 110 ff. See also Angulo, *Hist. del arte* **3**: pp. 43 ff., figs. 35–38, and pp. 69 ff., figs. 63–67, for some *artesonados* outside Antigua in Guatemala and El Salvador.
[25] The painted inscription reads as follows:
Line 1: [totally illegible]
Line 2: "su carpintero que
Line 3: "fue Estevan Alvarez
Line 4: "Siendo costeado Don Diego Yos
Line 5: "Se acavo esta arteson el dia 2 [?] de Junio del año 1788
Line 6: "Siendo cura el Sacerdote Don Manuel de Corze . . . [?] Por la Gracia de Dios."
See also Angulo, *op. cit.* **3**: p. 44, who, unaware of the inscription, dates this *artesón* on stylistic grounds in the seventeenth century.

5. *Vaults and Domes*

ASIDE from such factors as availability of skilled workmen and sufficient capital, it is noteworthy that in the sixteenth century the regular clergy of the monastic orders were directly engaged in church building in the country at large, especially in the Indian pueblos. Furthermore, the construction of vaulted churches was specifically prohibited in an action taken in the first Provincial Chapter of the Franciscan order held in 1567. It is not known, however, whether the Dominicans were of a like mind or not. Except for the *capilla mayor* where the main altar was located, Franciscan churches were not to be roofed with vaults, and their convents were to be of the simplest sort.[26] Whether these instructions were idealistic lip service or not is not the point. But in view of the scant population of the country, the lack of mineral wealth, and the almost complete absence of skilled workmen who could handle the geometry of vaults, it is not surprising that so few vaulted structures were built in the sixteenth century. The absence of a monumental architecture is to be expected, for this was a time when even the securing of food for the new Spanish settlers was a serious problem. This was finally solved only when a number of Indian towns were established in the area around Antigua to insure the production of a stable food supply.[27] There is but one sixteenth-century example of a barrel vault, that in the Dominican church of Tecpatán, Chiapas, but which may have been built soon after 1564.[28] The domed fountain of Chiapa de Corzo, also of mid-sixteenth-century date and still standing today, was considered remarkable even by the end of that century, impressing Ponce when he passed through there about 1586.[29]

It was not until well into the seventeenth century that vaulted buildings in appreciable numbers began to be built, and at that only in Antigua, the most important as such being the church of Santo Domingo, the Cathedral, and San Pedro. In general, most buildings continued to be roofed with wood and tile and, if vaulted at all, were so only over the *capilla mayor*, as in the church of San Francisco. There are many references to this mixture of roofing in the contemporary literature, not surprising since church plans were conceived of as having three distinct parts, not all of which were necessarily built at the same time. It frequently happened that a tile-roofed building was enlarged by the addition of a vaulted *capilla mayor* set at the east end, as in the case of San Sebastián where a sixteenth-century wood and tile building was enlarged by means of a vaulted *capilla mayor* in the seventeenth century. The churches of the towns of Asunción Mita, Chiquimula, Esquipulas, and San Cristóbal Acasaguastlán are but a few of the many examples noted in the contemporary literature of the seventeenth century.[30] Even so important a conventual church as that of San Francisco in Antigua still had this same combination of roofing as late as 1673, at which time the wood part was renovated.[31] This wood roof had been constructed about one hundred years before over the nave, the walls of which were tamped earth. But the *capilla mayor*, built at the same time, was roofed with a brick vault of some sort, probably domical in shape.[32]

In view of the fact that so little brick and stone was employed in the sixteenth and early seventeenth centuries, it is not surprising that the few vaulted structures should be cause for comment on the part of contemporary travelers and writers. Gage, who was stationed in the town of Amatitlán for a number of years during the second quarter of the seventeenth century, was particularly proud of the fact that he superintended the building of a two story cloister about 1635 with stone arcades which he considered as well finished as any in Guatemala.[33] It is hardly likely, as a nineteenth-century author believed, that the arcade was also vaulted, for Gage was dealing with inexperienced Indian workmen and he himself most likely could not have designed such.[34]

There are some few examples of churches of seventeenth-century construction, however, which had barrel vaults over the nave; for example, the Dominican church in San Cristóbal de las Casas still standing today, Santa Catalina, San Agustín, San Pedro, and Santo Domingo in Antigua. The latter structure was roofed with a rather curious arrangement of four barrel vaults which caved in when the dome over the crossing crashed down on them in the earthquake of 1717.[35] Apparently the nave, the *capilla mayor*, and the transepts were roofed with barrel vaults which converged on the crossing and formed the supports for a dome on pendentives. These vaults were completed about 1648, to judge from the date of the contract for the construction of the *capilla mayor* described as being roofed with a barrel vault.[36]

[26] Vásquez, **1**: p. 181.

[27] *AGG*, A 1.17.1 (1740) 5002–210, see also *BAGG* **1** (1935): pp. 5 ff., and Juarros, **1**: p. 63.

[28] Ximénez, **2**: p. 31; see also Olvera, *Joyas de la arquitectura*, pp. 14 ff., and Berlin, *Convento de Tecpatán*, pp. 5 ff.

[29] Ponce, *Relación*, p. 481.

[30] Fuentes y Guzmán, **2**: pp. 195 f., 197, 198, 242, and 248.

[31] Vásquez, **4**: p. 329; *AGG*, A 1.20 (1673) 476–10, also *BAGG* **10** (1945): p. 131.

[32] Ponce, *Relación*, pp. 410 ff.

[33] Gage, ch. xx, pp. 406 ff.

[34] García Peláez, **1**: p. 231.

[35] Ximénez, **3**: p. 354.

[36] *AGG*, A 1.20 (1648) 694–668, also *BAGG* **10** (1945): pp. 102 ff.

The church as a whole was finally completed in 1666.[37] Even by the end of the century the use of stone vaulting was considered the wonder of wonders. The church of Tecpán was described as not having even one finger of wood in all its construction even though it was built before 1590.[38] The latter statement is obviously an exaggeration, for the present church of Tecpán, though very old indeed and renovated various times, is roofed with wood and tile. It was never totally vaulted, and in 1586 was still an adobe-and-thatch structure, though a new church of brick was already in construction.[39] It may, however, have had a vaulted *capilla mayor* in conformity with the regulations relative to church construction emitted by the Franciscan order under whose ecclesiastical jurisdiction the town of Tecpán fell. The principal church of the Franciscans and the seat of the religious province served by that order in Antigua was roofed exactly in the same fashion. Only a century or so later, in 1684, a two-story infirmary was built with arches and vaults, and about 1695 the two-story cloister with barrel-vaulted arcades.[40]

According to Vásquez, the first church of Santa Catalina in Antigua, built in 1613, was roofed with a barrel vault until it was destroyed in 1630.[41] Actually this statement is wrong, for the contract let to construct a new church building is dated 1626, and was to be roofed with wood and tile.[42] The church of San Pedro in Antigua, built about 1662, was roofed with a barrel vault supported on ten rib arches which covered the total length of the building including the nave and *capilla mayor.*[43] It collapsed in the earthquake of 1717 according to Arana who inspected the ruin of the city on behalf of the civil authorities since it had "ni estrivos ni bestiones," that is, it lacked buttresses.[44] The church of San Agustín also had a barrel vault, but only over the nave, probably built in the seventeenth century, and still partly in place today. It would seem that, in general, vaulted construction was rarely attempted before the second half of the seventeenth century and that barrel vaults were preferred to domical construction with the exception, perhaps, of the fountain of Chiapa de Corzo mentioned above and also possibly the Capilla de Loreto in the church of San Francisco in Antigua described as being "de media bóveda."[45] This chapel must have been constructed toward the end of the seventeenth century when the façade as a whole was being remodeled.[46]

Not until after the earthquake of 1717 did vaulted construction become general in Antigua and were all churches built from the very first with vaults, as were additions to older buildings also. The church plan tended to grow in both directions on the longitudinal axis when a new façade with vaulted choir bays was added at the front end and a domed crossing and vaulted *capilla mayor* at the other. This process occurred in Los Remedios and Santa Cruz, to cite two examples in Antigua. (FIGS. 24, 131.) In the eighteenth century barrel vaults were rarely if ever employed, except in one or two instances in cloister corridors.

By far the most common type of vaulting employed in eighteenth-century Antigua was the pendentive dome, a variation of which was used to roof churches laid out with oblong bays. The earliest use of domical roofs is found in the Cathedral, completed in 1680. The entire roof, of which very little still remains, consisted of individual domes over each bay of the plan. After 1717 domical construction was almost universally employed in every church built or repaired in Antigua. When employed over oblong bays the resultant shape of these domical vaults might best be described as half ellipsoids cut through by four vertical planes thus forming a pendentive "half-watermelon" dome. (FIG. 1.) This type of dome was probably derived from the normal pendentive dome employed over square bays and described by Fray Lorenzo as "*bóveda vaída*" which he advises using where the height of the building must be kept down.[47] The external appearance of the soffits of the ellipsoid type developed in Antigua belies their true construction. Very frequently by means of the addition of plaster ribs these ellipsoidal domes simulate the structural effect of the intersecting of two barrel vaults of different sizes. The diagonal arches in the planes of intersection are treated as massive projecting moldings simulating ribs but which are only skin deep. Quite often these false ribs are frankly treated nonstructurally and do not meet at the common keystone or point of intersection, but rather radiate from a circular medallion or other decorative element on the crown. Actually, only the transverse semicircular arches resting on the piers of the side walls are structural, while the smaller lateral arches are false. The archivolts of the latter are corbelled out from the walls between piers in each bay. The springing of the vault proper is actually imbedded in the wall itself. The treatment of arches with plaster moldings is a practice

[37] Molina, *Memorias*, p. 117.
[38] Fuentes y Guzmán, **1**: p. 386.
[39] Ponce, *Relación*, p. 449.
[40] Vásquez, **4**: pp. 330 and 390 ff.
[41] *Ibid.*, p. 369.
[42] *AGG*, A 1.20 (1626) 757, also *BAGG* **10** (1945): pp. 221 ff.
[43] Vásquez, **4**: p. 381; Fuentes y Guzmán, **3**: p. 356.
[44] See Ximénez, **3**: p. 355.
[45] Vásquez, **4**: pp. 219 ff.
[46] *AGG*, A 1.20 (1675) 477–32, see also *BAGG* **10** (1945): pp. 133 ff., and Vásquez, **4**: pp. 71 ff., where the text is transcribed.
[47] *Arte*, pt. 1, ch XLVII, pp. 124 ff.

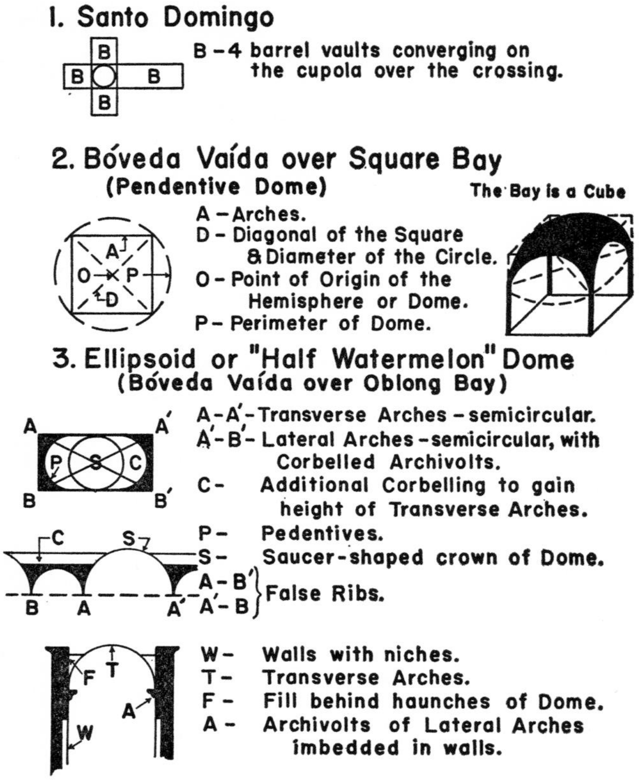

1. Santo Domingo

B - 4 barrel vaults converging on the cupola over the crossing.

2. Bóveda Vaída over Square Bay
(Pendentive Dome) The Bay is a Cube

A - Arches.
D - Diagonal of the Square & Diameter of the Circle.
O - Point of Origin of the Hemisphere or Dome.
P - Perimeter of Dome.

3. Ellipsoid or "Half Watermelon" Dome
(Bóveda Vaída over Oblong Bay)

A-A′ - Transverse Arches - semicircular.
A-B′ - Lateral Arches - semicircular, with Corbelled Archivolts.
C - Additional Corbelling to gain height of Transverse Arches.
P - Pedentives.
S - Saucer-shaped crown of Dome.
A-B′, A′-B False Ribs.

W - Walls with niches.
T - Transverse Arches.
F - Fill behind haunches of Dome.
A - Archivolts of Lateral Arches imbedded in walls.

FIG. 1. Typical Antiguan vaults and domes.

known in Spanish *mudéjar* workmanship as far back as the fifteenth century.[48] But here in Antigua not only the groin arches but the lateral ones too are only skin deep and have no organic relationship whatever to structural needs. The transverse arches, however, which limit each bay and cross the nave from wall to wall, it should be borne in mind, are indeed functional.

In effect the roof is worked as if it were to be covered with a barrel vault. The transverse arches are, therefore, laid out first as described in the construction of San Pedro.[49] A half ellipsoid pendentive dome is then constructed in each bay, that is, a *bóveda vaída*, which has the virtue of being low, a necessary precaution because of the earthquake danger. On examination of the ruins of San José, the Cathedral, Las Capuchinas, Santa Rosa, El Carmen, and others in Antigua, it is immediately apparent that the masons actually improvised a system of vaulting based on a modification of the true pendentive dome.

It must, furthermore, be borne in mind that even in the eighteenth century the master masons were not accustomed to working from detailed plans on paper drawn by professional architects or engineers. Plans in general, even when prepared by such well-known figures as Diez de Navarro, were of the most general sort and never included the complicated geometry for laying out the bays or constructing the vaults above. The workmen were not accustomed to seeing the building in complete detail on paper before translating it into brick and stone.[50] As a result, solutions to unforeseen problems had to be improvised as work progressed. This was especially true when the mason was dealing with a form of construction in which he had had little if any experience as happened, for example, in the construction of the vaulted cloister corridor of Las Capuchinas.[51] Such a workman builds directly, and when through some miscalculation parts of the plan seem impossible to carry out since he cannot erase walls as the draftsman would lines on paper, he must perforce improvise a solution or lose all the labor and material expended up to the moment of discovery of the error.

In order to force the pendentive dome to fit into an oblong bay, the mason first constructed more or less true spherical pendentives between the haunches of the arches. (FIG. 1.) But since the lateral and transverse semicircular arches were of unequal heights, he then raised a section, also more or less spherical and as a continuation of the pendentives, up above the lower crowns of the lateral arches. This additional masonry reached the height of the crowns of the wider and higher transverse arches. In this fashion an oval opening was formed which served as the outline of the lip of the dome proper. A true pendentive dome, *bóveda vaída*, is normally constructed over a square bay, for the pendentives then provide a circular opening. Here instead, an oval opening is formed, into which the mason then fitted a shallow flat saucerlike ellipsoidal crown which was usually the first part shaken loose in earthquakes. (FIGS. 147, 148.) By means of the plaster coating applied to the soffit which was also treated with false ribs, he gave the ellipsoidal dome the appearance of intersecting barrel vaults. In some cases, extra ribs were added to simulate six- or eight-part Gothic ribbed vaults, as for example in San Francisco. (FIG. 51-h.)

It is interesting to note that, in general, the courses of brickwork in Antiguan vaults and domes do not radiate from a single fixed point of origin. (FIG. 151.) The four corners between the haunches of the arches are filled independently of each other. Discrepancies in the curvature of the vault soffit are corrected by means of the finish coat of plaster. Normally only the transverse arches and the soffit of the vault are built of brick which is laid in a mortar almost

[48] Bevan, *Spanish Architecture*, p. 106.
[49] See fn. 43 above.
[50] See this chapter, no. 6, "Drawings and Architectural Handbooks," below.
[51] See Las Capuchinas, ch. XIV, no. 7, below.

as thick as the brick itself. The brick of the soffit is more like a veneer and probably served as a form laid directly on the centering and as an aid to hold the rest of the material of the vault in place during construction. The space behind the haunches is filled with a mixture of whole and broken brick, stones, and mortar, so that only the very crown of the vault is exposed on the roof above. In a sense, because of the great proportion of mortar and the random nonradiating brickwork, the vault must really be conceived of as a monolith. Vaults of this type do not function in a dynamic fashion exerting the lateral thrust normal in arcuated construction. That this type of workmanship was faulty is an understatement. Except in some few instances, none of the vaults are intact today and even those still *in situ* are so badly cracked that it is dangerous for one to remain for long inside those buildings where they can still be seen.

Pendentive domes on square bays, or nearly square bays, were employed in the corridors of Las Capuchinas. But here too improvisation was necessary. Because of a miscalculation in laying out the bays on the ground, the domes, perforce, resulted in being not geometrically true. The only examples of more or less perfect pendentive domes exist in the cloister corridors of the Escuela de Cristo from about 1740, the Ayuntamiento from about 1743 and the east end of the Capitanía from about 1769.[52] (FIGS. 162, 159, 206.) It is more likely than not that the builders here profited from the experience gained in the construction of Las Capuchinas.

Cupolas supported on spherical pendentives became quite common in the eighteenth century and were employed over the crossings of churches. Only two examples have survived, that of Escuela de Cristo and of La Merced. It was customary for the cupola to soar above the roof of the nave and of the *capilla mayor* with their lower saucer-shaped ellipsoid domes, as seen today in La Merced. (FIG. 71.) Another type of vault was that used over the chapel of the Seminario Tridentino. It is a long barrel vault pierced by windows at regular intervals in a manner reminiscent of, but structurally distinct from, the method known in Spain from the seventeenth century on.[53]

6. *Drawings and Architects' Handbooks*

CONSIDERING the number of buildings extant in Antigua and the rest of Central America, one is struck by the fact that only of a few are drawings or plans to be found among the countless documents in the archives of Guatemala City and Seville. Those few plans which have come to hand and the lesser number which have been published are almost all of eighteenth-century date.[54] This seeming scarcity of plans in the archives should not be taken to imply that plans were not drawn in the first place, for many may exist unknown to modern investigators and remain undiscovered in monastic and church archives. But in view of the minor character of so many of the buildings, especially parish churches, it may be assumed with some probability that the drawings were not considered important enough to have been preserved. Those drawings which do exist in the archives of Guatemala and Seville are mainly of buildings with which the crown was concerned and on which money from the royal treasury was expended, such as cathedrals and government buildings. But even in the case of the latter, the total number of buildings represented in the plans known to exist in the Archivo General del Gobierno in Guatemala City and the Archivo de Indias in Seville, is perhaps even less than twenty-five. That there are no plans for buildings of sixteenth-century date is not surprising in view of the fact that most buildings then were of nondescript character and constructed of ephemeral materials. It should also be borne in mind that there were no professional architects, except two perhaps, who had spent some time in Guatemala in the sixteenth century.[55] Another consideration which might explain the absence of sixteenth-century plans is a lack of skilled artisans in Central America. Furthermore, the religious orders had not yet prospered and were, in fact, represented by very few monks and nuns who were frequently obliged to build their churches and convents rapidly of whatever materials happened to be at hand in order to maintain their hold on the recently gathered and baptized indigenes.

It is not unreasonable to expect that plans were sent from Spain for the more important civic and religious buildings. In fact, town plans were drawn according to specifications required by royal authority. In some few instances, architects were actually sent from Spain to supervise the laying out of towns, as happened in the case of Antigua in the sixteenth century and of Guatemala City at the end of the eighteenth.[56] Quite often the instructions from Spain were

[52] See ch. XIV, nos. 8, 9, and ch. XV, no. 6, below for a complete description of these buildings.

[53] Kubler, *Arquitectura*, p. 39.

[54] See Angulo, *Planos*, for reproductions. Mencos, *Arquitectura*, who literally read and studied all the known documents dealing with Guatemala in the Archivo de Indias, did not come upon more than a scant few beyond those published by Torres Lanzas and Angulo.

[55] See Antonelli and Gárnica in ch. VI, "Brief Notices, etc.," below.

[56] Fuentes y Guzmán, I: pp. 130 ff.; González Mateos, *Marcos Ibáñez*, pp. 53 and 65 ff.

arbitrary and did not take into consideration local conditions. For example, the architect Francisco Sabatini who chose Marcos Ibáñez to direct the work of building the new capital in 1776, criticized Diez de Navarro's plan, the streets of which were oriented north-south and east-west so that the prevailing winds would supposedly blow down the streets. Quoting a principle from Vitruvius (!), he felt it would have been better to lay out the gridiron at an angle off the north-south orientation so that the prevailing winds would be broken. Actually the prevailing winds of Guatemala City are in no way similar to those of the Italian peninsula of Vitruvius' time, blowing northeast to southwest, so that Diez de Navarro's plan unwittingly took Vitruvius' principle into account. But even though he had spent many years in Guatemala, Diez de Navarro did not take it into consideration that the city is located about 15° north latitude and that even during the summer solstice the sun never swings far enough to the north to bring some warmth and light into buildings with a northern exposure. During the dry season from November to March when the mean temperature is the lowest it is actually more comfortable outdoors than indoors where the warming sun never reaches. The point is that the discussion between Ibáñez and Diez de Navarro was a purely academic one

At any rate, such discussions and royal orders are to be expected when dealing with buildings considered to be under royal prerogative, but hardly in connection with the smaller parish churches which were under the ecclesiastical jurisdiction of the monastic orders before the middle of the eighteenth century and of no direct concern of the crown. In view of the foregoing comments it is, therefore, not surprising that the majority of building plans which have come to light are of late eighteenth-century date and for the most part of structures over which the crown had a voice.

Yet, with the change in architectural character of building works by mid-seventeenth century from nondescript to permanent, one must expect that this new state of affairs was guided at least by some basic or rudimentary understanding of architectural matters which was doubtlessly common knowledge at the time. It is true, for example, that during the seventeenth century some architectural handbooks such as that of Fr. Lorenzo de San Nicolás were published in Spain. One cannot prove or disprove the presence of copies of such books in Guatemala. Yet time and time again, an examination of the construction technique of the actual buildings reveals many parallels described in the handbooks, particularly that of Fr. Lorenzo de San Nicolás which appeared in successive editions in Spain, some dating from the very end of the eighteenth century, and which must have enjoyed some measure of the same popularity in Central America. That such architectural handbooks could possibly have been published in Central America must not be categorically denied, though this is quite unlikely in view of the fact that it was not until 1662 that the first printing press in Central America was set up in Antigua. In none of the various catalogues and bibliographies of the books published in Guatemala during the time of the colony is such a handbook listed.[57]

Even if it were certain that such handbooks were either imported or published in Guatemala, it is highly problematical that they were actually used by the builders of the first permanent buildings in the seventeenth century. Most craftsmen were hardly literate judging by the fact that on the few contracts which have come to light in the archives they either signed with a mark or in a hand hardly that of someone accustomed to writing. Craftsmen, then as now, learned their trade by doing and not by reading books. Many illiterate craftsmen can and do draw plans. Even in contemporary Guatemala it more frequently than not happens that the architect makes a simple sketch plan with only the most essential details shown and then turns the actual job of construction over to the *maestro de obras* who, from practical experience, already knows how to lay out the most complicated structures of reinforced concrete, determining the load-bearing capacities of beams and columns, and instructing the workmen as to how many rods, stirrups, and tension bars to use in each member. In the seventeenth century and later in the colonial period, construction work was carried on in very much the same manner. It was a master stone mason, Francisco Hernández de Fuentes, who designed the church of Santa Catalina, yet he could not write his name and a witness signed the contract for him! And José de Porras who carried on the construction of the Cathedral after the Spanish architect left the job could not sign the necessary reports required by the authorities.[58] One

[57] See Bandelier, *Notes*; Beristain, *Biblioteca hispano-americana*; Brasseur de Bourbourg, *Bibliothèque*; Gavarrete, *Colección*; Medina, *Biblioteca*, also his *La Imprenta*; O'Ryan, *Bibliografía*; Rodríguez Beteta, *Nuestra bibliografía*; Sánchez, G. Daniel, *Catálogo*; and Villacorta, *Bibliografía*. No copies of architectural handbooks were found in the Biblioteca Nacional de Guatemala either, or in the Archivo General del Gobierno. For an article on the architectural handbooks which found their way to Hispano-America see Torre Revello, *Tratados de arquitectura*, who does not even mention Guatemala. It is possible, however, that those architectural handbooks which are known to have existed in Mexico may have also been brought to Guatemala. That this was so cannot be proved since none have been found there to date.

[58] *AGG*, A 1.20 (1626) 757, also *BAGG* **10** (1945): pp. 221 ff., for the contract. For Porras see biographical references in ch. VI, below, "Brief Notices, etc."

Martín de Ugalde, a master stonecutter, took the contract to finish the *capilla mayor* of Santo Domingo in 1648,[59] and in 1675 one Ramón de Autillo, also a master stonecutter, signed the contract for the *portada* of the church of San Francisco.[60] These three men were all master craftsmen who on their own responsibility undertook construction work over which they had absolute control and without the benefit of a professional architect, one of whom actually also drew the plan yet could not sign his name. There are other instances, especially in the eighteenth century, where building trade craftsmen and even retable builders undertook contracts for building construction.

7. *Earthquake Resistance*

MODERN critics have rightfully assumed that the architectural style of Antigua Guatemala might properly be called "earthquake baroque," a term which implies a conscious attempt by the builders to perfect a system of construction resistant to earthquakes.[61] But this is only partly true and specifically connected with the employment of squat massive supports in the cloisters of Las Capuchinas, Santa Clara, Santa Teresa, the University, the Seminario Tridentino, and the arcaded façades of the Ayuntamiento and Capitanía, all dating after the 1717 catastrophe. The exaggerated stubby proportions reflect an aesthetic change and were probably also resorted to because of the fear of earthquakes.[62] It is noteworthy, however, that the proportions of church façades do not change after this date though the elements of the decorative treatment do. Even a partial catalogue of the earthquakes which caused the capital city of Antigua to fall into ruins repeated times is demonstration enough that the builders were aware of the dangers inherent in their construction methods.[63] In fact, it was largely the fear of earthquakes which restrained the inhabitants of Antigua from constructing buildings of any great cost during the early part of the seventeenth century.[64]

Not before the eighteenth century, nevertheless, was it recognized that there was some physical connection between building methods and earthquake resistance. For example, Vásquez writing early in the eighteenth century noted that buildings fell no matter how deep their foundations, while many smaller structures with no foundations at all remained standing; that the strongest walls built of stone and lime mortar cracked and fell as if struck by rays of lightning, since they most strongly oppose the forces of the earth movements, whereas walls without foundations do not oppose them.[65] Ximénez, speaking of the earthquake of 1717, says that many of the buildings fell as a result not so much of the violent earth movements, as of bad building construction in general; that many walls had been perforated for doors and windows and then bricked up and that many of the roof timbers were old and rotted.[66] It is interesting that almost all churches in Antigua of post-1717 construction are roofed with vaults, a method not too common in the seventeenth century. It may not be too farfetched to assume, though there is no proof for this conclusion at all, that vaulted roofs were considered more earthquake-resistant than wood. As mentioned above, the extant cloister arcades are uniformly low and supported on very massive piers. All are post-1717 in date, except possibly the cloisters of La Merced, La Recolección, and San Francisco of which few vestiges remain. Some architectural historians in discussing the Guatemalan style have assumed that in the eighteenth century the fear of earthquakes brought about the custom of thickening church walls as a measure against destruction.[67] This is actually erroneous, for in that century the opposite occurred. The walls were reduced to screens between the piers supporting the vaulting above. The load-bearing quality of walls was minimized after 1717; they are almost all uniformly 1.50 m. thick, and further reduced to about 0.80 m. where niches are introduced. Earlier constructions follow no standard scheme but are generally thicker, in some cases actually 2.00 m.[68]

It was not before the eighteenth century that the civil authorities began to consider it a matter of public concern to prevent the continuing destruction of the city. After the earthquake of 1717 the *ayuntamiento* passed an ordinance, as a safety measure, limiting the height of private houses to one story.[69] This custom of building low, one-story houses prevailed during the eighteenth century,[70] and as a matter of fact is still quite common throughout the whole of Central America, even in the capital cities of the various republics.

[59] *AGG*, A 1.20 (1648) 694–668, also *BAGG* **10** (1945): pp. 102 ff.
[60] *AGG*, A 1.20 (1675) 477–32, also *BAGG* **10** (1945): pp. 133 ff.
[61] Kelemen, *Baroque*, pp. 122 ff.
[62] Angulo, *Hist. del arte* **3**: pp. 2 ff.
[63] See Vásquez, **1**: p. 183, for an account of sixteenth-century earthquakes, and **4**: p. 331, for seventeenth-century earthquakes; González Bustillo, *Razón particular*, for eighteenth-century earthquakes; *Efem*, pp. 23, 28, 29, 50, 61, 62, 107, 129, 145, 246, 251, for earthquakes in all three centuries; Fuentes y Guzmán, **2**: pp. 440 ff., for a seventeenth-century account of all earthquakes up to 1690, and Juarros, **1**: pp. 161 ff., for an account of all up to about 1800. See ch. I, fn. 78, above, for a bibliography of "earthquake literature."
[64] Vásquez, **1**: p. 264.

[65] *Ibid.*
[66] Ximénez, **3**: p. 353.
[67] See Angulo, *Historia del arte* **3**: p. 3; Kubler and Soria, *Art and Architecture*, p. 83.
[68] See no. 2, "Walls," in this chapter above, and "El Carmen," "Las Capuchinas," "San José," in ch. XIV, nos. 2, 7, and 10, below.
[69] *Efem*, p. 148.
[70] Cortés y Larraz, fol. 8.

Only in the twentieth century with the introduction of reinforced concrete earthquake-resistant construction was this custom changed, though private houses are for the most part still one story high. The need for safe building methods was recognized more and more as the eighteenth century advanced, and in 1751 the *maestro mayor de obras*, Juan de Dios Estrada, requested that the *ayuntamiento* actually control the construction of private houses much in the same way modern municipal building departments do.[71] But it was only in the newly transplanted capital of Guatemala City near the end of the eighteenth century that more rigorous instructions were announced as to how private houses were required to be built in order to avoid exposing them to the danger of earthquakes. At the same time regulations on materials and methods of construction were also made public.[72]

It would seem that not until the end of the colonial period were quasi-scientific or empirical earthquake-resistant methods of construction suggested, which began to depart from the traditional building methods. Bernardo Ramírez, *maestro mayor de obras* of the city, wrote a report in which the need for reorganizing the whole brickmason's craft was proposed. He took into account the problem of wages and cost of materials in view of the repairs constantly required as a result of earthquakes.[73] Also in 1799 an order was published indicating what special precautions in construction work were to be taken in order to avoid exposing the buildings to destruction in earthquakes.[74] Only as late as 1802 was an order given that no public building be begun until the plan had been submitted to and approved by the royal authorities.[75] But not until the invention of reinforced concrete was the only successful method of making buildings earthquake resistant devised, so that not before the twentieth century was there an actual complete break with the traditional colonial materials and methods of construction. This came about not as an internal historical development or outgrowth of local experience, but rather through outside extraneous influence.

8. *Rebuilding, Restoration, Use of Second-hand Materials, Afterlife of Colonial Building Methods*

THERE is scarcely a single colonial public or religious building in the whole of Central America that has not at some time or other been repaired or rebuilt. This is especially true in the city of Antigua which abounds with numerous instances where older structures were either torn down and the material reused, or else were repaired, altered, and enlarged. This practice of patching was forced on the builders of Antigua because of the constant destruction by earthquakes. The utilization of the ruined parts of buildings was frequently a matter of dire economic necessity. Even as far back as the early part of the seventeenth century, as an example of this permanently prevailing situation, the contract for the construction of the conventual church of Santa Catalina, referred to so frequently above, included the provision allowing the use of some of the materials from the former structure.[76] The construction and constant reconstruction, the altering and repairing of buildings was a constant process. As the seventeenth century advanced, it became more and more customary to replace structures of less permanent materials with others of brick and stone. Besides, as many of the early nondescript buildings were destroyed in earthquakes, it was deemed a good opportunity to rebuild in a more permanent fashion. There are literally hundreds of documents dating from the seventeenth through the eighteenth centuries in the Archivo General del Gobierno in Guatemala City, under the general classification A 1.10.3 in *legajos* 4046, 4047, 4048, and 4049, with petitions from towns all over Guatemala asking for help to rebuild, repair, or enlarge churches. The material condition of most provincial churches after the earthquake of 1751 was bad enough to prompt an investigation to verify the exact state they were in. Both the president of the *audiencia* and the various bishops were requested to report to the crown and indicate from what sources money to repair them could be expected.[77]

The use of second-hand materials in new constructions was taken as a matter of course probably because it was always felt that at best the new building would probably come down in the next earthquake. The sixteenth-century cathedral of Antigua was built with materials brought from the ruins of the first one in Ciudad Vieja,[78] and the 1680 cathedral with materials from the sixteenth-century one. In 1773 when the city of Antigua was laid waste again and it was decided to move the capital to the new site in the Valle de las Vacas, materials were carried off wholesale to the extent that many of the buildings which had been only partially ruined were now almost completely demolished. This was done with the explicit permission of the civil authori-

71 *AGG*, A 1.10.2 (1751) 18769–2447.

72 Larreinaga, *Prontuario*, pp. 107 ff.; and *AGG*, A 1.10 (1776) 1548–56.

73 *AGG*, A 1.16 (1773) 2830–148; also *Efem*, p. 245.

74 Larreinaga, *Prontuario*, pp. 107 ff.

75 *AGG*, A 1.1 (1802) 660–22.

76 *AGG*, A 1.20 (1626) 757, also *BAGG* **10** (1945): pp. 221 ff.

77 Larreinaga, *Prontuario*, p. 123.

78 Juarros, **2**: pp. 235 ff.

ties as a means of enforcing their will on those who were opposed to moving.[79]

When methods of construction are employed as a criterion for establishing the chronological sequence of the development of the architectural style of Antigua and the rest of Central America, as is frequently necessary in the absence of historical data, another important fact must be taken into account if erroneous conclusions are to be avoided; namely, the persistence and afterlife of colonial practices long after all political connections with Spain had been severed. The changes in the political sphere had no effect on the well-established and traditional building methods implanted in the course of colonial history, and were not reflected in the brickmason's or carpenter's craft in the nineteenth century. The long hallowed methods of Spanish origin continued regardless of political independence. Evidence of the persistence of colonial practices is found in the various buildings in Antigua which were restored and put to use again during the nineteenth century. The actual work of restoration was noted by a French traveler who marveled at what was being done to restore the ruined structures, largely through efforts of the *corregidor* of the department of Sacatepéquez, Juan María Palomo y Montúfar.[80] Among others, La Merced, the arch of Santa Catalina, and the church of San Pedro were rebuilt. On the other hand, the Santuario de Guadalupe (FIG. 37), the façade of which is quite colonial in style, was actually constructed in 1874 for the first time.

Even in the present century, restoration in the colonial manner has continued in Antigua and elsewhere in Guatemala. For example, the University was repaired in 1949 and 1950 and a new floor laid in the cloister corridors and courtyard. The square thin colonial-type bricks were laboriously cut by hand one at a time into a clover-leaf shape, the edges of each rubbed smooth with a piece of brick. These were then laid in lime mortar, each individually adjusted to make a close fit. Not only was the style taken from the past, but the very methods by which the floor was laid were no different from those of the colonial period. Another striking example of the employment of colonial methods of construction was noted in the restoration work carried out about the same time on the church and convent of San Juan del Obispo near Antigua. An addition to the church was built off the crossing in which the colonial style was carried out in the colonial materials and methods of construction. The walls were plastered in the traditional manner and architectural details worked in stucco, so that the addition is indistinguishable from the colonial parts of the building. The cloister was also restored and a new fountain built which looks no different now from the fountains of Antigua of supposedly colonial manufacture.

The afterlife of the colonial building practices, therefore, is not a phenomenon that can be simply classified as imitative or derivative and rooted in a nostalgia for the past. The truth is that the past is still quite current in parts of Guatemala. Appearances in Guatemalan architecture are more than skin deep, for the very methods by which colonial buildings were restored and even built anew in the nineteenth century were no different from those which had been employed in creating the prototypes they supposedly imitate.

[79] *AGG*, A 1.10.2 (1774) 1642–66; A 1.10 (1775) 1543–56; A 1.10.2 (1777) 1650–67, 4523–69, 4524–69; A 1.10.3 (1777) 4571–76; and A 1.10.1 (1779) 6458–307, being a ledger in which the total amount of material carried off from Antigua to Guatemala City is listed.

[80] Brasseur de Bourbourg, *Antigüedades*, pp. 164 ff.

IV LABOR AND THE BUILDING TRADES

ASIDE from other causes such as the deficient economic development, the scant Indian population not yet wholly converted to Christianity, and the small number of Spanish settlers, one may justifiably conclude that the general nondescript character of sixteenth- and early seventeenth-century architecture was in part also the result of a lack of skilled building-trade craftsmen. As building activity accelerated from the middle of the seventeenth century on, opportunities for learning a trade became increasingly possible for all elements of the population, not only for Spaniards born in Guatemala who, nevertheless, rarely engaged in manual labor, but also for Indians, mestizoes, mulattoes, and even Negro slaves. Any discussion of the question of the introduction and development of European crafts among that population in colonial times must perforce take into account that of the demography, with regard to both total numbers of inhabitants at any given time and the various races or castes into which they were classified by law and custom.

Building craftsmen comprised the smallest group represented among the first settlers of Ciudad Vieja. Except for carpenters, none of the building trades were sufficiently represented to form individual companies or groups according to craft-guild membership in order to take part as a body in the customary religious processions. The problem of protocol was finally settled in 1539 when, it is interesting to note, the carpenters were placed sixth in line, being preceded by the armorers, silversmiths, merchants, barbers, and tailors. They were ahead only of the blacksmiths and shoemakers.[1] Since no mention is made of bricklayers, stonecutters, or of *ensambladores*, one may suppose that, if any of these trades were represented among the scant one hundred and fifty Spanish settlers in Ciudad Vieja, they were not numerous enough to form separate groups but just trailed behind the others.[2] That the convenient supply of building lumber should have been one of the strongest arguments in favor of choosing the site for the first settlement seems natural in view of the preponderance of carpenters among the building-trade mechanics represented among the first settlers of Guatemala.

[1] Remesal, **1**: p. 52.

[2] *Ibid.*, p. 41.

The *encomienda* system whereby Indians were forcibly compelled to work for Spaniards had the effect of practically stifling any desire upon the part of the Spanish craftsmen from personally following their trades.[3] In fact, from the very first in 1529 the craftsmen demanded that they also be given *encomiendas*, refusing to work at their trades until Indians were assigned them. This was finally done as a matter of equity. After their demands had been met, few actually saw any purpose in practicing their crafts since they had, in effect, become landed proprietors. The authorities were thereupon constrained to take issue with them, and in 1534 passed an ordinance saying that unless the Spaniards worked at their trades the *encomiendas* of Indians would be taken away from them.[4] A similar measure also had to be taken in San Salvador because of like conditions.[5] Matters did not improve as the century progressed; in fact, the Spaniards became more and more reluctant to engage in manual labor, refusing not only to practice their trades, but even to engage in agriculture. A rather dim view of this situation was taken by the crown and a royal order was emitted in 1553 instructing the Spaniards to engage either in some trade or in agriculture or be sent back to Spain.[6] To combat the evident demoralization of the settlers, the *audiencia* suggested that land in the valley surrounding Antigua be distributed for the purpose of stimulating the Spaniards to work it and thus wipe out the evil of indolence among them.[7] The lack of craftsmen of any kind, Indian or Spaniard, in mid-sixteenth century, and of bricklayers in particular, was one of the reasons why the adobe or tamped-earth fences, required as a mark of having taken possession around the building lots distributed in November of 1541, were not built immediately. The time limit specified was six months ending in May, 1542, which had to be extended a

[3] García Peláez, **2**: p. 15. [4] Remesal, **1**: p. 248. [5] *Ibid.* **2**: p. 272.

[6] *AGG*, A 1.23 (1553) 1511–193; also Pardo, *Prontuario*, p. 58.

[7] *AGG*, A 1.23 (1559) 1512–270; also Pardo, *op. cit.*, p. 58.

full year. And even then the requirement was not fully complied with in some cases even as late as 1555.[8] Probably as a means of encouraging the practice of a craft, the *ayuntamiento* was ordered in 1548 to establish standard or official pay scales for artisans, implying that wages had been on a hit-or-miss fashion before then and not gauged to local conditions.[9]

Whether or not matters had taken a turn for the better by the end of the sixteenth century, that is, whether the older Spanish settlers finally followed their crafts or whether by then native Guatemalan born Spaniards had learned crafts, is difficult to ascertain. Yet it would seem that the Spaniard in general shunned manual labor as degrading, an occupation fit only for the inferior castes. López de Velasco who chronicles the period between 1571 and 1574 noted that there were five hundred Spanish heads of families, *vecinos*, in Antigua of whom seventy were *encomenderos* and the rest just ". . . pobladores y tratantes . . .," that is, residents and men engaged in commercial enterprises.[10] Ponce, who was there about a decade later, relates that the town was of medium size and inhabited by many noble but not very rich people.[11] Neither López de Velasco nor Ponce mention artisans of any kind, implying that Spaniards who engaged in crafts were so few in number as to have gone unnoticed. In the last quarter of the sixteenth century in most of the other regions of Central America, as for example in all of Honduras, in Nicaragua, particularly in the cities of Granada and León, also in San Miguel, El Salvador, and in Huehuetán, Chiapas, at least half and sometimes even the majority of the Spanish heads of families were *encomenderos.*[12] In Sonsonate, El Salvador, about the same time all one hundred and thirty *vecinos* were engaged in commercial pursuits.[13] Though the lack of craftsmen in the building trades was great in sixteenth-century Antigua, it was even more acute elsewhere, so that the few artisans resident there sometimes took work in the provinces, as happened in the construction of the church of Tequecistlán.[14] At the time Cerrato freed the Indian slaves in 1549,[15] there was a great fear among the inhabitants, voiced by none other than Bernal Díaz del Castillo, that if the Indians were given their liberty all building construction and all agriculture would cease since "los españoles oficiales son pocos y los demás no tienen oficio," that is, there are very few craftsmen among the Spaniards while the rest are without a trade. The Spaniards, he continues, are not artisans, nor is it fitting that they should be.[16]

One may safely conclude that, in general, Spanish craftsmen were never an important labor force in the sixteenth and early seventeenth centuries. Nevertheless, though few in number, they introduced the building trades to the other racial segments of the population by whom they were always outnumbered even from the very first. The Spaniard's point of view and the frame of reference from which he viewed manual labor are revealed by the custom of symbolical physical participation in pious works such as the erection of churches. Frequently Spaniards, as members of the highest caste, and even monks and other ecclesiastics would put on the guise of humble workmen to demonstrate a self-imposed humiliation and take part in the construction of religious buildings, but only symbolically and never realistically enough to produce something. This type of thinking is best represented in those passages of Vásquez' writings where he relates how the monks of the Tercer Orden in 1647 decided to do manual labor in the construction of El Calvario in Antigua and thus set an example for the noble gentlemen of the city who consequently not only donated money for workmen and laborers, but who also humbled themselves to the point of helping carry lumber, adobe, and tile, and even took up the spade and dug trenches for the foundations.[17] Even the spade(!), an agricultural implement used by the lowly Indian. Or else, by way of hyperbole, he relates how Fr. Gonzalo Méndez, who died in 1566, was the master craftsman and first *peon* in the work of building churches and founding towns.[18] The physical work alluded to is no more than a figure of speech.

It is a fact, however, that in many of the remote towns, the monks had no other alternative but to build their churches and convents with their own hands, thereby probably explaining the nondescript and temporary character of so many sixteenth-century religious buildings. Frequently the monks not only gathered the Indians into towns which they laid out with streets and fences, but also built the churches on which they personally labored in fact, having no alternative since the Indians did not yet know how to build in the Spanish manner.[19] Remesal, though lamenting the sixteenth-century Spaniard's abhorrence of manual labor,[20] points out as a mark almost of martyrdom the fact that Dominican monks frequently occupied their time mak-

[8] Remesal, 1: p. 45; *Efem*, pp. 7, 14.

[9] *AGG*, A 1.2.4 (1548) 2196–122; also *Libro de actas*, pp. 63, 107.

[10] *Geografía*, pp. 286 ff.

[11] *Relación*, pp. 410 ff.

[12] López de Velasco, *Geografía*, pp. 306, 317, 321 ff., 318, 297, 302, 303.

[13] Ponce, *Relación*, p. 403.

[14] Pineda, *Descripción*, p. 346. See ch. II, fn. 87, above.

[15] Ximénez, 1: p. 479.

[16] Zavala, *Contribución a la historia*, pp. 223 and 227.

[17] Vásquez, 4: p. 429.

[18] *Ibid.* 4: p. 20. See also Remesal, 2: pp. 244 ff. and 247, for a description of how Dominicans did manual labor of the meanest sort when necessary.

[19] Vásquez, 1: p. 107.

[20] Remesal, 1: p. 248.

ing adobes,[21] or quarrying stone,[22] and praises one in particular, Fr. Francisco de la Cruz who died in 1567 or 1568, who worked with his own hands on the church and convent at Copanaguastla, Chiapas.[23] He has the greatest admiration for Fr. Matías de Paz, who died in 1579. He built the Dominican convent in Antigua remaining on the job to take a hand in the actual building operations. It was a marvel, according to Remesal, to see this monk working all day long making adobes, laying them up, giving orders to the Indians as to what had to be done, and using mason's tools himself![24] Not to be outdone by the feats of the Dominicans, Vásquez gives a number of examples of Franciscans who engaged in manual labor in the construction of relgious buildings, as for example, the the construction of the church of San Francisco in Antigua rebuilt from the ground up in 1582. One monk was such an expert bricklayer that he actually constructed the vault over the chapel of Santa Ana. Unfortunately Vásquez does not give his name. Or another example, that of Fr. Blas de Morales who took his vows in 1607 and who worked as a *peon ca.* 1634 on the chapel of the Tercer Orden, thus inspiring the other monks to do likewise.[25]

If it was not the Spaniards who engaged in the building trades throughout the colonial period, one must conclude that it must have been members of other castes who did, the most numerous being the Indian and the mestizo. But the contribution of the Indian, that is, as an Indian apart from European culture, must not be assumed to have been as profound as on other Hispano-American styles. Nor should his contribution to the building arts be taken for granted to have been comparable to that in other areas of Latin America, particularly in Mexico and Peru. The role of the Indian and his culture in the architectural tradition of Guatemala and Central America has still to be delineated requiring as a prolegomenon a careful study of the population of Central America in general and the Indian element in particular. Though exact numbers are difficult to ascertain, it is reasonable to expect that the indigenous population at the time of the conquest was very scant indeed, and that the controversy ignited by Las Casas relative to the destruction of the Indians was almost pointless as regards Central America, though true enough for the West Indies, since there were few Indians in the first place to be killed off in highland Guatemala and the rest of tropical Central America.[26]

The question of the Indian contribution to the building arts of Guatemala and Central America cannot be answered simply through a search for supposedly aboriginal decorative motifs which, in fact, are absent in the existing colonial monuments.[27] To deal with the problem wholly in Indian terms is most misleading, especially since the Negro is as an important labor element, if not even more important, in the population in general and building trades in particular. On such a basis, one might just as misguidedly look for Negro art forms, especially in Antigua where Negroes began to be employed in building construction from as far back as the beginning of the seventeenth century.[28] And even in the sixteenth century in most of the towns of the Pacific coast of Costa Rica, for example, the Negroes had already been fused with the native population,[29] while in San Salvador by the middle of the seventeenth century the population of mixed Spanish and Negro ancestry, mulattoes, were numerous enough to form a company of militia of their own.[30] Mulattoes were so numerous in Antigua by the third quarter of the sixteenth century that a special order was emitted requiring them to pay taxes, *tributo*, as had already been required of the Indians. Also included in this order were Negro slaves as well as the newly arrived and as yet untrained Negroes, *bozales*.[31] By the middle of the eighteenth century, the Spanish, Indian, and Negro elements of the population were so intermixed that the former nomenclature for the castes resulting from mixtures of Spaniard with Indian or with Negro—*mestizo* or *mulato*—no longer had any significance, and a new term, *pardo*, was commonly employed.[32] The census of Antigua taken in 1740 disclosed that the mulattoes were more numerous than even Spaniards and mestizos. There were 2,240 Spaniards including women and children, 2,570 mulattoes and but 1,800 mestizos, presumably heads of families whose wives and children were not counted. Not included in these figures were the Negro and mulatto slaves or servants who were probably Indians. In

[21] *Ibid.* **2**: p. 421. [22] *Ibid.* p. 155. [23] *Ibid.* p. 478.

[24] *Ibid.* pp. 493 ff. [25] Vásquez, **1**: p. 245.

[26] For some works dealing with the population of preconquest America see the following: Bancroft, *The Native Races*; Baron Castro, *La Población*; Hanke, *Bartolomé de las Casas*; Jeffreys, *Pre-Columbian Negroes*; de las Casas, *Brevísima relación*; Kroeber, *Native American Populations*; Kubler, *Population Movements in Mexico*; McBryde, *Cultural and Historical Geography*; Poole, *The Spanish Conquest*; Rosenblatt, *El desarrollo de la población*; Sapper, *Die Zahl*; Seminario Centroamericano; Termer, *La habitación*; Viñas y Mey, *El estatuto del obrero*; Zavala, *Contribución a la historia*.

[27] See Neumeyer, *The Indian Contribution*, *passim*, who does exactly that. See ch. XI, fn. 1. For a theoretical study of the problem of the acculturation process in indigenous art forms, see Kubler, *Colonial Extinction*.

[28] García Peláez, **2**: p. 29.

[29] Cabrera, *Guanacaste*, p. 233.

[30] Díez de la Calle, *Memorial*, p. 120.

[31] *AGG*, A 1.23 (1574) 1512–447; Pardo, *Prontuario*, pp. 102, 103.

[32] See for example, *Efem*, p. 191, also *AGG*, A 1.10.3 (1705) 31. 280–4047, for some uses of the term.

the towns nearby the mulattoes also outnumbered the other groups as well.[33]

The Indian's contribution to building construction in the sixteenth century was largely that of a common day laborer. If he had any direct hand in actual building operations, it was only in the case of the temporary thatch-roofed wattle-and-daub structures which characterized the first settlements of Guatemala, Chiapas, and other regions of Central America. The Indians were thus employed in Ciudad Vieja from the very first days of the early settlement, and were even worked on Sundays in order to provide shelter for the Spanish inhabitants. In 1534 a regulation was passed imposing a fine of three *pesos* on those settlers who required their Indians to work on Sundays and holidays.[34] They were often treated more like beasts of burden carrying building materials on their backs, a great many of them being so employed after the destruction of Ciudad Vieja in 1541 and the construction of the new capital in Antigua. A city ordinance was finally passed limiting their loads to two *arrobas*, about fifty pounds.[35] Throughout the whole of colonial history the Indian was primarily a common laborer who was customarily obliged to leave his village and come to the towns for stipulated periods of time to serve his term of forced labor by way of paying his taxes.[36] It was common practice to employ Indians in the construction of churches and convents under the same system of *repartimiento* which was finally prohibited in 1558 when the crown advised that economic help be given for these ends and not the free labor of Indians.[37]

The question arises as to how many of the Indian slaves who were freed in 1559 were actually craftsmen in the first place, and how many were mechanics in the building trades. Bernal Díaz del Castillo's complaint was that the Indians did not yet know how to make constructions fit for a city, and that those who had been taught a trade would run off to the hills if freed, preferring not to work in the cities.[38] One of the four categories of slaves distinguished in 1551 included those owned by Spanish craftsmen and artisans from whom a clamor arose that, since they had taught these slaves crafts, they wished some recompense for the instruction and time expended.[39] In a sense, these Indian slaves had been apprentices who were now in debt to the master craftsmen from whom they had learned their trades. But exactly how many of the freed Indians had been slave apprentices or skilled workmen, and what proportion they represented of the total labor force in building construction is impossible to ascertain. After being freed all former slaves were required to pay taxes which most frequently consisted of labor in lieu of money. As a result, they were ordinarily employed as common day laborers on building works, quite frequently against their will, a custom which continued all through the colonial period even for the building of Guatemala City.[40] But by the end of the eighteenth century the custom was dying out, or at least the Indians from the villages and towns who had been called to work on the new capital city were now bold enough to register their opposition refusing to comply, saying they had to work in their fields.[41]

One may safely conclude that, in general, the Indians, except those few living in towns in the sixteenth century, were not craftsmen and that even fewer had become skilled mechanics in the building trades. It is true, however, that the monks sometimes taught crafts to the Indians. But this they did more out of necessity than charity since so few Spaniards deigned to do manual labor, though later writers considered it an altruistic act of benevolence on the part of the friars.[42] It is interesting that by the end of the sixteenth century some Indians had begun to build churches in their towns on their own initiative and to make contracts for the work directly with Spanish craftsmen. Such procedures were thereupon specifically prohibited unless license from the *audiencia* had been previously obtained. To have employed such craftsmen in church building works without permission from the *audiencia* authorities was considered an impertinence, not so much on the part of the Indians themselves it would seem, but of the ecclesiastical authorities administering these Indian towns. At any rate, in 1591 Indians were strictly prohibited from employing silversmiths, embroiderers, brickmasons, or carpenters to work on their churches.[43] Apart from any policy to control church building on the part of

[33] *AGG*, A 1.17.1 (1740) 5002–210, also *BAGG* **1** (1935/36): pp. 5 ff.

[34] Remesal, **1**: p. 46 ff.

[35] *Ibid.* **2**: pp. 46, 50; *Efem*, p. 7.

[36] Larrazabal, *Apuntamientos*, p. 101.

[37] *AGG*, A 1.23 (1558) 1511–238; also Pardo, *Prontuario*, pp. 20, 101.

[38] Zavala, *op. cit.*, p. 227.

[39] *Ibid.*, p. 223.

[40] *Ibid.*, pp. 250, 251, 254.

[41] See the following documents: *AGG*, A 1.10 (1777) 4473–62, Sololá; 4474–62, Quetzaltenango; 4475–62, Patzún; 4476–63, San Pedro Huertas; 4477–63, San Andrés Izapa; 4479–63, Izapa; 4480–63, Santiago Mataesquintla; 4481–63, San Pedro Pinula; 4482–63, Santa Apolonia, Chimaltenango; 4483–63, San Raymundo de las Casillas, Sacatepéquez; 4485–63, Petapa; 4486–63, Chimaltenango; 4487–63, Chimaltenango, being an order to capture Indians who refuse to work; (1778) 4488–63, Sololá; 4489–63, Sololá; 4490–63, Cubulco; (1779) 4501–63, permission given to use arms to enforce *repartimientos*.

[42] Vásquez, **1**: p. 107.

[43] *AGG*, A 1.16 (1591) 1751—fol. 26 v.

the *audiencia*, it would seem that there were so few craftsmen of any kind, either Indian or Spaniard, at the end of the sixteenth century that their services had to be strictly rationed. This policy also served another purpose too, that of keeping a stricter watch over affairs in the far-flung small Indian towns, the greater number of which were under the ecclesiastical jurisdiction of the monastic orders.

Pineda who visited Antigua about the same time mentions the existence of a neighborhood inhabited by Indians who had formerly been slaves (they must have been very old by that time) who were all craftsmen of all sorts.[44] Though he does not mention any trade specifically, some were doubtlessly also engaged in the building trades. In the more remote regions of Central America, particularly in parts of Honduras and Nicaragua, the majority of the Indians had not yet been converted to Christianity, let alone introduced to European culture, even as late as the seventeenth century. They still lived by hunting and fishing and could hardly be expected to have already learned trades.[45] The facts do not bear out García Peláez who believed that in the sixteenth century the Indians were already skilled craftsmen of all kinds. This may very well have been true for the city of Antigua only, but not generally so for the rest of Central America.[46]

The question of Indians as craftsmen, or of Negroes and mulattoes as craftsmen, is of course related to the problem of the development of the caste system in colonial Central America and may only be partially explained in terms of the acculturation process by means of which individuals learn an imported trade foreign to their native tradition. Granted that the Spaniards were the teachers and the Indians and Negroes the pupils, this teacher-pupil relationship could only have been true during the sixteenth century when Spanish craftsmen arrived in Ciudad Vieja along with the first settlers. The fact is that the Indian, the Negro, the mestizo, or the mulatto who engaged in one craft or another, did so, not as a member of his race, but as a craftsman. If he contributed anything to the building arts, he did so as a craftsman and not as an Indian, mestizo, or mulatto. The crafts they had learned were the traditional Spanish crafts, not Indian or Negro. In fact, by the seventeenth century, the acculturation process had been sufficiently completed for the Indian craftsman to think no longer as an Indian, but as a Guatemalan. The same is true for the Negro. The crafts which the different mechanics pursued were neither Negro nor Indian, despite the fact that the workmen themselves are referred to as belonging to one or another racial caste. For example, the neighborhood of La Candelaria in Antigua was inhabited by carpenters, masons, coppersmiths, tinsmiths, and shoemakers who still spoke the Pipil language at the end of the seventeenth century, while those in the neighborhood of Santa Ana were all butchers who spoke Cakchiquel.[47] On the other hand, the neighborhoods of Espíritu Santo and San Jerónimo were inhabited mainly by saddlers and cobblers who are referred to as "*indios ladinos*," that is, Europeanized Indians.[48] Early in the seventeenth century when the intermingling of races was perhaps not yet so complicated, Gage[49] noted that some Indians had the same trades as Spaniards, being smiths, tailors, carpenters, masons, and shoemakers, a list noted later at the end of the century by Fuentes y Guzmán.[50] One Blas Marín, described as an "*indio ladino*," took the contract to complete the church of Santa Cruz in Antigua in 1662.[51] Apparently, by the middle of the seventeenth century, members of the inferior castes were admitted to the craft guilds.[52] This munificence was really a necessity in Antigua where so few people engaged in skilled crafts of any kind; for, as mentioned in another connection above, when new stone quarries were discovered on the hills of San Cristóbal and San Felipe in 1686, the lack of skilled quarrymen prevented their being exploited.[53]

In the smaller towns, even in those exclusively Indian, there were always a few craftsmen among the inhabitants who worked in the building trades, as for example in San Mateo Ixtatán where native carpenters built the *artesón* for the local church,[54] or in Mixco where Gage had some trouble building a vault over one chapel. He employed only Indians whose craftsmanship won the admiration even of Spanish artisans.[55] Stonecutters from the towns of Santa María de Jesús, Amatitlán, Itzapa, San Cristóbal el Bajo, and Jocotenango were impressed for the work on the Franciscan church in Antigua in 1693.[56] It is questionable whether all the above workmen were really Indians, for the towns of Itzapa and Amatitlán had large Negro populations, whereas

44 *Descripción*, p. 330.

45 Vásquez, **3**: p. ix, and **4**: pp. 109, 79, 83, 164, 32.

46 García Peláez, **1**: p. 229.

47 Fuentes y Guzmán, **1**: pp. 401 and 403.

48 *Ibid.*, p. 390.

49 Gage, ch. XIX, p. 329.

50 See fnn. 47, 48, above.

51 *Efem*, p. 71.

52 García Peláez, **2**: p. 29. See Wethey, *Colonial Architecture*, p. 122, and Harth-Terré, *Artífices*, pp. 223–231, for similar developments in Peru and Bolivia.

53 Fuentes y Guzmán, **1**: p. 135.

54 Fuentes y Guzmán, **3**: pp. 86 ff.

55 Gage, ch. XIX, p. 329.

56 Vásquez, **4**: p. 390, fn. 2; also *AGG*, A 1.10.3 (1693) 31.272–4046.

only Santa María de Jesús is completely Indian to this very day. The inhabitants of the town of Jocotenango, on the outskirts of Antigua, were primarily brickmasons,[57] whose descendants continued in this craft even in the nineteenth century, establishing themselves in a village of the same name on the outskirts of Guatemala City after the removal of the capital from Antigua.[58] Jocotenango was founded by Pedro de Alvarado who settled his Tlascalan and Mexican auxiliaries there after the conquest. In general, the Indians of the small villages and the countryside were not craftsmen at all but primarily farmers, as they had been since preconquest times and still are to this very day.[59] When called on, the Indian would do whatever work was required of him, but he was not prone to specialization except in rare instances, as for example, in seventeenth-century Santiago Atitlán where, in addition to having a specialized occupation, each was also a farmer.[60]

In the building trade, since it was an urban occupation, one might expect to find greater numbers of Negro and mulatto workers than Indian. The Negro slave was employed more frequently in the cities than in agriculture, though it should be borne in mind that the majority of sugar mills and indigo plantations were operated with Negro slave labor. If García Peláez is correct, of the eighty-or-so master craftsmen in Antigua counted in the census of 1604, only thirty were Spaniards and the remaining fifty Negroes and mulattoes, among whom were one carpenter, one stonemason, and one brickmason, the latter a Negro.[61] The surprising fact is that so few individuals were reported as being engaged in the building trades, even taking into account the very scant building activity noted in the early seventeenth century. The castes into which García Peláez distributes the eighty artisans of 1604 are open to question since no such classification appears in the document to which he refers. He deduces caste membership from the way the names of the individuals are written! Such reasoning is unacceptable. Furthermore, it was only by the middle of the seventeenth century that members of the non-Spanish castes could enter craft guilds. One of the bricklayers who worked under José de Porras in 1669 on the Cathedral in Antigua was a mulatto, another a Spaniard, while the remaining twelve were Indians.[62] But there can be no doubt that by the eighteenth century mulattoes represented a force to be accounted for in the building trades as well as other crafts. They were numerous enough to have been included in the order prohibiting the payment of wages in advance.[63] Among the seven hundred slaves, probably Negroes and mulattoes, employed on the plantation owned by the Dominican order near Salamá at the end of the third quarter of the eighteenth century there were numerous craftsmen including carpenters and stonemasons.[64] The silversmiths' guild which had excluded Indians, mestizos, and mulattoes in 1745, emitted a new order in 1776 admitting them since the most skillful workmen in this trade belonged to these castes.[65] That the majority of workmen in the building trades were mulattoes was noted in 1775 by Martín de Mayorga, the president of the *audiencia*, who expressed the opinion that though they were skilled and practical mechanics they still had no formal training in architecture, and that an architect, Diego de Ochoa, should be put in charge of the construction work on the new capital.[66] Bernardo Ramírez, *maestro mayor de obras*, one of the most prominent architects and hydraulic engineers in Guatemala who had undertaken work of great responsibility in establishing the new capital at Guatemala City, was actually of mixed race and was denied the rights and privileges normally accorded the caste of Spaniards in which he wished to be classified, despite the fact that his service to the city merited great rewards.[67]

Of the approximately one million inhabitants in the whole of Central America at the end of the colonial period in 1810, approximately 313,000, or one third, were *pardos*, including some Negroes. Only 40,000 were considered white, that is Spaniards of pure European extraction, and about 650,000 were Indians. The Indian was still primarily an agriculturalist, the Spaniard a merchant or planter owning vast extensions of land, while the *pardos* were mainly artisans such as painters, sculptors, silversmiths, carpenters, weavers, tailors, shoemakers, blacksmiths, and others.[68]

The institution of the craft guild, *gremio*, which had been introduced in Guatemala by the first settlers went through a gradual transformation in the course of time so that by the eighteenth century it no longer had the meaning or importance for the various crafts as it had had at first in Guatemala or contemporary Spain. Aside from other well-known so-

57 Juarros, **2**: pp. 221 ff.

58 Irisarri, *Cristiano errante* **9**: p. 251.

59 García Peláez, **3**: p. 145.

60 Vásquez, **4**: pp. 46 ff., the editor quoting a document in the Archivo Arzobispal de Guatemala, A 4.5–2.

61 García Peláez, **2**: p. 29; also Villacorta, *Historia*, pp. 175 ff. See also *AGG*, A 1.2–6 (1604) 11810–1804, apparently the document on which García Peláez based his conclusions.

62 Mencos, *Arquitectos*, p. 171.

63 *Efem*, p. 236.

64 Cortés y Larraz, fol. 129 v.

65 Carrera Stampa, *Ordenanzas*, pp. 97 ff.

66 García Peláez, **3**: p. 93.

67 Mencos, *op. cit.*, p. 187.

68 Larrazabal, *op. cit.*, p. 87 ff.

cial and economic pressures, the changes in the significance of the craft guilds, originally medieval institutions, and their eventual disappearance came about because of the special conditions regnant in Central America where from the very first there were extremely few Spanish craftsmen, that is, guild members trained in the Iberian peninsula according to the traditional precepts and methods. Besides, very few individuals of Spanish extraction born in the New World learned and carried on as members of the craft guilds imported from Spain. The guild institution in Central America died out and lost its meaning, but for special reasons quite apart from those which caused its demise in Europe. The first guild members shunned manual labor and actually employed Indian slaves, some of whom they taught the craft. When the slaves were freed the Spanish master craftsmen turned to members of all the non-Spanish castes, including mestizos and mulattoes, to do the actual manual labor. This is especially true of the building trades where large numbers of unskilled workmen are normally required during construction. By the middle of the seventeenth century there are numerous instances where members of the non-Spanish castes take contracts to construct various buildings, presumably as independent masters and members of craft guilds to which they had of necessity begun to be admitted. Before that time Indian, mestizo, Negro, and mulatto craftsmen could work only under the responsibility of a Spanish master craftsman and official guild member. After 1650 members of these inferior castes were admitted to the guilds which formerly had been reserved for Spaniards only.[69]

It is quite natural to expect a greater preponderance of members of these non-Spanish castes to be attracted to the building trades because far more unskilled workers are needed in the course of construction than in other occupations. A single master brickmason could very well direct the manual labor of a multitude of day laborers and semiskilled bricklayers, as happened when José de Porras worked on the Cathedral in 1669 laying out the work and supervising the raising of the building. In time the unskilled and semiskilled workmen would learn the craft and apply for membership in the guild, wishing to be independent masters with the right to undertake contracts. The other crafts, those of silversmith, weaver, embroiderer, metal founder, and the like, which entailed the maintenance of a shop where the goods produced could be sold, were probably not opened to the inferior castes until much later. In fact, the silversmiths' guild was opened to the non-Spanish classes only in the last quarter of the eighteenth century, and only then on the basis of the fact that they were the best craftsmen available in the first place.[70] In such crafts as that of bell and cannon founders, *fundidores*, whose products were not needed with as great frequency as of the other trades, non-Spanish craftsmen were probably not admitted at all during the colonial period. For example, a Franciscan monk of the Tercer Orden, Fr. Tomás de Morales, probably a Spaniard, requested in 1747 that no one else in Antigua be allowed to follow his craft of bell and cannon founder.[71] Another individual of the same trade, presumably a Spaniard too, Juan Tiburcio de Paz, *fundidor*, requested payment from the *ayuntamiento* in 1739 for a statue he had cast.[72]

Another reason why members of the non-Spanish castes could be admitted to the building trade guilds without too much objection from Spaniards is that in almost all cases contracts for building works were for the supervision of construction only, *maestranza*, and rarely included both labor and materials. The most common procedure was for the master craftsman to undertake the contract for his supervision only, and possibly for drawing the plan too, the actual day laborers and materials being supplied by the other legal entity party to the contract, as for example, in the case of building the *portada* of the church of San Francisco in 1673 and 1675, the *altar mayor* of Santo Domingo in 1636, the completion of the *capilla mayor* in the same church in 1648, and the construction of the church of Santa Catalina in 1626.[73] In other words, the master craftsmen required no personal outlay of money to complete the job, nor did he require any interim capital for materials or salaries as would be necessary in crafts such as the weaver's, baker's, silversmith's, and others who maintained shops where the product of their labor was also sold as merchandise.

The question of competence of the master craftsman and the workmen he supervised can best be answered by an examination of the materials and the methods employed in the actual buildings still extant. Aside from political, economic, social, and religious factors which may have a direct influence on general architectural character, the quality of workmanship is never of the highly sophisticated or skilled type one expects to find in the churches of the principal cities of Spain. Regardless of the low or high degree of competence of the Guatemalan builders, his works were, nevertheless, thrown to the ground and shattered over and ove

[69] García Peláez, **2**: p. 29. See also fn. 52 above.

[70] Carrera Stampa, *op. cit.*, pp. 97 ff.

[71] *AGG*, A 1.16.4 (1747) 2811–148.

[72] *Efem*, p. 186.

[73] *AGG*, A 1.20 (1673) 476–10; (1675) 477–32; (1636) 690–69 (1648) 694–668; and (1626) 757.

again by earthquakes. It is a matter of great wonder that even the scant ruins one sees still standing today have not been completely obliterated. There were complaints registered for the first time during the eighteenth century about the skill of workmen, as for example, in the case of the construction of La Merced where some errors had been committed and an investigation ordered in 1749 to see whether the brickmasons who were working on the building really had titles of master craftsmen.[74] Apparently many building mechanics who had not received official license continued to be employed, for in 1752 the *maestro mayor de obras*, Juan de Dios Estrada, informed the *ayuntamiento* that those who had not been examined should be prohibited from working in building construction.[75] The year of the great earthquake, in 1773, Bernardo Ramírez, *maestro mayor de obras* of the city of Antigua, requested the *ayuntamiento* to set up the rules for the organization of the brickmasons' guild, just a few days before the first earth movements destroyed the city.[76] But by then the older notion of the craft guild as a private organization administered by its own members who supervised the instruction, the quality of work, and decided on its own membership no longer had any meaning. What Ramírez really proposed was to set up the regulations or a code concerning the actual building operations and the supervision of the workmen who hired themselves out as individuals. He was not concerned with the medieval craft guild member whose place was to be taken by the building contractor who is primarily an entrepreneur and not a craftsman. The regulations Ramírez proposed were to originate from the civil authorities and not from within the guild proper. This point of view permeates the code which Ramírez eventually wrote five years later in the new capital presenting a document with the rules covering the masonry trade to the president of the *audiencia* in 1782.[77] He lists all the workmen by name, the total number being 194 divided as follows: master bricklayers, *ladinos*, in Guatemala City, 44; *medios cucharas*, that is journeymen, in Guatemala City and from the towns of Jocotenango and Santa Ana, 101; stonemasons, *canteros*, from San Cristóbal, 28; and stonemasons from Jocotenango, Santa Ana, and San Gaspar, 8; *tapieros*, that is, those who work in adobe, 13 from Parramos, Chimaltenango, Jocotenango, and Ysapam, probably Izapa. And at last, under the impetus of the Sociedad Económica in the new capital and in the new scientific spirit of the times, all the various crafts were reorganized and their regulations published in 1798.[78]

[74] *Efem*, p. 202.

[75] *AGG*, A 1.69.3 (1752) 48142–5556.

[76] *Ibid.* A 1.16 (1773) 2830–148; also *Efem*, p. 245.

[77] Mencos, *op. cit.*, pp. 199–209, transcribing *AGI*, Guatemala, 466.

[78] *Regiamento de artesanos.* For a general survey of the reorganization of the craft guilds in the eighteenth century see Samoyoa Guevara, *Reorganización gremial.* For a history of the craft guilds in Guatemala, see also his *Los gremios*, especially ch. XIX, "La Corporación gremial y las castas coloniales," pp. 177–182.

V ARCHITECTS

THE history of the extant architectural monuments of Antigua is, for the most part, well documented in both the contemporary literature and the archives, yet in only a few rare instances are the names of architects associated with specific buildings mentioned. Rarer still are the cases where contemporary plans or drawings associated with the extant buildings have been found in the archives. In fact, from before the eighteenth century no plans at all have come to light. That this should be the state of affairs is not surprising when one realizes that only two formally trained architects are known to have worked in Central America in the sixteenth century; namely, Juan Bautista Antonelli who took part in choosing the site for the city of Antigua and who may have drawn the town plan soon after 1541, and Rodrigo Martínez de Gárnica who began the construction of the Cathedral, the plan of which he probably drew.[1] The apparent lack of plans is also explained by the fact that even with regard to master craftsmen and *maestros de obras* who might have had sufficient theoretical knowledge to draw simple plans and practical experience to direct construction work, only three are known from the sixteenth century. Francisco de Porras, or Porres, whose name appears in connection with the construction of the first church in all Central America, at Ciudad Vieja sometime between 1529 and 1537, was either a carpenter or engaged in some other building trade, but was hardly a formally trained architect. A little later in the sixteenth century another Spaniard, Francisco Tirado, a master bricklayer and stonemason, actually drew the plan and supervised the construction of the Franciscan convent in Ciudad Vieja. A contemporary of Tirado, the third and last known practical building craftsman from the sixteenth century, one Diego Felipe, sometimes spelled Phelipe, was a stonemason whose racial origin is unknown, though probably Indian or mestizo.

The scant number of formally trained architects and building mechanics is not surprising and is reflected in the generally nondescript character of sixteenth-century architecture which hardly required more than rough sketches to guide the workmen. It is known that many of the monks who were responsible for the evangelization of the natives whom they literally gathered from their widely scattered cornfields and established in towns, were also responsible for the design and construction of the first churches and even the town plans. The monks had, perforce, to be town planners and architects, if such they can be dubbed, for what they designed and built was largely of wattle and daub, adobe, and wood. The procedure the monks followed to bring the Indians together and establish them in towns in the sixteenth century is vividly described by Remesal,[2] who says:

> Ellos eran los que tiraban los cordeles, medían las calles, daban sitio a las casas, trazaban las iglesias, procuraban los materiales, y sin ser oficiales de arquitectura, salían maestros aventajadísimos de edificar. Cortaban los hoces de caña por sus manos, formaban los adobes, labraban los maderos, asentaban los ladrillos, encendían el horno de cal, y a ningún ejercicio por bajo que fuese se dejaban de acomodar.

That is to say, the monks stretched the lines and laid out the streets and the building plots, laid out the churches, procured the materials, and, though they were far from being architectural mechanics, they nevertheless became the most excellent masters of construction. They cut the cane with their own hands, made the adobes, dressed the lumber, laid the bricks, and lit the lime kilns, no work being too menial for them to undertake.

That the monks, especially the Dominicans, if Remesal's word is to be taken, were amateur architects of necessity in the sixteenth century there can be no doubt. He singles out three in particular, Fr. Pedro de la Cruz, Fr. Alonso de Villava, and Fr. Pedro de Barrientos as being adept at building-construction.[3] In addition to Remesal's three, six other

[1] For all names mentioned in this chapter, see ch. VI following "Brief Notices of Architects, etc." where they are arranged in alphabetical order with pertinent data given for each.

[2] Remesal, **2**: pp. 244 ff. and 247. For similar conditions and developments in Peru, see Wethey, *Colonial Architecture*, pp. 6 ff., 11, 29.

[3] Remesal, **2**: pp. 422, 536.

monks who were also amateur architects or builders are known from the sixteenth century: Rodrigo de León, Matías de Paz, Melchor de los Reyes, Agustín de Salablanca, Francisco de Santa Marta, and Benito de Villacañas.

In the seventeenth century the number of references to amateur architects and builders among the religious orders decreases, only two appearing in the contemporary literature. The number of practical builders, however, both specialized craftsmen and *maestros de obras*, increases from only three in the sixteenth to twenty in this century. On the other hand, the name of but one lone formally trained architect appears, that of Martín de Andújar who designed and began the construction of the second cathedral of Antigua, the one still standing in part today. Another interesting phenomenon in the seventeenth century was that many of the practical building craftsmen, especially masons and carpenters, more often than not designed as well as supervised the construction of both public and religious structures. The colonial archives of Guatemala City contain a number of contracts for additions to and alterations of buildings, the construction of which was undertaken by practical men who were, in some cases, responsible not only for the supervision of the actual work, but also for the plans; as for example: Martín de Autillo, Ramón de Autillo, Nicolás and Juan López, Blas Marín, José de Porras, Damián Rodríguez, Martín Ugalde, and Juan Bautista Vallejo. One in particular, José de Porras, a mason by trade, took over the supervision of the construction of the Cathedral from Martín de Andújar, the Spanish architect, and brought the building to completion.

From the eighteenth century Luis Diez de Navarro, a Spanish military engineer and architect who had spent about ten years in Mexico before coming to Guatemala, is outstanding. It is doubtful, however, whether he did much more than inspect the work in progress during construction of the various public buildings he had a hand in designing, to see that the plans and specifications were carried out, for he represented the interests of the civil authorities. The supervision of the actual construction work was usually turned over to a *maestro de obras*. The greater number of his jobs consisted of remodeling older structures. The Capitanía, the University, San Jerónimo, and the Beaterio de Indias, for example, were all altered or repaired under his supervision, but not designed from the ground up by him.

During the eighteenth century until after the destruction and removal of the capital in 1773, Diez de Navarro is the only architect in Antigua known to have had any formal education and training in architecture. But on the other hand, there are a number of men of practical cast whose personalities dominate much of the architecture of Antigua during that century which witnessed the most intensive building activity of the whole colonial period. The most prominent and gifted among these practical men were Bernardo Ramírez, Juan de Dios Estrada, Francisco de Estrada, Francisco Javier Gálvez, Diego de Porras, and José Manuel Ramírez. None had had any formal architectural education, learning their craft on the job and eventually rising to positions of responsibility as *maestros de obras*.

It was only after the destruction of Antigua in 1773 and the building of the new capital city that formally trained architects, all from Spain, appeared in Guatemala in appreciable numbers. Almost all came for the specific purpose of directing the construction of the cathedral in Guatemala City, succeeding each other in the post, so that there was never more than one or two present at the same time. When it was finally decided to begin construction of the public and private buildings of the new capital in 1775, the question arose concerning the need for an architect to supervise the work. The president of the *audiencia*, Martín de Mayorga, felt that a responsible architect should be put in charge since the few craftsmen in Guatemala were mainly mulattoes who had had no formal training in schools and did not know the fundamental principles of architecture. He, therefore, requested that one Diego de Ochoa be made responsible for supervising the construction of the new city. An opposite view was taken a few years later by the then president of the *audiencia*, Matías de Gálvez, who believed that the local workmen did not need such supervision since the buildings being constructed were so nondescript in character that a formally trained architect would only be superfluous.[4] But once the new cathedral was undertaken in 1774, since this was to be a building of monumental character, no less than seven architects were sent at various times to do the plans and supervise the construction. But so ingrained was the traditional practice of relying on practical builders rather than schooled architects, that the plan submitted by Ibáñez was first studied and approved by Bernardo Ramírez who had never had any formal architectural schooling but had had considerable experience in building construction and public works.

One may safely conclude that the majority of the civil and religious structures of Guatemala, and especially those of Antigua, were only in the rarest instances designed by formally trained architects, and that in equally as rare instances did architects supervise the actual construction of the buildings. Until the end of the eighteenth century not a

[4] García Peláez, **3**: p. 93.

single Spanish first-rate architect of even modest abilities, except perhaps Bautista Antonelli, had ever worked in Guatemala. Those Spanish architects who came were usually military engineers who worked on various fortifications in Central America, primarily in Honduras and Panama.[5]

Many of the master craftsmen who undertook the supervision of construction work, especially in the seventeenth century, were not able to sign the contracts though, in some instances, they actually had drawn the plans submitted with the specifications. Many of these men began as simple workmen, learning their craft as time went by until ultimately they emerged as *maestros de obras*, as for example, one Bernabé Carlos who in 1677 was a brickmason but by 1703 applied to the *ayuntamiento* of Antigua to be examined for the license of *maestro mayor de arquitectura y albañilería*. The training, then, of those in charge of building construction, either as craftsmen or as *maestros de obras*, was of a practical nature. They were entrusted with the responsibility of building works solely on the basis of their reputations as mechanics rather than because of any formal architectural training as draftsmen under the guidance of an older, established architect. This may in part explain the improvised character of so much of Guatemalan architecture, especially with regard to the construction of masonry vaults and domes.[6]

The activities of the building trades were not altogether anarchic, but subject to municipal control in the form of the licensing of craftsmen. Yet to a Spaniard of the late eighteenth century like Matías de Gálvez, the best of the Guatemalan buildings designed and constructed by these licensed practical builders scarcely deserved classification as architecture. Though the guild system as known in Spain really never took root in Guatemala, which might have insured higher standards, the civil authorities, nevertheless, undertook to regulate the various crafts and actually examined applicants who wished to practice a craft on their own reponsibility as masters. From the mid-seventeenth century on, more and more notices appear of craftsmen examined for license to practice one or another specialized building trade as well as to be *maestro de obras*. In 1687 a proposal was brought before the *ayuntamiento* that those who wished to engage in architecture as builders should be required to pass an examination.[7] Whether this regulation was immediately instituted or not is not known, but such an order was definitely published in 1723 requiring those who wished to practice architecture to take an examination for the license and the title of *maestro en arquitectura*.[8] But the enforcement of this regulation must have been lax, for in 1752 Juan de Dios Estrada, *maestro mayor de obras públicas*, reported to the *ayuntamiento* that in his opinion those who had not passed the examination in architecture should be prohibited from supervising building construction.[9]

Just what the examination was like is not known, but it must have been of a practical nature and in oral rather than written form. Judging by the popularity of the handbook of Fr. Lorenzo de San Nicolás, first published in 1633 and reissued in 1667, 1736, and even as late as 1796, one may surmise that his precepts for the education of the architect, of extremely practical character, may have also served as the basis for the training of the practical builders of Guatemala. It is hardly likely, however, that the first two of Fray Lorenzo's treatises, those dealing with arithmetic and geometry, could have meant much to the unlettered craftsmen of Antigua and the rest of Central America. The third treatise, dealing with building and following Vitruvius in great part, might have struck an harmonious chord of understanding in the minds of the practical builders of Guatemala. But all the foregoing is purely speculative; for there is no way of knowing just how many copies of his books found their way to Central America and whether the practical builders and craftsmen could themselves read, or have his works read aloud to them if they were illiterate. The indubitable fact is, nonetheless, that many of the building methods he describes, especially with regard to foundations, plastering, and brickwork have their counterpart in the extant colonial architecture of Guatemala. But caution must be exercised in drawing conclusions as to the influence of his handbooks, for the methods Fray Lorenzo describes were not invented by him, but were common practice before and during his time and could very well have been known to the craftsmen of Guatemala without recourse to books.

The founding of a formal school for architects in Guatemala City was proposed, on paper at least, at the very end of the eighteenth century. The Spanish architect Josef Sierra, who was employed on the cathedral of Guatemala City in 1794, submitted a plan to the Sociedad Económica de Amigos del País de Guatemala for the organization of a school which he called "Academia de Matemáticas." In the plan of studies he outlined, one course was devoted to architecture.[10]

[5] See Llaguno y Amírola, *Noticias de los arquitectos* for an early nineteenth-century account of the most renowned architects in Spanish history.

[6] See for example, Las Capuchinas and Santa Cruz in ch. XIV, no. 7, below, and ch. III, no. 5, "Vaults and Domes," above.

[7] *Efem*, p. 105.

[8] *AGG*, A 1.69.3 (1723) 48141–5556.

[9] *Ibid*. A 1.69.3 (1752) 48142–5556.

[10] Mencos, *Arquitectos*, pp. 192 ff., quotes "Plan de Ynstrucción que se ofrece dar en la Academia de Matemáticas, dispuesto por el Capitán y Yngeniero Ordinario de los Reales Exércitos don Josef de Sierra," *AGI*, Guatemala, 529.

But nothing was done to carry out Sierra's proposal; instead in 1795 the Sociedad proposed to form an academy of fine arts.[11] The crown thought that a school of painting, sculpture, and architecture was not necessary at the moment, suggesting that a school of drawing and mathematics be established first.[12] The Sociedad Económica finally opened the doors of a school of fine arts which was actually more a school of drawing, as had been realistically suggested by the crown in 1797. Seventy-seven students were enrolled and the architect Garci-Aguirre, who had come to work on the cathedral, was the director.[13] But there is no indication that architecture as a formal study was part of the curriculum, or that any classes were held in more than a most perfunctory manner. The school existed more on paper than in fact, there not being sufficient personnel with the proper training and education to staff the faculty.[14] In 1801 the pupils of Garci-Aguirre exhibited their work, this being the only notice of any activity of the school. It would seem that a course of study, the object of which was to train architects as outlined by Sierra in 1794, was never realized.

[11] Mencos, *op. cit.*, p. 195, quoting *AGI*, Guatemala, 529.
[12] *AGG*, A 1 1 (1796) 24902–2817; García Peláez, **3**: p. 199.
[13] Salazar, *Desenvolvimiento*, pp. 268 ff.
[14] Mencos, *op. cit.*, p. 194.

VI BRIEF NOTICES OF ARCHITECTS, *MAESTROS DE OBRAS* AND OTHER BUILDING-TRADE CRAFTSMEN

THE men listed in this chapter, whose names have been culled from the contemporary literature and archival documents, are all identified with the architecture of Guatemala and Central America either as professional architects, practical craftsmen responsible for building operations, or amateurs. This is no more than a random sample and should not be construed as being the sum total of all those who may have been so engaged during the course of almost two hundred and fifty years of building activity. Mentioned are but four formally trained architects, all of Spanish origin, who are known to have worked in Guatemala before 1773 and who probably represent the actual total. The remaining eleven listed all arrived after the destruction of Antigua and are connected with work in the new capital, La Nueva Guatemala. With regard to the practical builders, the number included below must in truth be only a fraction of all those who engaged in the building trades, not only in the city of Antigua but most certainly in the other cities of Central America of which very few notices have come to light. The total number of entries including all categories is seventy-one (table 2).

TABLE 2

	16th Cent.	17th Cent.	18th Cent.	After 1773	Totals
Spanish Architects—formally trained	2	1	1	11	15
Practical builders—*maestros de obras*	3	20	14	7	44
Amateur architects and builders	9	2	1	0	12
Totals	14	23	16	18	71

1. AMPUDIA Y VALDEZ, JUAN. Architect. After 1773.
In 1783 he did the plan of the Hospital de San Juan de Dios in Comayagua, Honduras.

Torres Lanzas, *Planos*, nos. 250 and 251 referring to *AGI*, 101-1-2 (1); also Angulo, *Planos*, pl. 173; Mencos, *Arquitectura*, append. XLIII, a transcription of *AGI*, Guatemala, 571, dated 6 VIII 1783.

2. ANDUJAR, MARTIN DE. Architect. Seventeenth century.
A Spaniard, who began the Cathedral in Antigua in 1669 and was also probably responsible for the plan. He was replaced by José de Porras in 1677 as supervisor of the construction. In 1677 both he and the master brickmason, Bernabé Carlos inspected the university buildings.

Juarros, **2**: p. 240; *Efem*, p. 86; Mencos, *Arquitectos*, p. 171; and Castellanos, *Relación sintética*, pp. 74 ff. See also Mencos, *Arquitectura*, ch. IV and append. IX, quoting *AGI*, Guatemala, 166, dated 19 XI 1677.

3. ANTONELLI, JUAN BAUTISTA. Architect. Sixteenth century.
He arrived in Guatemala after the destruction of Ciudad Vieja in 1541 and inspected the valleys of Jalapa, Las Vacas, Tianguecillo, and Panchoy, deciding on the latter as the best location for the new city. He probably remained in the New World until the end of the century asking for passage back to Spain in 1599. He may have been responsible for the original plan of the city of Antigua.

Aguirre Matheu, *Descripción*, pp. 73 ff.; Juarros, **2**: pp. 178 ff.; Fuentes y Guzmán, **1**: pp. 130 ff.; Schaeffer, *Indice*, II, no. 3582; Angulo, *Bautista Antonelli*, for a biographical account. An architect by the same name is mentioned by Llaguno, *Noticias de los arquitectos*, III, pp. 10 ff., who arrived in Spain in 1559. If this is actually the case, the Antonelli referred to must be another person than the one known to have been in Guatemala in 1541. It is quite possible, however, that Llaguno's information may not be altogether correct.

4. ARCE, FR. JOSE DE. Amateur architect and builder. Seventeenth century.
A Dominican monk who rebuilt the bridge at Sacapulas after it had been destroyed *ca.* 1616. He moved its loca-

tion to another site and designed it with six massive stone piers which supported a wooden roadway above. The date of his death is given as 1691.

Ximénez, **2**: p. 491.

5. Aristondo, Juan de Dios. Practical builder and architect. Eighteenth century.
On October 27, 1741, he was named *maestro mayor de arquitectura* of Antigua. He was also a captain of infantry of "*pardos.*"

Efem, p. 191.

6. Autillo, Martin de. Stonemason. Seventeenth century.
Together with Juan Bautista Vallejo he worked on the church of Santo Domingo in Antigua in 1636. His name is also given as Utillo or Utilla.

AGG, A 1.20 (1636) 690–69, also *BAGG* **10** (1945): pp. 101 ff.; Mencos, *Arquitectos*, p. 175.

7. Autillo, Ramon de. Stonemason. Seventeenth century.
He contracted to do the *portada* of the church of San Francisco in Antigua in 1675. He may possibly be of the same family as Martín de Autillo.

AGG, A 1.20 (1675) 477–32, also *BAGG* **10** (1945): p. 133.

8. Arroyo, Jose. Architect or draftsman. After 1773.
Working on the cathedral of Guatemala City until his death in 1788 as director of the construction, he was responsible for some of the plans.

AGG, A 1.10.2 (1788) 1670–68.

9. Barrientos, Fr. Pedro de. Amateur architect. Sixteenth century.
A Dominican monk who laid out and repaired the convent and church of his order in Chiapa de Corzo, Chiapas, sometime after 1562.

Remesal, **2**: p. 536.

10. Barrientos, Pedro de. Brickmason. Seventeenth century.
He contracted to do the vault of the *capilla mayor* of the church of San Sebastián in Antigua in 1668.

Efem, p. 79; Mencos, *Arquitectos*, p. 175.

11. Barruncho, Manuel Jesus. Stonecutter. Eighteenth century.
He worked on the fountain in the Plaza Mayor of La Nueva Guatemala under the direction of its designer, the architect Bernasconi. When the latter died in 1785, Barruncho was put in sole charge of the job which he completed soon after 1788.

Díaz, *Bellas artes*, p. 122; Mencos, *Arquitectura*, ch. viii, fn. 58, quoting *AGI*, Guatemala, 529, dated 14 xii 1785, and for another copy of the same, *AGI*, Guatemala, 470.

12. Bernasconi, Antonio. Architect and draftsman. After 1773.
He worked on the cathedral of Guatemala City until his death in 1788. In a document dated 1777 ordering reimbursement for his expenses in coming to Guatemala from Spain, he is referred to as a draftsman under the architect Marcos Ibáñez. He drew the plans for the Palacio Arzobispal adjacent to the cathedral as well as the fountain of Carlos iii which once stood in the main plaza in front of the cathedral and is now located in the Plazuela de España, Guatemala City. He also made a survey of the Maya site of Palenque in Chiapas, Mexico.

González Mateos, *Marcos Ibáñez*, p. 54; Castellanos, *op. cit.*, nos. 253, 254, 261, 262; Angulo, *Planos*, pls. 153, 171, 172; Mencos, *Arquitectura*, ch. ix, fn. 58, quoting *AGI*, Guatemala, 529, dated 14 xii 1785. For drawings of Palenque, see Torres Lanzas, *op. cit.*, nos. 256, 257, 258, 259, 260, dated 1784 and 1785, also Angulo, *op. cit.*, **3**: p. 391, pls. 133–138. These, however, were not drawn by him.

13. Bodega, Manuel de la. *Maestro de obras*(?) or architect(?). After 1773.
He was named superintendent of the cathedral in Guatemala City in 1786.

AGG, A 1.10.2 (1786) 1669–68.

14. Bonilla, Juan de. *Maestro de obras*. Eighteenth century.
In 1703 he replaced José de Porras who had died as *maestro de obras* of the city of Antigua.

Efem, p. 130.

15. Campo y Rivas, Manuel del. *Maestro de obras*(?) or architect (?). After 1773.
He presented a claim in 1797 saying he should be the superintendent on the work of the cathedral in Guatemala City.

AGG, A 1.10.2 (1797) 1672–68.

16. Carcamo, Agustin de. Brickmason. Seventeenth century.
In 1688 along with José de Porras (ii), Bernabé Carlos

and Andrés de Illescas, he carried out an inspection of the church of San Sebastián.

Mencos, *Arquitectura*, ch. IV, fn. 125, quoting *AGI*, Guatemala, 180.

17. CARCAMO, NICOLAS DE. Brickmason. Seventeenth century.
In 1665 he helped José de Porras (II) render a report on the condition of the church and hospital of San Pedro on the construction of which they were employed as bricklayers. Only Porras' name is followed by the title *maestro de albañilería*. Neither Cárcamo nor the other two mechanics employed on the job, nor Porras himself, could sign their names to the document.

Mencos, *op. cit.*, ch. V, append. XVIII, a transcription of *AGI*, Contaduría, 883 A, dated 20 X 1665, also 983 A, same date.

18. CARLOS, BERNABE. Brickmason and architect. Seventeenth century.
In 1677 his name appears as the brickmason who with Martín de Andújar inspected the university buildings. Along with José de Porras (II), Andrés de Illescas, and Agustín de Cárcamo in 1688, he inspected San Sebastián. He must have been a very young man when his opinions on construction were first sought, for twenty-six years later in 1703 he applied to the *ayuntamiento* to be examined for the title of *maestro mayor de arquitectura y albañilería*.

Efem, pp. 86, 130 and Andújar above; Mencos, *Arquitectos*, p. 174, also *Arquitectura*, ch. IV, fn. 125, quoting *AGI*, Guatemala, 180.

19. CRUZ, VICENTE. *Maestro de obras*. After 1773.
He reported to the *ayuntamiento* in 1777 on the availability of second-hand materials from various ruined public buildings in Antigua which might be serviceable in the new constructions in Guatemala City.

Efem, p. 260.

20. CRUZ, PEDRO DE LA. Amateur architect and builder Sixteenth century.
A Dominican monk who worked on the fountain of Zinacantán Chiapas, Mexico, in 1562 and who also built a church in a nearby Indian town. He also designed and built a caracol stair in Chamula, Chiapas, as good as any done by Spanish mechanics, though he had never had any previous training.

Remesal, **2**: p. 422.

21. DIEZ DE NAVARRO, LUIS. Architect and military engineer. Eighteenth century.
A Spaniard, he arrived in Guatemala in 1741 after having worked during the previous nine years or so in Mexico. He is the first formally trained architect to have worked in Guatemala for an extended period of time. Some of his drawings still exist in the Archivo de Indias in Seville. His first charge in Guatemala was the repair of the Cathedral and the general reconstruction and enlargement of the Capitanía. He also did considerable work in the provinces, building the fortifications of Omoa, Puerto Caballos, and Trujillo, Roatán, and elsewhere. As the official responsible for all constructions which fell under the royal authority, he had much to do with the building of the Beaterio de Indias, the University, and the water supply system of Antigua, as well as the town plan of Guatemala City after the destruction of Antigua in 1773. He was the first in Guatemala to understand that the flooding of the Pensativo River resulted from the cutting of the timber on the highlands surrounding Antigua as well as from the deplorable practice of farming on land of excessive slope, and to suggest remedies in a rational and scientific spirit. He served as a general factotum for the *audiencia* making a geographical survey of the whole of the Reino de Guatemala in 1756, and was still active in 1775 and 1776 when he drew some new plans for the layout of Guatemala City as well as for various religious and civil buildings to be erected there. For a short time he also served as governor of the province of Costa Rica. After 1776 his son, Manuel Diez de Navarro, applied for the title of military engineer before a commission on which his father served.

Berlin, *El Ingeniero Diez de Navarro*, pp. 89 ff.; Calderón Quijano, *Ingenieros*, pp. 40–47; *Efem*, pp. 203, 213, 217, 222, 223, 258; Castellanos, *op. cit.*, pp. 74 ff.; Mencos, *Arquitectos*, pp. 178 ff. and *Arquitectura*, ch. VI, append. XXX, a transcription of *AGI*, Guatemala, 657, dated 22 IV 1760 and 14 II 1769, also append. XXXII, a transcription of *AGI*, Guatemala, 316, dated 14 X 1757 and dealing with the Casa de Moneda. *AGG*, A 1.17.3 (1744) 17508–2335; A 1.1 (1760) 24871–2817; A 1.1 (1758) 3–156; A 1.17.3 (1756) 38302–4501; A 1.10.3 (1775) 4536–74; Torres Lanzas, *Planos*, no. 69, plan of San Jerónimo, dated 1767, also Angulo, *Planos*, pl. 163 for same; Torres Lanzas, *op. cit.*, no. 214 and Angulo, *op. cit.*, pl. 156 for provisional convent and church of Santa Clara in Nueva Guatemala, dated 1775; Torres Lanzas, *op. cit.*, no. 216 and Angulo, *op. cit.*, pl. 155, dated 1775, for provisional convent of San Francisco; Torres Lanzas, *op. cit.*, no. 220, for a map of Guatemala City still in construction and dated 1776, and no. 225, a map of the whole Reino de Guatemala, dated 1776; see also Calderón Quijano, *op. cit.*, p. 46, fn. 87 for a listing of his geographical reports, as well as the one published in Guatemala City, 1850, titled *Relación sobre el antiguo Reino de Guatemala . . . en 1745*, being the same as *AGG*: A 1.17.3 (1744) 17508–2335. For his Mexican works, see Angulo, *Historia del arte*, **2**: pp. 542 ff.

22. DOLORES, MANUEL. Master brickmason. 1788.

See José Marcelino below.

23. DORANTES, ANTONIO. Master brickmason. Eighteenth century.
He paved a stretch of street in Antigua from the Arco de Matasano to the Guarda de Ánimas between March 10 and July 13, 1764. In 1772 his name appears in connection with the inspection of the church of the Beaterio del Rosario (Beatas Indias) saying it was not necessary to repair it.

Efem, p. 226; Angulo, *Planos*, **4**: p. 413.

24. ESTRADA, FRANCISCO DE. *Maestro de obras*. Eighteenth century.
In 1755 he was named *maestro mayor de obras* of the city of Antigua replacing Juan de Dios Estrada. In 1762, together with Diez de Navarro, he was sent to investigate the causes of the flooding of the Pensativo River. Four years later he presented a project and the estimate of costs for the construction of a building adjacent to and continuing the same façade style as the Ayuntamiento, a proposal which was never carried out. He remained in his post until 1770 when he was replaced by Bernardo Ramírez.

Efem, pp. 212, 223, 231, 239; *AGG*, A 1.10.2 (1766) 18771–2447.

25. ESTRADA, JUAN DE DIOS. *Maestro de obras*. Eighteenth century.
In 1749 he reported to the *ayuntamiento* on the proposed construction of a dome supported on four columns for the Cruz de Piedra in the Calle Ancha de Jocotenango. And again in that same year he registered a vigorous complaint on the very bad workmanship on the part of the brickmasons employed in the construction of the church of La Merced then in progress. In 1752 he completed the construction of a hospital. The same year he also submitted a proposal to the *ayuntamiento* that those who had not been previously examined be prevented from directing construction work. He died in 1755 and his post of *maestro mayor de obras* was assumed by Francisco de Estrada.

Efem, pp. 202, 212; *AGG*, A 1.7 (1752) 1296–52, and A 1.69.3 (1752) 48142–5556; Mencos, *Arquitectos*, p. 183.

26. FELIPE, DIEGO (PHELIPE). Stonemason. Sixteenth century.
The first bridge at Los Esclavos built in 1592 has been attributed to him. Working along with him was Francisco Tirado who may have been the chief craftsman on the job under whom Felipe worked. It is tempting to conclude that Felipe may not have been a Spaniard, but was either an Indian or a mestizo and possibly even a mulatto, though there is absolutely no proof for this conclusion.

Fuentes y Guzmán, **2**: p. 130; Villacorta, *Historia*, p. 177; Mencos, *op. cit.*, p. 166.

27. GALVEZ, ANTONIO DE. Master carpenter. Eighteenth century.
In 1734 he helped Diego de Porras in the inspection of Santa Clara. The title appended to his name is given as *maestro de carpintería y baluartes*.

Mencos, *Arquitectura*, ch. v, fn. 26, append. XXVII, a transcription of *AGI*, Guatemala, 229, dated 22 v 1734.

28. GALVEZ, FRANCISCO JAVIER. Carpenter and architect. Eighteenth century.
On March 1, 1757, the *ayuntamiento* granted him the title of *maestro mayor de arquitectura civil y de carpintería*. Three years later, together with Diez de Navarro, he made an inspection of the Capitanía in order to estimate the cost of repairing the building. A year later the work began under his direction.

Efem, pp. 214, 217, 219; Castellanos, *op. cit.*, pp. 74 ff.

29. GAMUNDI, SEBASTIAN. Architect(?). After 1773.
He was temporary director of the construction of the cathedral in Guatemala City until he died in 1788.

AGG, A 1.10.2 (1788) 1670–68.

30. GARCIA, SALVADOR. Stonemason. Seventeenth century.
His name appears in 1641 in a document in which he requests to be examined for the grade of master stonemason(?).

AGG, A 1.16.22 (1641) 38298–4500.

31. GARCI-AGUIRRE, PEDRO. Architect. After 1773.
In charge of the construction of the cathedral in Guatemala City for about one year in 1802. He did other work in Guatemala City both for private people as well

as religious groups, and directed the construction of the new convent of Santo Domingo where he was buried when he died in 1809. He was probably a Spaniard who came to Guatemala late in the eighteenth century and was for a time director of a school of fine arts in Guatemala City.

AGG, A 1.10.2 (1802) 1678–68 and A 1.10.1 (1804) 1484–65; Mencos, *Arquitectos*, pp. 193 ff.

32. GARNICA, RODRIGO MARTINEZ DE. Architect. Sixteenth century.
In 1542 he undertook to build the cathedral in Antigua for which he was paid 400 pesos, according to one source and 1,200 according to another. This building was completely demolished in the seventeenth century and its place taken by the present Cathedral completed in 1680.

Efem, pp. 7 ff.; Fuentes y Guzmán, **2**: p. 407; Mencos, *op. cit.*, p. 164; Castellanos, *Relación sintética*, pp. 74 ff.

33. GONZALEZ BATRES, JUAN JOSE. Amateur architect. Eighteenth century.
As a member of the city council, *ayuntamiento*, he was in charge of the construction of the Ayuntamiento (building) in Antigua in 1743. It is unlikely that he did more than supervise the work, nor was he responsible for the plan.

Batres Jareguí, *América Central*, **2**: p. 526; *Efem*, p. 195.

34. GUERRERO, PEDRO. Architect. After 1773.
He was responsible for the plans, drawn in 1784, of the building where the *audiencia* was housed in San Salvador.

Torres Lanzas, *op. cit.*, no. 225, also Angulo, *op. cit.*, pl. 175.

35. HERNANDEZ DE FUENTES, FRANCISCO. Stonemason. Seventeenth century.
He took the contract to build the church of Santa Catalina in Antigua in 1626 drawing the plans and constructing the building as well.

AGG, A 1.20 (1626) 757, also *BAGG* **10** (1945): pp. 221 ff.

36. IBAÑEZ, MARCOS. Architect. After 1773.
He arrived in Guatemala in 1777 along with Antonio Bernasconi who worked with him as draftsman. The express purpose for which he came was to work on the city plan as well as the cathedral. His plan for the city was preferred over that of Diez de Navarro but was, in fact, based on Diez de Navarro's original scheme of a north-south east-west orientation which had been criticized by Sabatini. Ibáñez's plan for the city is dated 1778 while that of the cathedral is 1782, implying that he did not direct the latter work very long since he left Guatemala in 1783.

Torres Lanzas, *op. cit.*, nos. 234, and 247; also Angulo, *op. cit.*, pls. 147, 148; *AGG*, A 1.10 (1777) 1575–59; González Mateos, *Marcos Ibáñez*, pp. 49–55.

37. ILLESCAS, ANDRES DE. Brickmason. Seventeenth century.
In company with José de Porras (II), he inspected Los Remedios in 1678. He was also part of the group of experts including Porras (II), Agustín de Cárcamo and Bernabé Carlos who rendered a report on San Sebastián in 1688.

Mencos, *Arquitectura*, ch. IV, append. XV, a transcription of *AGI*, Guatemala, 30, dated 16 II 1678; also *ibid.*, ch.IV, fn. 125, quoting *AGI*, Guatemala, 180.

38. LEON, FR. RODRIGO DE. Amateur architect and builder. Sixteenth century.
A Dominican friar who designed the fountain in the plaza of Chiapa de Corzo, Chiapas, Mexico, which was completed by a Spanish craftsman in 1562.

Remesal, **2**: p. 422.

39, 40. LOPEZ, NICOLAS and JUAN. Master carpenters. Seventeenth century.
In 1673 and 1674 the brothers López took the contract to rebuild the roof of the conventual church of San Francisco in Antigua which had rotted out.

Efem, pp. 83, 84; *AGG*, A 1.20 (1673) 476–10, also *BAGG* **10** (1945): p. 131; Vásquez, **4**: pp. 67 ff.

41. LORENZO, CRISTOBAL. Master brickmason. Seventeenth century.
Together with Andrés Serrano, he inspected the convent of La Concepción in Antigua after the earthquake of October 9, 1607.

Mencos, *Arquitectura*, ch. III, quoting *AGI*, Guatemala, 176, dated 25 and 30 X 1607, also append. IV where the document is transcribed.

42. MARCELINO, JOSE. Master carpenter. 1788.
He and Manuel Dolores, a brickmason, drew the plan

of the work to be done on the convent of La Concepción in San Cristóbal de las Casas, Chiapas, Mexico.

Angulo, *Planos* pl. 303, **2**: p. 321; **4**: pp. 637 ff.

43. MARIN, BLAS. Master carpenter. Seventeenth century.
Described as an "indio ladino," he took the contract to complete the church of Santa Cruz in Antigua in 1662.

Ch. XIV, no. 5, below; *Efem*, p. 71.

44. MARQUI, SANTIAGO. Architect. After 1773.
Born in Madrid in 1767, he arrived in Guatemala in 1804 where he undertook the supervision of the construction of the cathedral in Guatemala City. On various occasions, in 1813 particularly, he requested permission to return to Spain permanently. This was denied since he was the only trained architect in the country competent to direct the work on the cathedral. He died there about 1820 or so. He also worked on a school building, Educatorio de Indias, for which he requested payment in 1810. He is the last of the Spanish architects who worked on the cathedral which was still not quite finished at the time of the independence from Spain in 1821. What remained to be done, however, was of minor character.

Mencos, *Arquitectos*, pp. 195 ff.; *AGG*, A 1.1 (1806) 702–23, A 1.1 (1810) 5212–221 and 5213–221, A 1.1 (1820) 922–30.

45. MATA, FR. FELIX DE. Amateur architect and builder. Seventeenth century.
He was responsible for parts of the Dominican convent in Antigua, the *portada* of the church, and especially the fountain done in 1618. (FIGS. 92–94.) The latter was very graphically described by Gage who was there early in the seventeenth century. Mata was born about 1595 in Ocaña in the province of Castilla, Spain, and died in Guatemala in 1634.

Mencos, *op. cit.*, pp. 166 ff.; Ximénez, **2**: pp. 233 ff.; Castellanos, *op. cit.*, pp. 74 ff.; Gage, ch. XVIII, pp. 283 ff.

46. MEDINA, DIEGO DE. *Maestro de obras*. Eighteenth century.
For a short time he was in charge of the construction of the hermitage of El Calvario in Antigua in 1720. In the same year he helped Diego de Porras report on San Agustín. He is referred to as a carpenter.

Efem, p. 150; Mencos, *Arquitectura*, ch. VI, fn. 108, referring to *AGI*, Guatemala, 309, dated 23 XI 1720.

47. MONZON, NICOLAS. *Maestro de obras*. After 1773.
Supervised the construction of the church of the town of Xenacoj, Guatemala, in 1797.

AGG, A 1.10.3 (1797) 18827–2448.

48. NAJERA, DIEGO DE. Carpenter. After 1773.
In 1810 he presented an estimate for the carpentry work to be done on a house in Guatemala City which once belonged to one Juan Ramírez.

AGG, A 1.1 (1810) 18009–2377.

49. NUÑEZ, AGUSTIN. Architect, sculptor, and retable builder. Seventeenth century.
In 1687 he was granted the title of *maestro mayor de arquitectura* by the *ayuntamiento*, and in 1689 that of *maestro mayor de las artes de escultura y ensambladura*, implying that he apparently practiced all three crafts at the same time. In 1689 he took the contract to build a retable, including the sculptures, for the conventual church of La Concepción. He is heard from again in 1706 when he delivered two retables for the church of the Compañía de Jesús in Antigua. No notices have appeared indicating whether he actually worked in building construction, though he may have done so for private parties, which may explain the lack of documents recording building contracts.

Efem, pp. 105, 108; *AGG*, A 1.62.2 (1706) 48139–5556.

50. OCHOA, DIEGO DE. Architect. After 1773.
He was mentioned as the architect who should be put in charge of the actual construction work in the new capital of Guatemala City in 1775.

García Peláez, **3**: p. 93.

51. PAZ, FR. MATIAS DE. Amateur architect. Sixteenth century.
A Dominican monk, he came to Guatemala in 1539 and worked in connection with the construction of the Dominican convent in Antigua after 1541. He remained in Guatemala until 1551.

Mencos, *Arquitectos*, pp. 164 ff.

52. PORRAS, DIEGO DE. Practical builder and architect. Eighteenth century.
In 1703 he was granted the title of *maestro mayor de arquitectura* by the *ayuntamiento* of Antigua. His name,

sometimes spelled Porres, has been associated with the construction of the church and convent of La Recolección which he may very well have designed and which were completed in 1717 when he was about thirty-seven years old. In connection with a report he made in 1720 on the Escuela de Cristo he gives his age as forty. His name is also associated with the plans of the Casa de Moneda, begun in 1733, and part of the Capitanía. (FIG. 197.) After the earthquake of 1717 he was commissioned by the *ayuntamiento*, as *maestro mayor de obras*, to take an inventory of the damage. In 1734 in company with Antonio de Gálvez he reported on Santa Clara. In 1739, in this same official capacity, he inspected the site where the Mercedarian order wished to build the Colegio de San Jerónimo. In that same year he was paid the sum of one hundred *pesos* and given a testimonial letter of praise for the work he had done on the fountain in the main plaza of Antigua. He died a year later and his post was taken over by Juan de Dios Aristondo.

García Peláez, **3**: p. 27; Díaz Durán, *Casa Moneda*, p. 216; *Efem*, pp. 161, 186, 191; Mencos, *Arquitectos*, pp. 170 ff., and *Arquitectura*, ch. IV, append. XVII, quoting *AGI*, Guatemala, 309, dated 9 XII 1719, a report on the condition of Los Remedios and also ch. V, fn. 26, append. XXVII, a transcription of *AGI*, Guatemala, 229, dated 22 V 1734, and append. XXVIII, a transcription of *AGI*, Guatemala, 309, dated 11 VI 1720, also append. XXXI, a transcription of *AGI*, Guatemala, 314, dated 18 VIII 1738.

53. PORRAS, FRANCISCO DE. Building craftsman. Sixteenth century.
His name appears as having constructed the first Central American cathedral in Ciudad Vieja some time between 1529 and 1537.

Juarros, **2**: pp. 234 ff.; Castellanos, *op. cit.*, pp. 74 ff.

54. PORRAS, JOSE DE (I). Building craftsman. Seventeenth century.
Mentioned in connection with the building of the earlier seventeenth-century church and convent of La Compañía de Jesús in Antigua completed in 1626. There is no evidence which might indicate that he was related to the Francisco de Porras who worked on the first cathedral in Ciudad Vieja almost one hundred years before, or to the José de Porras who worked on the Cathedral of Antigua between 1669 and 1680. He could very well have been the father or grandfather of the latter and a descendant of the former, but of this there is no proof. The similarity of names and the building trades with which these men were associated does not make it farfetched to expect that they were all members of the same family.

Efem. p. 48.

55. PORRAS, JOSE DE (II). Master brickmason and architect. Seventeenth century.
He may have been the father or the uncle or even the grandfather of Diego de Porras. He worked on the church and hospital of San Pedro from 1662 to 1665 rendering a report on the state of the building in the latter year. From this document it is learned that he was twenty-six years old. Though entrusted with directing the job, he still could not sign his name. José de Porras began as a master brickmason on the Cathedral of Antigua in 1669, taking over the direction of the work from Martín de Andújar in 1677 and completing the job in 1680. In 1678 he rendered a report on Los Remedios. In 1688, along with Andrés de Illescas, he inspected San Sebastián. He may also have worked on the church of Santa Teresa and the conventual hospital of Belén, but this is not certain. His name also appears in connection with inspections he made on behalf of the *ayuntamiento* of the church of San Sebastián in 1689 and the hospital of San Alejo in 1693. He died in 1703 and his post of *maestro mayor de obras* was taken over by Juan de Bonilla.

Castellanos, *op. cit.*, pp. 74 ff.; Juarros, **2**: p. 274; *Efem*, p. 130; *AGG*, A 1.10.3 (1672) 31258–4046; Mencos, *Arquitectos*, pp. 170 ff., and *Arquitectura*, ch. IV, also append. IX, a transcription of *AGI*, Guatemala, 166, dated 19 XI 1677, as well as append. XV, a transcription of *AGI*, Guatemala, 30, dated 16 II 1678. See also *ibid.*, fn. 125 quoting *AGI*, Guatemala, 180, and ch. V, append. XVIII, a transcription of *AGI*, Contaduría, 883A, dated 20 X 1665. See also, ch. XIII, no. 3, fn. 16, below.

56. QUIROZ, GREGORIO NACIANCENO. Architect. Eighteenth century.
The architect of the cathedral of Tegucigalpa, Honduras, begun in 1756.

Ypsilanti, *Monografía*, p. 15.

57. RAMIREZ, BERNARDO. *Maestro de obras* and architect. Eighteenth century.
Of a family of practical builders in Antigua where he was born in 1741, he was in 1770 appointed *maestro mayor de obras* of the city continuing to occupy the same post when the capital was transferred to Guatemala City where he was responsible for constructing the water supply system. When Ibáñez, a trained professional

Spanish architect, presented the plans for the cathedral in Guatemala City, the archbishop sought the advice of Ramírez whom he asked to look them over and give his opinion. Ramírez also worked on many of the public and religious buildings of the new capital, actually making the plans for the convents of Las Capuchinas, La Recolección, Santa Catalina, Beaterio de Santa Rosa, and Santa Teresa. In 1782 he was put in charge of formulating the new regulations of the brickmasons' guild after having indicated about nine years previously that this was urgently needed. In 1798 he was in charge of the actual construction work on the cathedral. This would imply that his resignation as *maestro mayor de obras* which he submitted to the *ayuntamiento* in 1790 had not been accepted. He refused the honor of being named *oficial de pardos*, that is, an officer in charge of a military unit of soldiers of mixed ancestry, petitioning instead that he be classified as a Spaniard. This was denied him since he was apparently of mixed ancestry. The last notice relative to him dates from 1803 when he presented a bill for forty pesos to the *ayuntamiento*.

Efem, pp. 239, 245; *AGG*, A 1.16 (1773) 2830–148, A 1.10.2 (1783) 1660–68, A 1.2 (1790) 15833–2213, A 1.10.2 (1798) 1673–68, A 1.2 (1803) 15857–2214; Mencos, *Arquitectos*, pp. 174 ff.; *AGI*, Guatemala, 471 and 643; Villacorta, *op. cit.*, p. 313.

58. Ramirez, Jose Manuel or Manuel Jose. Master brickmason and *maestro de obras*. Eighteenth century.
Born in Antigua in 1703, he was the father of Bernardo Ramírez. In his capacity as *maestro mayor de obras* of Antigua, he reported to the city council in 1747 that the vaults and the bell tower of the Ayuntamiento were in danger of ruin because of dampness. Together with the carpenter Manuel de Santa Cruz he made an inspection of the old university building in 1763 in order to appraise the property as well as to make estimates of the repairs which would be required to make the structure serviceable. His name appears again as a building lot appraiser in 1780; if this refers to the same person, he would have been at that time seventy-seven years of age.

Efem, p. 199; *AGG*, A 1.3 (1763) 1157–45, also *ASGH* **17** (1941/42): pp. 376 ff.; *AGG*, A 1.1 (1780) 17990–2374, also *BAGG* **5** (1939/40): p. 367; Mencos, *op. cit.*, p. 184.

59. Rodriguez, Damian. Master carpenter. Seventeenth century.
On November 9, 1629, he was given the contract for setting the wooden pillars for the corridor of the Ayuntamiento in Antigua.

Efem, p. 49.

60. Reyes, Fr. Melchor de los. Practical builder. Sixteenth century.
He arrived in Guatemala early in the sixteenth century as a lay brother of the Dominican order. He may have died either in 1557 or 1559 at which time his passing was considered a great loss for the order. He reputedly worked so fast setting stone that six Indian helpers could hardly keep him supplied with material. He was, apparently, one of the few skilled craftsmen in Central America in the first days after the conquest who advised the monks on their construction work. According to Remesal, for want of his advice after his passing, many of the subsequent buildings were poorly constructed.

Remesal, **2**: pp. 248, 329.

61. Salablanca, Fr. Agustin. *Maestro de obras* (?). Sixteenth century.
He worked during the second half of the sixteenth century in Guatemala and may possibly have been commissioned to build the church at Tecpán which was under the doctrine of the Franciscan order. This is rather unusual, if true, for Salablanca was a member of the Dominican order. In the sixteenth century the church at Tecpán was of the most rudimentary type of construction. One may assume, therefore, that Salablanca had very little to do with it, if anything at all. His name is also mentioned in connection with the construction of the Dominican convent in Ciudad Vieja earlier in the century.

Fuentes y Guzmán, **1**: pp. 386 ff., Angulo, *Historia del arte*, **2**: p. 47.

62. Salazar, Jose Nicholas. Master brickmason. Eighteenth century.
In 1776 he reported on the work accomplished on the cathedral of León, Nicaragua, by Bishop Juan Carlos Vilches y Cabrera. In the same document he gives a more or less detailed account of the plan and the condition in which the bishop found the building as well as the progress made to date.

Mencos, *Arquitectura*, append. XLV, a transcription of *AGI*, Guatemala, 606, dated 27 IX 1776.

63. SANTA CRUZ, MANUEL DE. Master carpenter. Eighteenth century.
In 1763, together with José Manuel Ramírez, he made an inspection of the university building and also prepared an estimate of the cost to put it into serviceable condition.

AGG, A 1.3 (1763) 1157–45, also *ASGH* **17** (1941/42): pp. 376 ff.

64. SANTA MARTA, FR. FRANCESCO DE. Amateur builder. Sixteenth century.
After the earthquake of 1575 he was put in charge of reconstructing the Franciscan convent in Ciudad Vieja, Guatemala, taking six years to complete the job.

Vásquez, **2**: p. 238; Mencos, *Arquitectos*, p. 165.

65. SERRANO, ANDRES. Master brickmason. Seventeenth century.
See Cristóbal Lorenzo above.

66. SIERRA, JOSEF DE. Architect and engineer. After 1773.
He arrived in Guatemala City about 1788 and took over the direction of the building works of the cathedral. After making a study of the already existing parts, he changed the plans to include a vaulted roof rather than one of wood and tile. He remained in charge of construction for about ten years, but apparently also found time to work on the gunpowder factory in Antigua for which he presented his bill in 1798. In 1800 he carried out an inspection of the cathedral of Granada, Nicaragua, then in construction. He proposed to found a school to train architects in Guatemala which was then sorely needed, for there was not a single native Guatemalan engaged in the building trade there who had received the proper professional training. In view of the general lack of skill on the part of the brickmasons and carpenters, he proposed that a school for practical mechanics be attached to the school of architecture. He also did some work in military architecture, particularly on the forts of San Carlos and of Trujillo. In 1802 he asked to have his salary for the work on the cathedral paid him since he planned to return to Spain shortly.

Torres Lanzas, *Planos*, nos. 267 and 268 and Angulo, *Planos*, pls. 151 and 152; *AGG*, A 1.10.2 (1788) 1670–68, A 1.1 (1798) 514–18, A 1.10.2 (1800) 1674–68, A 1.10.2 (1802) 1677–68; *AGI*, Guatemala, 529; Mencos, *op. cit.*, pp. 190 ff.

67. TIRADO, FRANCISCO. Stone- and brickmason. Sixteenth century.
Probably born in Málaga, Spain, about 1520. Sometime before 1581 he did the plan and supervised the actual construction of the Franciscan church and convent in Ciudad Vieja. In 1586, bearing the title *maestro mayor de cantería y albañilería*, he carried out an inspection of the Franciscan convent and church in Antigua. Along with Diego Felipe, he built the bridge at Los Esclavos in 1592. He probably died not long afterward.

Fuentes y Guzmán, **2**: p. 130; Juarros, **2**: p. 74; Villacorta, *op. cit.*, p. 177; Mencos, *Arquitectos*, pp. 165 ff., also *Arquitectura*, ch. II, fn. 57 quoting *AGI*, Guatemala, 10, a letter to the crown dated 24 IX 1581, and his append. I for a transcription of *AGI*, Guatemala, 966, dated 30 IV 1581.

68. UGALDE, MARTIN. Stonemason. Seventeenth century.
In 1648 he took the contract to complete the work on the *capilla mayor* of the church of Santo Domingo in Antigua.

Efem, p. 59; *AGG*, A 1.20 (1648) 694–688, also *BAGG* **10** (1945): pp. 102 ff.

69. VALLEJO, JUAN BAUTISTA. Stonemason. Seventeenth century.
Together with Martín Autillo, he worked on the church of Santo Domingo in Antigua in 1636.

Efem, p. 52; *AGG*, A 1.20 (1636) 690–69, also *BAGG* **10** (1945): pp. 101 ff.; Mencos, *Arquitectos*, p. 175.

70. VILLACAÑAS, FR. BENITO DE. Amateur architect and builder. Sixteenth century.
He built a stone bridge in Sacapulas, Guatemala, in 1570 which was destroyed and rebuilt in 1590 and again in 1616. According to the stories told, he had a hut built on the river bank so that he could live right on the job and supervise the Indian workmen.

Remesal, **1**: pp. 338, 606; Ximénez, **2**: p. 48.

71. VILLAVA, FR. ALONSO DE. Amateur architect and builder. Sixteenth century.
A Dominican monk, he worked on and completed the church of Tecpatán, Chiapas, about 1562.

Remesal, **2**: p. 422.

VII BUILDING ACTIVITY

THE number of churches, monastic establishments, and government buildings constructed in Central America during the colonial period is related to the number of inhabitants, to the measure of success in the program of the Christianization of the indigenes, and to the economic development of the region. Such matters must at least be clarified somewhat, if not answered completely, before any conclusions as to building activity can be reached.

It may be safely stated that the number of buildings constructed in the sixteenth century was not very great, notwithstanding the fact that these were largely of nondescript architectural character. Not such obvious and facile deductions are possible with regard to questions of population. Throughout the entire colonial period the total number of inhabitants in all of Central America, both Spanish and indigenous, was scant indeed when compared to that of contemporary Mexico. At the outset of Spanish domination in the sixteenth century, Central America must have been extremely sparsely inhabited considering that some two centuries later, in the census taken in 1778, the total population numbered about 800,000.[1] Just what were the total figures for the entire population during the sixteenth and seventeenth centuries it is impossible to tell. Only those Indians who had been Christianized, who lived in established towns, and who were heads of families were included in official tabulations. Rarely were individuals as such counted, except for Spaniards, since the reasons for taking a census were largely for tax purposes. When the civil authorities took a count, only heads of families were included. When the church was the interested party, only those who were old enough to be considered communicants were numbered in the census. Be the sum total of the counted plus the not-counted population what it may, the only figure which has any import at all as regards building activity is that which represents the inhabitants who could be and were actually counted. Only the countable population, then, is the factor that determines the rate of building activity. Though the figures derived in any one of the various censuses taken may be inaccurate as regards the actual total number of inhabitants in Central America, these figures do at least represent the population accessible in cities, villages, and other localities where religious and civil buildings existed. The unbaptized Indians who were still living as they had been in preconquest times were not a factor in building activity. In the sixteenth century, for example, the mere handful of Spanish settlers more directly influenced building activity than the hordes of Indians only recently Christianized. In all the province of Guatemala in 1586, including Chiapas and El Salvador, there were but four Spanish cities; namely, San Cristóbal de las Casas, San Salvador, San Miguel, and Antigua, as well as one *villa*, Sonsonate.[2] It is no wonder, then, that practically no buildings of sixteenth-century date are extant today, so few having ever been built in the first place.

From the very first the impetus to building activity, especially of churches and monastic establishments, was greatly strengthened because quite frequently the crown contributed in some part toward the construction and maintenance of church property. In the eighteenth century, after the secularization of the doctrine and the ousting of the monastic orders from parish administration, the crown's financial contributions were of even greater importance for new construction as well as for the repair of churches damaged in the frequent earthquakes. But before this administrative change had come about, the Dominicans and Franciscans who had been most active in the evangelization of the native population were naturally also the most active in the construction of churches and monastic buildings, especially after a royal *cédula* emitted in 1538 made it legal to use the income from tithes for church and convent construction.[3] So fervent was the missionary activity and so strong the desire to impress the native population, that the friars very often undertook the construction of churches and convents out of all proportion to the towns of miserable mud and wattle and daub thatched huts. There were churches in Chiapas, according

[1] Juarros, **1**: p. 14; *Gaceta de Guatemala* **6**: 256 (1802): p. 100; Baron Castro, *La Población*, pl. LVIII, pp. 225 ff.

[2] Ponce, *Relación*, pp. 383 ff.

[3] Remesal, **1**: p. 73; Fuentes y Guzmán, **3**: p. 218.

to Fr. Tomás de la Torre, which were more extensive than those found in Spanish villages of twice and three times the population.[4] This phenomenon was not unique in Central America where the ecclesiastical establishments were never as numerous, imposing, or grand as those of Mexico. The impression is frequently given that during the whole of the colonial period in Mexico nothing was done except the construction of myriads of churches, supposedly some eight thousand or more in the eighteenth century alone.[5] Even allowing for the most extreme exaggeration, no such numbers were ever remotely equaled in Central America.

The greatest building activity throughout the colonial period would naturally be expected to occur in the capital city of Antigua, in the towns of the immediate surrounding region, as well as highland Guatemala and Chiapas where the native population was densest. The country toward the Atlantic coast, including the present-day Departments of Petén, Alta Verapaz, and Baja Verapaz in Guatemala, northern Honduras, and northern Nicaragua, is still quite depopulated today. Missionary activity of any intensity began rather late there and only after the highlands and the Pacific coastal region from the Soconusco to, and including, present-day El Salvador and Nicaragua had been consolidated politically. The outlying regions in the areas now comprising the greater part of Honduras, Nicaragua, and Costa Rica were active fields for missionary activity throughout the whole of the seventeenth and eighteenth centuries. In parts of Costa Rica, as well as the Petén and Verapaz in Guatemala itself, the work of the missionaries continued well into the nineteenth century. In view of the foregoing, one would hardly expect to find buildings in any number or of any architectural importance in the regions so scantily populated, and by Indians not yet Christianized at that.

The principal cities of Central America, those founded by and for Spanish settlers as distinguished from the *pueblos* established for Indians exclusively, are the *loci* where the architecture of Central America developed. Because Antigua was the capital and most populous of all, it is not an unexpected phenomenon that the Antiguan style is the prototype of the architecture which developed in the remaining cities of Central America—San Cristóbal de las Casas in Chiapas, Cartago in Costa Rica, Comayagua in Honduras, Granada and León in Nicaragua, and San Salvador, Sonsonate, and San Miguel in El Salvador. Even the smaller and less important towns, not only those reserved for Spaniards but for other segments of the population as well, also looked to the capital city for the style to follow in the construction of civil and religious buildings. In the outlying regions not yet Christianized, buildings which could be classified as architecture were largely nonexistent. The colonial archives of Guatemala City are replete with documents attesting to the fact that in the remote areas of Central America the basic problem was ever one of the conversion and the Europeanization of the native population rather than church building.[6] In Honduras as late as the year 1814 missionary activity was still being carried on, especially on the north coast.[7] In 1818 there were still 122 *reducciones* there, that is, converted groups of Indians, of which forty-five were older while the remainder had been organized during the term of office of the then present governor of the province.[8]

By mid-eighteenth century the architectural monuments of the city of Antigua had reached the maximum number built, for with the great earthquake in 1773 and the removal of the capital soon afterward all construction activity ceased.

[4] Blom, *Tomás de la Torre*, p. 195.

[5] Atl, *Iglesias de México* **3**: p. 5, gives 4,000 or so as the number built. In another place, **4**: p. 87, he states that 8,000 were built in the eighteenth century alone. See also, Guido, *Redescubrimiento*, p. 228, who gives the number as 10,000 in eighteenth-century Mexico, and Sanford, *Architecture in Mexico*, p. 156, who says that within seventy-five years after the conquest 400 monasteries were built by the Franciscans, Dominicans, and Augustinians.

[6] See the following documents in *AGG* under the general classification A 1.23 for missionary activity dating from the sixteenth to the beginning of the nineteenth century: (1594) 1513–10065, Taguzgalpa, Honduras; (1607) 1514–107, same; (1609) *Libro de reales cédulas*, 600–1615–Leg. 1514 fol. 147, Nicaragua; (1610) *Lib. rea. ced.*, 1514 fol. 160 in Honduras; (1610) 10069–200, Nicaragua; (1613) 10069–232–1514, Costa Rica and Nicaragua; (1643) 100–71–1516, in the Petén, Guatemala; (1656) 10073-1518, Verapaz; (1664) 10075–1520–137, Honduras; (1671) 10075–1520–13, Nicaragua; (1680) 10076–213, general missionary work; (1686) 10077–1522–140, general; (1686) 10077–1522–189, Verapaz; (1702) 10079–1523–53, Petén; (1709) 10080–1525–44, unidentified region; (1709) 10080–1525–46, Costa Rica; (1709) 10080–1525–59, general; (1709) 10080–1525–57, Petén; (1713) 10080–1525–254, general; (1713) 10080–1525–262, Petén; (1714) 10080–1525–317, Petén; (1715) 10081–1526–15, Petén; (1716) 10081–1526–60, Petén; (1717) 10081–1526–68, Petén; (1759) 10083–1528–138, general; (1764) 10083–1528, region unidentified; all the foregoing are reproduced in *BAGG* **5**: (1939/40), the entire volume devoted to documents of this sort.

See also the following for later missionary activity in outlying regions under the classification A 1.12: (1768) 506–50, region unidentified; (1768) 2472–117, Honduras, Nicaragua, and Costa Rica; (1768) 4826–119, Nicaragua; (1768) 4826–129, Nicaragua; (1769) 2474–117, Nicaragua; (1769) 2475–117, Nicaragua; (1771) 4831–119, Nicaragua; (1775) 2477–117, Costa Rica; (1785) 514–50, Honduras; (1788) 2482–118, Nicaragua; (1795) 2486–118, Chiapas; (1813) 2489–118, general missionary activity.

The above documents are not listed in the catalogue appended to the bibliography.

[7] *BAGG* **7** (1941/42): pp. 146 ff.

[8] *AGG*, A 1.37 (1818) 17517–2335, also *BAGG* **7** (1941/42): pp. 175 ff.

The public and ecclesiastical buildings were listed in a geographical report in 1740 as follows:[9]

1) Cathedral.
2) Parish church of San Sebastián.
3) Hospital Real de Santiago, for clergymen.
   Convents for men, a total of six.
4) Santo Domingo.
5) San Francisco.
6) San Agustín.
7) Compañía de Jesús and its Colegio Borja.
8) San Juan de Dios and its hospital.
9) Belén and its hospital.
10) Colegio de Nuestra Señora de la Asunción.
   Convents for nuns, a total of five.
11) La Concepción.
12) Santa Catalina.
13) Santa Clara.
14) Santa Teresa.
15) Las Capuchinas.
16) Casa y Colegio de Niñas.
17) Casa de Recogidas.
18) El Calvario and its chapels of the Via Crucis.
19) Chapel (hermitage) of Santa Lucía.
20) Chapel (hermitage) of Santa Cruz.
21) Chapel (hermitage) of San Lázaro.
22) Beaterio de Santa Rosa, for Spanish women and widows.
23) Hermitage of Cruz del Milagro.
24) Hermitage of Ntra. Sra. de los Dolores del Cerro.
25) Hermitage of Ntra. Sra. de los Dolores de Abajo.
26) Hermitage of Ntra. Sra. de los Dolores del Manchén.
27) Chapel Espinosa.

The churches on the outskirts of the town such as Jocotenango, San Cristóbal el Bajo, Santa Ana, and Santa Isabel do not figure in this list. The Beaterio del Rosario which was built later does not appear either, nor do any of the other buildings such as the Seminario Tridentino and the University, nor the public buildings such as the Ayuntamiento and the Capitanía, all of which date after 1740.

A more complete listing of buildings was given by the archbishop Cortés y Larraz a few years before the catastrophe of 1773 in his report of a personal inspection he had made of his diocese. The area under his jurisdiction included most of present-day Guatemala and part of El Salvador as well, numbering a total of one hundred and twenty-three parish seats or *cabeceras de curatos*.[10] One would suppose that there was at least one church in every parish seat and an indeterminate number in the little villages, *visitas*, under the ecclesiastical administration of each parish. In another place he contradicts himself and gives the number of parishes as 115 including four in Antigua, but gives no information in any case as to the number of church buildings in these parishes.[11] More complete information is derived from another source, a document dated in 1777, where the number of churches in the same region, that is, Guatemala and the larger part of present-day El Salvador, is given as 419.[12] Not included in this figure are the churches in the regions comprising present-day Chiapas, Nicaragua, Alta Verapaz, Baja Verapaz, Honduras, and Costa Rica. The most nearly complete and accurate figures are given by Juarros who depends on an actual census taken in 1778 in which the total number of churches in the whole Reino de Guatemala, that is, the whole of Central America from Chiapas to and including Costa Rica, is given as 759.[13] The total number of churches in the bishopric of Guatemala alone including part of present-day El Salvador is given as 424 and thus corroborates the information given in the document of 1777. The remaining 335 churches were located as follows: in the bishopric of León, Nicaragua—88; in that of Chiapas, Ciudad Real (San Cristóbal de las Casas)—102; in that of Comayagua, Honduras—145. Juarros does not give any figures for Costa Rica, but does list the towns there under the heading of *Provincia de Costa Rica* as follows: four principal towns with six smaller ones attached, as well as five small places which he calls *doctrinas*. The latter were probably villages where recently baptized Indians had been gathered.[14] One may assume that in all of Costa Rica there were hardly more than a dozen or so churches in 1778.

Juarros gives the population at the end of the eighteenth century of the entire region of Central America as being a total of 797,214.[15] When his individual figures for each region are added together, the total comes to 829,093. In the latter figure he probably included those towns still under the jurisdiction of the regular clergy, thus accounting in part for the discrepancy.[16] According to a contemporary author, who corroborates the figures given by Juarros, the number of churches in Honduras at this time was 145 in all.[17] A further corroboration of some of Juarros' figures comes from an early nineteenth-century source dated in 1810 in which the number of towns in the whole Reino de Guatemala is given as 774. This would make Juarros' figure of 759

[9] *AGG*, A 1.17.1 (1740) 5002–210, also *BAGG* **I** (1935): pp. 5 ff.

[10] Cortés y Larraz, fol. 279 v. ff.

[11] *Ibid.*, fol. 6 ff.

[12] Larreinaga, *Prontuario*, pp. 56 ff.

[13] Juarros, **I**: pp. 14 and 72.

[14] *Ibid.*, p. 73.

[15] *Ibid.*, p. 14.

[16] *Ibid.*, p. 72.

[17] Goicoechea, *Relación*, p. 314.

churches appear very near the truth.[18] By the early nineteenth century in all of Costa Rica there were but 39 parishes and 3 missions with a total of 88 churches,[19] and in Honduras 137 towns were divided into 35 parishes with a total of 145 churches, the same figure as given by Juarros for 1778.[20]

To sum up then, at the very end of the colonial period there were fewer than 800 churches in all of Central America, of which at least 450 were located in present-day Guatemala and in part of El Salvador. Hardly the impressive figures reported for Mexico, but still considerable in proportion to the total population of Central America which numbered about 800,000 at the most at the end of the eighteenth century.

[18] Larrazabal, *Apuntamientos*, p. 107.

[19] Larrazabal, *Bosquejo*, p. 123.

[20] Larrazabal, *ibid.* pp. 124 ff

PART II

The Antiguan Style: Analysis of Elements

VIII PLANS

1. *Church Plans*

NOT until well into the seventeenth century when the use of masonry materials became common were buildings constructed with any desire for accuracy. The mid-sixteenth-century Dominican establishment in San Cristóbal de las Casas was of such extremely humble and nondescript character, for example, that the church, cloister, dormitory, and other offices were laid out by pacing off the plan on the ground, stakes and mason's lines not being used at all.[1] Such primitive methods would be normally employed for temporary structures. For more permanent buildings, it is more likely than not that even in the sixteenth century such methods as those later described by Fray Lorenzo de San Nicolás were employed, methods which are still good building practice to this very day where optical instruments are not employed in laying out buildings on the construction site.[2] He advises that the lines be stretched using a square, the bigger the better, and that once the foundation trenches are filled the lines be rechecked to see whether they have not been moved. After this has been done, the foundation is then leveled off before proceeding with the raising of the walls.

The cruciform type of church plan with projecting transepts described by San Nicolás[3] is not common in Antigua. Except for two or three of the more important churches, all were actually single nave in plan and not unlike one of the common types enumerated by Brizguz y Bru, an eighteenth-century author of an architectural handbook.[4] The single-nave church plan was also quite popular in seventeenth- and eighteenth-century Andalucía.[5] It is noteworthy that the optimum proportions suggested by Brizguz y Bru have their counterpart in some of the contemporary constructions in Antigua; namely, a proportion of about 1:4½ or 1:5 for single-nave churches with a crossing. Another interesting detail Brizguz y Bru advises on is the location of the sacristy (*sagrario*). It should be to one side of the presbytery if it cannot be placed at the front of the church. In almost all Antiguan examples, the sacristy is at the front of the church projecting to one side of the façade.

Except for these few generalities with regard to the layout of the building, *antigüeño* church plans follow a pattern quite their own and tend to be rather long and narrow. This elongated character may be present either as a result of additions, as in the case of Santa Cruz and Los Remedios, or it may have been so designed from the very first, as in El Carmen.[6] (FIGS. 131, 24, 116.) Plans with a wide central nave and side aisles are not too common in Antigua or in the rest of Central America. The Cathedral, the church of the Compañía de Jesús, La Recolección, La Merced, and possibly the no-longer-existing church of Santo Domingo are the only examples of triple-aisled churches in Antigua. Only the Cathedral, Escuela de Cristo, and La Merced had visibly projecting transepts. (FIGS. 40, 86, 105, 62, 161.) The other three-aisled churches were no different in exterior appearance from the single-nave barnlike buildings. The transept bays which are necessarily of equal length as the crossing are lined up with and are as wide as the side aisles. Furthermore, they are not differentiated on the exterior for their roofs are of the same height as the side aisles. In a strict sense, clerestories as such are absent because the vaulting of the nave and side aisles are also equal in height. In other words, the *cajón*, that is, the barnlike or boxlike structure of the single-nave church was extended in width and the interior divided by the addition of two rows of supports to form a central nave and two side aisles. Only in the Cathedral and La Merced are the nave roofs raised above the side aisles to form true clerestories. The single-nave barnlike church is not only

[1] Remesal, **2**: p. 147.

[2] *Arte*, pt. I, ch. XXIV, pp. 52 ff.

[3] *Ibid.*, ch. XIX, pp. 42 ff.

[4] Brizguz y Bru, *Escuela*, pp. 97 ff.

[5] Kubler, *Arquitectura*, pp. 26 and 36. See Wethey, *Colonial Architecture*, pp. 11, 29, for the same type of plan in Peru derived from the Andalusian *mudéjar* church; also Sancho Corbacho, *Arquitectura, passim*, for some examples in Andalucía.

[6] For details of buildings in Antigua referred to in this chapter as well as the others of pt. II, see the specific buildings described in detail in pt. III, chs. XII through XV below.

common in Guatemala, but is also widespread in Mexico. Buildings of this type are uncomplicated to plan and are easily constructed. If any architectural interest or dramatic accent appears on the exterior, it is placed on the façade only. The main emphasis was actually concentrated on the interior space by way of decorative architectural details and furnishings such as retables.[7]

The most common layout or scheme of the *antigüeño* church plan may be divided as follows: The first bay behind the façade is the choir. In most cases this part of the plan is divided vertically by a mezzanine floor, either framed of wood or vaulted, which separates the lower and upper choir, *coro bajo* and *coro alto*. An organ is usually placed in the upper floor. The second part of the plan is the nave, sometimes referred to as the *cuerpo*. It is divided into well-marked bays if vaulted. If roofed with wood and tile, the nave consists of a long narrow hall, sometimes with two rows of wooden posts set on stone bases to support the tie beams of the *artesón* above. The third element of the plan is the crossing. In almost all cases it is square, and in the eighteenth century was frequently surmounted by a cupola consisting of a lantern, dome, and drum supported on pendentives. Finally, at the opposite end of the church the *capilla mayor* or presbytery is located. This dependency is sometimes also referred to as the *altar mayor*. Quite often it is a step or two up from the floor level of the nave and crossing. An added feature of the plan is the sacristy (*sagrario*) which, in most cases, is a separate unit abutting on and projecting from one side of the façade. In some examples a symmetrical structure is built on the opposite side of the façade. The two extensions thus form the lower stories of twin towers which frame the central retable portion of the façade, referred to as the *portada*.

The complete church was conceived of as consisting of the above parts which could be, and were in fact, built separately as conditions warranted. It was quite common, especially during the sixteenth and early seventeenth centuries in small poor towns, to build the *capilla mayor* first so that mass could be celebrated. Or, in some instances, a small nondescript barnlike building without any architectural interest would be reformed by the addition of a *capilla mayor* at the far end and a choir bay and a new façade at the front end. This actually happened in Los Remedios, Santa Cruz, and even in the case of as important a church as San Francisco. (FIGS. 24, 131, 51.) Only in the eighteenth century did it become more and more common to build all the parts of the church plan at once, as happened in San José el Viejo and El Carmen. (FIGS. 167, 116.) Churches were also frequently enlarged as a matter of course after having been damaged in earthquakes, so that few remain which were constructed in a single building period, Las Capuchinas being perhaps a unique example. (FIG. 145.)

The custom of building the *capilla mayor* first probably goes back to the sixteenth century when the Indians were being hurriedly gathered into towns and converted, so that places of worship were needed almost overnight. In the Yucatan peninsula there were many such temporary thatch structures, known as *ramadas*, sometimes built in connection with open chapels where the main altar was located.[8] The thatch *ramada* served as a sort of nave for the congregants. A special chapel for Indians built in the sixteenth century and located next to the convent of Santo Domingo in Antigua was still in use in the early seventeenth.[9] It is doubtful, however, whether it was like the open chapels of Mexico or whether the nave was in the form of a *ramada*.[10] The contemporary literature of the sixteenth and seventeenth centuries contains many references to churches which were being altered or enlarged, in which the nave, *el cuerpo*, of the church was roofed with wood and tile while the *capilla mayor* was vaulted.[11] There are also instances in the outlying regions where the *capilla mayor* was roofed with wood and tile and the nave with thatch, as for example, that of San Pedro Necta even as late as the end of the seventeenth century.[12] Even so important a building as the pre-1680 Cathedral of Antigua had a *capilla mayor* roofed with wood and tile.[13]

The church building itself was always spoken of as being composed of different parts. The *cuerpo* is always referred to as separate and distinct from the *capilla mayor*. For example, in the contract for the construction of Santa Catalina in Antigua the choir, nave, *capilla mayor*, and sacristy are described as independent units.[14] The use of different materials for the nave and the *capilla mayor* is frequently an indication that the two parts were built at different times. Even simple barnlike churches entirely roofed with wood and tile are,

[7] See Angulo, *Eighteenth Century Church Fronts*, pp. 27 ff., for some Mexican examples.

[8] Roys, *Conquest Sites*, pp. 145 ff.

[9] Remesal, **2**: pp. 152 and 246; also Ximénez, **1**: p. 485.

[10] See Toledo Palomo, *Capilla abierta*, pp. 40 ff. For open chapels in Peru, see Wethey, *op. cit.*, pp. 13, 52, 127, 171. See McAndrew, *Open Air Churches*, for a monumental work on the open chapels in Mexico. He is of the opinion that this type of structure did not appear in Guatemala except for the two known from the contemporary literature in Antigua. See his pp. 348–349, also p. 283, fn. 9, for *posas*.

[11] Ponce, *Relación*, pp. 410 ff. and 439 ff.

[12] Fuentes y Guzmán, **3**: p. 72.

[13] *Ibid.*, p. 356.

[14] *AGG*, A 1.20 (1626) 757, also *BAGG* **10** (1945): pp. 221 ff.

nevertheless, always distinguished by a difference in roof levels over different parts, that of the *capilla mayor* rising above the rest of the building. In the manuscript of Cortés y Larraz, especially on the maps where he locates towns and draws the church buildings, the rear end of the tile-roofed churches depicted, presumably the *capilla mayor*, is invariably shown raised above the main part of the structure.[15] The same feature, a raised portion square in plan at the rear end of the boxlike nave, can still be seen in any number of little churches all over Guatemala today. None, of course, are altogether in their pristine colonial state. The churches in the towns of Patzicía, Sumpango, San Bartolomé Milpas Altas, to mention but three as examples, look as if constructed of two boxes, one of which is set upright at the far end to serve as the *capilla mayor*.

The church building was always integrated into the surrounding area where it was located and never considered as an independent feature unrelated to the site. In the city of Antigua the street plan determined the setting for the church, usually, if possible, on a plaza. Or if not, then a small area in front was always left open so that the church façade was never actually on the same building line as the adjacent structures on the street. The open space in front of the church, referred to as the atrium, very frequently also served as the main square in many small towns. In Mexico, especially during the sixteenth century, the atrium was walled and served as a place of defense.[16] The atrium in Guatemala was primarily for processions and spectacles of various sorts connected with divine services, as was also true in Mexico. Sometimes a little chapel, *posa*, likewise as in Mexico, was located in each of the four corners of the more or less square open area. Remains of these little chapels still exist in the towns of Palín (San Cristóbal Amatitlán of colonial times), San Antonio Aguas Calientes, and perhaps in one or two other places in Guatemala. As a matter of fact, the atrium with its little chapels in front of the church in Malacatancito was referred to as the "cuadro processional," that is, the processional square.[17] Not all the spectacles which took place in the open area in front of the church were necessarily religious, to judge by the fact that in 1615, during the festival of Corpus Christi, the City council of Antigua permitted the staging of theatrical performances in the atrium of the Cathedral, that is the raised platform or *lonja* approached by steps from the Plaza Mayor below.[18] (FIGS. 40–42.)

Frequently the church was built on a platform which raised the building above the general level of the plaza and which was referred to as the *lonja*. The *lonja* formed part of the façade ensemble. That churches should have *lonjas* was considered necessary and important enough to have warranted special considerations on the part of the *ayuntamiento* of Antigua who granted a needed extra bit of land to the church of Santa Lucia in 1712 for just such a purpose.[19] Definite instructions were promulgated by the crown in 1573 requiring that cathedrals be raised above the general ground level and be approached by steps so that they would stand higher than the other buildings in the square.[20] This device of raising the church building above the general ground level became quite common in many Guatemalan towns, even in those with but a single church, let alone a cathedral. Low-stepped *lonjas*, however, are the rule not only within the city limits proper of Antigua, but also in the *barrios* just on the outskirts of town; as for example, Santa Cruz, Santa Ana, Jocotenango, and San Cristóbal el Bajo. (FIGS. 132, 176, 209, 38.) But on the other hand, small parish churches approached by high, imposing flights of steps giving access to the *lonja* immediately in front of the façade can be seen in Palín, Patzún, Patzicía, Sumpango, and Chichicastenango, these being but a few examples that come to mind.

2. *Convent Plans*

THE first three monastic orders which had been established in Central America during the second quarter of the sixteenth century were, of necessity, so engrossed in the work of the evangelization of the native population on the one hand, and were represented by so few monks on the other, that they had neither the interest nor the means to build conventual houses of more than temporary nature. For example, the Dominican convent in Zinacantán, Chiapas, including a cloister, dormitories, cells, and other offices was built in the space of three days in 1546. This would imply, of course, that the structure was no different from the native type and probably of wattle and daub and roofed with thatch. A more formal plan was conceived for the convent of the same order in 1550 in nearby San Cristóbal de las Casas, but it is hardly likely that it was carried out in any but the simplest building materials. The plan included a church, a residence for the monks, a school, an infirmary, a guest

[15] See *Descripción* for a map of the parishes Cortés y Larraz visited as well as the individual maps of each parish.

[16] Atl, *Iglesias de México* **6**: p. 119; also Toussaint, *Arte colonial*, pp. 49 and 78. [17] Fuentes y Guzmán, **3**: p. 175. [18] *Efem*, p. 43.

[19] *AGG*, A 1.10.3 (1712) 16543–2280.

[20] *Fundación de pueblos*, pp. 331–360. See also *Archivo Nacional*, Madrid, MS. 3017, "Bulas y cédulas para el gobierno de las Indias," an English translation in *Hispanic American Historical Review* **4** (1921): pp. 743–758 and **5** (1922): pp. 249–254.

room, a kitchen, a dining hall, other necessary rooms or offices, and a garden as well as a corral for animals.[21] The types of dependencies listed are interesting, being no different from those normally found in the conventual establishments in later centuries in Antigua. Temporary type of construction, as has been pointed out, began to be abandoned by the end of the sixteenth century, so that by the early seventeenth century the monastic orders were housed in permanent and formally planned structures. Even discounting the exaggerations of Friar Thomas Gage, the convents of Santo Domingo and San Francisco he describes in the early seventeenth century were already somewhat sumptuous and planned so as to provide more than the mere amenities and comforts of communal monastic life.[22] Yet there are instances of less opulent orders, as for example, the Augustinians who established themselves in 1610 in some private houses where they continued living until they were able to build a formal convent later in the middle of the seventeenth century.[23] The custom of converting private houses for use as convents may be cited also, among others, in the case of Santa Catalina in that same century, and as a purely temporary measure in the case of Santa Clara even as late as the eighteenth century.

All monastic establishments, those for monks as well as those for nuns, shared a common type of plan dominated by the church and a formal cloister which was usually two stories high and surrounded on all sides by arcading. (FIGS. 78, 144, 152, 165.) This is understandable, for in fact in Mexico by 1550 monastic architecture had been standardized on order of the first viceroy who imposed a plan which was to become traditional.[24] The church was always adjacent to the main cloister and usually arranged so that access was had both from the exterior for the public and through the convent for the resident members. (FIGS. 51, 62, 105, 145.) In the case of the nunneries of La Concepción, Santa Clara, and Santa Catalina, the main façade of the church lies within the convent walls. (FIGS. 127, 141, 22.) The lay public entered by means of a door on one of the long sides facing the street. This special type of orientation reserved for nunneries is paralleled in Mexico where it also was customary to locate the church longitudinally along the street with the principal door opening from one of the long sides directly into the nave. The intention behind this scheme was to have the church and convent behind form an integral part with the neighboring private houses in order to emphasize the monastic character of the buildings.[25] The arrangement of the other two nunneries in Antigua, Santa Teresa, and Las Capuchinas is no different from that employed for convents for monks. The main church façade faces the street from which it is set back, the space forming the atrium. (FIGS. 76, 146.) The lay public enters, as is normal with monastic churches of monks, directly into the choir bay of the church. In all instances, regardless of whether the main façade is within the convent walls or on a street, the church shares one wall with the cloister.

[21] Remesal, **2**: p. 128.

[22] Gage, ch. XVIII, pp. 283 ff.

[23] Remesal, **2**: p. 353; Juarros, **1**: pp. 126 ff.

[24] Atl, *op. cit.* **6**: p. 17.

3. *Arcaded Cloisters and Façades*

BY the eighteenth century all monastic establishments in Antigua had cloisters two stories high, all of which are presently in ruins. The cloisters, still extant in part and of which enough remains to reveal the original plan, are La Merced, Santa Teresa, Las Capuchinas, Santa Clara, and Escuela de Cristo. (FIGS. 63, 78, 143, 165.) Extremely few or no traces at all remain of the cloisters of Santo Domingo, Santa Catalina, La Concepción, San Jerónimo, and San Agustín, and only fragments of the cloisters of San Francisco, La Recolección, and La Compañía de Jesús. The Palacio Arzobispal also had a cloister, only a small part of which is still intact. (FIG. 102.) The University, the Seminario Tridentino (FIGS. 184, 193), and the Colegio de Indias all had one-story cloisters which are for the most part still standing, though repaired at various times in the last 150 years or so. The cloister of the Colegio de Indias, in the same block as the latter two but probably part of the same building complex, has been altered, the building now serving as a public school. The two principal governmental buildings, the Ayuntamiento and the Capitanía are distinguished by their graceful two-story arcaded façades which are still largely intact, though about half the length of the latter was rebuilt during the late nineteenth century. (FIGS. 156, 202.)

In all cases, except in that of the cloister of the Palacio Arzobispal, the supporting piers or columns of the arcades are of exceedingly squat and massive proportions. All connecting arches employed are most frequently based on a semicircle. In some cases, however, the arches are described as secants of circles or are based on three separate centers. On the other hand, the arches of the Seminario Tridentino and the University are mixtilinear in outline. (FIGS. 184, 193.) The arcades, including both the supports and arches, are invariably built of the usual brick and stucco, except those of the Ayuntamiento and the Capitanía where stone

[25] *Ibid.*, p. 103.

is used. Aside from the supposed desire for greater solidity as a measure to insure resistance to earthquakes, if such was intended, the massive character of the supporting members of arcades is in part due to their brick construction which prevented them from being built in proportions approximating the smaller ratio of diameter to height possible with stone. Yet, it is interesting that in the two examples where stone is employed, the same squat proportions characterize the columnar supports of the arcades.

The cloister of the Palacio Arzobispal is the only example extant which differs somewhat from the rest. (FIG. 102.) The arches of the arcade are rather flat, being based on three centers, while the columns are noticeably thinner than those commonly employed after 1717. These are the classical Tuscan type and about four diameters high. The archivolts of the arches are left plain and an entablature, simple in profile, is set just above the crowns of the arches. Except for the painted decoration, the scheme as a whole is very reserved indeed. The same type of Tuscan order is employed in the cloisters of Las Capuchinas and Santa Teresa, both of which, however, are characterized by very squat unfluted columns supporting arches based on secants of circles. (FIGS. 152, 153, 78.) These also have undecorated archivolts and simple architraves. Each bay of the corridors is conceived of as a cube into which the arched opening is fitted. The same proportions, but with some small variations, obtain in the case of the columns of the arcaded façades of the Ayuntamiento and the Capitanía, both of which are later in date and built of stone. (FIGS. 155, 156, 206.) Here the Tuscan columns are equally as squat in proportion, but are fluted. The arches, now full half circles, have specially accented archivolts. In the Ayuntamiento, the soffits of the arches are also half-round in section and on their surface the fluting from the column shafts continues. (FIG. 160.) The bays at spandrel level are marked by small slightly projecting pilasters which rest on the abaci of the arcade columns and support the entablature above. The latter is reduced to a cornice member only. The treatment of the elevation of the central bays of the Capitanía is more classical in inspiration. A small pilaster is engaged to the fluted column. (FIGS. 203, 204.) Another pilaster above the capital sets off the elevation of each bay at the spandrels in which the arched opening is fitted. The columns of the remaining bays to either side on the arcade are like those of the Ayuntamiento across the plaza. On the second story the spandrels are bounded by small pilasters which rest on the abaci of the capitals. (FIG. 201.) These are nonarchitectonic and frankly decorative, consisting of elongated triangular shapes tapered toward the bottom with breaks or indentations on the long sides.

In lieu of massive squat columns, square piers are employed in the cloister of Santa Clara, in the Casa de Moneda located in part of the Capitanía, in the extant parts of the cloister of Santa Catalina, and in that of San Francisco. The pilaster treatment of the piers which support the arches of the Santa Clara cloister to mark off each bay are close to classical prototypes. (FIGS. 143, 144.) Square piers are also employed in the cloister of the University, but in combination with mixtilinear arches of very complicated profile. (FIGS. 189, 192, 194.) The pilasters engaged to the piers framing each bay are extremely ornate and quite frankly nonarchitectonic. In the case of the Seminario Tridentino, however, though the arches are also mixtilinear in outline approximating an ogee shape, squat Tuscan columns with plain and unfluted shafts lacking diminution are employed. (FIG. 184.) These provide a note of reserve in an otherwise exuberant scheme. The exterior elevation of the arcade is picked out with some plaster ornament accenting the archivolts, while the soffits of the arches are fluted. Of rather unusual and even ornate character are the supports of the semicircular arches of the cloister of the Escuela de Cristo, being an extravagant development from a pier normally square in section. (FIGS. 162, 166.) The process of elaboration might be described as follows: first, the corners of the square are indented with square chamfers; then, half circles are added to each of the four sides. (FIG. 11.) The same shape is also frequently employed for window openings in the eighteenth century and even for basins of fountains. Another unusual type of pier is employed in La Merced cloister. It is octagonal in section. The soffits of the connecting arches continue the same three-fasciae profile as that of the sides of the pier facing the opening. (FIGS. 63, 75.) In other words, the pier in section is based on a square which has been chamfered to form an eight-sided plan. The edges of the archivolts of the arches above are also chamfered to form three fasciae.

IX RETABLE-FAÇADES

THE outstanding feature of the Guatemalan church is the treatment of the main façade. The space behind this frontispiece, so to speak, is of little architectural interest and hardly more than a large box, be it roofed with wood and tile or vaults. Not unlike so many of the eighteenth-century churches of Mexico City, those of Antigua are quite remarkable with regard to the absence of any dramatic or striking effects in the interior, displaying a preference instead for a concentration of decorative features on the façades.[1] The greatest efforts and imagination were expended in adorning the façade which belies the simple barnlike structure behind. More frequently than not, the façade as a whole is but a screen wall and not necessarily related to the interior plan of the church. La Merced, for example, though three aisles in plan has, nevertheless, but one central door, the side aisles ending in blank walls. (FIGS. 62, 64.) Or where flanking towers are employed, these frequently project to either side far beyond the narrow width of the nave behind and from which they are quite isolated in plan. (FIGS. 167, 168.) The façade or frontispiece, *portada*, is conceived of as an independent part of the church attached to the nave, *cuerpo*, but not necessarily integrated with it in a unified design. In fact, as has been pointed out in connection with church plans above, the façade was not always constructed in the same building operation as the rest of the church. There are many instances where façades and choir bays represent later additions to an earlier and simpler church. The same is also frequently true with crossings and presbyteries, also added at different times to the *cuerpo* or nave of a basic boxlike plan.

The church façades in Antigua and in the rest of Guatemala and Central America may be classified into three types: two-tower façades; single-tower façades; and façades without towers. In all three types the most important element is always the central portion, the *portada* which is treated like a retable, and is usually of the same width as the nave behind. The door is the central element and the very focus of the symmetrical design. The façade is really a screen, its surface quite variegated and broken by niches, moldings, windows, and applied orders all integrated in a balanced composition of which the door is the pivotal feature. When present, the towers act as a frame for the central retable as a whole.

In those instances where there is but one tower or none at all, the main front of the church appears as if it were a freestanding retable. In fact, in many seventeenth-century examples the single tower, set at one side of the façade and stepped back somewhat, was not at all conceived of as an auxiliary feature to frame the central retable. When single towers are present they frequently house the *sagrario*, sacristy. In some cases sacristies were actually built independently of the façade in a separate building operation. Fuentes y Guzmán, though not always a reliable reporter as to the appearance of buildings, does take special note of the presence or absence of towers, mentioning, among others, that the churches of the towns of San Pedro Solomá near Huehuetenango, San Mateo Ixtatán in the same region, Malacatancito, and Santa Eulalia all had single towers. Curiously, the last one mentioned though still roofed with thatch had a formal façade with a bell tower.[2] Antiguan examples of the single-tower façade, that is, where the tower is an independent unit projecting to one side, are Los Remedios and San Sebastián. (FIGS. 24, 25, 79, 81.) Other examples of the same type outside Antigua may be seen in the towns of Patzún, Alotenango, Escuintla, Palín, Patzicía, and San Juan del Obispo.

Twin tower façades are not as common in the seventeenth century, the outstanding examples being Belén, the Cathedral, the church of San Francisco, all in Antigua, and the Franciscan church in Ciudad Vieja which was drastically remodeled in the nineteenth century. (FIGS. 35, 42, 53.) In the above four examples, the towers are massive fortresslike constructions which project beyond the width of the nave and actually enclose usable space. On the other hand and in contrast to the latter examples, the façades of Santa Cruz, San José el Viejo, and Santa Ana have twin

[1] Angulo, *Eighteenth Century Church Fronts*, pp. 27 ff.

[2] Fuentes y Guzmán, **3**: pp. 84, 87, 88, and 175.

towers that are employed as decorative, nonfunctional, flanking, symmetrical elements to frame the central retable and which project to either side beyond the width of the nave. (FIGS. 131, 132, 167, 168, 176). Furthermore, these towers are of very reduced size and just large enough to enclose a caracole stair leading to the belfry and roof over the choir. In Santa Cruz only one tower is actually free-standing, the other being integrated into the construction of the sacristy. The façades of the Compañía de Jesús and La Merced show massive towers to either side of the central retable, but these are actually lined up with the side aisles and are contained within the width of the three-aisled body of the church behind. (FIGS. 83, 62, 64.)

In general, the vogue for massive projecting towers, sometimes containing the sacristy, is by and large a seventeenth-century phenomenon. By the eighteenth century the tower is treated more decoratively, as in the cases already mentioned of San José, Santa Cruz, and Santa Ana. Or else, towers are absent altogether, and the central retable is framed by piers, as for example: Santa Teresa, Nuestra Señora de Dolores del Cerro, El Carmen, Santa Clara, Las Capuchinas, Santa Rosa, and Santa Isabel, to mention some examples in Antigua. (FIGS. 76, 100, 118, 140, 146, 178.)

The retable-like design of the central portion of the façade is not surprising for the reason that many "architects" in seventeenth-century Antigua were actually retable builders by trade, *ensambladores*, as for example, one Diego Méndez who in 1661 took the contract to finish the church in Amatitlán, and Agustín Núñez who began as an architect passing his licensing examination in 1687 and that of sculptor and *ensamblador* in 1689.[3] It was not uncommon in seventeenth-century Seville for *ensambladores* to practice architecture.[4] In fact, in seventeenth-century Guatemala the façade of the church is sometimes actually referred to as a retable.[5] This is in keeping with the needs of the popular outdoor religious rites and processions more common in colonial times than now. The façade is part of the general setting, the open area or atrium fronting the church with chapels in the four corners and a stone cross in the center where religious spectacles were enacted. Within this setting and frame of reference, the church façade does indeed serve as a gigantic retable, the main door of which gives a direct longitudinal view to the altar at the far end of the building and the final stopping place of the procession which winds around the atrium before disappearing into the building itself. Even today, in many of the small towns of Guatemala during special festivals and holidays, processions wind through the streets ending at the church front which acts as a monumental ceremonial backdrop where the whole pageant comes to a dramatic culmination.

The similarity between the interior retable and the exterior *portada* is most striking. In fact, very frequently some of the church retables are of such great size as to equal that of actual architectural constructions. Núñez, the very year he was granted license as an *ensamblador*, designed a retable for the conventual church of La Concepción twelve *varas* high by nine *varas* wide, that is, approximately 36×27 feet.[6] Reading the contract signed by the *ensamblador* Vicente de la Parra in 1690 for a retable to be installed in the church of Santa Catalina, one is struck by the close similarity even as to terminology with actual church façades. He agreed to do the work according to the ". . . diseño y modelo, dibujo y mapa, que para ello hice . . .," that is, according to the plan and model which he made. The height was to be eight and three-quarter *varas* measured from the altar to the finial, and the width seven *varas*, that is, about 26×21 feet. The layout was to consist of three bays, *calles*, and three stories, *cuerpos*. On the first story the Ionic order was to be employed, on the second the Corinthian, and on the third story which he calls the finial, *remate*, the composite order. All the columns were to be solomonic and decorated according to his plans. It is interesting that he could not sign the contract himself, for he did not know how to write, yet he could design the retable and make a plan and model of the proposed work![7]

The three by three division described in the contract is typical of *antigüeño* façades, even to the detail of calling the third story the *remate*, that is, the finial. The *remate* is usually confined to the central bay only and acts as a pinnacle or terminating feature of the design. The proportions alluded to are also the ones most commonly found in the retable façades of Antigua, that is, almost a square, if not actually square, excluding the finial. (FIGS. 8, 79.) In the central bay, usually wider than the lateral bays, the door is located at ground level while a large niche-window with a statue occupies the second story. The third story of the central bay, the *remate*, juts above the roof line and has a niche with statuary flanked by mixtilinear half-pediments which carry the eye down to the side bays. The side bays almost always begin with high podia, though in the case of Los Remedios they are relatively low, on which the applied orders rest. (FIG. 25.) Engaged columns of various types

[3] *Efem*, pp. 70, 105, and 108.

[4] See Saenz de la Calzada, *El Retablo*, p. 212.

[5] Fuentes y Guzmán, **3**: p. 175.

[6] *Efem*, p. 108.

[7] *AGG*, A 1.20 (1690) 695–119, also *BAGG* **10** (1945): pp. 224 ff.

mark off the bays and frame niches in which statuary, bonded with the wall behind, is placed. (FIG. 29.) Entablatures divide the first and the second stories as well as the second and the third or finial. The second-story columns usually rest on a podium or low Roman attic of shorter height than that of the first story and which is sometimes broken by the large niche window in the central bay. Niches are recessed between the columns of each of the side bays repeating the scheme of the first story. The use of podia of different heights in each story makes possible a more harmonious proportion, for the upper-story columns would otherwise have to be greatly elongated.

In the case of the Cathedral, a colossal order is employed because of the greater height and width of the three-nave building. (FIGS. 42, 43.) Had a proportionately higher podium been employed here, the monumental effect, apparently sought by means of the tall columns, would have been lost. The same type of colossal order is also employed on the church of the nearly town of San Juan del Obispo, more or less contemporary with the Cathedral. In general, colossal orders two stories in height are avoided by the *antigüeño* builders, since they would have been out of place in the relatively small overall dimensions of the façades.

The strict adherence to the three-by-three scheme is broken in only four cases in Antigua, those of San Pedro and Santa Teresa, where the lower story is treated more like a triumphal arch, in Espíritu Santo, and in San José el Viejo. (FIGS. 33, 76, 99, 168.) The second story of the latter retable-façade is actually the *remate*, there being no third story. This results in a very low, wide proportion for the façade as a whole, an effect which is carried even further by the projecting twin towers. Simplified retable-façades are frequently employed for modest churches in some of the smaller towns of Guatemala, as for example, in San Miguel Escobar near Ciudad Vieja and in San Miguel Morazán, known in colonial times as San Miguel Tejar.[8]

[8]For some provincial examples see Kelemen, *Some Church Façades*, pp. 113–126. There are some examples of striking similarity in Peru which have been likened to gilded wood retables transposed into stone. See Wethey, *Colonial Architecture*, pp. 18 ff., figs. 110, 111, 177, 179, 182, 186, 187. The examples known from Mexico are not as close in appearance, see Kubler and Soria, *Art and Architecture*, p. 76; also Wethey, *op. cit.*, p. 138, quoting Atl, *Iglesias de México 3*: p. 187; and Angulo, *Historia del arte 2*: pp. 673 ff., figs. 622, 625, pp. 626–630 and pl. XXIX, for some retable-façades in Oaxaca, Mexico. The retable-façade is not uncommon in contemporary Spain, though rarely indeed does it occupy the total width of the church front as it normally does in Antigua. Parallels to the Antiguan employment of a retable as a complete church front (apart from the style of the retable itself) are to be found in Valladolid on the church of Nuestra Señora de las Angustias, dated 1597–1606, see Schubert, *El Barroco*, pp. 166 ff., fig. 50; near Jerez de la Frontera on the Cartuja, dated 1667, see Kubler, *Arquitectura*, pp. 75 ff., fig. 101; in Valencia on the church of San Miguel, where there are flanking towers also, dated 1632–1644, see Kubler, *op. cit.*, pp. 75 ff., fig. 98; and in Ubeda on the church of El Salvador, dated after 1536, see Chueca Goitía, *Arquitectura*, pp. 250 ff., fig. 217.

X FAÇADE TREATMENT

THE architectonic quality of the applied orders employed in the ornamentation of façades was really never violated. Even when the fanciful columns and pilasters, often pure decorative inventions, appear in the eighteenth century, they are employed in conformity with structural logic. The scheme of the classical orders—stylobate, column, and entablature—is always retained. By the eighteenth century it also became customary to decorate the plane surfaces of the retable-façade with both geometric and floral patterns in plaster, *ataurique*. In some cases the lacelike ornamentation also runs over the surfaces of the column shafts in a manner not unlike that common on the interior altar retables of the same period.

1. *Pilasters and Engaged Columns*

i. Tuscan and Ionic

Engaged columns of Vitruvian inspiration are not uncommonly employed for the applied orders. Unlike the ancient prototypes, the shafts are sometimes covered with plaster ornament. More severely classical and unadorned shafts are seen on the façades of Los Remedios, San Agustín, San Cristóbal el Bajo, the Cathedral, and San Sebastián. (FIGS. 26, 29, 38, 43, 82.) Yet even these are on occasion decorated with painted ornament as those of La Compañía de Jesús. (FIG. 83.) The façade of the church in the nearby town of San Juan el Obispo is very much like that of the Cathedral in Antigua. In both instances the proportions of the Tuscan columns are long and slender. They are really colossal orders two stories high. All the retable-façades with unadorned column shafts enumerated above date from the period between 1650 to about 1700. When the Tuscan and Ionic orders are decorated with floral or other decorations in plaster, as for example on the façades of the churches of La Merced, Espíritu Santo, and El Carmen they date from the eighteenth century. (FIGS. 65, 67, 99, 118.)

ii. Solomonic Columns.

There are not too many examples of retable-façades in Antigua treated with solomonic or twisted columns. Those employed on the façades of San Francisco and the side door of San Pedro have plain unadorned shafts, whereas the solomonic columns of the convent entrance of La Merced and the church façade of La Candelaria are richly overlaid with *ataurique* ornament looking almost like interior altar retables. (FIGS. 54, 74, 123.) All of the above examples noted here date from near the end of the seventeenth century.

iii. Invented or Free Forms.

A number of invented "free form" or nonarchitectonic pilaster shafts were introduced in the eighteenth century. These might be classified as distinctive species or variations of the *estípite*. But the Antiguan examples are different in appearance being composed of elements quite remote from those seen in the usual *estípite*, that is, inverted obelisks or inverted elongated truncated pyramids. The inverted pyramid type of *estípite*, considered the hallmark of the baroque style in Europe, appears in Mexico in the second or perhaps third quarter of the eighteenth century.[1] An earlier variation is employed in 1712, but the inverted pyramid type actually becomes common only after 1749.[2]

The opinion has been held that the Mexican *estípite* influenced the Antiguan types.[3] This is hardly likely for these variations of the nonarchitectonic pilaster appeared independently in Antigua as a local expression of the widespread vogue in the Hispanic world for the pyramidal *estípite*. The stylistic affinities of the Guatemalan *estípite*, if thus it should be dubbed, are more likely than not to be found in the comparable examples in the province and city of Seville, Spain.[4]

[1] See Villegas, *El gran signo*, *passim*, for an historical survey of the *estípite*, especially chs. XIII and XIV dealing with the baroque type in Spain and Mexico.

[2] *Ibid.*, pp. 154 ff., for the example of 1712 and Kubler and Soria, *Art and Architecture*, pp. 79 ff., fn. 49, for the example of 1749 and later ones too, also p. 369, fn. 28, for a definition of the *estípite*. For some discussions of the *estípite* and its Mexican development see Angulo, *Eighteenth Century Church Fronts*, p. 30, fn. 12; Baird, *Style in 18th Century Mexico*, p. 262, fn. 7, and also his *Ornamental Niche-Pilaster*, pp. 5 ff.

[3] Buschiazzo, *Estudios*, p. 45.

[4] Some comparable examples in Seville and its province of nonarchitectonic pilasters, *estípites*, which do not employ the inverted truncated pyramid element, are as follows: the church of the convent

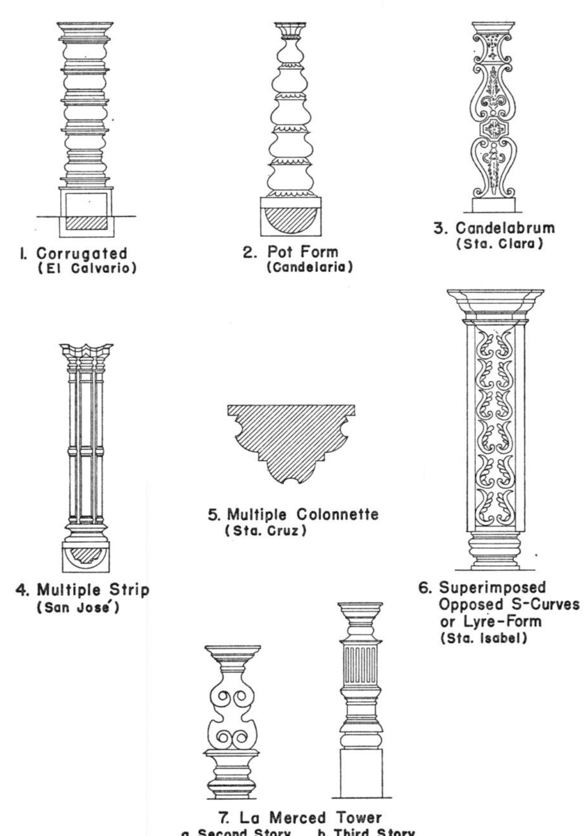

FIG. 2. Nonarchitectonic pilaster shafts.

And as in the case of the Andalusian types, perhaps the closest point of reference with regard to the possible origin of the *antigüeño* nonarchitectonic pilaster shaft, is the Hispanic retable. In fact, such an origin has been suggested for the very Andalusian *estípite*, the Peruvian, and the *antigüeño* as well.[5]

The distinctive nonarchitectonic pilaster shaft was introduced in Antigua during or just before the second quarter of the eighteenth century, that is about the same time that the *estípite* first appears in Mexico. The *antigüeño* types must, therefore, be considered as independent developments, for in a few instances some antedate the comparable Andalusian examples, most of which are from the second half of the eighteenth century.[6]

The seven invented pilaster shaft types are: (FIG. 2.)

a. Corrugated Shafts. This type is composed of a number of superimposed squared amphoras or squared Attic bases and is first seen on the gate house of El Calvario, dated 1720. Corrugated shaft pilasters are also employed on the church façades of Santa Ana and Santa Rosa. (FIGS. 110, 173, 177.) The type appears on the side entrance of San Jerónimo, on the small pilasters which frame the niches of the retable-façade of San José el Viejo, and on the topmost story of the north tower of San Francisco. (FIGS. 182, 169, 53, 54.)

b. Pot-form Pilaster Shafts. This is really a variation of the corrugated type. Only one example is known in Antigua, and is on the door of the ruined structure abutting on the south side of the main façade of the church of La Candelaria, dated in the early eighteenth century. (FIG. 124.) Each unit of the shaft is circular in section and shaped like a squat full-bellied amphora in elevation. The sharp divisions between the three main members of the Attic base (a large lower torus, a scotia, and a smaller upper torus) are blunted so that the whole is blended into a continuous S-curve.[7]

c. Multiple Strip Shafts. This type is first seen on the façade of San José el Viejo dated soon after 1740. (FIG. 169.) It is also employed to frame the side bays of the main door of the Seminario Tridentino dated 1758. (FIG. 183.) The shaft as a whole is half-round in section and is composed of a number of small, narrow, ribbon-like vertical strips each of which is divided by a horizontal astragal molding about the middle of its length.

---

of Los Terceros in Seville, dated 1690–1713; the church of La Vera Cruz in El Arahal, of uncertain date, but doubtless eighteenth-century; the tower of the parish church in Bollullos Par del Condado, dated 1775–1779; the tower of the convent of Las Marroquíes in Ecija, dated 1760; the third story of the façade of the church of San Juan de Dios in Seville, dating probably from near the end of the eighteenth century; on the patio door of the Albornoz house in Ecija, of late eighteenth-century date; the door of house no. 7, Calle Fernando Llera in Fuentes de Andalucía, dated 1753. For published photographs and documentation of the above examples see in the same order as given above Sancho Corbacho, *Arquitectura*, pls. 65, 138, 166, 207, 308, and 316; also pp. 20–25, fig. 1, for a discussion of the origin and development of the *estípite sevillano*. The dissimilarity between the Mexican and the Antiguan *estípites* is recognized by Angulo, *Historia del arte* **3**: p. 31, and by Kubler and Soria, *op. cit.*, p. 84.

[5] Angulo, *op. cit.*, **3**: p. 31, sees the origin of the Antiguan *estípite* in the Hispanic retable. See also Wethey, *Colonial Architecture*, pp. 18 and 239, who believes the *estípite* originates in eighteenth-century Andalucía and was employed in Mexico about the same time, its origin being in the retable builders' craft. Sancho Corbacho, *op. cit.*, p. 22, also states that the origin of the *estípite sevillano* is to be found in the contemporary retable.

[6] In discussing the façades of the churches of Los Terceros in Seville and of La Vera Paz in El Arahal, Sancho Corbacho, *op. cit.*, pp. 140 and 199, pls. 65 and 138, says of the former that one has the sensation of standing before a monument from the Spanish colonies, and that the latter could have been built by some master from overseas. He goes on to say, regarding the façade of Los Terceros, that in analyzing the architectonic elements and decorative motifs, one realizes, however, that there is an indubitable relation between these two façades and the other Sevillian façades of the period.

[7] It is very close to the type on the church of La Vera Paz in El Arahal. See fn. 6 above.

d. Multiple Colonnette Shafts. There is but one example of this type of shaft, that on the retable-façade of Santa Cruz, dated 1731. (Fig. 133.) In effect, a cluster of small columns rises from a base shaped like a squat open-mouthed, large-bellied pot or amphora. Between the colonnettes moldings which are triangular in section project and thus form sharp arrises. Each colonnette is in turn deeply grooved with a concave incision or flute. Described otherwise, the continuous spherical surface of the shaft is broken up into the smaller spherical units, the colonnettes. The spherical surface of each of these is then further disturbed by the alternating plane surfaces of the protruding arrises and receding flutes.

e. Candelabra Shafts. This shaft is made up of two pairs of opposed S-curves or scrolls, the larger spirals meeting at the center. The type is seen on two buildings only, Santa Clara and Escuela de Cristo, both dating from the third decade of the eighteenth century. (Figs. 140, 141, 163, 164.)

f. Superimposed S-curves or Lyre-form Shafts. This is a unique type and seen only on the retable-façade of Santa Isabel, probably dated sometime about the middle of the eighteenth century. (Figs. 178, 179.) It consists of a number of pairs of superimposed opposed S-curves, each pair flaring toward the bottom to outline a lyre. The same lyre shape appears in the metopes of the frieze of the courtyard arcade of the University and also as an ornament on the low parapet wall on the exterior of the Seminario Tridentino. (Figs. 194–196, 183.)

g. Pilasters from the Towers of La Merced. These are purely fanciful forms and completely unrelated even to carpentry treatment representing an expression of the free imagination of the plasterer who created them. (Figs. 66, 72.) Some of the moldings of the one on the third story recall the corrugated pilaster shaft, but the main feature is a form reminiscent of a triglyph or an iron grill. The pilaster employed on the second story is reminiscent of a spiral form and bears even less resemblance to carpentry work.

2. *Door Openings*

i. Semicircular Door in Concentric Niche.

The most common door opening is that spanned by a semicircular arch and set within a recess or niche with a concentric semicircular header. This type is seen in Antigua on the gatehouse of El Calvario, the doors in the lateral bays of the Cathedral façade, Los Remedios, San Agustín, San Cristóbal el Bajo, San Sebastián, Santa Cruz, and Escuela de Cristo. (Figs. 26, 29, 38, 42, 81, 115, 132, 161.) The same door treatment is employed in the churches of the towns of San Antonio Aguas Calientes, San Pedro Huertas, Palín, Patzún, and Tecpán.

The door jambs are usually surmounted by small pilaster capitals from which the semicircular arch springs. Frequently, both the jambs and the archivolt of the door are decorated with some ornamental pattern. When the door is of brick, then the ornament is worked in plaster as on the church of Santa Cruz. Otherwise, when stone is employed for the door, the pattern is worked in low relief, as in the case of San Cristóbal el Bajo. (Figs. 136, 39.)

ii. Semicircular Door in Niche with Semicircular Stilted Arch.

The second most commonly employed door opening is a variation on the first in that both door and niche arches are semicircular. The niche arch, however, is not concentric, but is rather stilted springing from a higher level to form a tympanum above the crown of the door arch. The springing of the niche arch lies at about the height of the crown of the door arch and is marked by a horizontal molding which spans the opening to form the half-circle tympanum. A niche with a statue is usually placed in the tympanum. This type of treatment is employed for the main door in the central bay of the Cathedral and also on the façades of the churches of Belén, San Francisco, Santa Teresa, Compañía de Jesús, and in a modified form at Jocotenango. (Figs. 35, 43, 53, 76, 83, 208.) It also appears on the parish church in Escuintla, a town near the Pacific coast.

In both of the above two door types, the treatment of the reveals of the niche jambs and the soffits of the niche arches usually is quite plain. In the Compañía de Jesús, on the other hand, the simple square-edged niche jamb is rejected in favor of a more complicated and ornate arrangement in which both the reveals and the arch soffit are variegated in section being composed of a number of moldings of different profiles. (Fig. 85.) This type of multiple surface treatment of jamb reveals and arch soffits is normally employed when mixtilinear arches span the door niches.

iii. Mixtilinear Niche Headers with Semicircular Doors.

The ogee arch was sometimes employed for the door niche in the eighteenth century. Yet the header of the door opening itself is always semicircular and not unlike the first type with regard to the square edge treatment of door jambs and archivolt. Mixtilinear niche headers describe a general ogee outline, each individual example varying from all others, no two being exactly alike, as in El Carmen, La Candelaria, and Santa Ana. (Figs. 121, 123, 176.) The reveals of the jambs and the soffits of the arches are composed of a number of moldings complicated in profile which add to the ornate quality of the recess in which the simple semicircular door is set. Mixtilinear arches of this type are also employed for the cloister arcades of the Seminario Tridentino and the University. (FIGS. 184, 189.) On the basis of the dates of the buildings mentioned above, it appears that the mixtilinear arch was not employed in Guatemala before the second or third decade of the eighteenth century.

iv. Trumpet-shaped Door Niche, "Abocinado."

There is but one example of this type in Antigua, the church of El Calvario dated about 1720. (FIG. 115.) The door is set in a very deep niche with splayed reveals and an arch with a trumpet-shaped soffit overhead. This is decorated with elongated petals or flutes to give a seashell effect. The same treatment is seen in the case of the niche-window on the façade of the church at Patzún.

v. Rectilinear Shallow Door Niche.

There is but one example of this type of door treatment in Antigua; namely, in Santa Rosa dated 1750. (FIG. 171.) The niche itself which has a horizontal header is hardly recessed at all, while the actual door opening is spanned by a semicircular arch. The effect is that of a rectilinear panel framing the semicircular opening like a modified *alfiz* motif known in Mexico and probably derived from *mudéjar* prototypes.[8] This type of opening also appears in the little churches of eighteenth-century date in the nearby towns of San Gaspar Vivar, San Lorenzo el Cubo, and Santa María de Jesús. It is also employed in the clerestory and on the inside face of the niche-window in the choir of the Cathedral which now lies open to the sky above the present vaults constructed in the nineteenth century. (FIGS. 48, 50.)

3. *Niches*

RETABLE-FAÇADES are invariably treated with niches which are always placed between the pairs of engaged columns or pilasters in the side bays of both stories as well as in the central bay of the third or finial. Sometimes niches are also introduced in the space to either side of the large niche-window of the second-story central bay. In such cases, two small niches, one above the other, are set on each side of the window as seen in the Cathedral, La Compañía de Jesús, and Santa Rosa. (FIG. 42, 83, 84, 172.) The retable-façade of El Carmen is the only church in all Antigua devoid of niches with statuary, the space for niches in each side bay actually being taken up by an extra pair of columns set on a podium projecting from the general plane of the façade. (FIG. 118.)

Almost invariably the niches are apsidal shaped, that is, semicircular in plan and covered with quarter domes. Generally, only one niche is employed in each of the side bays, except in the Cathedral, San Francisco, Santa Teresa, and Compañía de Jesús where two niches are set one above the other to fill the space between the engaged columns. (FIGS. 42, 54, 76, 83.) A similar arrangement is also seen on the churches of the nearby towns of San Miguel Escobar and San Juan del Obispo.

There is great variety in the framing treatment of niche openings with regard to the type of pedestals, pilasters, and pediments employed. Triangular pediments are more common in the seventeenth century and are seen on the churches of Los Remedios, San Agustín, Belén, the Cathedral, San Sebastián, and La Compañía de Jesús. (FIGS. 27, 31, 35, 42, 82, 83.) But this type is also used later, as for example, on La Merced, Santa Cruz, and Santa Clara. (FIGS. 69, 133, 140.) In the latter, however, the niche header is an ogee arch. Segmental pediments over niches are employed along with the triangular type on the same monuments, very frequently in the upper story. Broken pediments, some with raking cornices ending in scrolls are seen, among others, on San Sebastián, La Candelaria, Santa Isabel, and over the main door of Santa Cruz. (FIGS. 82, 125, 178, 132.) Some of the broken pediments vary slightly in that the upper ends of the raking cornices are joined by a swagged concave section which continues the molding profiles of the cornices, as on La Merced and Santa Isabel. (FIGS. 69, 178.) Sometimes the raking cornices are slightly swagged and joined at the top by a short horizontal member, as in San José and over the door of San Jerónimo. Ogee headers are employed for the niche openings of both Santa Ana and San José el Viejo. (FIGS. 169, 177, 182.)

It can be stated that, in general, simpler forms are employed in the seventeenth century and more ornate forms later.

4. *Niche-windows*

THE second focal point of interest of the retable-façade is the large window set in the second-story central bay on center with the main door. The most common type consists of an opening with a trumpet-shaped semicircular header not unlike a niche, except that the wall behind is perforated and glazed. The reveals and soffits of niche-windows are very markedly splayed providing, thereby, a deep wedgelike space for a free-standing statue on the sill or floor. The surface of both the reveals and soffits is usually decorated with plaster ornament, some very simply and

[8] Toussaint, *Arte mudéjar*, p. 56.

others very ornately. Normally, the niche-windows repeat the general outline of the main door in the lower story, as for example, in Los Remedios, San Francisco, La Merced, Santa Teresa, San Sebastián, and Santa Rosa. (FIGS. 25, 54, 68, 76, 80, 172.) The same type of stilted arch employed in the main door niche of the Cathedral is repeated for the window niche above. Square-headed rectangular niche-windows are to be seen in the churches of Santa Cruz and Escuela de Cristo. (FIGS. 42, 132, 161.)

Octagonal windows, so common on the façades in Mexican churches of the eighteenth century,[9] are represented in only four examples in Antigua: San Cristóbal el Bajo, El Carmen, Santa Ana, and Santa Isabel, though this shape is used quite frequently elsewhere in the church building. (FIGS. 38, 120, 176, 178.)

5. *Entablatures*

CLASSICAL prototypes seem to have inspired most of the entablatures employed in the seventeenth century and even later. Architraves are usually made of three fasciae and friezes are quite plain. Cornices consist of a few simple moldings, but with an exaggerated projection. Widely projecting cornices are quite necessary in order to carry the rain water away from the façades, this being an especially important function in the tropics where torrential downpours are a daily occurrence from March through October. Triglyph and metope friezes are employed on the façades of the churches of San Agustín, San Pedro, the Cathedral, and Santa Teresa. (FIGS. 29, 33, 43, 76.) Except for the façade of La Candelaria (FIGS. 123, 124), broken entablatures are avoided in Antigua, though extremely ornate entablatures sometimes appear in the interior of churches, as for example, La Concepción and El Carmen, the latter a rippling or undulating variety.[10] (FIGS. 130, 119.) Pulvinated friezes are employed only on the façades of Santa Cruz, San José el Viejo, and Santa Rosa, though such a frieze appears on the interior of the chapel of the Seminario Tridentino. (FIGS. 135, 168, 171, 186.) In general, the architectonic quality of the entablatures is maintained, even when pilasters of the order are absolutely free of any structural architectural restraint.

6. *Plaster Decoration and Ornament*

PROBABLY the earliest use of plaster decoration, known as *ataurique*, is seen on the interior piers and on the pendentives of the domes of the Cathedral. (FIGS. 47–49.) Other early examples where decoration of this sort is employed are in the niches of the façade of Belén and in the spaces between columns on the façade of San Sebastián. (FIGS. 35, 82.) Painted ornament is employed on the façade of the Compañía de Jesús, as it also is in the cloister of the Palacio Arzobispal. (FIGS. 83, 103.) But in general, plaster ornament does not become common until after the earthquake of 1717, when such buildings as El Carmen, La Candelaria, Santa Cruz, and Santa Clara are so treated. (FIGS. 118, 125, 135, 140.) Yet there are examples of eighteenth-century church façades such as Las Capuchinas and Escuela de Cristo which are entirely devoid of such ornament. (FIGS. 146, 161.) The finest example of over-all ornament of this type appears on the façade of La Merced where it was applied to an older structure which was remodeled about 1763 and probably repaired again in the nineteenth century. (FIGS. 67, 74.) Even so ornate a façade as that of San José el Viejo does not have any over-all plaster decoration, the applied orders being extravagant enough. This is also true in the case of Santa Rosa. The zenith of the plasterer's craft was reached not in the decoration of façades, but in the cloister of the University and on the ceiling of the chapel in the Seminario Tridentino where the delicacy of workmanship and ingenuity of design have no equal anywhere in Guatemala. (FIGS. 185, 187, 195, 196.)

Because of the general similarities of the craft, this type of plaster ornament, *ataurique*, falls within the continuing *mudéjar* tradition of contemporary Spain. But it should be borne in mind that the repertoire of ornaments employed in Guatemala is not as variegated as that in the Iberian peninsula. The execution of details tends to be flatter and in lower relief.[11] Plaster ornament became common in Guatemala only toward the end of the seventeenth century, while in Mexico it was already known at the beginning of that century.[12] Furthermore, the style of the Guatemalan *atau-*

[9] Angulo, *Eighteenth Century Church Fronts*, p. 28.

[10] The rippling or undulating frieze of El Carmen is the only known example in Antigua and is without doubt related to the type so popular in eighteenth-century Andalucía, particularly in the province of Seville, as for example: on one of the *espadañas* and on the cupola of the church of San Pablo in Seville, dated 1697; on the cupola of the church of San Pedro in Carmona, dated 1760; on the cupola of the church of El Salvador, also in Carmona, dated 1700–1720; on the cupola of the parish church of La Campana, dated 1784; on the cupola of the church of San Francisco in Fuentes de Andalucía. These are but a few that come to mind. For published photographs of the above, see Sancho Corbacho, *op. cit.*, pls. 10, 19, 82, 91, and 185.

[11] See Bevan, *Spanish Architecture*, pp. 105 and 110; also Kubler and Soria, *op. cit.*, p. 78, for a statement on the nature of Spanish, Mexican, and Guatemalan stucco decoration.

[12] Toussaint, *op. cit.*, pp. 10 and 42 ff.

rique cannot be even remotely compared with that of the sixteenth century in Santo Domingo and the Antilles where it is closely related to plateresque motifs derived from Spain.[13]

In addition to its use for surface ornament, plaster was the principal material employed in Guatemala for the modeling of architectural details, such as the applied orders including podia, columns, and entablatures. Even the sculptures in the niches of the retable-façades were constructed of brick bonded to the structural wall behind and finished in plaster. The practice of rendering architectural details in plaster simulating a functional appearance is universal in all church interiors. The lateral arches which seemingly support the nave vaults really consist of brick corbelling with a plaster finish, while the vaults themselves actually spring from and are incorporated with the walls. Likewise, the ribs employed on the soffits of vaults have no direct relationship to function or structure. They are built of single courses of thin square bricks laid on a diagonal so that one corner protrudes from the vault soffit to form the rib outline. These protruding arrises then form the core on which the molding profiles are finished in plaster to give the appearance of functional ribs. All interior pilasters with their fluted shafts, capitals, bases, and entablatures are all worked in brick and stucco. In one instance only, that of Las Capuchinas, are the pilaster shafts of stone. (FIG. 149.) But it is a stone veneer in lieu of a plaster veneer, and like plaster is nonfunctional and merely decorative.

[13] Angulo, *El gótico*, pp. 35 ff., pls. 24, 25, and 45–54.

XI THE ANTIGUAN STYLE

1. *Introductory Observations*

THE dating of each individual building in Antigua by means of objective external evidence, regardless of similarities in appearance to analogous architectural monuments elsewhere in the Hispanic world, is perhaps the most pressing problem which must be solved first, and is a necessary prelude to the threefold task of discovering the origins, tracing the development, and establishing the stylistic terminology of the colonial architecture of Antigua, and, by corollary, of contemporary colonial Central America as well. Once the date of each building is firmly established, an absolute chronology for the stages or periods of development of the architecture as a whole may be delineated, a chronology proper to it regardless of its apparent Iberian and Hispano-American stylistic affinities. On the basis of information derived from contemporary literary and documentary sources and from a direct physical study of the buildings themselves, in fine, by the employment of archaeological methods, objective evidence may be brought to bear on the problems of chronology, stylistic origins or connections, and terminology.

The dating of buildings in colonial Central America solely by means of stylistic comparisons with Iberian prototypes is most hazardous and presupposes a close synchronous development on both sides of the Atlantic which may or may not be true, and which also remains to be demonstrated. It is obvious, nevertheless, that the architectural style of Antigua is an import, but that it went through a stylistic development independent of its Spanish origins is not altogether patent, yet is to be expected. The process of this stylistic transformation, as a phenomenon apart from the question of influences from the Iberian peninsula, can only be ascertained with any measure of accuracy by means of the external evidence alluded to above.

There are many reasons why transformations and adaptations of the style of origin should have occurred. In the first place, the greater majority of the extant buildings were designed and constructed by practical builders rather than formally trained architects. Nor were the abilities of the building craftsmen of the caliber of that common among workmen in contemporary Spain. The lack of experienced and competent craftsmen in Guatemala resulted frequently in the emergence of forms that seem strange and illogical; for example, the many vaults and domes erected without adequate knowledge of the geometry involved. This is equally true for sculptured decoration where a so-called "denaturalization" or "primitization" of the European vocabulary took place and which has been incorrectly adduced to the hand of the Indian.[1]

A second reason why stylistic comparisons with Spain may not be employed as an altogether safe criterion for dating is the afterlife of colonial building methods into the nineteenth and even the twentieth centuries. Coupled with the constant destruction due to earthquake activity, all buildings have invariably been altered and repaired in the traditional manner so as to mislead even the most astute and knowledgeable critics. The same bricks, the same lime mortar, the same mason's tools, including the same type of

[1] Neumeyer, *Indian Contribution*, pp. 112 ff., believes such a process actually took place in Mexico and Peru and even in Guatemala. He cites the Casa de los Leones in Antigua as a prime example of how the baroque forms were transformed back into an early medieval status by the *native* mode of perception. But there is no way of knowing the race of the craftsman or craftsmen who executed the door, whether they were Spaniards, mestizos, Negroes, or mulattoes. The one certain fact is that the workmanship is not very good, and that the rampant lions to either side of the door are probably from the hand of a stonemason and not a sculptor. The style of these lions does not necessarily represent an abandonment of optical methods of representation in favor of conceptual methods because of the Indian mentality of the workman. It would be closer to the facts to recognize in the door of the Casa de los Leones a stereotyped composition imported from Spain and copied by an unskilled workman who tended to flatten the figure using the profile or two-dimensional view because he did not have the experience or skill to render it in a more plastic or three-dimensional manner. This lack of technical skill and the resultant "primitive" appearance which characterizes all sculptured decoration in Central America, be it in stucco or stone, has been misconstrued as Indian. See fn. 13, below, for further discussion of this problem.

plumb bob, were and are still in use in Guatemala since early in the colonial period.[2]

A third reason why the Spanish stylistic vocabulary must be adjusted to local conditions in Guatemala is that there is much overlapping in the chronology of the different regional Iberian styles from which the Antiguan is derived. It seems as if the builders of Antigua selected styles from a catalogue—the catalogue of Spanish architectural history, particularly that of the city and province of Seville.[3]

A fourth reason why the Guatemalan style must be analyzed in its own terms before any stylistic connections with the Iberian peninsula can be clarified is that in Central America, as elsewhere in Hispano-America, building activity progressed independent of direct supervision from Spain, except in the few instances when Spanish architects arrived in Guatemala for the express purpose of designing and constructing specific buildings. More frequently than not, buildings were erected first and official plans drawn afterwards which were then sent to Spain as part of progress reports on construction, as happened for the Capitanía.[4]

2. *The Four Periods of the Antiguan Style*

THE fourteen years which elapsed between the establishment of the first capital at Ciudad Vieja in 1527 and its destruction in 1541 are not represented even by any ruins ascribable to that period, let alone buildings in some stage of preservation. The architecture of Ciudad Vieja during this time was, at best, of the most rudimentary and nondescript character. A similar condition also prevailed even in Antigua during the first century of its existence from about 1543 to 1650.[5] The few major constructions which were attempted in the sixteenth and the first half of the seventeenth centuries have all but disappeared in the later reconstructions and enlargements. Some buildings were actually dismantled and new ones begun from the ground up on the same sites. The architecture of the first half of the seventeenth century, then, is not represented in Antigua at all, but possibly elsewhere.[6] According to some eighteenth-century authors, from about 1590 to 1650 the earthquakes which racked the city of Antigua were so frequent and devastating that the inhabitants did not dare undertake building construction of any value.[7] This observation is, in truth, borne out by the historical evidence and the architectural remains, thus corroborating the fact that only after the middle of the seventeenth century did any accelerated building activity begin.

It was, therefore, fully more than one hundred years after the conquest that the first truly architectural style appeared in Antigua and in the rest of Guatemala and Central America. The architecture which had existed before that time, taking into consideration the various factors of population, materials of construction, and others, if it must be given a stylistic classification, may best be described as "marginal" or "pre-architectural." Beginning then shortly before 1650 and continuing until the final abandonment of the city in 1773, the *antigüeño* style is encompassed within a period of about one hundred and twenty-five years or so during which time it served as the prototype for the architecture of the whole of Central America.[8]

On the basis of the contemporary literary and documentary evidence and from a physical examination of the actual architectural remains, the buildings of Antigua have been found to fall into four periods as follows: the first, from about the second quarter or middle of the seventeenth century to 1680 culminating in the Cathedral; the second, from 1680 to the destructive earthquake of 1717; the third, from 1717 to 1751, witnessing the flowering of a truly native style proper to Antigua; and finally, the fourth period, from 1751 to 1773 when this native style ends abruptly with the destruction and abandonment of the city.[9]

3. *Terminology and Stylistic Connections*

THOSE few authors who have ever dealt with the Antiguan style at all have employed a terminology ranging through the whole gamut of the Spanish stylistic vocabulary more or less contemporary with the colonial period.[10] Two Do-

[2] See ch. III above, no. 8, "Rebuilding, Restoration, etc." Also the case of San Agustín, ch. XII, no. 3, fn. 20, where the seventeenth-century form of the façade was intentionally retained in the eighteenth-century repairs.

[3] See ch. IX above, fn. 8, for the retable-façade in Spain; also, ch. X above, fnn. 4, 5, 6, 7, for the nonarchitectonic pilaster in the province and city of Seville; and here, below, fnn. 20, 21, 22, 23, 24, 25, and 27, for other architectural details of Spanish origin.

[4] See Guido, *Redescubrimiento*, pp. 233 ff., for a discussion of this problem as related to Hispano-American architecture as a whole.

[5] See Kubler and Soria, *Art and Architecture*, p. 83, who also noted this fact.

[6] The Dominican church at Cobán may date from this period. But it has been so drastically reformed that little of its original character can be derived. One or two buildings in Chiapas, the Dominican establishments at Tecpatán and Copanaguastla, are the only other examples of buildings still existing from this early period. See fn. 11 below for bibliographical references to these two buildings.

[7] González Bustillo, *Razón particular*; Vásquez, **1**: p. 264.

[8] Angulo, *Historia del arte* **2**: p. 44, recognized the singular position of Antigua considering it the cradle of all Central American art.

[9] For stylistic summaries of each period see the introductory sections to chs. XII, XIII, XIV, and XV below.

[10] McAndrew, *Relationship of Mexican Architecture*, p. 32, points out the confusion in the terminology employed for Hispano-American architecture as a whole. Villacorta, *Historia*, pp. 306 ff., employs

minican conventual churches in Chiapas, one in Copanaguastla and the other in Tecpatán, supposedly dating from the end of the sixteenth century, have been dubbed plateresque on the analogy of the contemporary Mexican examples.[11] Those Antiguan façades decorated with nonarchitectonic pilasters have been labeled churrigueresque.[12] The Antiguan style has also been seen as a phenomenon demonstrating the "fusión arquitectural hispano-indígena."[13] With somewhat more felicity and nearer the actual fact, some critics have recognized indubitable *mudéjar* traits in the architecture of Antigua.[14]

In searching among the Spanish regional styles more or less contemporary with the colonial period, the nearest to the Antiguan is that to be found in the former Reino de Sevilla, which includes the modern provinces of Seville, Huelva, Cádiz, and part of Málaga.[15] But this comparable stylistic mood is not to be found in the monumental churches of the capital cities of this part of Spain, rather in the small towns of the countryside. The little villages to the west of Seville in the region known as El Aljarafe and also those to the south toward Jerez de la Frontera and Cádiz are dotted with churches where "disembodied" architectural elements, taken out of the context of the buildings on which they appear, have their counterpart in Antigua. As one travels through these small towns, some with ancient churches dating from as far back as the thirteenth and fourteenth centuries, there unfolds before one's eyes the evolution of the artistic tradition in which the Moslem past, the *mudéjar*, acts as the unifying element through the vagaries of the changing architectural styles in the sixteenth, seventeenth, and eighteenth centuries. Direct counterparts of the ancient *mudéjar* churches of the Reino de Sevilla are of course totally lacking in the New World, but the craft tradition which produced them is not. The deeply rooted *mudéjar* craft traditions had transformed the nonindigenous styles of Andalucía into a local regional native architectural idiom which, as the very speech of Andalucía, was in turn exported to Guatemala. That is to say, the Italianate *renacentista* and the baroque styles came to Antigua as a veneer on the "timeless" *mudéjar*.

Mudéjar traits are indeed quite apparent in the architecture of Antigua. In fact, the *mudéjar* is the one Iberian style which predominates and underlies all the other recognizable styles from which the Antiguan style is derived. As in Andalucía, the *mudéjar* tradition is a constant which runs through all of Antiguan and Central American architecture. Though not always overt, it is, nevertheless, the basic core on which the other imported Iberian styles appear as an accretion. In this respect, the architectural tradition of Antigua is but an extension of that of the Reino de Sevilla where the *mudéjar* style is not confined to a single stylistic period, but one which lies submerged in the nonindigenous styles such as the Gothic, Renaissance, and baroque. The same process of assimilation and acculturation of architectural styles seems to have taken place in Antigua. Though the repertoire of styles of Antiguan building tradition is itself an import in the first place, the *mudéjar* root stock of Andalusian Spain was transplanted early so that all subsequent stylistic imports followed a phenomenological process of transformation and development not unlike that which had transpired and was transpiring in Andalucía itself.

But the question of terminology cannot be answered by the uncritical acceptance of patent stylistic similarities. Conclusions based on appearances alone have led to confusion in the classification of Hispano-American architecture in general, and that of Antigua and Central America in particular. A more objective method is that of meticulously noting the incidence of "disembodied" architectural elements in both Spanish and Antiguan architecture, that is, by means of comparative "anatomical analyses," and by collating the dates of the monuments from which these elements are extracted.

What are the specific elements which, when taken out of context, point to a relationship between *sevillano* and *anti-*

---

a terminology for the building tradition of Antigua derived from Mexican and South American styles, and even includes the "mission style" from southern California.

The terminology appropriate to Mexican architecture is not generally applicable to that of Guatemala and Central America. The distinctive lines of development between the two, from as far back as the sixteenth century, have been recognized by Kubler, *Mexican Architecture* **2**: pp. 281 ff.

[11] See Olvera, *Copanaguastla*, pp 115–136, and his *Joyas de la arquitectura*, pp. 14–17, 29; also Berlin, *Convento de Tecpatán*, pp. 5–13.

[12] Pardo, *Guía*, p. 20.

[13] Villacorta, *op. cit.*, pp. 306 ff. See also fn. 1 above.

Wethey, *Colonial Architecture*, pp. 8, 20 ff., and 156, uses the tern "mestizo style" for the building tradition which flourished in the remote areas of the Andean region which were almost wholly Indian in population. He goes on to say that the chronology and development of this style are still obscure for lack of documentary information, and also because some of the early examples have been destroyed. According to him, the peak of mestizo art in Peru was reached about 1750. He follows the reasoning of Neumeyer, *Indian Contribution*, *passim*, seeing in the "primitivism of simplified geometric design" the naïve hand of the Indian. For objections to the use of the term "mestizo style" see Kubler and Soria, *Art and Architecture*, pp. 91 ff.

Indian modes of expression may be patent in the art and architecture of Peru and Mexico, but for Guatemala and Central America, the data on population, races, and labor in the building trades do not warrant any such conclusion. See ch. IV, above, "Labor and the Building Trades."

[14] Toussaint, *Arte mudéjar*, pp. 55 ff.

[15] See Sancho Corbacho, *Arquitectura*, pp. 7 ff.

güeño architecture? But to search for direct copies of Spanish prototypes in Antigua would be a fruitless endeavor indeed, and would also be tantamount to negating the existence of the phenomenon of the morphological alteration of those very prototypes in Antigua. A line of reasoning which might be followed with some profit has been aptly set forth by Chueca Goitía who says in part, "... no existe en el terreno de la cultura una transmisión que no lleve consigo alguna alteración."[16] Though no exact replicas can be cited, even a cursory examination of certain "disembodied" structural and decorative details in Antigua results in the inevitable conclusion that these are related to the architecture of Seville and the surrounding region.

First of all, the similarity of craftsmanship and materials of construction indicate that these are part of the same building tradition. One need but examine the walling methods, the manner of setting foundations, the methods of plastering, and the finishing of walls to realize that no further documentation is needed to validate the conclusion that Antiguan craft practices were an extension of the Andalusian, or better still, the *mudéjar*. In fact, brick construction is universal in Antigua and is also the most common type in contemporary Seville even for buildings of major importance and of monumental scale.[17]

Another point of contact with Andalucía and the *mudéjar* tradition is seen in church plans. It should be borne in mind, however, that most of the buildings in Antigua were frequently altered and changed during the colonial period, especially in the eighteenth century. Like the Sevillian *mudéjar* prototype, the most common Guatemalan church consists of a single-nave plan roofed with an *artesón* and a presbytery covered with a brick vault. In the more humble churches the presbytery, the *capilla mayor*, is square in plan and is also roofed with wood and tile. (In fact, the first cathedral building in Antigua was roofed in this manner.) But the main ridge of the nave roof abuts on the higher elevated wall of the *capilla mayor*. A hip roof of pyramidal shape distinguishes the presbytery from the rest of the building since the presbytery walls are higher in elevation than those of the nave. The interior gable end of the nave roof actually abuts on the presbytery wall facing the main axis of the church.[18] The object of this arrangement is to distinguish clearly on the exterior the two principal parts of the church, the nave and the *capilla mayor*, for the floor plan is of the same width throughout both parts. This type of plan was common throughout the whole of the colonial period in the small towns of Guatemala and the rest of Central America. In Antigua itself, however, there are numerous examples of churches which in the course of time were ultimately covered with wood and tile nave roofs and vaulted presbyteries. The latter may be a short barrel vault or, in later examples, a small pendentive dome, *bóveda vaída*. This type of arrangement represents a modification of the traditional *mudéjar* plan of Andalucía: the nave roofed with an *artesón* and the *capilla mayor* with a ribbed Gothic vault.[19]

Mudéjar traits are also reflected in the use of mixtilinear arches for door and niche headers and cloister arcades. These appear in the third period of the Antiguan style, that is, after the earthquake of 1717.[20] A comparable phenomenon is to be noted in the eighteenth-century baroque architecture in and around Seville, specific details of which were duplicated almost exactly in Antigua during the same century.[21]

[16] Chueca Goitía, *Arquitectura del siglo XVI*, p. 13.

[17] See ch. III, above, "Methods of Construction," nos. 1, 2, and 3. The vogue for the use of brick even extended to monumental construction in Andalucía especially during the seventeenth and eighteenth centuries; see Kubler, *Arquitectura*, pp. 98 ff., and Sancho Corbacho, *op. cit.*, pp. 15 ff.

[18] See ch. III, above, no. 4, "Wood and Tile Roofs."

[19] The sixteenth-century Franciscan church in Antigua might have had a Gothic ribbed vault in the presbytery, though this is not actually stated. See ch. XIII below, no. 1, "San Francisco." This type of plan, of *mudéjar* origin, was also introduced in Peru according to Wethey, *op. cit.*, pp. 11, 29. In Andalucía the simple *mudéjar* boxlike churches of the vernacular tradition from the second quarter of the seventeenth century onward began to be decorated with stucco ornament in patterns of Italian origin, according to Kubler and Soria, *op. cit.*, p. 11. See also Schubert, *El barroco*, pp. 144 ff., who believes Juan Martínez introduced this type of decoration in the churches of Seville.

[20] See ch. X, above, "Façade Treatment," no. 2, iii, "Mixtilinear Niche Headers etc." See also ch. XV, nos. 4 and 5, below, "Seminario Tridentino" and "Universidad."

[21] See the following examples: (1) The ogee or keel arch on the tower of the parish church in Manzanilla, Spain, dated 1760. A variation of the same type appears in the door to one side of the main façade of La Candelaria, the façade of San José, Santa Ana, and Santa Clara in Antigua. The same type also appears in the parish church of La Luisiana, Spain, dated late eighteenth century, where it is used for the arch, *arco toral*, dividing nave and presbytery. (2) The stepped trefoil arch in the arcade of the patio of house no. 5, Calle San Roque in El Arahal, Spain, of mid-eighteenth-century date. An involved variation appears in the cloister of the University in Antigua. The same type of arch is also employed in the patio of the house of D. Fernando de Quintanilla in Lora del Río, Spain, of approximately the same date; in the patio of the house of the Condes de Valverde in Ecija, Spain, dated late eighteenth century; and also on the upper part of the *portada* of the house of the Albornoz family in Ecija, Spain, also of the same date. This type of arch is used in the niches of the façade of Santa Ana, in Antigua. A variation is also employed in the niches of the façades of Santa Clara and San José.

These are but a few examples of mixtilinear arches which have counterparts in contemporary Antigua. For published photographs of the above Spanish examples see in the same order, Sancho Corbacho, *op. cit.*, pls. 78, 140, 283, 286, 296, and 306.

By far the most commonly employed mixtilinear arch in Seville is the trefoil. One of the earliest examples of this type is seen on the

Vault construction is another clue which may be followed in tracing stylistic connections. It is interesting that the earliest type built in Guatemala was the simple barrel vault. This was usually pierced by windows at intervals directly in the haunches and lacked lunettes which were ordinarily formed by the intersection of vaults of smaller diameter. The latter device was quite common in seventeenth-century Spain. Barrel-vaulted churches are represented by at most four examples in Antigua and by one in Chiapas. The barrel vault, but with the differences mentioned here, was introduced to Guatemala in the same century. But it had little acceptance there, the builders preferring the simple *mudéjar* wood and tile roofs.[22] With the construction of the Cathedral in Antigua, completed in 1680, the pendentive dome, *bóveda vaída*, was introduced in Central America. But this type of vault construction did not achieve widespread acceptance until after the earthquake of 1717, and only then to replace the old weakened wood and tile roofs. *Bóvedas vaídas* were also used in new construction work, the *mudéjar* wood and tile roofing being no longer trusted to withstand earthquakes. Low pendentive domes were also employed over oblong bays in the naves of churches as well as in the normal fashion over square bays in arcaded cloister corridors.

It is interesting to note that the *bóveda vaída*, which had been introduced in Andaluíca in the third quarter of the sixteenth century, came to Guatemala for the first time a full century later, and only about one hundred and fifty years later did it find any widespread acceptance there. This tardy use of the *bóveda vaída* cannot be ascribed to an unwitting time lag in the passage of this construction technique from Spain to the New World. The more or less simultaneous appearance of the barrel vault disproves such an assumption. The fact is that Antigua did not lag far behind Seville, for the *bóveda vaída* also reappears in Andalucía in the eighteenth century and just about the time of its popularity in Antigua.[23] The manner of employment of the *bóveda vaída* in eighteenth-century Seville and the surrounding region is remarkably similar to the way it is used in Antigua. The pendentive dome, meant to cover a square plan, is forced into the oblong bays of the nave. It is built of the same materials, brick, rubble, and lime mortar.[24]

The development of the Antiguan style and the characteristics which set it apart as a native creation independent of its Iberian origins, are best revealed in the retable-façade. The *antigüeño* retable-façade is more than a mere variation of the well-known Spanish type which appears in many places and is not confined to Seville. The retable-façade was not an invention of the *antigüeño* builders, for it is well known in the Iberian peninsula from the sixteenth century on.[25] The Spanish retable-façade rarely if ever serves as the total frontispiece or as a complete church front, as does the *antigüeño* from its very inception in the seventeenth century.[26]

---

north door of the tower of Don Fadrique, built right after the Reconquest in 1252. But during the eighteenth century the trefoil arch became quite common; see for example: the *portada* of the chapel of the church of San Pablo in Seville, dated 1725; the façade of the chapel of San José in Seville, dated 1747; the same on the church of El Buen Suceso in Seville, dated 1690–1730; on the façade of the church of Jesús Nazareno in Lora del Río, dated 1733–1743. The trefoil was not too popular in Antigua and appears only on the façades of the churches of Espíritu Santo and Santa Rosa, dated after 1702 and *ca.* 1750 respectively. For photographs of the Spanish examples of the trefoil arch see Sancho Corbacho, *op. cit.*, pls. 14, 92, 96, and 116.

[22] See ch. III, above, no. 5, "Vaults and Domes." For the treatment of windows in lunettes, formed by the intersection of smaller barrel vaults with the larger longitudinal vault of the nave, see Kubler, *Arquitectura*, p. 39, who believes this to be a device originated by Michelangelo in the Sistine Chapel in the Vatican in Rome. San Agustín, San Pedro, Santa Catalina, and Santo Domingo in Antigua were barrel vaulted. Santo Domingo in San Cristóbal de las Casas is also roofed with a barrel vault with windows piercing the haunches directly.

[23] For the introduction of the *bóveda vaída* in sixteenth-century Andalucía see Chueca Goitía, *op. cit.*, pp. 247 ff., 250, and 258, who attributes its introduction to Andrés de Valdevira, 1509–1575, citing as examples the church of the Hospital de la Sangre in Seville, the church of El Salvador in Ubeda, and the cathedral of Jaen. The *bóveda vaída* is also employed in La Colegiata in Osuna, *ibid.*, p. 197, and in numerous other sixteenth-century Andalusian churches.

[24] For examples of the *bóveda vaída* in the eighteenth century in Seville and the nearby region, the following may be cited: the parish church of Santa María Magdalena in El Arahal, employed over the oblong bays of the nave; the parish church of Nuestra Señora de la Encarnación in Bormujos, where it is employed to roof the *capilla mayor*; the parish church of San Juan Bautista in Las Cabezas de San Juan, also employed over oblong nave bays; and in the parish church of La Concepción in Castilleja de la Cuesta, over the *capilla mayor*. For published data, drawings, and photographs of the above examples see Hernández Díaz, *Catálogo arqueológico* **1**: pp. 165 ff., 229 ff., **2**: pp. 3 ff., 296 ff.

[25] See ch. IX, fn. 8 above, for a discussion of the origin of the retable-façade. For some sixteenth-century retable-façades from Seville and the surrounding region see Chueca Goitía, *op. cit.*, pp. 197 ff., who refers specifically to the *portada* of the crossing of the principal church in Puerto de Santa María, the church of Santa María in Utrera; the church of the Hospital de la Sangre in Seville, p. 266, fig. 241; the church of El Salvador in Ubeda, p. 250, fig. 217; and outside Andalucía, the *portada* of La Anunciación of the Cathedral of Orihuela, p. 272, fig. 250.

[26] See ch. IX, above, "Retable Façades," and ch. XIII, below, "The First Period: The Seventeenth Century to 1680," for some examples in Antigua. The church of Nuestra Señora de las Angustias in Valladolid, dated 1597–1604, is the nearest analogy where the total width of the façade is occupied by the retable, see Schubert, *El barroco*, pp. 116 ff., fig. 50, and Kubler and Soria, *op. cit.*, p. 17, fig. 12, for some discussion of this church.

Some retable-façades other than in the area of Seville in which the

The general conception of the Antiguan church plan is one where the façade is directly related to the interior retable. The exterior retable is but a prelude to that behind the main altar at the far end of the building. In this respect the Antiguan church is comparable to the many small-town churches of the Sevillian countryside, where there too the *portada*, the entrance door and its decoration, is the main accent on the otherwise plain façade. The difference is that in Antigua, except when there are flanking towers, the retable-façade occupies the total width of the church front and is not confined to the area immediately surrounding the door opening.[27]

The earliest extant examples of the retable-façade in Antigua date from the first period, that is, from about 1650 to 1680. Invariably these are all rather reserved and Italianate in design. Furthermore, they are devoid of any ornament which might detract from the tectonic quality of both the architectural orders and the façade wall to which these are applied. It is only in the second period, 1680 to about 1717, that painted designs, stucco ornament, and solomonic columns, heralds of the baroque style, appear. The stucco and painted ornament are sometimes applied to the surface of both the orders and the façade wall too. Stucco ornament in designs of Italian origin had appeared in Andalucía about fifty years before.[28]

The third period is marked by the introduction of the use of nonarchitectonic pilasters in applied architectural orders.[29] Mixtilinear arches, as already noted above, also appear for the first time in the third period which, by the fourth period, become quite profuse and involved in design. The nonarchitectonic pilasters, *estípites*, though not exact replicas of the Andalusian, are nevertheless of the same generic types. The Antiguan *estípites* are frankly expressive of the plaster material in which they are carried out and remote from the wood architectural forms on retables from which they originated. Despite the addition of these nonarchitectonic members, the structural quality of the façade as a whole is never abrogated. The *estípites* are employed according to the canons of the normal architectural orders. In other words, the Italianate Tuscan column gives way to the nonarchitectonic support. The active restless façades with undulating or curved surfaces so common in the late baroque or the rococo are totally absent in Antigua. The façade wall is always built in rectilinear planes to which the orders are added. Elevations of niches, however, may be quite curvilinear in outline, but podia, entablatures, and pediments are interrupted by right-angle mitering in and out.

The overall appearance of the retable-façade has with good reason led some critics to consider the architecture of Antigua as falling wholly into the baroque style. And on the analogy with the emergence of the Spanish baroque, the Antiguan was thought to have been preceded by an Italianate or *renacentista* phase.[30] The churches of Los Remedios, San Agustín, San Cristóbal el Bajo, and the Cathedral, all from the first period, have retable-façades which are perhaps closest to the style which might, with some serious reservations, be called *renacentista*. There can be no doubt that this type of retable-façade is the prototype or seed from which the later baroque type develops, that is, the core on which the extravagant decorative elements are added in the eighteenth century.

The *antigüeño* builders developed an independent regional or local style; but the continued stylistic influence from Seville in the late seventeenth and all through the eighteenth century is a fact that is quite patent, and hence cannot be put aside. It cannot be mere coincidence that the development of the Sevillian baroque coincides more or less with the *floruit* of this style in Antigua, that is, from the last two decades of the seventeenth century to the destruction of 1773.[31] It also cannot be mere coincidence that the reappearance of

organization of the elements of the design are somewhat akin to the *antigüeño*, but unrelated otherwise, might be cited as follows: the *portada* of the church of Nuestra Señora de las Angustias in Granada, dated about 1664–1671, Kubler, *Arquitectura*, pp. 89 ff.; the façade of San Juan de Dios also in Granada, dated 1737–1759, *ibid.*, p. 313; the *portada* of the church of Santa María in Alicante, dated 1721–1724, *ibid.*, p. 317, fig. 409.

27 The connection between the *portada* and the interior main altar with its retable was noted by Kubler, *op. cit.*, pp. 40 ff.

28 See fn. 19 above for the introduction of stucco ornament in Seville in the second quarter of the seventeenth century.

29 For the relation of the nonarchitectonic pilaster with that of contemporary Seville see ch. x, above, "Façade Treatment," no. 1, iii, and especially fn. 4.

30 Angulo, *Historia del arte* **2**: pp. 56 ff., says that the whole of Antigua is baroque. On the analogy of the emergence of this style in Spain, he believes one must look elsewhere for examples of the Renaissance style from which the Antigua baroque is derived. As an example of the earlier style, he cites the church in Rabinal which he believes would date from about 1600 were it located in Spain. The one example of the Renaissance style from Antigua which Angulo mentions as being the possible prelude to the eighteenth-century baroque, the church of San Pedro, was almost totally destroyed in 1773 and rebuilt and altered in the nineteenth century. One must proceed, therefore, with some caution in making stylistic comparisons with this building.

31 The Sevillian baroque extends from the last two decades of the seventeenth century to about 1780, according to Sancho Corbacho, *op. cit.*, p. 11. As a matter of fact, the colonial architecture of *La Nueva Guatemala*, the new capital and present-day Guatemala City, is almost wholly neoclassical. The baroque style disappeared with the destruction of Antigua in 1773.

the *bóveda vaída* in Andalucía and its popularity in Antigua, that the vogue in both Andalucía and Antigua for decorative baroque elements such as stucco ornament, mixtilinear arches, and nonarchitectonic pilasters, should have occurred more or less contemporaneously. The great innovator of the Sevillian baroque was Leonardo de Figueroa whose work came to a culmination in the years from about 1680 to 1730. Features of his decorative innovations spread far afield in the Reino de Sevilla and were introduced into the *antigüeño* vocabulary after 1717.[32] The very alphabet of architectural details which obtained in contemporary Seville seems to have also been used in part in Antigua. Some of the mixtilinear arches, window openings, and nonarchitectonic pilaster shafts of Antigua seem to have been extracted from the catalogue current among the *sevillano* builders.[33]

The term "earthquake baroque" has had some acceptance among those few architectural historians who have treated the architecture of Antigua.[34] In view of the long history of destructions by earthquakes, it has been assumed that the changes which took place in the style, particularly in the eighteenth century, were in answer to the need to make the buildings more earthquake-resistant. Unfortunately, though overt appearances might seem to indicate this to be a fact, that is, the squat proportions of the supports of arcaded corridors in cloisters and on façades, the evidence—architectural, documentary, and literary—does bear out such a conclusion. It is just as valid to assume that the massive character of the eighteenth-century arcades resulted from the introduction of a new aesthetic ideal, as it is to conclude that the new proportions were functional and for the purpose of earthquake resistance.

Here again a point of contact with the Iberian peninsula is illuminating. This type of massive squat support for arcading was not unknown in Spain from as far back as the sixteenth century.[35] That the builders of Antigua were perplexed as to what means to employ in order to strengthen buildings, there can be no doubt. As a matter of fact, after the earthquake of 1717, the introduction of the *bóveda vaída*, though seemingly childish now because of the well-known futility of such devices, was probably thought of as a means of providing a more substantial roofing than the usual wood and tile which would come down with the slightest tremor.[36] It has been stated that in the eighteenth century in Antigua walls were thickened and heights shortened in order to cope with the ever-present threat of earthquake. In fact, however, it was during this century that the walls were greatly reduced in thickness and treated like screens between the piers and buttresses which supported the vaulting above. These screens were made even thinner by recessing niches into them.[37]

The use of twin towers as an adjunct to the retable-façade does not have any direct parallels in Andalucía. But as is true of the retable-façade itself, the use of towers to frame façades seems to be not an unknown phenomenon in the Iberian peninsula. But of quite different inspiration are the well-known towers which pierce the sky of the former Reino de Sevilla in Ecija, Carmona, Estepa, Palma del Condado, Osuna, Manzanilla, and many other towns in this region, all from after the earthquake of 1755 and all inspired by the famous Giralda of Seville. The towers, in all the cases here mentioned, are not integral parts of the façade, but each appears as an isolated element not necessarily directly related to the church *portada*.[38] The contrary is the case in Antigua where the twin towers are usually designed and

---

[32] See Sancho Corbacho, *op. cit.*, pp. 46 ff., and also his *Leonardo de Figueroa*, pp. 341 ff.

[33] See Sancho Corbacho, *Arquitectura*, p. 26, fig. 2, where some typical lobulated and mixtilinear arches are illustrated. Though not direct counterparts, many in Antigua are variations of the *sevillano* types. They can be added to the list Sancho Corbacho gives and would not be out of place there. The same is true with regard to window openings illustrated in his fig. 20, two of which are quite common in Antigua. The plaster ornament of Antigua is not as rich or as variegated as that of contemporary Seville, but there too, certain motifs, especially the arrow point used to decorate pendentives, are to be found; see his fig. 21.

[34] The term "earthquake baroque" was coined by Kelemen, *Baroque*, pp. 122 ff.; see also Angulo, *Historia del arte* **3**: pp. 1 ff.

[35] A sixteenth-century example exists in the former garden of the Palace of the Dukes of Tarifa, in Bornos, in the province of Seville, where a sculpture gallery is built with very short massive piers that are square in plan. Also from the sixteenth century, one may cite the arcaded façade of the Palace of the Pizarro family in Trujillo, Badajoz. Eighteenth-century examples of similar supports such as those in Antigua may be cited as follows: in the cloister of the Franciscan monastery of La Rábida, near Palos de Moguer in Huelva (this is not to be confused with the *mudéjar* cloister there); in one of the cloisters of the Convento de Loreto in Espartinas near Seville; supports of heavy proportions but not as exaggerated as the above, may be seen on the arcaded façade of the former royal stables located in the Plaza de los Reyes in Jerez de la Frontera, dated 1768; and also in some of the arcades of the Plaza Mayor of Ecija. See also, ch. VIII, above, no. 3, "Arcaded Cloisters and Façades."

[36] See ch. III, above, no. 7, "Earthquake Resistance."

[37] See ch. III, above, no. 2, "Walls." Kubler and Soria, *op. cit.*, p. 83, in speaking of the diffusion of Italianate and Granadine elements in façade treatment in Quito, believe that this could not take place in Antigua because ". . . Guatemala was soon forced to change in order to meet seismic threat . . ." and that as a result walls were thickened and heights shortened. Evidence to the contrary is presented in the architectural data on the monuments described in chs. XIV and XV below.

[38] For the relation of the towers of the region to the Giralda of Seville, see Kubler, *Arquitectura*, pp. 278, 281, also Sancho Corbacho, *Arquitectura*, *passim*, for references to the towers of the towns mentioned above. The tower façades, such as San Miguel in Jerez de la

built with the rest of the façade in a single building operation. In fact, in the very few examples of three-aisled churches, the towers are lined up with the side aisles while the retable itself occupies only the area of the width of the central aisle or nave. But in the more common single-nave churches, where flanking towers are employed to frame the retable, these actually project to each side of the nave, adding extra width to the façade as a whole. These towers are then unencumbered on three sides and are attached to the nave of the church by a party wall. There are, however, examples of seventeenth-century date with but one tower, usually of very square massive proportions, which is set slightly back from the main plane of the façade. These single towers usually house the *sagrario* or some special chapel and are not always contemporary with the construction of the rest of the façade. But in the eighteenth century, symmetrical towers are commonly employed to frame the central retable. These are invariably of very reduced plan looking more like oversized buttresses. Twin-tower façades are also quite common in eighteenth-century Mexico and to a lesser extent elsewhere in Hispano-America. The Antiguan twin-towered façade is part of a widespread style in Hispano-America, whose immediate origins are closer to home in the New World.[39]

Frontera, the parish church in Constantina, or San Pedro in Arcos de la Frontera are unrelated to the *antigüeño* types.

[39] See Angulo, *Historia del arte* **2**, *passim*; especially chs. XIII, XVI, in Puebla and nearby towns; ch. XVII for Oaxaca; ch. XVIII for Michoacán, Jalisco, and Querétaro; ch. XIX for Guanajuato; ch. XX for some other states too. For South American examples see Kubler and Soria, *op. cit.*: San Francisco, in Quito, fig. 40a; La Compañía, in Cuzco, fig. 43b; San Francisco in Lima, fig. 44b; the cathedral of Córdoba, Argentina, fig. 52a. For Peruvian examples, see Wethey, *op, cit.*: in Cuzco, the cathedral, pl. 62, la Compañía, pl. 67, San Pedro, pl. 68, Belén, pl. 89, San Sebastián, pl. 70; in Lima, the cathedral, pl. 103, San Francisco, pl. 108; in Pisco, La Compañía, pl. 117; in Nazca, San José, pl. 118, San Xavier, pl. 131; in Ayacucho, Sta. Teresa, pl. 145, San Francisco, pl. 146, Sta. Ana, pl. 148; and elsewhere.

PART III

The Monuments of Antigua: Historical and Architectural Data, Chronological Synthesis of the Style

XII THE FIRST PERIOD: THE SEVENTEENTH CENTURY TO 1680

PRIOR to the completion of the Cathedral in 1680, the conventual church of the principal house of the order of Santo Domingo, located in Antigua and of which not a single trace remains, very likely was the most imposing and most important architectural monument in all Central America, judging by descriptions of it in contemporary literature and archival documents. Such a conclusion is warranted on the analogy of one of the lesser Dominican churches, still well preserved, in San Cristóbal de las Casas which in plan and construction in great part resembles the Antiguan church known only from verbal descriptions. Vaulted roofs were rarely employed in this first period, the barrel vault being the most common type. Four such vaults roofed the Dominican church in Antigua, completed about 1666. Remains of a barrel vault are still extant in the church of San Agustín and barrel vaults also once roofed Santa Catalina and San Pedro.

Perhaps even more characteristic of the period is the sobriety of the retable-façades which are treated with a great deal of reserve. Surfaces are left plain with no plaster ornament at all to soften the architectonic quality of the applied orders of classical inspiration. Massive twin towers flank the façades in two cases, the church of Belén and the Cathedral, so that the central retable is dwarfed. The façade of Belén is quite plain and now lacks applied architectural orders. But the building has undergone so many repairs and alterations, some even as late as the 1940's, that it is quite impossible to tell whether applied orders too had not been once present on the façade. In general, towers are of reduced size and frame the central retable. Actually the retable is the most dominant feature of the façade, as for example in San Agustín and Los Remedios, while the towers, if functional at all, are just large enough to allow room for a caracole stairway, as in San Cristóbal el Bajo.

Doors are spanned by semicircular arches and are set in recesses or niches with concentric headers. A variation is noted, however, in the Cathedral where the niche jambs are considerably taller, thus forming a semicircular tympanum between the crown of the door arch and the intrados of the niche arch. A diminutive niche with a statue is located in the tympanum. The same type of door is also used on the church of Belén. The church of Santa Catalina is oriented so that the main façade is within the convent grounds. The building is entered through two doors on the long side facing the street. This type of orientation is employed in all but two of the other nunneries in Antigua.

Most of the buildings constructed in this first period have histories going back to an earlier date. Some were also reconstructed and altered, the most notable example being that of Los Remedios whose walls of tamped earth and brick clearly point to a number of additions ranging from the early seventeenth to well into the eighteenth century. The church of San Pedro is an example of the afterlife of colonial building methods. It was rebuilt in the nineteenth century on the ruins of an earlier structure. The same is true of the Santuario de Guadalupe of the nineteenth century, done in the colonial style utilizing parts of an older building.

The period comes to a culmination with the construction of the Cathedral, completed in 1680. It was repaired and remodeled in the nineteenth century. Its three-nave cruciform plan is not too common. The use of a colossal order on the façade is unique in Antigua, there being only one other example of any note, that in the little town of San Juan del Obispo nearby of more or less contemporary date. The arrangement of niches in the retable-façades, one above the other between engaged columns in each of the side bays as well as to either side of the niche-window in the second-story central bay, is a pattern which was to be repeated in the following periods on many buildings.

The Cathedral was roofed with domical vaults, marking a departure from the previous limited vogue for barrel vaults and a practice which was also to become quite common in the following periods. Likewise, the plaster decoration, *ataurique*, the hallmark of so much Antiguan architecture, appears on the necking of the nave piers, around windows, and on the pendentives of the cupola of the crossing.

1. *Santa Catalina: Church and Convent, 1626 and 1647*

(FIGS. 3, 22, 23.)

Historical Data

THIS convent, the official title of which was "Convento de Nuestra Señora Santa Catharina Virgen y Mártir," was founded December 27, 1604, by four nuns from the convent of La Concepción in Antigua.[1] The founding date of 1604 refers only to the legal institution of the new monastic community and not to the construction of an actual physical establishment where it was to be housed. Not until 1609 or 1610 did the four nuns actually move from La Concepción and enter a convent building of their own, apparently in some private houses donated by Don Francisco González and located on the site where the church and convent of San Agustín now stands.[2] In addition to this real property, one Don Miguel Muñoz donated the sum of 10,000 *pesos* as a financial endowment making the establishment of the convent possible. By 1631 the original number of four had grown to fifty-two and by 1697 had again increased to one hundred and ten nuns and six novices. Apparently by mid-eighteenth century the religious community had markedly declined in size, for there were but forty-nine nuns and one novice resident there.[3]

The remodeled private houses soon proved inadequate, and in 1615 a new structure was built in a new location, the site where the present ruined church which had been built later on is to be seen today. The abandoned, makeshift, first conventual building was turned over to the newly arrived monks of San Agustín.[4]

In 1609 the *cabildo ecclesiástico* had legalized the new convent according to canonical law. Official license from the civil authorities was still lacking, however.[5] The existence of the convent came to the attention of the crown later, for in 1614 an inquiry was made why this convent had been founded without previous permission.[6] It would seem, therefore, that the original quarters were intended to be provisional only while awaiting official license. When approval was finally granted, possibly in the same year, a more formal convent building was constructed on the new location. According to Vásquez, the new church and convent had already been occupied by the nuns in 1613, apparently before being completed. The church was roofed with a barrel vault. But these buildings were destroyed in 1630.[7] Therefore, Vásquez' description of the church building must refer to one built later in the seventeenth century. What the first formal church and convent of 1614/1615 were like is impossible to tell. But we are fortunate enough to have the contract for the construction of a second church on the new site dated September 2, 1626, from which we learn not only about this particular building, but also in general about construction methods of the time.[8] It is noteworthy that a master stonemason was the general contractor who drew the plan of the church as well as undertook the actual construction.

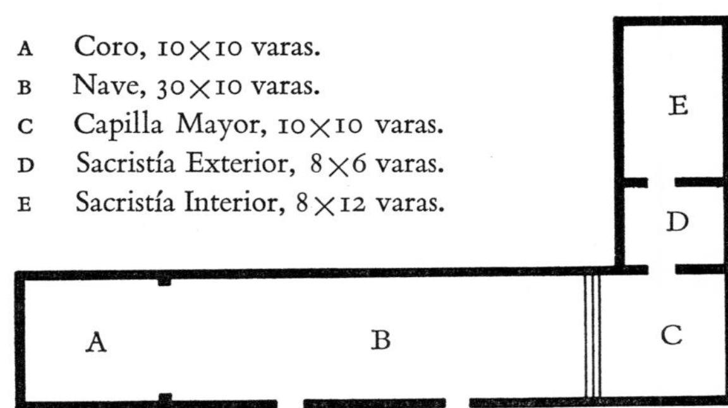

FIG. 3. Plan of Santa Catalina, 1626, as described by Francisco de Hernández de Fuentes.

The contract also gives the exact specifications. Hernández, the contractor, agreed to construct a wall fifty *varas* long (the *vara* is approximately 33 inches) against which he was to build the upper and lower choir ten *varas* square with a single window in both the east and west walls if this be necessary. (FIG. 3.) Both interior and exterior walls, according to the contract, were to be constructed up to ground level of lime, stone, and sand, the mix to be according to the proportions customary in Antigua namely, two parts sand, two parts earth, and one of lime. The nave, the *capilla mayor*, and the choir were to be roofed by a framework of trusses,

[1] *AGG*, A 1.18 (1740) 07/5022–211, see also *BAGG* **1**: 2 (1936): pp. 127 ff., and **10**: 3 (1945): pp. 225 ff.

[2] Díaz, *Romántica ciudad*, p. 53; Pardo, *Guía*, p. 126; Fuentes y Guzmán, **3**: p. 134; *AGG*, *loc. cit.*; and *Efem*, pp. 38, 39.

[3] *AGG*, *loc. cit.*

[4] Remesal, **2**: p. 354; Vásquez, **4**: p. 368; Juarros, **1**: p. 133; and *Efem*, p. 41, give the date as 1613.

[5] *Efem*, p. 38.

[6] Larreinaga, *Prontuario*, p. 206.

[7] Vásquez, **4**: pp. 368 ff.

[8] *AGG*, A 1.20 (1626) 757, also *BAGG* **10**, 3 (1945): pp. 221 ff. The contract reads in part as follows: "Francisco Hernández de Fuentes, maestro del arte de cantería, residente en la dicha ciudad . . . a tratado de que se haga iglesia nueva en el dicho convento por la necesidad de ella ay y yo el dicho Francisco Hernández de Fuentes e ofrecido hacerla y dada acabada de todo punto, tomándola a destajo, conforme a la planta que hize, con que se me den por mi trabajo, materiales y demas gastos de yndios y oficiales, seys mill tostones de a quatro reales de plata cada uno, pagados los dos mill luego de contado y dos mill en todo el mes de febrero del año que viene de seyscientos y veinte y siete y los otros dos mill para el dia que diere acabada la obra, la cual daré acabada de oy en un año. . . ."

armadura de par y nudillo, running the total length of the building to the gables which were to be sheathed with boards in the manner customary in the city. The exterior walls were to be finished with dark plaster on which bands of white plaster were to be applied so as to simulate coursed ashlar masonry. The parts of the walls between the brick and stone sections, of which there were to be eight in all, were to be adobe reinforced with timber. He also agreed to build some wooden steps set in brick leading to the main altar like those in Los Remedios, and to place two door openings in the east wall of the nave, one of stone and the other of brick. Both were to fit the wooden doors already in existence. He also agreed to build four recesses for confessionals, each with an exterior door on the convent side of the building.

From the details given in the above contract it appears that the roof was of wood and tile, and not a barrel vault as stated by Vásquez. Other details in the contract are worthy of note. The walls consisted of stretches of masonry, *rafas*, about six feet long, interspersed with sections built of adobe and wood reinforcing, *zoquete*. The walls of Los Remedios, the building he refers to with regard to the altar steps, present a similar combination of masonry and adobe, and must be accounted a method common to the early seventeenth century and one by which major economy of cost could be effected.

The mortar mix Hernández specified as being the one commonly employed in Antigua consisted of two parts sand, two parts earth, and one part lime—a rather strange procedure which includes earth. On the basis of a comparison with present-day building practices in Antigua proper and in the whole of Guatemala, one may conclude that the sand referred to is river sand, and the earth is that which is known as *arena amarilla*, a yellow sand of volcanic origin. This yellow sand is used today with lime for mortar to lay bricks and also for the scratch and brown coats in plastering walls. The lime is the common mason's unslaked lime universally used where portland cement is not available. It was only in the second quarter of the twentieth century that portland cement began to be produced in Guatemala. Prior to that time it was imported and used only on those buildings where reinforced concrete was employed. Cement mortar is not commonly employed for laying brick even today. The mortar that Hernández agreed to provide is then a 4:1 mix and not as poor as would seem, for the river sand is very sharp and the *arena amarilla*, or *tierra* as he calls it, is actually inorganic and very light and also makes a very good aggregate. Another detail is also of interest, the finish of the exterior wall, presumably the one on the street, the *lienzo pared* as he calls it, of a dark plaster interspersed with ribbons of white plaster to give the appearance of masonry.

The church he designed occupies the very same site where a second church, now ruined, stands. He divided the *lienzo pared* fifty *varas* long (between 137 and 138 feet counting each *vara* as 33 inches) into three main parts: ten *varas* for the choir; thirty for the nave; and ten for the *capilla mayor*. He also agreed to build an exterior sacristy eight by six *varas* and an interior sacristy eight by twelve *varas*. These two sacristies abutted on some part of the church, probably to the east off the *capilla mayor* and into the convent property. (Fig. 3.) The remains of these two sacristies are not in evidence among the present ruins for some private houses are now located in the area to the north blocking off the church wall there. The present church, occupying exactly the same plot as that built by Hernández in 1626 is approximately the same length. The modern houses abutting on the north actually face the plaza of La Merced at the corner of the block. The convent property probably did not extend to the plaza. On this basis one may conclude, therefore, that the exterior sacristy abutted on the *capilla mayor* to the east, for it could not project to the north, and that the interior sacristy continued beyond in the same direction, thus making a more logical arrangement. In this fashion not only are the convent grounds enclosed by buildings, but a more convenient access for the nuns is also provided directly from the interior of the convent too, as shown in the reconstruction of the plan in figure 3.

The present church, the third one on the site, was inaugurated September 15, 1647.[9] Cristóbal de Salazar began the church with funds donated by one Alonso de Cuellar, a former resident of Mexico who died in Guatemala and bequeathed his fortune to be expended in pious works making Salazar his executor. Since Cuellar had not left enough money, Salazar's wife used some of her own to complete the work after her husband's death, a total of 36,000 *pesos* being expended.[10] The church previously built by Hernández had cost 6,000 *tostones*, that is 3,000 *pesos*, and must have been a much simpler structure than the one in ruins today. Of course, the 36,000 *pesos* included the cost of the convent too. Even if the church represented but one-fourth of the total cost, it was, nevertheless, a much more expensive building than the 1626 church.

[9] Juarros, **1**: p. 134.

[10] Antonio de Molina, *Memorias*, p. 36. He refutes González Dávila who in his *Teatro de Indias* gives the cost as only 20,000 *pesos*. This account is repeated almost *verbatim* by Ximénez, **2**: p. 244, who was one of the editors of the memoirs of Molina, and is also repeated by Vásquez, **4**: p. 368.

The number of nuns and pupils in residence apparently once again increased very markedly after the completion of the 1647 church and convent, for in 1683 permission was sought from the *ayuntamiento* to close off the street to the south on which the conventual buildings faced so that some houses on the other side might be incorporated into the convent.[11] This petition was denied, and the order then acquired some houses on the street running north-south directly opposite the convent and church. In 1693 the conventual authorities wished to build an arch across this street in order to connect both parts so that the nuns and pupils would not be obliged to leave the convent in going from one part to the other. Over the protests of some of the neighbors, permission was granted and an arch constructed which was completed in June of 1694.[12]

The church, convent, and arch which had apparently withstood the earthquakes of 1717 and 1751, were destroyed at last in the cataclysms of 1773.[13] Some minor damage to the barrel vaults of the church is mentioned as having occurred in the earthquake of 1717,[14] indicating that the 1647 church was not roofed with wood and tile as the 1626 church had been. The arch which spans the street was repaired and rebuilt in the middle of the nineteenth century and still stands today. (FIG. 18.) But there is no way of knowing whether the original plan had been followed in its reconstruction.[15] It was still standing after the earthquake of 1773, but in very precarious condition and was a hazard to passersby.[16]

Architectural Data

NOTHING remains of the convent today and the site is now built up with houses constructed in the nineteenth century.[17] The private buildings immediately to the south of the church may still preserve some of the original features of the convent and of one of its cloisters, all of which have been so modified that it is well nigh impossible to work out the original plan. The part of the convent across the street connected by the arch is completely gone, the area now occupied by some private houses and commercial establishments.

All that remains of the 1647 church are the four exterior walls, while the interior is heaped high with the debris of the vaults from the destruction of 1773. Both of the doors (FIG. 22) which face on the street have been bricked up and only a small opening is left in the southernmost through which one may enter the church up a flight of rough stairs cut in the debris still piled about two meters above the street level. Some squatters, occupying a small thatch hut inside, cultivate a small vegetable garden in the nave. They were still there in 1957.

The church is oriented north-south with the long east wall on the street, a device common also in the nunneries of Santa Clara, La Concepción, and in many examples in Mexico. The main façade of the church to the south, that which encloses the choir, is incorporated into the convent area. The lay public entered by doors in the exterior long wall facing the street.

The two doors in the long east wall are not exactly alike, though both have rusticated stone jambs and headers. That nearest the choir, the southernmost, has a blazon with the royal Spanish coat-of-arms immediately above the lintel. (FIG. 23.) The north door lacks the coat-of-arms and has slightly different rustication on the two jambs and the header. The jambs of both doors are treated like pilasters with bases, shafts, and capitals. The bases are quite simple, but the shafts are rusticated. The jambs of the south door are divided into three elongated panels simulating three rusticated blocks with deeply recessed joints. The north door shafts are treated with rustication too, but here the blocks are laid in eight courses with a vertical joint on center in every other course. The lintels repeat the patterns common to the jambs of each door. The actual joints of the stone blocks do not coincide with the rustication patterns. The lintels above are straight arches composed of two voussoirs and a keystone. The rustication pattern is designed to obscure the arcuated character of the lintel and is visible now because the keystones have slipped slightly revealing their wedge-shaped outlines. Each lintel is surmounted by a stone entablature consisting of a narrow frieze band and cornice supported on small consoles or brackets pendant on each side of the lintel.

Much remodeling has been carried out on the surface of the wall facing the street, yet still visible on portions just over the doors are the remains of a triglyph and metope frieze worked in stucco. (FIG. 22.) Directly above this frieze a small cornice molding divides the height of the wall at about three-fourths of the way up. Another cornice molding of slightly wider projection acts as a coping for the top of the wall. The space between the lower and upper cor-

[11] *Efem*, pp. 96 ff.; Pardo, *Guía*, pp. 128 ff.

[12] *AGG*, A 1.2–2 (1693) 11778–1784, and *BAGG* **8**, 1 (1943): p. 77; also *Efem*, pp. 113 ff., and Pardo, *op. cit.*, pp. 128 ff.

[13] Díaz, *Romántica ciudad*, p. 53; Pardo, *op. cit.*, pp. 128 ff.; Juarros, **1**: pp. 133 ff.

[14] Ximénez, **3**: p. 356, and González Bustillo, *Razón puntual*, also *Ciudad mártir*, pp. 77 ff.; also González Bustillo, *Extracto*, also *Ciudad mártir*, p. 130. See also Mencos, *Arquitectura*, append. VII, transcribing *AGI*, Guatemala, 305, "Reconocimiento del convento de monjas de Santa Catalina de Santiago, en Guatemala a 18 x 1717."

[15] Pardo, *Guía*, p. 129.

[16] González Bustillo, *Razón particular*, also *Ciudad mártir*, p. 102.

[17] Díaz, *op. cit*, p. 153.

nices corresponds to the haunches of the interior vaulting. Some water spouts are still *in situ* just under the coping.

2. *Nuestra Señora de los Remedios: Church ca. 1650 and 1687*

(Figs. 24–27.)

Historical Data

This church was first founded early in the sixteenth century in Ciudad Vieja on land which Pedro de Alvarado himself had allotted for the building of an oratory.[1] Later, when this city was destroyed and the site abandoned in 1541, the church was removed to the new location in Antigua. According to Vásquez, who refers to a *cédula* of that time, one Baltazar Estévez began work on a new church building in 1575 converting a thatch structure to one roofed with wood and tile,[2] confirmed by a document referred to which still exists in the colonial archives in Guatemala City.[3] It would seem, then, that for about the first thirty-five years after its removal to Antigua, the church of Los Remedios was hardly more than a thatch hut, and that not until about 1587 was the more formal building with the tile roof completed. In that same year, a royal order had arrived instructing the local authorities to give economic help in the construction and outfitting of the church, now referred to as an *ermita*, that is, an hermitage.[4] The bishop Francisco Gómez de Córdova, who died in Los Remedios in 1598, did some work on the building toward the end of the sixteenth century.[5] In 1594 a movement was started to raise the hermitage to a parish.[6]

The church building constructed during the last quarter of the sixteenth century must have been of modest character, though it probably had either a bell tower or belfry where a bell inscribed with the date July 10, 1610, donated by one F. P. J. M. Gutierres, once hung. In 1957 this bell was lying on the floor of one of the corridors of the cloister of La Escuela de Cristo. The hermitage was not raised to a parish at once in 1594 owing to certain changes in both the political and ecclesiastical administrations which had occurred about that time.[7] In the meantime, the church building deteriorated, as did the house adjacent to it, and in 1625 these were offered to the order of San Agustín, which did not accept.[8] It was only then that the rebuilding of the church was undertaken and in 1641 was finally raised to a parish, at which time, or not long after, the new building was probably completed.[9] Vásquez, writing *ca.* 1695, says that the church was very nicely furnished and built from the ground up a matter of forty-five years before, which would make the date for the final completion about 1650. It is, however, possible that the majority of the work on the structure had been done during the episcopate of the bishop Agustín de Ugarte y Sarabia, 1632–1641.[10] Actually the building which Vásquez knew had been radically altered about a decade or so before the time he was writing.

In 1676 this building had fallen into a bad state of repair, especially the *artesón*.[11] Two years later José de Porres and Andrés de Illescas rendered a report on its condition as well as an estimate of the cost of the repairs necessary.[12] In 1679 license was granted to José Aguilar y Revolledo, then *alcalde* of the city, and to Francisco Antonio de Fuentes y Guzmán to collect alms to be used for the rebuilding of the church of Nuestra Señora de los Remedios.[13] Work began soon after and was carried on rather haltingly because of the lack of funds and the extreme poverty of the parishioners.[14] In 1687 the repairs and alterations were finally completed. Included among the changes in the original plan was the addition of the present façade, the tower, and the vaulted *capilla mayor*.[15] Some minor details still remained incompleted. For example, one José de Arria, who had contracted to cast a bell for the church, was being sued in 1710 by Andrés Ruiz de la Cota, *mayordomo* of the building works of Los Remedios, be-

[1] Fuentes y Guzmán, 1: pp. 153 ff.

[2] Vásquez, 4: pp. 382 ff. For documentary confirmation of Vásquez' report, see fn. 3 below and also Mencos, *Arquitectura*, ch. IV, n. 84, quoting *AGI*, Guatemala, 171, "Información de Baltasar Estévez, hermitaño de Nuestra Señora de los Remedios sobre que le haga una limosna, en Guatemala a 23 I 1585."

[3] *AGG*, A 1.23 (1586) 1513–660; see also *Efem*, p. 28. Also the document in *AGI* cited immediately above.

[4] *AGG*, A 1.2–4 (1587) 2195–247; see also Pardo, *Prontuario*, p. 57.

[5] Vásquez, *loc. cit.*

[6] *AGG*, A 1.23 (1594) 5113–750; see also Pardo, *op. cit.*, p. 57, *Efem*, p. 33, and Vásquez, *loc. cit.* All the foregoing information is also given by Juarros, 1: pp. 147 ff., who repeats the statements of Vásquez almost verbatim. For the *cédula* raising Los Remedios to a parish church see Mencos, *Arquitectura*, ch. IV, n. 87, quoting *AGI*, Guatemala, 286, Leg. 22, fol. 141, dated in Madrid 29 V 1594.

[7] Vásquez, *loc. cit.*

[8] See San Agustín below, no. 3; also Mencos, *Arquitectura*, ch. IV, n. 89, quoting *AGI*, Guatemala, 177, "Instancia hecha de los frailes de San Agustín sobre la Ermita de los Remedios, en Guatemala, año de 1636."

[9] *Efem*, p. 55; see also Juarros, 1: p. 68.

[10] Villacorta, *Historia*, p. 218.

[11] Mencos, *Arquitectura*, ch. IV, n. 92, quoting *AGI*, Guatemala, 179, "Memorial de D. José de Lara en Guatemala a 13 X 1676."

[12] *Ibid.*, append. XV, transcribing *AGI*, Guatemala, 30, "Reconocimiento de los Remedios practicado por los alarifes José de Porres y Andrés de Illescas, en Guatemala a 16 II 1678."

[13] *AGG*, A 1.10–3 (1679) 31.267–4046.

[14] Mencos, *op. cit.*, ch. IV, n. 93, quoting *AGI*, Guatemala, 78, "Carta de la Audiencia de Guatemala al Rey, en Guatemala a 30 IV 1678."

[15] *Ibid.*, ch. IV, n. 106, quoting *AGI*, Guatemala, Contaduría 983A, "Testimonio de la yglesia de Nuestra Señora de los Remedios, en Guatemala a 22 V 1700."

cause the bell had not yet been delivered.[16] The building works, begun in the last quarter of the seventeenth century, came to a final conclusion by 1716 when the city council ceded an allotment of water for the use of the church.[17] Just how the building fared in the earthquake of 1717 is difficult to ascertain because of conflicting reports.[18] However, in 1762 it was slightly damaged by the flooding of the Pensativo.[19]

It would seem from the foregoing that the first formal sixteenth-century permanent wood-and-tile-roofed structure of *ca.* 1587 was largely obliterated in the alterations carried out between 1625 and about 1650; and that, furthermore, the 1625–1650 building was altered or rebuilt during the last quarter of that same century. In 1773 the building was destroyed, the cupola and the *artesón* coming to the ground.[20]

Architectural Data

THE mixture of walling material is most curious and reflects the constructional history of the church. (FIG. 24.) The façade, the choir, the east end of the church enclosing a small section of the nave, and the whole of the *capilla mayor* are of masonry. The nave walls are of tamped earth interspersed with sections of stone and brick, no doubt representing the portion of the church completed about the middle of the seventeenth century. There are no buttresses on the exterior of any of the walls, an unusual condition since one would expect the tamped-earth walls to be reinforced as was ordinarily done elsewhere where walls of this material were employed.[21]

The present plan of the church falls into three main parts: the choir bay just behind the façade; the nave which is divided into an indeterminate number of bays (probably five to judge by the niches); and finally the *capilla mayor*. The masonry walls of the tower with the chapel, *sagrario*, abutting on the north side of the choir and entered from it, are bonded to the façade and choir walls. Remains of the cross wall which once separated the choir from the nave are still visible inside. This cross wall probably represents the original mid-seventeenth-century façade. It was altered to support the mezzanine floor of the upper choir when the new façade, choir bay, and sacristy were added in the later construction work begun after 1679. A long vertical crack in the north nave wall just inside the ruined cross wall is evidence that the present façade and choir bay represent a later addition to the original church. It divides the older, tamped-earth nave from the later masonry choir.

The mezzanine floor which divided the lower choir from the upper was originally supported on wooden beams, the emplacements of which are clearly seen on the inside of the façade wall. It is not possible to ascertain whether the cross wall between choir and nave was pierced by an arch as shown on the plan (Fig. 24), or whether a heavy timber girder spanned the opening on which the nave ends of the mezzanine beams rested.

The original length of the nave of the mid-seventeenth-century building extended to just short of the west or nave side of the *capilla mayor*, for here a change from tamped-earth to masonry walling is noted. Like the present plan which includes the later additions, the original was a single-nave church. Two rows of wooden posts or columns probably supported the transverse tie beams of the timber-and-tile pitched roof. Unlike most of the single-nave churches of Antigua built in the eighteenth century, there are no engaged piers or pilasters to divide the nave into definite bays. However, five pairs of niches are arranged in both long walls. The first two pairs were bricked up in the later construction leaving three niches each on the north and south tamped-earth walls. The last two niches of the north wall were pierced through and converted into arched openings leading to the priest's house abutting on this wall. It is quite possible that these last two arched openings are part of the reconstruction work carried out between 1625 and 1650.

In a strict sense, the church lacks a crossing, for the last bay, once roofed with a cupola, the pendentives of which are still in view, was the probable location of the *capilla mayor*. Normally the cupola is set over the next to last bay, the crossing. This last bay of the existing plan with its cupola was added in the later reconstruction, to judge by the difference in walling. Directly on the longitudinal axis of the building and resting in part on a low buttressing wall which extends across the entire width of the rear of the church, a small niche protrudes slightly from the *capilla mayor*. Abutting on the north side of the nave are the remains of what may once have been two separate constructions, no doubt

[16] *AGG*, A 1.11 (1710) 2044–94.

[17] *AGG*, A 1.2–6 (1716) 16584–2284.

[18] Ximénez, **3**: p. 356. According to a report of Diego de Porres, some arches and the façade were cracked. The *artesón* also was in danger of falling; see Mencos, *op. cit.*, append. XVII, transcribing *AGI*, Guatemala, 319, "Vista de ojos y reconocimiento de la yglesia de Ntra. Señora de los Remedios, por el Maestro Mayor de Obras Diego de Porres, en Guatemala a 9 XII 1719."

[19] Juarros, **1**: p. 165.

[20] González Bustillo, *Razón particular*, also *Ciudad mártir*, p. 92.

[21] According to the information quoted in the document cited in fn. 15 above, there were six buttresses eleven *varas* in height on each side of the building. No such buttresses or vestiges of any remain today. It would seem that the report, an account of the money spent in construction, and written thirteen years after completion, was not based on a physical inspection of the building then standing.

the house for the curate and its dependencies. These structures, now in total ruins, were built of tamped earth, of which hardly more than some of the perimeter walls can be made out.

The overall plan of the church is of rather elongated proportions, greater than usual in Antigua, and no doubt the result of the alterations and additions alluded to above. The total width of the façade, including the tower housing the *sagrario* which has the appearance of an independent unit but which is attached to the façade, is 20.61 m. This measurement also includes the pier projecting to the south of the retable-façade. The dimensions of the pier are 1.85 m. on the front face by 1.35 m. deep by 1.40 m. on the inside return face where it joins the long wall of the nave. Like the tower to the north, it protrudes beyond the width of the nave.

The total exterior length of the building measured from the façade to the protruding niche of the *capilla mayor* is 55.70 m. The exterior width of the nave is 13.45 m., making an actual proportion to length of about 1:4 instead of the apparent proportion of slightly more than 1:2½ when compared with the width of the façade as a whole. The interior length of the nave measured from the cross wall of the choir to the inside transverse arch of the *capilla mayor* is 34.50 m. Its width of 10.65 m. makes a proportion of just a little more than 1:3. These dimensions and proportions are probably very close to those of the 1650 church plan.

The tamped-earth nave walls are approximately 1.40 m. thick, while the masonry walls of the façade and choir are about 1.50 m. The projection of the applied orders are not included in this dimension. The second story of the façade is not set back, as is common in most churches in Antigua, but is of uniform thickness throughout its height. Another peculiarity of the wall construction is the treatment of the nave niches, all of which are shallow including those which had been bricked up. The latter were originally about 3.30 m. wide while the rest, including the two with door openings, are about 4.10 m. wide. Another difference in the thickness of the walling is to be noted in the *capilla mayor*, the walls of which are a little less than 1.70 m. and hence much thicker than those of the rest of the church. The inside dimensions of this section of the plan are about 10.20 m. square. The *arco toral*, that is, the arch which spanned the side of the bay bounded by the nave is missing. Remnants only of the lower portions of the supporting piers are still visible. Parts of the pendentives which once supported a dome or a cupola are still in evidence in the far corners of the room against the rear wall.

The exterior dimensions of the tower are as follows: 5.40 m. on the façade side by 6.58 m. deep by 5.30 m. on the rear or return wall. The *sagrario* inside is 4.45 m. square and is roofed with a dome on pendentives built between the niche arches, all still *in situ*. The diameter of the dome is approximately 3.90 m. The three freestanding walls have niches, while the part of the wall abutting on the church adjacent to the choir has a door 3.15 m. wide.

The nave and the choir were once covered with a pitched roof of timber and tile, the outline of which is still discernable on the inside of the façade wall. One may suppose that the roof was finished as an *artesón* with tie beams across the nave supported on wooden posts, a system which remained unchanged when the masonry parts of the structure were added after 1679.

The façade design is of the typical retable-like frontispieces common in Antigua, but is very reserved and devoid of any surface ornament. (FIG. 25.) The applied orders which decorate the three-storied surface adhere strictly to logical or structural forms with no nonarchitectonic departures or free innovations for purely decorative ends except for the pulvinated frieze in the second story. The total width of the façade, as has already been stated, is 20.61 m. and is divided as follows: 5.40 m., the width of the tower; 13.46 m., the central portion or retable; and 1.85 m., the projecting pier. The total height of the façade at the highest point is approximately 15.00 m. making a proportion of about 4:3, that is, wider than high. The tower, however, is treated as a separate unit attached to the north end of the façade, thereby altering the low squat proportions of the retable if viewed separately. The retable proper is 13.46 m. wide, or slightly less than the height of about 15.00 m., thus making the proportions of this central portion of façade about 4:5, which is still rather square except for the statue or the cross which no doubt once surmounted the pinnacle adding to the total height of the composition.

The retable-façade is laid out in the usual manner of three vertical bays and three horizontal stories or divisions. The widest is the central bay, in the first story of which the main door, set within a round-headed niche or recess, is located. (FIG. 26.) Directly over the door in the space between the extrados of the door arch and the intrados of the niche arch is a coat-of-arms worked in stucco. In the second story, immediately above in the same central bay, the wall is pierced by a niche-window with splayed reveals and spanned by a segmental arch, the soffit of which is also splayed. Framing the window are very simple pilasters and a high pediment, on the tympanum of which palm leaves flanking a rosette are worked in stucco relief. (FIG. 25.) Directly above, the third-story central bay consists of an *espadaña* or parapet wall with three small niches, the center one larger than the other two. This parapet wall, equal in width to the central bay and the only one on the third story, is flanked by low scroll or volute pediments which carry the eye down to the side bays. Broken scroll pediments are also set on top of the parapet wall to either side of a low pedestal or base where a statue or a cross once stood providing, thereby, a soaring pinnacle and focal point to the triangular composition of the story as a whole. The lateral bays are but two stories high and punctuated with small round-headed niches framed by pilasters and triangular pediments. The bays are marked off by engaged columns of very simple design, complete with bases and entablatures which break miters on column centers. The shafts are plain except for a projecting astragal molding about one-third of the way up. The order employed on the lower story is Tuscan, that on the upper Ionic. The columns which flank each side of the *espadaña* of the third story are more ornate having capitals which are a variation of the Tuscan with concave abaci. The first-story columns do not rest on a podium, as is so common in Antigua, but run the total height of the story. (FIG. 27.) As a result, the niches are set rather low leaving a blank space between the tops of their low triangular pediments and the entablature of the main order above which runs across the retable as a whole.

The tower abutting to the north of the façade is treated as an independent unit set slightly back and is an asymmetrical element in the overall design. (FIG. 25.) Its exterior elevation was originally laid out in three stories or horizontal divisions, the topmost of which is no longer extant. The first story is unadorned except for some moldings in the form of a cornice at the top at the level of the pediments of the niches in the second story of the retable-façade. Except for a single window near the top and just under the cornice, the surface is unbroken. Immediately above is the belfry proper with arched openings facing the three exterior sides. The pinnacle, now missing and possibly pyramidal or domical in shape, probably reached at least the height of the base of the statue or cross which once surmounted the central bay of the retable-façade.

3. *San Agustín: Church and Convent, 1657 and* ca. *1761*

(FIGS. 28–31.)

Historical Data

THE Augustinian order was never very important in Central America. In fact, the convent in Antigua was the only one ever established in Guatemala and was under the jurisdiction of the house in Mexico.[1] Nor did the Augustinians engage in any of the intensive missionizing activity which so characterized the Franciscan and Dominican orders.[2] Two members of the order arrived in Guatemala about 1610, but were hardly welcomed by the city authorities of Antigua. In a *cabildo* dated July 21, 1611, the city council stated quite categorically that it was against the establishment of the Augustinian order, giving the reason that the city was too small and poor to support still another monastic establishment.[3] Nevertheless, soon after their arrival, the two monks took over some houses which had formerly belonged to Lic. Diego de Paz y Quiñónez and in a sense founded the order on an informal basis. The protestations of the city council were apparently to no avail, for the convent was officially established in 1611 under the patronage of the captain Manuel Estévez who donated 24,000 *tostones* for that purpose.[4] Gage mentions in passing that the order owned a sugar mill in Petapa where twenty Negro slaves were employed.[5] This property probably represented the endowment given them by Estévez. The houses where they had established themselves soon after their arrival in 1610 were located near the convent of San Francisco. They did not live very amicably with their neighbors with whom they were soon engaged in controversies and arguments, as a result of which they had to move elsewhere to quarters which were considered inadequate. Yet there they remained until 1615. At this time ten additional monks arrived in Guatemala and the convent was transferred to some houses which had been vacated by the nuns of Santa Catalina.[6] By 1617 the order seems to have been well established and to have made some improvements on the conventual buildings, for in that year the fountain in the cloister was completed for which the city council granted a water allotment.[7]

The houses which they had taken over from Santa Catalina were hardly of monumental character, nor had they been intended as a convent in the first place. Furthermore, these had been damaged in an earthquake in 1607 and were patched up and used until 1615 when the nuns moved away to new quarters, making them available to the Augustinians.[8] These quarters though now supplied with water were at best makeshift, and so ten years later (in 1625) it was pro-

[1] Vásquez, **4**: p. 363.
[2] Fuentes y Guzmán, **2**: pp. 373 ff.
[3] *AGG*, A 1.2–2 (1611) 11.766–1772, see also *BAGG* **8**, 1 (1943): pp. 27 ff.
[4] *AGG, loc. cit.*; also *Efem*, p. 40.
[5] Ch. XVIII, p. 299.
[6] Remesal, **2**: p. 353.
[7] *Efem*, p. 44.
[8] Vásquez, **4**: pp. 324 and 368 ff.

posed to give them the church of Los Remedios which at that time was in rather deteriorated condition. This proposal was vetoed by the crown.[9] The monks must have remained where they were for some time, for some thirty-five years later, on January 28, 1657, a new church and convent were formally dedicated.[10]

The first notice concerning the building from the eighteenth century is a report that the damage suffered by the church and convent of San Agustín was only superficial.[11] This was probably not altogether accurate, for in a letter to the crown in August of 1720 the monks stated that they had been in bad financial straits since the earthquake, living only on the scant contributions gathered by its alms seekers as well as the payments they received from the performance of religious services such as masses and burials.[12] Repairs began soon afterward but came to a standstill in 1726. Some work had been done on the *capilla mayor* and some on the walls. Five bays and the *portada* still required work.[13] The city authorities had the same measure of enthusiasm for the order as they had had a century or so earlier, and did nothing to help in the reconstruction of the damaged buildings. Finally, in 1739 a *cédula* arrived in which the *ayuntamiento* was ordered to help in the reconstruction of the convent of San Agustín.[14] It may be assumed that the necessary repairs were carried out, probably on the façade and some of the vaults.

In the earthquake of 1751 the convent and church were again badly damaged, so seriously that the neighbors in the vicinity actually petitioned the city council that the church be torn down, especially since the order itself was indigent and lacking sufficient income to carry out the repairs.[15] Since nothing was done to repair the building, the *ayuntamiento* finally issued an order in 1755 to dismantle the ruinous parts of the wall of the *capilla mayor* down to the level of the windows.[16] It would seem, then, that the principal damage occasioned in the earthquake of 1751 was in the *capilla mayor* where the domical roof, as well as the supporting walls, must have been destroyed. If any repairs were carried out at all, they probably were done in the following decade, for a notice in 1761 records a petition of the monks of San Agustín requesting the masons' guild to cooperate and help with the reconstruction of the church.[17]

Some repair work must have been done on the church. At least the *capilla mayor* was certainly reconstructed for it exists to this very day. In the earthquake of 1773 the church and convent were again very badly damaged. The cupola over the church crossing remained intact until 1917 when in the earthquake of that year it was brought to the ground.[18] Of the convent not a single vestige remains, and the site where it once stood contiguous to the church is now occupied by private houses.

Architectural Data

THE church plan has but a single nave and an independent bell tower attached on the north side of the façade. (FIG. 28.) The exterior length of the church is 50.80 m. The width of the façade, exclusive of the bell tower, is 12.56 m., the exposed portion of the latter measuring 7.50 m. The lower story of the north wall of the tower has been utilized in the construction of the adjacent private house making it impossible to ascertain its exact dimensions. On the basis of the thickness of the other walls, it is estimated that the total width of the bell tower at ground level on the main façade was originally 8.10 m. The length measured through the doors is 7.60 m. The *lonja*, running across the whole front, including the façade and the tower, is 20.50 m. long by 4.50 m. wide.

The interior width of the church is 9.20 m. measured from wall to wall, and 7.30 m. between piers. The longitudinal interior measurement on pier centers is as follows: the choir or first bay, 7.93 m.; the second bay, 5.65 m.; the third bay, 5.45 m.; the fourth bay, about the same, 5.42 m.; the fifth bay, 6.10 m.; the crossing, 10.45 m.; the *capilla mayor*, 6.50 m. The unequal length of the nave bays is a puzzling feature, and must be the result of the subsequent alterations in the building during the eighteenth century. It would appear that the crossing with its cupola or dome on pendentives, the *capilla mayor*, and the fifth bay represent eighteenth-century construction which was not finished until after 1761 and which was added to the length of the older seventeenth-century structure.

Extensive repairs were also carried out on the whole of the south wall including that of the choir and the façade too. The walling on this side of the church is 1.50 m. thick and

[9] *Ibid.*, p. 383; Juarros, **1**: p. 127; see also Los Remedios above.

[10] Ximénez, **2**: p. 244, and *Efem*, p. 66.

[11] Ximénez, **3**: p. 356. An opposite view with regard to the extent of the damage is given in *AGI*, Guatemala, 309, dated 23 XI 1720, quoted by Mencos, *Arquitectura*, ch. VI, n. 108.

[12] *AGG*, A 1.11 (1720) 16779–2292.

[13] Mencos, *op. cit.*, ch. VI, n. 112, quoting *AGI*, Guatemala, 309, "Testimonio del estado de la obra del convento de San Agustín de Guatemala, en Guatemala a 29 I 1726." The work remaining is specified to be carried out in the same form as that already existing.

[14] *AGG*, A 1.10–3 (1739) 18806–2448.

[15] *Efem*, p. 204.

[16] *Ibid.*, p. 212.

[17] *AGG*, A 1.16.1 (1761) 17133–2312.

[18] Díaz, *Romántica ciudad*, p. 58.

considerably heavier than that on the north side opposite. A two-storied arcaded cloister once probably abutted on it. The buttresses of the street side are all uniform, and except for those of the crossing are more decorative than functional. Actually there are no buttresses to either side of the street door which cuts through the south wall in the third bay. The door is treated like a small projecting *portada* framed by pilasters attached to shallow piers which run to about half the height of the wall. The north wall is much thinner, 1.10 m. as opposed to 1.50 m. The buttresses of the thicker south wall are not all paired with those on the north, either in number or character. Another unusual feature which might indicate repairs carried out at different times, is the nature of the fill between the buttress on the heavier south wall. It is approximately 3.00 m. high where it abuts on the wall but pitches down about a meter so that it is approximately 2.00 m. on the exterior face. However, these dimensions vary between bays, possibly indicating that this fill was not laid in one building operation but at different times.

The lower-story wall of the bell tower is of an extremely exaggerated thickness where it abuts on the church, 2.00 m.; yet it is bonded to the thinner south wall. (FIG. 30.) Therefore, the bell tower must be considered an integral part of the original façade plan, that is, conceived as an asymmetrical composition, and not an afterthought. The exterior stairs on the west side of the tower leading to the second story are not original.

A barrel vault extends over the length of the nave from and including the first bay or choir to the crossing. The crossing had once been roofed with a cupola composed of a drum and hemispherical vault supported on pendentives. The conventual church of Santo Domingo in San Cristóbal de las Casas, Chiapas, Mexico, is the only other extant example of the combination of barrel vaults with a dome over the crossing.

The mezzanine floor which separates the first bay into the lower and upper choirs is a flat ellipsoidal pendentive dome, the soffit of which is treated with a pattern of false ribs. Domes of this type were first employed in the eighteenth century. A similar dome roofs the *capilla mayor*.

The piers, emerging as buttresses on the exterior and as pilasters on the interior from which the transverse arches of the vaulting spring, are not uniform throughout the church. The pairs which support the vaults of the lower choir, of the *capilla mayor*, and of the crossing are of one type, while those which support the transverse arches of the nave barrel-vault are smaller and much simpler in design. All, however, have bases of stone. The shafts and capitals of the pilasters are of brick bonded to the walls and finished with a coat of plaster.

The bell tower, though slightly off square on the exterior, has a perfectly square interior plan, 4.60 m. by 4.60 m. The doors, set on the same axis, are both 2.55 m. wide. (FIG. 28.) The bell tower probably served as an entrance to the convent giving access to the main cloister, one wall of which no doubt abutted on the thinner north church wall reinforcing it. The second story of the bell tower is set back slightly but follows the same plan as that of the first story. Two windows pierce the second-story wall both in front and in back, the lower being a smaller rectangular opening surmounted by a longer round-headed opening. A small dome with a lantern crowns the structure as a whole. Likewise, the floor between the first and second stories of the interior is supported by a dome on spherical pendentives. The exterior of the tower is worked with an order of flat pilasters without capitals which support triglyph and metope friezes.

The *portada* or central retable portion of the façade, flanked by the bell tower to the north, is framed to the south by a massive pier measuring in plan 1.70 m. on the front and 2.00 m. on the south return, extending slightly beyond the line of the projection of the buttresses lining the south wall. (FIGS. 29, 30.) It reaches the total height of the façade, but its topmost parts are ruined. Some sort of finial once probably surmounted the pier so that it framed the freestanding third-story central bay of the retable-façade. A small caracole stairway begins in the second story of the pier giving access from the upper choir to the roof. Tiny windows, one on the south side and another higher up on the main façade side, light the stair.

The layout of the *portada* or central retable of the façade is of the common type seen in Antigua, though its proportions are rather more square than usual and reminiscent of the design of Los Remedios. (FIG. 25.) Actually the three-by-three division is not as pronounced here in San Agustín, for the third central bay rises more like a roof comb above the second story.

The applied orders in both the first and second stories are quite simple and have almost classical Tuscan capitals, the echinuses of which are adorned with an egg-and-dart motif. (FIG. 29.) The entablatures break miters on column centers, but lack architraves. The friezes recall classical prototypes being divided into triglyphs and metopes. The triglyphs are rather pronounced and have six heavy tenon-shaped guttae directly below the taenia. The metopes are not quite square, being a little narrower than high. Each is decorated with some religious or purely decorative motif.

An unusual feature, one which again recalls the façade of

Los Remedios, is the very low podium with pedestals breaking miters on which the lower-story columns rest. A low podium or Roman attic also runs across the bottom of the second story, thus repeating the same motif.

The main door of the church, located in the central bay of the first story, is of the most common Antiguan type, namely, spanned with a semicircular header and set in a recess of the same outline. The niches in the lateral bays of both stories occupy almost the entire space between the columns and are framed with simple architectonic pilasters supporting triangular pediments in the first story and segmental ones in the second. The niches proper, that is, the recesses where some statues still are *in situ*, are semicircular or apsidal in plan, the soffits of the half domes being fluted. (FIG. 31.) Lighting the upper choir is a deep window with the same decorative treatment as the niches, but with a triangular pediment.

The third story, which is really a finial for the retable, is reduced to the central bay only and accented by very elongated pyramidal merlons to either side set on column centers. The bay itself, quite high in proportion to the second story below, is rather plain being flanked by single columns to either side which support an entablature with a triglyph and metope frieze. Surmounting the bay and unifying the composition is a broken triangular pediment. On center and lining up with the door and window in the first and second stories respectively, a framed niche with a segmental pediment breaks the otherwise plain surface of the third-story wall. Completing the composition are mixtilinear half-pediments sweeping down in undulating diagonals to the side bays.

The design and decoration of the main façade is most confusing with regard to stylistic attributions. (FIG. 30.) The observation made by Angulo that its design is rather archaic (he believed the *portada* might have been built *ca.* 1728) is most perceptive and right for the wrong reasons.[19] It is certain that the building was in a very bad state of repair after the earthquake of 1717 and that the façade was certainly repaired. But its original seventeenth-century form was specifically retained.[20] The rather severe architectonic quality of the façade adhering to strict structural principles in its applied orders puts it out of concert with some of the rather extravagant and ornate façades of eighteenth-century Antigua. The vaulting in the church is quite obviously of two types: the barrel vault of the nave, more commonly employed in the seventeenth century; and the half-ellipsoid vault of the choir, well known in the eighteenth century. The nave part of the plan is indubitably of seventeenth-century origin, while the choir, the crossing, and *capilla mayor* are from the eighteenth century. It is quite possible that the alterations to the choir also necessitated basic changes in the structure of the façade; yet stylistic comparisons with that of Los Remedios seem to point to a seventeenth-century date. The original seventeenth-century design of the façade was purposely preserved in the later reconstruction.

[19] *Historia del arte* 3: p. 19.

[20] See fn. 13 above.

4. *San Pedro: Hospital and Church, 1662, 1675, and 1869*

(FIGS. 32–34.)

Historical Data

THE present name of this church and hospital is "Hospital del Hermano Pedro," sometimes also referred to as "San Juan de Dios." It was rebuilt in the nineteenth century, and it is difficult to distinguish between its colonial and post-colonial elements.

About the middle of the seventeenth century, in 1646 to be exact, some houses which had belonged to one Lope Rodríguez de las Varrilas were acquired by Bishop Bartolomé González Soltero for the purpose of converting them into a hospital for clergymen.[1] It is unlikely that the hospital was built at that time since the bishop died before he could realize his plan.[2] It was not until 1654, under the impetus of his successor, Bishop Payo de Rivera, that work on the church and hospital actually began.[3] By 1662 the hospital was completed and only then did construction on the church begin; the cornerstone was laid in November of that year by Payo de Rivera himself.[4]

José de Porras, twenty-six years old at the time, was in charge of the job. The church was still not finished in 1665, and lacked the nave vaults and the cupola. In a letter to the crown dated October 29, Rivera, repeating the information in the report rendered by Porras, stated, in part, that ten of the arches forming the support of the three vaults of the nave and presbytery had been constructed, and that all that was lacking was to cover the church which should be completed within a year and a half, more or less.[5] Rivera's esti-

[1] Fuentes y Guzmán, 3: pp. 349 ff.; also Juarros, 1: p. 128.

[2] Pardo, *Guía*, p. 116.

[3] Fuentes y Guzmán, *loc. cit.*, and 3: p. 355; *Efem*, p. 64.

[4] Fuentes y Guzmán, 3: pp. 349, 355; *Efem*, pp. 69, 70; also Juarros, 1: p. 128.

[5] Fuentes y Guzmán, 3: pp. 356 ff. For an account of the work done by José de Porras between 1662 and 1665 see Mencos, *Arquitectura*, ch. V, also his append. XVIII, transcribing *AGI*, Contaduría, 883A, "Declaración de José de Porras, maestro de albañilería, y Ni-

mate must have been a little too optimistic for the church was finally inaugurated December 2, 1675,[6] serving as a temporary cathedral from the day of its dedication until 1680 when the Cathedral was finally finished.[7]

In the earthquake of 1717 both church and hospital were left in extremely ruinous state, and the vaults destroyed.[8] Ximénez, quoting Arana and as usual differing with him on the extent of the damage, informs us that according to Arana the church was in total ruins with only the *portada* left standing, and that the vaults had caved in because they had neither "estribos ni bestiones" to support them.[9] From this description, it would seem that the vaulting was supported directly on the walls which were not reinforced with buttresses, a rather unusual method of building even in the seventeenth century. Ximénez, furthermore, minimizes the extent of the ruin leading one to believe that the church was left almost intact,[10] a statement which obviously does not jibe with the facts reported as late as 1719 in which it is clearly stated that all the churches of Antigua are in use again except San Pedro because of the ruinous condition of its vaults.[11] It would seem that Arana's report was nearer the truth, for he gives even such small details as the fact that the sacrament had to be rescued from the ruined church which could be entered only from the door connecting the hospital,[12] concurred in by another anonymous contemporary source.[13] The church was finally repaired in 1730 when the ruined vaults were replaced with an *artesón*.[14]

Just what the original appearance of this seventeenth-century church was like may never be known. According to Arana it was a single-nave church and had a cupola over the *capilla mayor*, and the vaults were not reinforced by buttresses.[15] Furthermore, we have no information as to what reconstruction work was carried out on both church and hospital after the destruction of 1717 other than the installation of a wood and tile roof. There can be no doubt, however, that both were rebuilt, for the next reference that comes to hand is a document dated in 1775 in which it is reported that the demolition of the hospital is complete and that all serviceable material has been sent to the site of the new capital for use there,[16] implying that these buildings had been in use until the earthquake of 1773. The hospital of San Pedro was merged with that of San Juan de Dios and the statues transferred to a church in the new capital city.[17]

Architectural Data

THE present church and hospital are, by and large, nineteenth-century reconstructions in which older elements were no doubt employed. (FIG. 32.) The present façade of the church retains its original colonial retable-like character. (FIG. 33.) Its overall design and the style of the applied orders are close to that normally found in seventeenth-century churches in Antigua such as Los Remedios and San Agustín. The central bay of the lower story projects slightly and is surmounted by a low triangular pediment recalling the effect of a "triumphal arch." The same device appears on Santa Teresa, implying, thereby, that José de Porras (II) may have built that church also.[18] The hospital *portada* is better preserved than the church façade, and its original seventeenth-century character seems to have been left more or less intact in the later alterations. (FIG. 34.)

The door on the east side of the hospital building behind the church proper which has been bricked up, but which once gave access to the convent is quite different. The pair of solomonic columns framing the door opening are very close in style to those of the church of Jocotenango. All in all, the church façade, though rebuilt in 1869,[19] is still quite

colás de Cárcamo sobre el hospital de San Pedro de Santiago, en Guatemala a 20 x 1665." Also his append. XIX, transcribing *AGI*, Contaduría, 983A, "Testimonio de la casa del hospital de San Pedro de Santiago, en Guatemala a 20 x 1665."

[6] Juarros, **2**: p. 240.

[7] Fuentes y Guzmán, **3**: p. 359; Vásquez, **4**: p. 381; Pardo, *op. cit.*, p. 116. The dates for the completion of the church are somewhat confused by Juarros by the fact that the Cathedral was located there. In one place (**1**: p. 152) he says the building was dedicated December 2, 1663, and the cathedral activities installed there on that very day. But from Fuentes y Guzmán (**3**: pp. 359 ff.), who quotes Payo de Rivera, we learn that the church of San Pedro was not yet roofed. In another place (**2**: p. 240), Juarros gives the date December 2, 1675, for the inauguration of the church and its use as a temporary cathedral. The latter date conforms with the facts concerning the Cathedral building of 1680 given by him elsewhere and from other sources. The year 1663 must be an error on his part, or of the editors of his manuscript.

[8] *AGG*, A 1.2–5 (1719) 15776–71, also *BAGG* **8** (1943): p. 126. For the same information see Mencos, *op. cit.*, ch. V, n. 17, quoting *AGI*, Guatemala, 309, "Año de 1723. Testimonio de los autos fechos sobre lo que tendrá de costo la reedificación de la iglesia del ospital del Señor San Pedro de la ciudad de Guatemala, en Guatemala a 3 x 1722."

[9] Ximénez, **3**: p. 355.

[10] *Ibid.*, p. 368.

[11] *AGG*, *loc. cit.*; *Efem*, p. 149.

[12] Arana, *Relación*, also *ASGH* **17** (1941/42): pp. 156 and 232.

[13] Vásquez, **4**: p. 393, a later addition to ch. 27 of bk. V, tract. 2.

[14] Mencos, *op. cit.*, ch. V, citing *AGI*, Guatemala, 309, dated 20 x 1730.

[15] *ASGH* **17** (1941/42): p. 232; Arana, *loc. cit.*, also Ximénez, **3**: p. 355. This fits the description given by Porras in his report of 1665: see fn. 5 above.

[16] *Efem*, p. 257.

[17] Juarros, **1**: p. 151.

[18] See ch. XIII, no. 3, fn. 16, below; also ch. VI above, for biographical data on José Porras (II).

[19] Díaz, *Romántica ciudad*, p. 74; Pardo, *Guía*, p. 117.

colonial in flavor except for minor details such as the false rustication of some of the plane surfaces, and represents another incidence of the afterlife of colonial methods well on into the postcolonial period.

5. *Belén: Church and Convent, 1670*

(FIGS. 4, 35, 36.)

Documentary Data

THE Bethlehemite religious order was founded in Antigua by Rodrigo de Arias Maldonado soon after the death of Pedro de San José de Betancourt in 1667.[1] A hospital, including a small oratory and consisting of no more than a thatch hut,[2] was already in existence before this time, perhaps since 1653. The Beaterio de Belén with a hospital for women on the adjacent side of this little plaza was established not long after.[3] Arias Maldonado began and completed the church of Belén which was roofed with very beautiful vaults, the total cost of which was about 70,000 *pesos*. The hospital, built at the same time, stood as an independent structure surrounded by three small plazas which have since disappeared.[4] The church was supposedly one of the most expensive and beautiful in Guatemala.[5] The cost of the church was not met by Arias de Maldonado alone, for Fernando Francisco de Escobedo, who served as temporary governor from 1671 to 1687, also donated 55,000 *pesos* for the building.[6]

The church, for the most part, was completed in December of 1667 and inaugurated in February of 1668.[7] Pedro de Betancourt, the inspiration of Arias de Maldonado, had died on April 25, 1667, just prior to the time when the latter supposedly began work on the church, hospital, and convent buildings.[8] If the date of Betancourt's death is correct, it would mean that the construction took about eight months to complete. This is not likely since in September of 1670, almost two years after the inauguration, one José Bernal de Cabrejo donated the sum of 1,330 *pesos* for the completion of the church, probably for interior furnishings.[9] It is more likely than not that the church and hospital were already in construction before 1667 when Pedro de Betancourt died, perhaps in 1666, for in February of 1665 the *ayuntamiento* granted Betancourt a parcel of land he had requested in order to build a habitation for one monk.[10] In July of 1666 he was given an additional piece of land adjacent to the hospital which was mentioned as being already in construction.[11] Among the goods which Betancourt bequeathed in his will was a two-story room still in construction.[12] Rodrigo de Arias Maldonado had been with Pedro de Betancourt four years before the latter died, that is, since about 1663.[13] Therefore, it would be more logical to conclude that construction of the church, hospital, and convent began while Pedro was still alive, perhaps sometime after 1663; and that the church, at least, was complete enough to be inaugurated about a year after his death in 1667, but that details of interior decoration and furnishings still remained unfinished so that as late as 1670 donations were received for that purpose.

In the earthquake of 1717 some unspecified damage to the buildings occurred,[14] though Ximénez says there was none.[15] However, as late as 1721 the order asked help to repair the damage caused in 1717.[16] To judge by a request made even as late as 1725 for permission to use a certain building site for the making of adobe bricks for the convent building, it would seem that it had not yet been totally repaired.[17] Some damage was also caused in the earthquake of 1751.[18] The final destruction came in 1773 when the convent especially was badly hurt. Yet, to judge by the present state of the church building, the effect of the earthquake was light, for the interior vaulting is still *in situ*. The church and convent are now private property and have been converted into a hotel. The convent was drastically reformed while the church remains more or less as it was left in 1773.

Architectural Data

THE main façade which faces west on the little plaza is the most interesting feature of the church. (FIG. 35.) It is divided into three divisions of almost equal width. The central portion, where one would expect to find the retable, is devoid

[1] *ASGH* **2** (1925/26): pp. 318 ff.
[2] Juarros, **1**: pp. 129 ff.
[3] See Beaterio de Belén, no. 6, below.
[4] García de la Concepción, *Historia bethlemítica*, pp. 26 ff. and 29; also Villacorta, *Historia*, pp. 225 ff.
[5] Vásquez, **4**: p. 391, gives the cost as 55,000 *pesos*, while Juarros, *loc. cit.*, no doubt using García de la Concepción as his source, gives the cost as 70,000 *pesos*.
[6] García Peláez, **2**: p. 153. Fuentes y Guzmán, **1**: p. 142, claims that the same individual donated more than 80,000 *pesos* for the construction of the church as well as for the interior furnishings such as altars, lamps, organs, and bells.
[7] *AGG*, A 1.2–2 (1668) 11.775–1781, also *BAGG* **8**, 1 (1943): pp. 33 ff.
[8] See n. 1 above; also *Efem*, p. 76.
[9] *Efem*, p. 81.
[10] Vásquez, *loc. cit.*; *Efem*, p. 72.
[11] *Efem*, p. 73 ff.
[12] *ASGH* **2** (1925/26): pp. 324 ff.
[13] García de la Concepción, *loc. cit.*
[14] Vásquez, **4**: p. 393.
[15] Ximénez, **3**: p. 356.
[16] *Efem*, p. 152.
[17] *Ibid.*, p. 158.
[18] *Ibid.*, p. 205.

of any special features such as projections and indentations or vertical accents, applied orders being totally absent. The flanking towers are quite plain too, and except for some very shallow pilasters on the first story, the surfaces are unadorned. There is no way of determining, however, whether applied orders were ever existent on the façade, since it has been subject to some reconstruction work in the twentieth century and earlier too.

The central bay has no horizontal divisions except the moldings high up at the level where the finial once stood and which must have been freestanding rising above the roof behind. The only architectonic accents on this the central portion of the façade are the round-headed door flanked by a small niche on each side and the elongated narrow window above. The remains of a niche are visible immediately above the window which is not original.

The actual door opening is set within a deep niche with a semicircular header. The door arch describes a similar curve to that of the niche, but is set below the intrados of the niche arch with the result that a tympanum area is formed between the two. In other words, the reveals or jambs of the niche are taller than those of the door, but both are spanned by semicircular arches. A small niche flanked by little pilasters is located in the tympanum right under the soffit of the niche arch and slightly above the crown of the door arch.

The window above the door is oblong with a horizontal header. Both the header and the jambs are splayed receding at an angle so as to reduce the size of the opening in back. The window opening shows evidence of recent work. A semicircular niche is set immediately above the window, giving the ensemble an elongated shape.

The framing elements of the small niches to either side of the door below project only slightly from the wall, and consist of a horizontal molding which forms the floor of the niche, and two narrow pilaster-like uprights supporting a small triangular pediment. The niches, with statues of brick and plaster still in place, are semicircular in plan and covered with quarter-spheres, the interior surfaces of which are decorated with some geometric and floral patterns worked in plaster.

The towers, that is, the two side bays, are treated somewhat differently from the central bay being divided into three clearly marked stories. (FIG. 35.) The pinnacles of the belfries which once surmounted each tower are now gone. Only part of the north belfry exists today. Each tower at ground level is only slightly narrower than the central bay. Yet they are more accented since each is treated with three Ionic pilasters and a window. (FIG. 36.) This window, with a half-circle header, is framed in turn with pilasters and a broken triangular pediment, the upper ends of its raking cornices turned back to form small volutes.

The pilasters set at either extreme of the lower story of the towers rest on high, narrow pedestals with projecting crown moldings, their flat, undecorated, and slightly projecting shafts terminating in a necking band and a very simple Ionic capital. The volutes of the capitals emerge from horizontal moldings as tight little spirals and project noticeably beyond the vertical limits of the shafts below. The order continues with a two-faciae architrave surmounted by an unadorned frieze above which there is a boldly projecting cornice composed of a number of moldings, the principal one being a rather broad corona. The architrave and frieze surfaces are broken by mitered projections following the scheme of the pilasters below. The arrangement of the applied order has an architectonic quality following a logical functional scheme.

But the third pilaster in the center of the tower bay adds a nonarchitectonic and purely decorative accent. It rises from between the volute ends of the raking cornice of the broken pediment over the window, the shaft actually resting on a bracket. In height, the central pilaster is less than half that of those to either side. Yet it is crowned with a capital of equal size. The presence of the window below adds a disturbing note too, for it seems as if the central pilaster rests on a void.

The second story of the towers is unadorned except for three flat, vertical bands or pilasters set on the same centers as those below. These lack capitals or other architectonic features. The belfry of the north tower, the only one extant, is really a third story and lacks the pilaster treatment, the surface being left plain. The finial of the belfry is gone now.

The plan of the church is divided into a central nave three bays long, an equal number of side-aisle bays, a projecting rectangular presbytery, and an extra chapel which extends beyond the middle north side-aisle bay (FIG. 4.) Not all these individual elements are uniform throughout, implying that some represent changes and additions to the original plan which were carried out during the course of the building's history before its abandonment in 1773.

The three nave bays are square in plan and are the only ones that are uniform in size. They are roofed with true pendentive domes, *bóvedas vaídas*. The presbytery which projects beyond the main body of the church, wider than the nave, is rectangular in plan and approximately twice as long as any individual nave bay. It is roofed with a barrel vault. The side aisles are quite narrow and line up with the towers. They can be entered only from the nave, there being no doors in the towers which might otherwise give

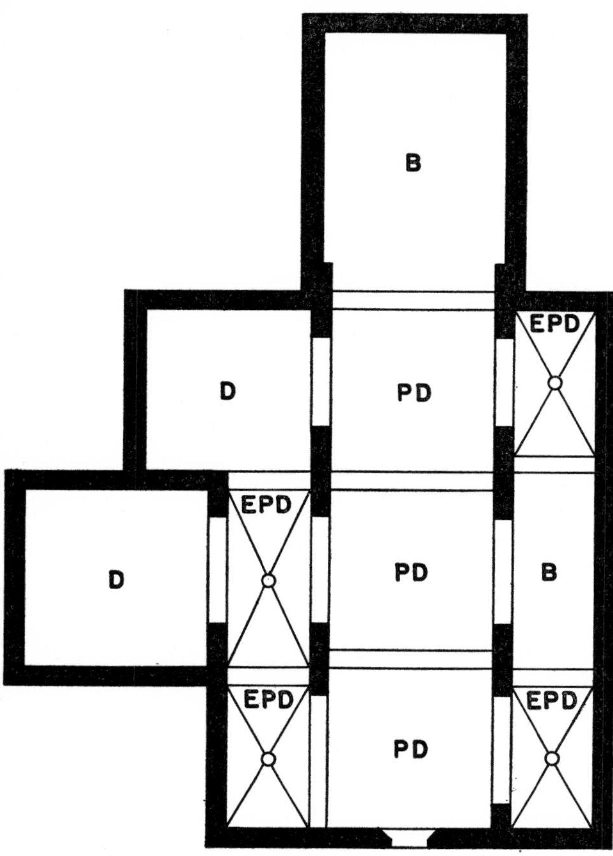

FIG. 4. Belén. Plan (not drawn to scale).

B Barrel vault
D Dome on pendentives over square bay
PD Pendentive dome over square bay (bóveda vaída)
EPD Ellipsoid pendentive dome over rectangular bay

access to them. The six side bays vary in length one from the other, and they do not have the same longitudinal dimension as the corresponding nave bay to which each is adjacent. The first and third south side-aisle bays and the first two north side-aisle bays are more or less of the same length and width, and are all roofed with half-ellipsoid pendentive domes, "half-watermelon" domes, fitted over the rectangular plan. The second or middle south side-aisle bay is somewhat longer and is roofed with a flat segmental vault. The third north side-aisle bay is square and roofed with a low dome on pendentives. The bay which projects beyond the second or middle north side-aisle is likewise square in plan and is also roofed with a dome on pendentives. The ensemble of side bays does not comprise true side aisles for purposes of circulation; rather the individual bays served as independent chapels and must have originally been furnished with retables. The extra bay off the second north side-aisle bay also probably served as a chapel, perhaps the *sagrario*.

The interior order is very plain. A simple classical entablature with a dentil course is supported on wide, unadorned pilasters. In lieu of massive piers, a structural or bearing wall is pierced with large semicircular-headed openings and the short remaining stretches adorned with pilasters. The wall continues above the entablature into which the longitudinal supporting arches of the vaulting are embedded rising above the height of the side-aisle roofs. In the lunettes formed by the longitudinal arches clerestory windows have been installed.

6. *Beaterio de Belén, 1670; Santuario de Guadalupe, 1874*

(FIG. 37.)

Historical Data

THE Beaterio de Belén was located adjacent to the convent and church Belén, and built soon after 1670.[1] Living quarters for the nuns (*beatas*), a hospital for women, and a small church comprised this monastic establishment for women. The hospital and nunnery were rebuilt in 1951 and converted into a hotel, so there is no way of knowing what the original plan was like.

The little church had probably been built toward the end of the seventeenth century, that is, not long after that of the Order of Belén diagonally across the square.[2] Exactly what the appearance of this building was when first built is hard to tell for it had been almost totally destroyed in the earthquake of 1773.[3] It was reconstructed in 1874 and rededicated as El Santuario de Guadalupe.[4]

Architectural Data

AN examination of the façade of this little church, now known as El Santuario de Guadalupe, reveals a startling conclusion, that the popular vernacular colonial style persisted in Guatemala right through the nineteenth century. (FIG. 37.) Two thin elongated towers treated like columns flank the façade. The narrow surface is indented with two broad, shallow flutes meeting in sharp arrises and bounded by heavy convex tubelike half-round moldings at the cor-

[1] Juarros, I: p. 138. See also García de la Concepción, *Historia bethlemítica*, p. 71.
[2] See Belén, no. 5, immediately above.
[3] González Bustillo, *Razón particular*.
[4] Díaz, *Romántica ciudad*, p. 68.

ners. A horizontal molding separates the shaft element from the arcaded belfries above. The space between the towers is treated like a retable. It is divided vertically into three bays and horizontally into two stories plus a third or finial. A large door with a half-circle header occupies the ground story of the central bay. A niche-window is set directly above in the second. The side bays have two superimposed niches in the first story and single niches in the second.

Details in the decorative treatment of the façade show unmistakable connections with many of the colonial churches of Antigua, particularly of the eighteenth century. For example, the engaged columns of the orders which mark off the vertical bays of each story were no doubt copied from the church of Santa Cruz but a block or two away on the other side of the river.[5] (FIGS. 132, 133.) The base which is set on a thin plinth, is composed of heavy torus surmounted by a peculiarly shaped pot form, the surface of which is treated with foliage worked in plaster. The door is set in a deep niche with the splayed reveals. The soffit of the arch above is trumpet-shaped (*abocinado*), diminishing in size toward the rear, and is fasciculated with an elongated Doric leaf and spine pattern somewhat like a shell in a manner very closely imitating the door of the church of El Calvario.[6] (FIG. 115.)

Along the west side of the building, the buttresses, one story high and set at irregular intervals, emerge as engaged columns terminating with capitals consisting of two cyma reversas separated by a broad fillet decorated with rosettes. This profile continues between buttresses and forms part of the coping cornice of the wall. Beneath the cornice, a molding imitating hanging drapery with tassels runs the length of the wall, the prototype of which may be seen on La Concepción.[7]

7. *San Cristóbal el Bajo: Church, Seventeenth Century*

(FIGS. 38, 39.)

Historical Data

THIS was an unimportant little church on the outskirts of the city of Antigua which, ironically enough, is one of the best preserved. Literary and documentary references are very scant, so that its history is not completely known.

The earliest mention of the church is in the form of a petition to the city council, *ayuntamiento*, dated April 19, 1607, in which Cristóbal de Calada and Mateo de Zúñiga, the *encomenderos* of the corn lands of San Cristóbal el Bajo, asked authorization to build a hermitage.[1] Whether or not this petition was immediately granted there is no way of knowing, but by the last quarter of the seventeenth century a church building was already standing there and is listed as being under the ecclesiastical jurisdiction of the convent of San Francisco of Antigua.[2] In 1689 this little church appears among those under the doctrine or ecclesiastical jurisdiction of the Franciscan convent of the nearby town of San Juan del Obispo, implying a change of administration and, furthermore, the number of confessants, whose native language is Cacchiquel, is given as 350.[3] According to Fuentes y Guzmán, the change over to the doctrine of San Juan del Obispo had taken place about 1668. He also states that about this time, *ca.* 1690, there were 350 parishioners, thus corroborating the information about this matter from other sources, and that the frontispiece of the church had a tower.[4] By the end of the eighteenth century, however, the town was listed in the parish of Los Remedios, indicating a third change of jurisdiction due to the secularization of the doctrine in mid-century.[5]

The only conclusions possible on the basis of this very scant information available are that the church of San Cristóbal el Bajo may have been built some time in the seventeenth century. Nothing is known of its history during the eighteenth century and later, though it is hardly likely that this little building remained intact during the great number of earthquakes which plagued the city and that it did not undergo any changes or repairs. It is still in use as a church today, which probably explains its well-preserved state.

Architectural Data

THE most interesting feature of this little church is the façade which may be compared to those of the churches of Los Remedios and San Agustín above. (FIGS. 38, 25, 30.) The layout of the whole façade is the usual retable type with three-by-three divisions flanked by twin towers which, in this case, are very small. The general proportions of the retable are somewhat squatter than usual in Antigua, but it is doubtful that this was so at first, for the finial, that is, the third-story central bay, is not original in all details. Its topmost portion is of recent construction, and probably sup-

[5] See Santa Cruz, ch. XIV, no. 5, below.
[6] See El Calvario, ch. XIV, no. 1, below.
[7] See La Concepción, ch. XIV, no. 4, below.

[1] *Efem*, p. 37.
[2] *AGG*, A 3.2 (1673) 15207–825, also *BAGG* **10** (1945): pp. 128 ff.
[3] Vásquez, **4**: pp. 37 ff., referring to *Archivo Arzobispal de Guatemala*, A 4.5–2.
[4] Fuentes y Guzmán, **1**: p. 379.
[5] Juarros, **1**: p. 79.

planted a pinnacle which had been better integrated to the parts below.

The design of the central bay differs somewhat with regard to the layout of the decorative elements and number of openings from the usual retable-façades of other contemporary churches. The door in the first story of the central bay with its double arch treatment, a semicircular-headed door set within a recess or niche with a concentric header, is typical, however. (FIG. 39.) Yet, an unusual feature is the use of stone for the archivolt and jambs of the door which are decorated with a repeated lozenge set within a depression or panel framed with moldings. The imposts are treated like Tuscan capitals and, as a matter of fact, the jambs are treated as pilasters.

An octagonal window with only slightly splayed reveals is set on center in the second-story central bay directly over the door. (FIG. 38.) This window does not run the total height of this bay, but is surmounted in the space above by a small deeply-recessed round-headed niche which actually breaks across the entablature dividing the second story from the third. The horizontal moldings of the entablature are not interrupted, but continue around the head of the niche forming the archivolt for the semicircular opening.

The third story, as noted above, has been modified somewhat. The remaining original parts consist of a low parapet wall with a segmental-headed niche flanked by short unadorned pilasters and scroll or volute half-pediments. Another member once surmounted this composition and has been replaced by a small bit of walling with concave sides on which a cross is now placed. Just what the original finial was like is hard to say, but it must have continued the design of undulating lines by means of volutes and scrolls terminating in a pinnacle considerably higher than the present modern reconstruction.

The side bays with the applied orders are very reserved and of a type exactly like those of Los Remedios above. (FIG. 26.) The first story begins with a low podium, the entire width of the side bay, which supports a pair of very simple Tuscan columns framing a niche with a triangular pediment. The columns have a very marked diminution and a slight entasis, or rather an attempt at entasis. A small astragal molding cuts across the shaft a little less than halfway up.

The entablature which runs across the total width of the retable, as is the case in Los Remedios and San Agustín above, lacks an architrave and breaks miters over column centers. (FIGS. 26, 29.) A triglyph is set over each column center and another over the center of the niche directly above the peak of its pediment. Curiously enough, there are no triglyphs in the central bay at all. The use of triglyphs is also seen on the façade of San Agustín above, where the entablature also consists of two members only. The rather wide blank spaces between the frankly decorative triglyphs can hardly be considered metopes. The second-story niches have segmental pediments, an arrangement also duplicated on San Agustín.

The second story is slightly shorter than the first, though the applied orders are exactly the same, except that the low podium or Roman attic is unbroken on column centers running across the whole front. The octagonal window of the central bay actually cuts through the podium resting directly on the cornice of the first-story entablature. The shorter second-story columns are correspondingly thinner than those of the first story and have a scarcely noticeable diminution. The frieze here lacks the triglyphs. Though this story is shorter than the one below, its cornice is actually much higher and of greater projection than that of the first story, repeating the same molding profiles. This larger cornice was necessary to protect the façade more efficiently from rainwater.

The third story which is largely confined to the central bay has already been described above. The scroll pediments to either side of the parapet wall overlap but half the width of the side bays. Small merlons are placed to each side and are set on center with the outside columns. It is not possible to ascertain whether these merlons are original or recent replacements.

The twin towers, set back slightly from the main central portion or retable, add to the width of the façade. They extend beyond the width of the nave behind and appear as freestanding when viewed from the rear. In appearance they are more like massive square buttresses about three-fourths the height of the central retable, the entablatures of which slightly overlap the front face of the towers. A belfry surmounted each tower; only the north one is still in place. A semicircular-headed opening pierces the three exterior sides. Merlons are set in each of the four corners as well as on the pinnacle of its rather low, pyramidal roof. Whether these merlons are original or not is difficult to say, for they are close to those of the Capitanía of late eighteenth-century date which were restored in the middle of the nineteenth.[6] (FIGS. 201, 202.)

The long sides of the church are accented with five massive buttresses which project to line up with the projection of the towers. They rise not quite the total height of the walls and are surmounted by mixtilinear scroll pediments

[6] See Capitanía, ch. xv, no. 6, below.

looking like gigantic consoles laid horizontally. Rectangular windows, set between the buttresses, light the single-nave interior.

The interior of the church has been subjected to much reconstruction and alteration. Along with the Cathedral, La Merced, San Lázaro, La Capitanía, and other buildings, this church must have been reconstructed in the nineteenth century and repaired on various occasions even in the twentieth.

The present roofing of the single-nave plan consists of whitewashed wood planking laid out to simulate a barrel vault indicating, perhaps, that a masonry vault of that shape once covered the building. The *capilla mayor* at the east end of the plan is roofed with a dome on pendentives. Four small windows, one over each pendentive, with semicircular headers pierce the haunches of the dome and appear as dormers on the exterior. The interior order is quite simple. A single deep inset square flute marks the pilasters which support an entablature with a plain and undecorated pulvinated frieze. Immediately above, a low section of wall is pierced by windows between pilasters to form a clerestory from which the barrel vault once presumably sprang. There is no clear evidence, however, that a barrel vault once existed, for a wood *artesón* can with equal reason be accepted as having originally roofed the nave.

8. *Cathedral, 1680*

(FIGS. 40–50.)

Historical Data

THE Cathedral extant today is the second one on the site, being preceded by a structure built in the sixteenth century. Land for the first cathedral in Antigua was assigned at the time the city was laid out right after the destruction of the former capital in nearby Ciudad Vieja. The foundations of the sixteenth-century structure were laid in 1542 by Rodrigo Martínez Gárnica, who probably worked but one year on this phase of the job.[1] Once the foundations were set, work progressed very slowly, mainly for lack of money.[2]

Then in 1545, at the request of the bishop Francisco Marroquín, funds were assigned and the work continued.[3] More funds were made available in 1547, when the income from the *encomiendas* of the *conquistador* Pedro de Alvarado and his wife Beatriz de la Cueva, which had been accumulating since their deaths, was now assigned to be used in the building operations.[4] In 1550 a royal *cédula* arrived ordering the completion of the cathedral, the cost of which was to be paid one-third from the royal treasury, one-third from the *encomiendas* and rents of the bishopric, and one-third by the citizens and Indians of the crown.[5] Apparently some changes in plan were made at this time, but the building was not completed even then, and more money was assigned in 1552.[6] Work must have progressed in a hit-or-miss fashion, for in 1560 the *ayuntamiento*, city council, decided to audit the accounts of the money spent to date on the cathedral.[7] Finally in 1576, thirty-four years after the foundations were begun, the crown ordered an investigation of the state of the building on which no work had been done for a number of years. It was revealed that those parts already completed had deteriorated in the meantime and were in bad condition because of long exposure to the weather.[8] Just when the first cathedral in Antigua was completed is not certain, but it must have been a patchwork affair when it finally was.

The foundations, and perhaps the lower courses of masonry of the walls, were built of the rubble from the ruined first capital city of Ciudad Vieja which Marroquín had bought for 1,600 *pesos* in 1545.[9] Furthermore, it had apparently been temporarily roofed with thatch which was in place until 1571.[10] Soon after, this must have been replaced with a wood *artesón*, for a report of almost a century later mentioned the fact that the roof timbers had rotted out and that it was necessary to rebuild the cathedral.[11] It is hardly likely that the sixteenth-century forerunner of the present Cathedral, inaugurated in 1680, had the same layout or was

[1] Juarros, **2**: pp. 235 ff.; Fuentes y Guzmán, **2**: p. 407; see also ch. VI above for biographical data on Martínez Gárnica. See also Mencos, *Arquitectura*, chs. III and IV for a compilation of all the documents in the Archivo de Indias relative to the history of both the sixteenth- and seventeenth-century cathedrals. For a brief summary of the probable architectural character of the sixteenth-century cathedral, see Luján Muñoz, *Noticia breve*, pp. 61–82.

[2] Mencos, *op. cit.*, ch. III, n. 17, quoting *AGI*, Guatemala, 156, a letter from the bishop to the king dated in San Pedro de Puerto Caballos (Honduras), 15 I 1543; and n. 18, quoting *AGI*, Guatemala, 156, another letter dated 15 I 1545; also n. 20, quoting *AGI*, Guatemala, 9, a letter from the *audiencia* to the king, dated in Gracias a Dios (Honduras), 30 XII 1545. Progress reports are given in each as well as an account of the straitened financial circumstances which impede carrying on construction in a normal manner.

[3] *Efem*, p. 10; also Juarros, *loc. cit.*

[4] *Efem*, p. 11; also Pardo, *Prontuario*, p. 27; and *AGG*, 1.2.4 (1547) 2196–153.

[5] Pardo, *op. cit.*, p. 27; and *AGG*, A 1.23 (1550) 1511–148.

[6] *Efem*, p. 12.

[7] *Ibid.*, p. 16.

[8] *Ibid.*, p. 22.

[9] Juarros, **2**: pp. 235 ff.

[10] Mencos, *op. cit.*, ch. III, n. 45, quoting *AGI*, Guatemala, 9, a letter from Dr. D. Antonio González to the king, dated in Guatemala, 15 III 1572 in which it is stated that the cathedral was still roofed with thatch in 1571.

[11] *AGG*, A 1.10.3 (1672) 31.255–4046.

of the same size, though Juarros asserts this to be true.[12] At any rate, the building continued in an incomplete state through the sixteenth century and even suffered damages from the earthquake of 1583. Parts of the roof fell in in 1600. No bells had yet been installed. Some repairs were finally undertaken soon after, which continued as late as 1616.[13]

Even the presbytery (*capilla mayor*) had been covered with a wood roof which by 1662 was in very bad condition as Bishop Payo de Rivera reported to the crown that year.[14] It is unusual that one of the principal churches of Central America should have been of such modest construction even as late as the mid-seventeenth century when even some of the humbler Guatemalan churches already had vaults or domes over the *capilla mayor*, while the naves, sometimes built separately, were roofed with wood *artesonados*, as for example, in Asunción Mita, Chiquimula, Esquipulas, San Cristóbal Acasaguastlán, and Zacapa.[15] The seventeenth-century church of San Francisco in Antigua, though roofed with wood and tile, had, nevertheless, a dome over the *capilla mayor*.[16]

By 1568 the wood and tile *capilla mayor* of the otherwise thatch-roofed cathedral was considered sufficiently complete to receive the remains of the *conquistador* Pedro de Alvarado which were brought there for burial,[17] but which were unfortunately lost about 1670 during the demolition necessary before beginning work on the present structure.[18] Needless excavations carried out in the 1930's, of course, failed to uncover any skeletal remains.[19]

Some chapels were added long after the building was supposedly complete, while others had to be built anew because of the extensive repairs required. The sacristy (*sagrario*) was not built until 1659.[20] Much of the interior furnishings and appurtenances still remained incomplete throughout the seventeenth century. A more permanent and fitting main altar was contracted for only in 1617.[21]

The workmanship was most shoddy, and during most of the seventeenth century the cathedral was constantly in need of repair. In 1662, as noted above, the roof of the *capilla mayor* was in an extremely bad state. Furthermore, many of the walls had been thrown off plumb in the earthquakes of 1651, 1663, and 1666 and were generally weakened.[22] It was finally decided that it would be better to demolish the building and begin from the foundations up, rather than attempt to repair the old sixteenth-century structure.[23] In 1669 as a spur to this decision, the *ayuntamiento* promised to donate the sum of 200 *pesos* annually for six years to be used toward the cost of the construction of the new cathedral, work on which was set to begin in 1670.[24] Two years later, in 1672, more money was allocated from the income of the vacant *encomiendas*.[25]

Demolition of the sixteenth-century *capilla mayor* and construction of the new edifice began almost at the same time in 1669, the building remaining in use part of the time. The old structure was razed to the ground bit by bit, while the new construction progressed piecemeal as the older parts were cleared away. By 1675 both demolition and construction had progressed to the point where services could no longer be held in the building so that on December 2 all activities of the cathedral were temporarily moved to the church of San Pedro, the very day the latter was inaugurated. Work must have progressed rapidly during the following year and the vaulting completed, for in 1676 the main altar was begun.[26]

The Cathedral of 1680

THE architect responsible for the design of the new cathedral was Martín de Andújar, who abandoned the job one year and a half later after a disagreement with his assistant, José de Porras, over the construction of the vaults. (FIG. 40.) Porras, a gifted but unlettered craftsman, formerly "*maestro menor*" became "*maestro mayor*" and brought the Cathedral to completion.[27] Not long after 1672 he petitioned for an increase in his daily wages in view of the fact that he was in sole charge of the masons and Indian helpers, starting work

[12] Juarros, *loc. cit.*

[13] Mencos, *op. cit.*, ch. III, n. 50, also append. II, transcribing *AGI*, Guatemala, 15, "Memoria de las necesidades de la yglesia catedral de Santiago, abril de 1601." Also his n. 66, where he states that Cristóbal Lorenzo was in charge of the work between 1614 and 1616.

[14] Fuentes y Guzmán, **3**: p. 356.

[15] *Ibid.* **2**: pp. 195, 197, 198, 242, 245.

[16] *AGG*, A 1.20 (1673) 476–10, also *BAGG* **10** (1945): p. 131; see San Francisco, ch. XIII, no. 1, below.

[17] *ASGH* **16** (1939/40): p. 406, quoting a document in *AGI*, Est. 2–Caj. 1–Leg. 2/19, tomo II, fol. 7.

[18] Díaz, *Romántica ciudad*, pp. 28 ff.

[19] Kelemen, *Baroque*, p. 40; see also Díaz, *Bellas artes*, p. 151.

[20] Juarros, **2**: pp. 236 ff., and **1**: p. 201; also Díaz, *op. cit.*, pp. 144 ff.

[21] *Efem*, p. 43.

[22] Juarros, **2**: p. 239.

[23] *AGG*, A 1.10.3 (1672) 31.255–4046.

[24] *Ibid.*, 1.2.2 (1669) 11.774–1780, also *BAGG* **8**, 1 (1943): pp. 29 ff. See also Lemoine Villacaña, *Historia*, pp. 418 ff., where 1669 is given as the year when actual work began.

[25] *AGG*, A 1.10.3 (1672) 31.256–4046.

[26] Juarros, **2**: p. 240.

[27] For a detailed description of the construction of the Cathedral, see Lemoine Villacaña, *op. cit.*, *passim*, who transcribes the copy of the report rendered by Gerónimo de Betanzos y Quiñones to the bishop Juan Ortega y Montañés who in answer to an inquiry of the crown made use of it in his official report. Betanzos' copy is in the Archivo General de la Nación, Mexico. The bishop's letter is in the Archivo de Indias in Seville, and is cited by Mencos, *op. cit.*, append. IX, transcribing, *AGI*, Guatemala, 166, "Carta del Obispo D. Juan

each day at four o'clock in the morning.[28] It is most likely that, in keeping with a custom prevalent in Guatemala even today, a general plan must have been drawn by de Andújar, but the actual construction work had been carried out under the direct supervision of Porras even before Andújar retired from the job. At any rate, about eleven years after the first work of demolition had been begun, the Cathedral was inaugurated on November 5, 1680.[29] It must have been the largest and most sumptuously furnished church in all Central America, and was raised to the rank of a metropolitan cathedral and the seat of an archbishop in 1743.[30]

There are many contemporary descriptions of the building which, except for some dimensions and minor details, generally give a picture consistent with the remains existing today. The cupola over the crossing was about 32 *varas* high, that is, approximately 21.00 m. The *capilla mayor* was located in the bay directly behind the crossing, and behind it was still another dependency, the *capilla real*. The plan of the Cathedral had a central nave with bays square in plan and side aisles. Eight chapels were located off either side aisle which added to the width of the church, sometimes described as being divided into five aisles or naves. The sacristy was located in the southeast corner and the *sagrario* in the southwest. Low pendentive domes once roofed the nave, side aisle, and chapel bays.[31] The final cost of the building was said to have been more than 150,000 *pesos*. A detailed account of the actual monies spent in the construction is preserved in a long and detailed document.[32] A description of the Cathedral contemporary with the time of its completion is given by Fuentes y Guzmán, details of which vary somewhat from that of later writers, especially Juarros.[33]

For a period of about forty years the building stood almost intact except for some slight damage in the earthquake of 1689.[34] Then in 1717 the catastrophic seismic activities which laid Antigua waste also caused rather extensive damage to the Cathedral. Except Ximénez, contemporaries who had lived through the earthquake all agree in stating that the cupola and the top story of the façade were badly damaged.[35] Ximénez, however, insists that neither were completely destroyed but were only slightly damaged.[36] It is certain, however, that in the earthquake of 1751 the cupola was destroyed,[37] and was rebuilt in 1754, but not as high as it had originally been.[38] Except for this slight departure from the original plan, the Cathedral remained unchanged until it was destroyed in the earthquake of 1773.

In the early nineteenth century part of the Cathedral was converted into the parish church of San José.[39] License to begin the work of remodeling was issued in 1820, and was probably finished not long after.[40] In order to facilitate the use of the parts designated in the new plan, much of the already ruined structure was torn down. The first two bays and the *sagrario* only were utilized for the new parish church, the rest of the building remaining in the ruinous state in which it had been left since the destruction of 1773. The interior, open to sky, was used as a burying ground when John Lloyd Stephens visited Antigua about 1839, the graves being shaded by a forest of dahlias and trees seventy to eighty feet high which rose above the walls.[41] (FIG. 45).

Architectural Data

THOUGH in part obscured by later reconstructions and alterations, the plan of the Cathedral is still clearly discernable today. (FIG. 40.) A *lonja*, considerably higher than the level of the plaza, fronts the building and the archbishop's palace continuing around the south side where a high monumental door also gives access to the interior. On the south side it terminates at the present house of the sacristan where many of the original seventeenth-century details still remain intact. The *lonja* as it exists today, broken by steps at the main entrance in front and again at the side door, may represent a nineteenth-century modification, to judge from a rough

---

Ortega y Montañés, sobre la obra de la cathedral de Santiago, al Rey, en Guatemala, a 19 de noviembre de 1677." See also ch. VI, above, for biographical data on José Porras (II) and Lemoine Villacaña, *op. cit.*, pp. 418 ff.; also Mencos, *loc. cit.*

[28] *AGG*, A 1.10.3 (1672) 31.258–4046.

[29] Juarros, **2**: p. 241.

[30] *Ibid.* **1**: p. 168.

[31] For dimensions and disposition of the plan see Lemoine Villacaña, *op. cit.*, pp. 426 ff., and Mencos, *op. cit.*, append. IX. Also Juarros, **1**: p. 64 and **2**: p. 241.

[32] *AGG*, A 1.10.3 (seventeenth century) 31.386–4051. González Bustillo, *Razón particular*, quoting a royal *cédula* of 1718 gives the cost as more than 200,000 *pesos*. See also the documents cited in fnn. 27 and 31 above.

[33] Fuentes y Guzmán, **1**: p. 139.

[34] Juarros, **2**: p. 247.

[35] Vásquez, **4**: p. 393; Arana, *Relación*, also *ASGH* **17** (1941/42): p. 155. Juarros, *loc. cit.*, repeats this information almost a century later.

[36] Ximénez, **2**: p. 353. González Bustillo, *Extracto*, says the damages to the Cathedral had amounted to 10,000 *pesos*.

[37] *AGG*, A 1.10.3 (1751) 4215–33, A 1.10.3 (1751) 31.349–4049, A 1.10.3 (1751) 127–8 fol. 72; see also Salazar, Fr. Juan José, *Piedra fundamental*.

[38] Juarros, *loc. cit.*

[39] *AGG*, A 1.2 (1813) 11.815–1805, also *BAGG* **8** (1943): pp. 199 ff.; also documents 11.818–1805, 11.819–1805, 11.820–1805, 11.821–1805.

[40] *AGG*, A 1.10.3 (1820) 18831–2448.

[41] Stephens, *Central America* **1**: p. 271.

sketch appended to a document dated 1784 which shows the *lonja* with a continuous set of steps extending to the archbishop's palace.[42] (FIG. 41.)

There are some differences in the appearance of the present-day façade and that depicted on the sketch of 1784. For example, the present façade lacks belfries. (FIG. 42.) The finial of the central bay is also quite different from that appearing in the sketch. The belfries, however, were still *in situ* when Catherwood did his drawing *ca.* 1840. (FIG. 19.) The archbishop's palace adjacent to the cathedral to the north is but one story high today, whereas in the sketch it is two stories high, the upper one shown as an arcade with an overhanging balcony which may very well have existed before the changes and demolitions of the early nineteenth century. The sketch also shows a small cupola directly over the *sagrario* consisting of a drum with circular or polygonal windows, a dome, and a lantern. In general, the layout of the orders on the façade is the same as exists today, except that the niches do not appear in the sketch. This may be due to the fact that it was not drawn to scale and the niches could not be fitted into the cramped space between the engaged columns. The sketch shows three steps directly in front of each door between the pedestals on which the columns rest, steps which lead up from the level of the *lonja* to the threshold. Such steps exist today. All in all, it seems that the builders changed the façade very little when the Cathedral was repaired for service as the parish church of San José.

New vaults were constructed, however, somewhat lower than originally, level with the top of the entablature of the first-story order of the façade and just under the open niche-window of the second-story central bay which once lighted the upper choir of the original church. (FIG. 50.) The window now is outside and above the present vaults, but was originally under a roof, a fact borne out by an examination of the ornamental treatment of the opening toward the former interior of the church. The splayed reveals of the window are treated with series of moldings of complicated profile. These are repeated on the soffit of the very flat arch above and covered with an intricate pattern of lace-like stucco decoration. The interior side of the opening is framed with an *alfiz*, an element above the arch made up of a horizontal molding as wide as the diameter arch itself and connected by two verticals which run to the imposts of the arch marking off the spandrels. The *alfiz*, as a whole, including the spandrels and the space above the crown are also treated with plaster decoration. Such a complicated and ornately decorated window could hardly have been meant to be above roof level as it is today. The same window design appears in the crossing and clerestory. (FIG. 48.) This roof is doubtless part of the nineteenth-century reconstruction when new vaults were built over the portion of the church just behind the façade and the orientation changed to north-south. The rest of the nave and side bays were left in their ruined state, in which they remain to this very day.

It may be safely concluded that, except for the absence of the belfries and the differences in the finial which in the 1784 sketch is shown as a single bay with a niche with mixtilinear half-pediments to either side, the present façade is essentially the same as it was after the repairs of 1717 and 1754. (FIG. 42.) The lower story of the façade, however, is very likely much the same as it was when completed in 1680.

The first-story central bay has the main door set in a deep niche with a stilted arch. (FIG. 43.) The door header and niche header are not concentric though both are semicircular, the latter being set considerably higher so that a space is left between the two. Directly over the crown of the door arch is a horizontal cornice supported on small brackets. In the semicircular tympanum a small niche is set. The doors of the side bays are recessed in even deeper niches with concentric headers and lack the intervening spaces or tympana as in the main door of the central bay. Directly above each side door there is a window so that in effect each side bay gives the appearance of being two stories high. Four pairs of rather long, thin Tuscan columns resting on high podia flank the three doors. The entablature which breaks miters over each column has a very small architrave, a triglyph and metope frieze, and a boldly projecting cornice to mark off the first story. The total effect is that of a colossal order framing doors and windows.

The second story is unusual in that it is reduced to the central bay only. Two pairs of Tuscan columns frame a tall, deep niche directly over the door below. The lower part of this niche is pierced by the window discussed above and which once lighted the upper choir. To either side and framing the central bay are mixtilinear half-pediments with scrolls which appear on the sketch of 1784 alluded to above. These formed connecting elements to the belfries, now missing.

The surface of the façade is left entirely plain. The decorative treatment is achieved solely by means of the niches which punctuate the surface between columns. Two niches, one above the other, are located between each pair of columns on both stories. Two additional niches are set in the space to each side of the second-story window. There is then a total of sixteen niches on the façade as a whole. There

42 *AGG*, A 1.10.3 (1784) 15.091–2123. The sketch of the façade was made about nine years after the destruction of Antigua and was appended to a survey of the area comprising the parish served by the *sagrario* of the Cathedral.

were twenty-three niches originally.[43] A statue is set in front of the window opening of the second story as well as in the sixteen niches. Unlike most of the façade statuary of Antigua which is built up on a core of brick bonded to the wall behind and finished with plaster, these are all free-standing sculptures.

The Cathedral is laid out in three aisles with salient transepts in a cruciform plan, a type rather uncommon in Antigua and the rest of Central America, where most churches are single nave and without projecting transepts. (FIG. 40.) Located off each bay of the side aisles are chapels, so that the area between the towers and the transepts forms a sort of second side aisle. The first two bays of the plan immediately behind the west front were utilized for the present parish church of San José.

Two rows of massive piers, cruciform in plan because of pilasters engaged on each of the four sides, looking somewhat like a cluster of piers, once supported the vaulting over the central nave and side aisles. (FIGS. 45, 46.) Pendentive domes, *bóvedas vaídas*,[44] roofed each of the bays of the higher central nave and the lower side aisles and chapels. Except for the domes over the *capilla mayor* and the *capilla real* directly behind it, all are now missing. A cupola on spherical pendentives once surmounted the crossing which came down in the earthquake of 1717, but was rebuilt immediately only to be totally destroyed again in 1773. All that remains are the pendentives with rather curious plaster reliefs, depicting the four evangelists. (FIGS. 48, 49.) Beneath each, standing on the capitals of the four piers where the transverse and lateral arches meet, are other long-robed figures of the same material in the round. The entire surface of each pendentive is covered with a lace-like pattern of intertwined ribbons.

The piers are marked by a great deal of reserve with a minimum of surface treatment so that their structural function is quite evident. (FIG. 46.) The four pilasters attached to each side of the square pier rest on rather high plinths and typical Attic bases made up of three members. The pilaster shafts are unadorned, a smooth troweled surface coating of plaster covering the brick. The capitals are rather complex and are not altogether uniform. Under each there is a necking band on which the chain pattern of the order of the Golden Fleece, *toisón de oro*, is worked in stucco. (FIG. 47.) A twisted rope molding marks off the necking from the shaft below. Immediately above the necking is a quarter-round molding, the echinus, on the surface of which a modified egg-and-dart pattern is worked. Above that is a flat unadorned fillet, the abacus. A small section of entablature, also laden with a floral pattern in stucco, surmounts each pier and forms the impost for the arches of the vaulting. In other words, the entablature does not continue across the lunettes of the clerestory to unite the individual piers.

The treatment of these short sections of entablature is most ornate. A plain, unadorned band surmounted by a projecting and heavily decorated molding forms the architrave. This molding, made up of a number of members of different profiles, has a scalloped pattern as part of its decoration. The frieze above is completely covered with an intertwined ribbon pattern in which some floral motives are also included. The cornice above is composed of a four-member molding, the most prominent being an ovolo on which a repeated oval shape pointed at both ends is incised. An added touch in the crossing and *capilla mayor* bays is that of a freestanding sculpture set at the corners on each capital facing the nave. Some of these are still *in situ*, the four in the crossing and one in the bay just in front of the crossing. (FIGS. 48, 49.)

In addition to the three doors which gave access from the Plaza Mayor, another gave access to the Palacio Arzobispal adjacent to the north wall of the Cathedral. (FIG. 40.) Two were also located at the east end of the building at the sides of the *capilla real* and opened on the street behind the cathedral. A rather large monumental door dominates the south side which one approaches from the street via the *lonja*. (FIG. 44.) The door opening has a semicircular header and splayed reveals. Framing the opening are two rather thin pilasters which support a Doric entablature with a triglyph and metope frieze. On center over the door axis a small round-headed niche with a small triangular pediment rises to the height of the side wall.

[43] Lemoine Villacaña, *op. cit.*, p. 428.

[44] The *bóvedas vaídas* of the Cathedral are, along with those in Belén, the earliest examples of this type of vaulting in Antigua. Only three were built in Belén, whereas the Cathedral was originally roofed with vaults of this type over each of the nave, side-aisle, and side-aisle-chapel bays. For Belén, see ch. XII, no. 5, above.

XIII THE SECOND PERIOD: 1680–1717

THE structural and decorative innovations introduced in the Cathedral mark a departure from the style which was in vogue during the seventeenth century. In a strong sense, the Cathedral bridges the first and second periods, representing, on the one hand, a culmination of the stylistic developments in the former and, on the other, serving as the prototype for the builders of the latter. As a matter of fact, the *bóveda vaída*, employed for the first time in Antigua in the Cathedral and in Belén, finds its greatest acceptance in the third and fourth periods after 1717.

The second period, which lasted but some thirty-seven years and is represented by nine buildings and some fountains, came to an end in the disheartening destruction of Antigua in 1717. Some of these structures had originally been built in the first period, but were altered, enlarged, or reconstructed after 1680. During this time four of the conventual organizations embarked upon their most ambitious building campaigns: La Merced, San Francisco, La Compañía de Jesús, and Santa Teresa. La Merced was basically reformed during this period, while the churches of La Compañía de Jesús and Santa Teresa were built from the ground up, the convent of the latter actually being rebuilt again in the following period.

The most remarkable changes noted in this period occur in the design and ornamentation of the retable-façades. Solomonic columns are used on the *portada* of San Francisco and the convent entrance of La Merced. On the other hand, even in those examples where the classical spirit of the applied orders is still maintained, these are now embellished with plaster ornament as seen on San Sebastián and La Merced. In the case of the latter, however, the ornament which decorates the flat surfaces of the façade and even encroaches on the columns may date from the fourth period. The ornate solomonic columns of the convent entrance, however, are definitely from about 1690. In lieu of plaster ornament, painted decoration is sometimes utilized, as for example, the façade of La Compañía de Jesús and the cloister columns of the Palacio Arzobispal.

A minor variation in the arrangement of the retable-façade appears on the churches of Santa Teresa and of Nuestra Señora de los Dolores del Cerro, where a scheme reminiscent of a triumphal arch frames the main door in the central bay and provides the most important accent on the façade as a whole. In the case of both San Francisco and La Merced, massive twin towers frame the central *portada*. Those of La Merced, though rebuilt or altered after 1750, are of late seventeenth-century origin, or possibly even from the first period before 1680. The towers of the church of San Francisco are not symmetrical. The original façade had but one tower dating from earlier in the seventeenth century. The second one, not an exact replica of the earlier, was added at the end of that century to complete the scheme of framing the central retable. The façade of San Sebastián has but one massive tower. Santa Teresa and Nuestra Señora de los Dolores del Cerro have no such arrangement, the central retable being flanked by narrow plain-surfaced buttresses or bastions more like those of Los Remedios from the preceding period.

The arrangement of niches in tiers or pairs in each of the side bays as well as to either side of the niche-window in the second-story central bay, introduced in the Cathedral, persists in San Francisco and La Compañía de Jesús. The other buildings adhere, however, to the simpler scheme of a single niche between columns in each side bay. The door type used in the Cathedral also predominates in this period, though the simpler type where the door niche follows the outline of the door opening is also employed.

The proportions of the retable-façades tend to be square overall, but in common with the Cathedral, side bays are quite narrow in comparison with the central bay. The central bay tends to be greater in width than that of both side bays together. This is true of San Francisco, La Merced, and La Compañía de Jesús. In the case of San Sebastián, the width of the central bay is almost exactly equal to the total of the two side bays, a proportion nearer that of the façades of San Agustín and Los Remedios from the preceding period.

Domical vaults over square bays, *bóvedas vaídas*, were introduced with the construction of the Cathedral. The vaulting of the church of San Francisco, now ruined, was based on the half ellipsoid over a rectangular bay, a type which was used later throughout the whole of the eighteenth century. In this period the system of reducing the walls to mere screens between massive piers on which the vaulting rests appears for the first time in San Francisco, La Compañía de Jesús, and Santa Teresa.

During this period the water-supply system was improved and enlarged and many fountains were built throughout the city, the most beautiful being that known as Pila de Campo completed just about the same time as the Cathedral. The dates of the other fountains are uncertain.

In general, this period and the previous one might be grouped together, except that the Cathedral seems to stand out as a unifying prototype for those constructions which follow until 1717. As was stated above, the main differences in this period from the one immediately preceding were the introduction of the solomonic column, the continued use of the simplified classical orders now ornate by means of plaster ornament, and the introduction of a new type of domical-vaulted construction as opposed to the barrel vault of the previous period. The retable-façade design is in part influenced by the proportions of the Cathedral, with narrow side bays in relation to the central bay. But a change to a proportion of 1:2, side bays to central bay, appears on San Sebastián and Espíritu Santo. The façade of Nuestra Señora de los Dolores del Cerro, though left plain is, nevertheless, in the spirit of the time, for the niche treatment is rather free and nonarchitectonic and the entablatures are broken in a manner which becomes quite common in the third period.

The proportions of the columns of the cloister corridor of the Palacio Arzobispal are rather slender when compared with those of San Francisco and La Merced. The simple Tuscan column is to be used again in the succeeding periods and to undergo a drastic change in proportions which result in giving it a massive, sturdy, and squat appearance.

1. *San Francisco: Church and Convent, 1675, 1690, and 1702*

(FIGS. 51–61.)

Historical Data

THE Franciscan order was one of the first to be established in Central America, about 1530 in Ciudad Vieja. There the principal convent remained until the removal of the capital to Antigua soon after 1541.[1] By 1542 a new convent building of sorts must have already been standing and the income from various Indian towns were assigned the order for its maintenance by the civil authorities.[2] The new building, obviously erected in a great hurry, could hardly have been more than a temporary structure. It was located on the same site where the Escuela de Cristo now stands, remaining there but three years. In 1544 the convent moved to its present site and the *ayuntamiento* assigned it a new water allotment.[3] That same year about twenty-five additional monks arrived in Antigua. By 1548 a formal convent building was already under construction of a size, no doubt, commensurate with the needs of the enlarged community.[4]

In the earthquake of 1565 the conventual buildings, in use only about twenty years, were damaged. The church fared so badly that it was on the verge of caving in, the roof having to be shored up for many years until the whole structure was rebuilt from the foundations up.[5] This was not done immediately, however, for as late as 1575 the order asked for help in the reconstruction which apparently had not yet begun.[6] In the meantime, the Franciscan province of Central America, "Provincia del Dulcísimo Nombre de Jesús," had been established about ten years before as a separate entity and independent of the province of Mexico.[7] Yet, nothing was done to build a convent commensurate with the position of importance of the principal house of the new administrative division until a royal *cédula* was emitted in 1576 explicitly ordering the reconstruction of the church and convent. But it was three years more before work was actually begun, in 1579, which was at long last completed in 1582.[8] Both the friars and the novices personally worked on this building and even the president of the *audiencia* joined in, so great was the enthusiasm and impatience of the people who wished to see the convent completed.[9] Vásquez describes the structure, saying in part that the *capilla mayor* was roofed with a vault which was unsatisfactory because it was very low. The unusual character of this vault, which was finally removed in 1692,[10] was also noted by Fray Alonso Ponce who visited Guatemala *ca.*

[1] Juarros, **1**: p. 123; also Vásquez, **1**: pp. 317 ff.
[2] *Efem*, p. 7.
[3] Vásquez, **1**: p. 164; *Efem*, p. 9.
[4] Juarros, *loc. cit.*; Vásquez, **1**: p. 165.
[5] Vásquez, **1**: pp. 154 and 183; also Fuentes y Guzmán, **2**: p. 441.
[6] *AGG*, A 1.23 (1575) 1512–474 and 1.23 (1575) 1512–475; also Larreinaga, *Prontuario*, p. 64; also Vásquez, **1**: pp. 43 and 227; and Fuentes y Guzmán, **3**: p. 460.
[7] Juarros, *loc. cit.*; Vásquez, **1**: pp. 317 ff.
[8] Vásquez, **1**: pp. 242 ff.
[9] *Ibid.*; also Juarros, **1**: p. 184.
[10] Vásquez, *loc. cit.*

1586. He said the convent had been originally built of tamped earth, *tierra*, and was then being rebuilt of the same material, but with many reinforcements of stone, brick, and lime. He goes on to say that the main chapel of the church was very strong and roofed with a brick vault.[11]

This late sixteenth-century church could have hardly been a very monumental structure, not only in view of the materials employed but also because of the lack of skilled craftsmen. One of the monks, who had learned the craft of mason probably only shortly before, served as master bricklayer and was responsible for the curious vault over the *capilla mayor*.[12] Nondescript though it was, this church served as the principal house of the Franciscan order in Central America for more than a century until it was damaged beyond repair in the earthquake of 1689. It was torn down in 1692 and a new structure was built from the ground up.[13]

Exactly what the façade of the late sixteenth-century church looked like is impossible to tell. It must have had niches with stone sculptures, however, to judge from a peripheral statement made by Vásquez in speaking of the Capilla de Loreto who says that this chapel was located on the spot where in 1601 some stone figures were being carved for the *portada* of the church.[14] Like the church, the walls of the convent were also mainly of tamped earth and masonry reinforcements. A chapel of the Tercer Orden was rebuilt in 1634 using the same sort of materials.[15] The sixteenth-century conventual structure was also replaced by one of more formal and permanent character after having been damaged in the earthquake of 1651.[16] Later on, January 1, 1680, a fire broke out in the wood roof of a storeroom where some charcoal was kept and threatened the whole convent with destruction, but fortunately only some cells were badly damaged.[17]

By the third quarter of the seventeenth century the Franciscan order had prospered and become one of the most influential in the whole of Central America. Under its ecclesiastical jurisdiction there were reputedly 120 towns with 17,985 Indian families representing a total population of about 55,000 souls administered by an unstated number of monks, of whom ninety resided in the convent in Antigua.[18] By the end of the century the number of monks in residence increased to about one hundred, about eighty remaining permanently in Antigua.[19] It was decided to rebuild the church in 1673 when it was found that roof timbers had rotted out, a fact noted in a contract signed on February 13 of that year by the conventual authorities and the carpenters Nicolás and Juan López in which it was stated ". . . por cuanto la iglesia del convento fundado en esta ciudad se pretende derribar, por hallarse las maderas de ella muy antiguas, podridas y pasadas y con amenaza de ruina. . . ."[20]

Vásquez gives an account of the building works carried out on the convent and church at the end of the seventeenth century based on his personal experience in the matter. His statements can be accepted with confidence; for, except for some precise dates, they are more frequently than not corroborated by documents still in existence in the Archivo General de Gobierno in Guatemala City.[21] The main problem confronting the investigator today is that of identifying the parts of the church and convent mentioned by Vásquez in the existing ruins.

In 1673 when Fr. Fernando de Espinoso was elected *provincial*, he undertook the repairs of the church which was about one hundred years old at that time. This is corroborated by the contract entered into with the brothers López mentioned above. The specifications of the work to be done give a vivid description of late seventeenth-century building methods and are worthy of note. The López brothers agreed to do the *artesón*, wooden ceiling, over exactly as it then was, "en la mesma forma que aora está . . . de par y nudillo, artesón y trese pares de tirantes en la mesma forma que va. . . ." The order agreed to furnish the materials and the scaffolding, while the carpenters were to supply the labor promising to complete the job within one year counting from September of 1673. Vásquez offers some further information concerning the materials, saying lumber for the new *artesón* was sought from a distance of up to twenty leagues, cedar, pine, and cypress being used for the various parts. The lumber was bought as standing timber, then felled, and carried to the job where it was cut to size as spe-

[11] Ponce, *Relación*, pp. 410 ff.

[12] Vásquez, *loc. cit.* The nave was roofed with wood and tile and the *capilla mayor* with a vault, "*bóveda*," a combination not uncommon in Guatemala during the whole period of the colony; see Markman, *JSAH* **15** (1956): p. 13, fn. 12. In view of the early date of this building and its roofing, it is not outside the realm of possibility that the vault alluded to was the Gothic ribbed type. If this was the case, then the church of San Francisco was not unlike the many *mudéjar* churches of the region around Seville roofed with *artesonados* over the nave and Gothic ribbed vaults over the presbytery.

[13] Vásquez, *loc. cit.*

[14] *Ibid.* **4**: p. 222.

[15] *Ibid.* **3**: p. 157.

[16] *Ibid.* **4**: p. 327.

[17] *Ibid.* p. 272.

[18] *AGG*, A 3.2 (1673) 15207–825, see also *BAGG* **10**, 2 (1945): p. 128.

[19] Vásquez, **4**: pp. 14 ff.; also **4**: p. 34, where a document in the Archivo Episcopal, A 4.5–2 (1689), is transcribed.

[20] *AGG*, A 1.20 (1673) 476–10; see also Vásquez, **4**: pp. 67 ff.

[21] Vásquez, **4**: pp. 329 ff., "Libro v, Tratado 2, Capitulo 27, De las obras que se han hecho en el convento de Nuestro Padre San Francisco de Guatemala."

cified by the *maestros de obra*, the López brothers, no doubt. One year was required to prepare the timber. At the end of the rainy season in 1674, which would probably be in November, the roof tile was also ready. When the old tile was removed, it was noted that the roof beams which were to be re-used to support the new *artesón* had rotted out. These would have collapsed had they not been replaced. The dates given by Vásquez at this juncture are a little confusing, for another contract to tear down the old roof, dismantle the *retablos*, remove the pictures, build the necessary scaffolding, and then replace the pictures and *retablos* had been signed by the convent authorities and the brothers López the twenty-first of June, 1674.[22] Work continued for months afterwards, according to Vásquez, and the church was finally inaugurated on February 5, 1675, ". . . de tan primorosa trabazón de lacería y artesón, remates y tirantes pintados y dorados, perfiles plateados de la forma del cordón de San Francisco, que de maderambe no ha habido otra semejante en este reino." It would seem then that the main structural parts of the late sixteenth-century church were not changed, that only a new roof and *artesón* were placed on the older walls, and that the old vaulted *capilla mayor* was not touched at this time.

Some changes not mentioned by Vásquez must, however, have also been carried out on the main façade of the church at this time, for in a contract dated March 2, 1675, one Ramón de Autillo agreed to make an arch of stone for the main door

> . . . un arco de piedra de cantería, hechura de frisco encojinado, con su base y capitel y humbral de abajo para la puerta principal de la yglesia del Convento de Nuestro Seráphico Padre San Francisco . . . y labrado ajustado a punto de pico en la forma referida, a la medida de dicha puerta . . . .

He agreed to complete the work by May of that same year for the sum of 200 *pesos*, supplying the labor only while the convent authorities were to provide all the materials as well as his helpers.[23] The description given in the contract seems to fit the door still *in situ*. It is, therefore, quite possible that it was incorporated into the later rebuilding at the end of the seventeenth and early part of the eighteenth centuries.

Vásquez also describes some work done on the convent, most of which is almost impossible to identify in the ruins today. (FIG. 51.) But his descriptions throw a great deal of light on the materials and methods commonly employed in Guatemala and Central America during that period. Soon after 1675 a new wood and tile roof was also built for the infirmary and a wood floor installed between the two stories of that dependency. These repairs must have proven inadequate and the infirmary was rebuilt after 1684 from the ground up as a three-story unit just south of the old one, for this was the only direction possible in which to make additions. The lower-floor plan consisted of three cells to either side of a dormitory, also repeated on the second story. But this time the floors were supported on arches and vaults. The third story had but four cells as well as a terrace with arcaded porches affording a pleasant view. Another dependency adjacent to this infirmary was also three stories high and likewise constructed with vaults and arches, the top story containing a little chapel oriented east-west. The convent was damaged in the earthquake of 1689, the repairs of the cells alone being estimated at 30,000 *pesos*. Some patchwork, mostly of wood, replaced some fallen barrel vaults.[24]

In 1692 it was decided to rebuild the church crossing and the *capilla mayor* which had not been touched in 110 years. The most recent earthquake had no doubt weakened the low vault or dome first seen by Ponce *ca.* 1586. Although the arches of the *capilla mayor* were not in bad condition, the dome and the lantern above, however, leaked so that the interior required whitewashing after each rainy season. Work began the twenty-fifth of June, 1692, and was still in progress at the time Vásquez was writing in 1695. The date of this operation is corroborated by a document dated 1693 in which the procurator of the convent requested that two stonecutters from the town of Santa María de Jesús be assigned to the work being done on the church.[25]

Some later additions to Vásquez's manuscript in a different hand and dating from about 1716 continue the story of the building works during the years 1695 to 1715.[26] Work on the crossing was finished about three years after it was begun, in 1695 or 1696. Building operations on the principal cloister, the one adjacent to the church, continued for about three years longer, perhaps until about 1698. Its four arcaded corridors were covered with barrel vaults. Between 1692 and 1698 some work was carried out on the sacristy (*sacristía*) which had a plan of three bays, *tres bernegales*, possibly meaning bays with pendentive domes shaped like inverted wine cups or goblets. At the same time the crossing, the *capilla mayor*, and the transepts were completed, all also referred to as having *bernegales* except the crossing which was roofed with a dome. These parts of the plan may be tentatively identified as follows on the plan (FIG. 51): the

[22] *Efem*, p. 84.

[23] *AGG*, A 1.20 (1675) 477–32, see also *BAGG* **10**, 2 (1945): p. 133.

[24] Vásquez, *loc. cit.*; also *Efem*, p. 107.

[25] *AGG*, A 1.10–3 (1693) 31.272–4046.

[26] Vásquez, **4**: p. 390 ff.

sacristy as *j*; the crossing as *c*; the *capilla mayor* or presbytery as *d*; and the transepts as *f* and *f′*.

The new *provincial*, elected in 1697, began the reconstruction of the nave part of the church, specifically from the crossing to the façade, matching the style and appearance of the cornices and windows on both sides, and dividing that section into six vaulted bays roofed with the same type of pendentive domes, referred to by the author as *bernegales*, supported on pilasters, four pairs of which were in the nave proper and two pairs in the choir. (FIGS. 55–58.) The new nave was built of masonry with pilasters equal in size to those already standing in the crossing and *capilla mayor*. Underneath each bay were subterranean chambers for burials. The work was not quite completed by 1700 when the *provincial* went out of office, the scaffolding still remaining in place. This description fits the present nave and choir of the church which is of exactly the same style as the crossing, transepts, and *capilla mayor*, *a*, *a′*, and *b* on the plan (FIG. 51).[27] Between 1700 and 1703 the *Aula Magna* was built next to the church in the lower cloister next to the main stairway and roofed with a barrel vault which may be identified on the plan as *l*. The nave and choir were finally completed in 1703 at which time the final touches were also made to the central retable portion of the façade. The scaffolding was removed from the interior of the church and the whole painted. An iron rail was installed in the choir and on the platform of the organ. All the windows of the church were glazed and fitted with wire grills. (FIG. 60.) The main altar retable, five stories high with statuary, and five other collateral retables as well were installed. The lamps over the main altar were of silver weighing four *arrobas*, approximately 100 pounds. All the cornices were adorned with railings in the form of arcades painted green and red and on each pilaster a painting of an angel in a gilt frame was hung. The church had been inaugurated a few months before final completion, the twenty-fifth of September, 1702, but was formally consecrated about twelve years later by the bishop Fr. Juan Bautista Alvarez de Toledo.[28]

The Capilla de Loreto, the foundation of which went back to 1600 when various families, the Medranos, Solórzanos, Alvarez de Vegas, and Aldanas donated money for this purpose, is difficult to identify in the present ruins, though Vásquez is at great pains to describe it. According to him, it was located in an oblique angle which had temporarily served as a workshop in 1601 for the stone cutters who were carving some sculptures for the *portada*. It was square in plan, measuring about 4.62 m. on each side, that is, twenty-two units of about 0.21 m. or the fourth part of a *vara* which varies from about 0.835 m. to slightly more.[29] The door on the north side gave access to the church not far from the main entrance, and another small door in the east wall faced in the general direction of the sacristy. From this description, one may conclude that the Capilla de Loreto was located on the site of or in the south tower of the present façade. (FIG. 53.) The oblique angle he refers to must have been an open area which was unoccupied before the entrance rooms to the convent were built after 1692. Vásquez also mentions a window which looked west. Remains of such an opening are still in evidence today. Therefore, we may tentatively accept the south tower as the earlier of the two. It went through many transformations both at the end of the seventeenth century when the church was rebuilt, and again in the eighteenth when earthquake damage was repaired after 1717 and 1751.

Vásquez continues his description going on to say that there is a room directly above the chapel and another above that where the monk who takes care of the clock and bells lives. The final story was a belfry which once had bells, but since another tower of similar design had been built on the other side of the *portada* some years later, it now serves as a clock tower. This can only mean that the Capilla de Loreto was located in one of the towers flanking the central retable of the façade which had been built originally as a bell tower. The disposition of the doors as mentioned by Vásquez points to the south tower as the location of the Capilla de Loreto, and hence the earlier of the two which now frame the façade, dating originally from the early seventeenth century.

The north tower is the later one, but it also originally dates from sometime in the seventeenth century. We may safely assume that changes in structure had taken place in both towers during the eighteenth century and even later due to earthquakes, a fact corroborated in part by the literary and documentary evidence. The south tower, the one where the Capilla de Loreto was probably located, had undergone some alterations even as far back as 1635 when it was hit by lightning and repaired the next year or so. The king, Felipe IV, donated 200 ducats for this purpose.[30] The north tower was repaired in the twentieth century, probably

[27] It is interesting to note that this is the first example in Antigua of the use of ellipsoidal pendentive domes over oblong bays, the *bernegales* mentioned by the author of the addition to Vásquez's manuscript. See also fn. 41 below.

[28] Villacorta, *Historia*, p. 219.

[29] Vásquez, **4**: pp. 219 ff. The dimensions he gives are twenty-two palms on each side.

[30] *Ibid.*, p. 232.

after the earthquake of 1917.[31] A photograph published in 1924 shows the entire northwest corner still out. The lower story is now complete, while the upper portion still shows a gaping hole on the corner.[32] (FIG. 53.)

The church and convent were not to remain intact for very long after completion and dedication, for in 1717 both were damaged by the earthquake as noted by the author of the addition to the Vásquez manuscript. The chapel of San Antonio of the infirmary and pharmacy, which had been been built in 1685, was completely destroyed and the ruins which were still left standing had to be demolished completely. The vaults of the church were cracked in various places and the bell tower, possibly the south, was also badly damaged.[33] This information is in part corroborated by another contemporary author who says the church vaulting was slightly damaged but the convent hardly at all.[34] The chapel of San Antonio was rebuilt about 1723, this time with buttresses to secure the walls.[35] In the earthquake of 1751 the church vaulting again was damaged. Repairs began sometime in July of 1752[36] and the church was rededicated in 1754.[37]

The earthquakes of 1773 finally left the church and convent in total ruins. The majority of the vaults are now gone and the nave is open to the sky. The convent which had been built for the most part of tamped earth is now a labyrinth of ruined walls almost impossible to identify.[37a] The plan of the main cloister which was built of brick is the only part of the convent that can be made out. The emplacement of the piers which once supported the corridor vaults is still visible. But nothing of the vaults remains except in the southeast corner. Some willful destruction of the convent was occasioned in the nineteenth century during the term of office of President Barrillas, when the ruins served as a source for cheap building materials.[38] Early in the nineteenth century the chapel of the Tercer Orden was rebuilt, the apse of which cuts into the north wall of the church nave, *o* on the plan (FIG. 51). This probably occurred sometime just before 1817 when the remains of Pedro de Betancourt were placed there.[39]

Within the convent grounds are a number of independent structures, some of masonry and some of tamped earth, all difficult to identify. (FIG. 51.) The girdling wall which surrounds the entire precinct of about two square blocks has been subject to some repairs, but is for the most part still intact. Two monumental entrance gates to the *atrium* or area to the west and north of the church are still in good repair. (FIG. 52.) Especially noteworthy is the gate to the west of the main façade which is more or less in the same style as the *portada*, and may be considered contemporary with it, namely, from the end of the seventeenth and beginning of the eighteenth centuries. (FIG. 53.)

Architectural Data

i. The Plan (FIG. 51.)

The church is single nave and cruciform in plan. The transepts project to either side of the crossing and the *capilla mayor* extends behind to form the cross. The towers at the west end of the church also project to either side of the nave.

The overall exterior dimensions are about 75.5 m. in length and 30.2 m. in width across the façade including the towers, and approximately 14.66 m. in width at the nave. The interior of the nave measures 11.66 m. in width, exclusive of the recesses for the niches. The width from niche face to niche face including the recessed portion is 13.2 m. The longitudinal measurements of the individual nave and choir bays are as follows: the first choir bay, *a*, from inside face of façade wall to pier center—6.60 m.; the second choir bay, *a′*, measured on pier centers—7.00 m.; the four nave bays, *b*, measured on pier centers—7.60 m. each; the crossing, *c*—about 11.5 m. on pier centers; the *capilla mayor*, *d* 11.45 m. measured from pier center to inside face of back wall; the open tower, *e*, 6.20 m. by 6.00 m.; the transepts *f* and *f′*—9.80 m. by 11.50 m.; the chapel adjacent to the north transept, *g*—8.00 by 11.87 m.; the rooms off the south transept, *h*, including the entry room which is 9.30 m. by 3.30 m., or the two bays measured together, a total of 16.90 m.; the long corridor, *i*—2.50 m. wide by an undetermined

[31] Kelemen, *Baroque*, p. 133.

[32] See *ASGH* **1** (1924/25): p. 326, for the photograph.

[33] Vásquez, **4**: p. 393.

[34] Ximénez, **3**: p. 355. Some corroborating information is also given in a document cited by Mencos, *Arquitectura*, ch. v, n. 161, where it is stated that all altars in San Francisco were functioning, implying that the damage had been slight. He quotes *AGI*, Guatemala, 305, "Testimonio de donde se celebran los divinos oficios, en Guatemala a 14 XII 1717."

[35] *Efem*, p. 154.

[36] Juarros, **2**: p. 248. See also Pardo, *ASGH* **24** (1949): p. 380, and *Efem*, p. 207.

[37] See a sermon preached on the occasion by Fr. Juan José Salazar, noted by Medina, J. Toribio, *Biblioteca hispano-americana*, pp. 123 ff., no. 246.

[37a] Work on the reconstruction of the church for the purpose of rededicating it (after a lapse of almost two hundred years) as a house of worship was well under way during the summer of 1963. The piers had been hollowed and filled with reinforced concrete columns. The nave was completely obstructed with wood scaffolding and forms prior to pouring the vaults, also with reinforced concrete.

[38] Díaz, *Romántica ciudad*, p. 76. For an account of the condition of San Francisco as a result of the earthquake of 1773, see Mencos, *Arquitectura*, ch. v, n. 163, citing *AGI*, Guatemala, 661.

[39] Pardo, *Guía*, p. 94; Díaz, *op. cit.*, p. 77.

length; the sacristy, *j*, including the entry room—5.10 m. by 8.45 m., and the main room which is divided into three bays—9.08 m. by 22.90 m. for the interior, and 12.00 m. by 24.50 m. on the exterior; the cloister, *k*—30.00 m. by 30.00 m. measured from wall to wall including the corridors which are approximately 5.00 m. wide; the entrance system to the convent, *l* and *m*—5.00 m. wide by approximately 20 m. long; the south tower, *n*—approximately 7.5 m. by 9.5 m. including the wall shared with the church; the north tower, *n'*—approximately 11.1 m. by 12.5 m.; the post-colonial chapel—approximately 11 m. by 25 m.; the conventual buildings, *q* and *r* occupying an area measuring approximately 100 m. by 100 m. to south and east of the church.

The church plan presents a unified scheme despite the fact that the different parts were built as separate operations to replace the older seventeenth- and in part sixteenth-century structure. The façade with its towers, however, may be conceived of as independent units structurally, for they preserve more of the older structure than do the nave, transepts, crossing, and *capilla mayor*. The cloister too, though occupying the same area as the older one, is structurally independent of the church on which it abuts.

Projecting from the east end of the church is a curious open tower, the upper stories of which are missing. It is not mentioned in any of the writings of Vásquez and, therefore, it may be assumed to have been added during the eighteenth century. The interior of the back wall of the *capilla mayor* on which the open tower abuts is punctuated by four semicircular-headed niches one directly above the other. (FIG. 56.) Sometime after 1951 the bottom niche was cut through, thereby providing access to the bottom story of the open tower. In other words, the tower was originally an independent unit attached to the church but not accessible from it. The first floor is about on a level with the threshold of the bottom niche. Massive piers are located on the outside corners and are reinforced by buttresses flaring out at a 45-degree angle, the top surfaces of which are treated as scrolls. A half-circle arch connects the two freestanding piers and similar arches return to tie them to the church wall and support a little pendentive dome with the floor above. There is no indication as to how access was gained to the first floor, nor are there stairs between floors. Just what the function of this structure was, is hard to surmise, probably either a clock or bell tower.

ii. Methods of Construction and Vaulting

The choir, nave, transepts, crossing, and *capilla mayor* are divided by means of a series of massive piers joined by both transverse and lateral arches on which the vaulting was once supported. (FIGS. 56–57.) In a strong sense the walls between piers are screens which are markedly reduced in thickness because of the deeply recessed niches. This fact was understood by the author of the addition to Vásquez's manuscript alluded to above in the historical data.[40]

The bays and/or vaulting of the church are referred to by Vásquez and his successor with the word "*bernegal*," probably meaning a bay covered with a pendentive dome, *bóveda vaída*.[41] The vaulting over all the nave and choir bays were originally half ellipsoids, or "half-watermelon" shaped domes, that is, ellipsoidal pendentive domes over oblong bays. Domes of this type became quite common in eighteenth-century Antigua. Those which once covered the nave bays were somewhat higher and more domical in appearance, while those still *in situ* over the two lower choir bays are extremely flat and shallow. (FIG. 58.) The nave and upper choir vaults are supported on transverse semicircular arches, the crowns of which are considerably higher than the lateral arches engaged to the side walls where lunettes pierced by windows at clerestory level are formed. (FIG. 60.)

The soffits of all the vaults were once ornamented with false ribs to give the appearance of intersecting barrel vaults. The crossing, square in plan, was once surmounted by a cupola of which only the supporting pendentives remain. The *capilla mayor* just behind also had a dome on pendentives which must have been lower in elevation than that over the crossing. (FIGS. 56, 57.)

iii. The Walling

The character of the walling in the different parts of the church varies, especially that of the north tower. As has been pointed out above, the nave and choir walls are really no more than screens. The walling of the church, exclusive of the piers, is uniformly 1.5 m., a dimension quite common in the eighteenth century. The niches are about 80 cm. deep thus reducing the wall thickness to about half. The contrary is the case with regard to the towers where the massiveness of the masonry is remarkable, especially that of the north tower which probably dates from later in the seventeenth century. From the historical data it may be concluded that the towers antedate the nave. At foundation level the walls of the north tower are about 3 m. thick. The walls of the south tower are about 2 m. thick at the foundation. This tower, as has already been pointed out above, may have

[40] Vásquez, **4**: pp. 390 ff.

[41] Angulo, *Historia del arte* **2**: pp. 65 ff., also believes this to be the meaning of word "*bernegal*." See fn. 27 above.

been the location of the Capilla de Loreto and is older than the north tower, both, however, dating from the seventeenth century. It is in very bad condition now and shows unmistakable evidence of repairs and alterations which must have been carried out at the end of the seventeenth and beginning of the eighteenth century when the additions to the cloister were built, and again after the earthquakes of 1717 and 1751.

The party wall shared by the cloister and the nave is approximately 2 m. thick and represents two separate constructions. The cloister remodeling was begun *ca.* 1695 and continued to *ca.* 1698. The church nave was constructed anew from *ca.* 1697 to *ca.* 1700. It is interesting that the portion of the wall facing the cloister corridor is built of thin square brick approximately 4 cm.×20 cm.×20 cm. and is not bonded to the church wall at all. Its total thickness is about 50 cm. The church wall, on the other hand, is built of a more modern looking brick which measures approximately 75 cm.×11 cm.×20 cm. The total thickness of this portion of the wall is 1.5 m. or the same as the rest of the church walling.

Another feature, not uncommon elsewhere in Antigua, is the fact that exterior buttresses are employed only on the north nave wall, while corresponding buttresses are absent on the 2 m. south party wall. But buttresses are employed on the uppermost section of the latter emerging above the former roof line of the second-story cloister corridor at the clerestory level of the nave. The south wall of the church was apparently reinforced by the extra girth of the party wall, the buttresses above being more decorative than functional. Also, only the north transept, the *capilla mayor*, and the sacristy have exterior buttresses. The interior units of the plan abutting on other structures lack buttresses.

iv. The Applied Orders and the Decoration of the Church Interior and Convent

As has already been stated, the construction of the main body of the church behind the façade is based on a series of massive piers connected by arches which support the vaulting. (FIG. 55.) All the piers are uniformly treated with pilasters and an entablature which runs around the entire interior of the church. The total projection of the piers including the pilasters, measured at the base, is 1.00 m. beyond the wall. Each pilaster rests on a stone base consisting of an extremely high plinth, a torus, a scotia, and smaller torus to form the ensemble. (FIGS. 56, 57.) The materials employed for the pilaster shafts are no different from the rest of the church walling; namely, brick and mortar with a plaster veneer and finished with a single deeply inset channel or flute. The capitals which surmount the shafts are an extremely simple Tuscan type.

A three-member entablature completes the order. (FIG. 55.) The architrave has but two fasciae separated by a molding of simple profile. The frieze is a flat continuous band broken at widely-spaced regular intervals by a vertical member in the form of a bracket set under each dentil of the cornice above, the total effect of which is not too far different from that of a triglyph and metope frieze. The vertical member of the bracket is as wide as the dentil above and projects only slightly from the plane of the frieze. The head of the bracket consists of a volute with fluting, in appearance very much like the scroll or peg box of a violin. The bracket as a whole is a simplified version of the type seen on the Capilla de San Isidro of the church of San Andrés, Madrid, dated 1657–1689.[42] Completing the entablature, the cornice is made up of a few simple moldings, the total height of which is not great, but is of pronounced projection so that the dentils below really serve as corbels. The entablature as a whole breaks miters over the piers and pilasters. The semicircular headers of the niches between pilasters in each bay cut through the bottom fascia of the architrave.

Above the entablature in the lunette of each bay, windows with splayed jambs and segmental headers are located. (FIG. 60.) Both jambs and header are treated as a series of complicated moldings with two major concave channels. The windows are centered directly below the slightly projecting corbelled archivolts of the lateral arches. In the space just above the extrados of the segmental window arch and unattached to the moldings which frame the window, a segmental pediment is worked in even higher relief than the archivolt of the lunette. These same independent segmental pediments are repeated in each bay on the exterior of the building between buttresses.

The interior of the church was once very ornately decorated with panel paintings, sculpture, and with relief deoration in plaster worked directly on some of the walls. A small fragment of *ataurique* still exists on the spandrels of the arch separating the choir from the nave. (FIG. 61.) It consists mainly of continuous scroll leaf and floral patterns which cover the entire surface leaving no empty space and was once painted with bright colors which have all but disappeared in the course of time.

There are also some remains of mural painting in the first long room as one enters the convent. (FIG. 59.) Apparently the walls of the two rooms just west of the main cloister were also once decorated with frescoes. Only a small por-

42 See Kubler, *Arquitectura*, fig. 93, for a photograph.

tion exists today on the wall shared with the south tower on which about four or five male figures in ecclesiastical garb are still visible. Most of the original colors have faded. The workmanship seems to be quite stereotyped and not of extraordinary quality.

v. The Main Façade

The exact dates of the towers are not known. The central portion, the *portada*, is designed like a retable and dates between 1675 and 1703. (FIGS. 53, 54.) The organization of the elements of the design, however, is remarkably like that of the church of La Compañía de Jesús finished about 1698. (FIGS. 83, 84.)

The proportions of height to width of the façade, including the towers, are rather squat; yet the central retable by itself tends more to the vertical than is common in Antigua in both the seventeenth and eighteenth centuries. The central bay is relatively wider than usual, and the lateral bays narrower. (FIG. 54.) Dominating the lower story of the central bay is a deep recess with a semicircular header with extra-long jambs in which the door with a similar header but shorter jambs is set. The door was constructed and no doubt completed in 1675 by Ramón Autillo, as attested to by the contract mentioned above. The niche in which the door is set is so wide that almost no space intervenes between the jambs and the solomonic orders of the lateral bays. The crown of the niche arch, because of the longer jambs, reaches to a level just under the entablature which runs across the whole *portada*. An empty space results between the intrados of the niche arch and the extrados of the stone door arch to form a semicircular tympanum, a device also seen on the main door of the Cathedral and La Compañía de Jesús. (FIGS. 43, 85.) A horizontal molding is corbelled at the spring line of the niche arch to outline a semicircular tympanum or blind lunette. In the tympanum there is located a small niche with a statue still *in situ*. This little niche is apsidal shaped, that is, semicircular in plan and roofed with a half dome. It is framed by solomonic columns and an entablature, but lacks a pediment.

The second story of the central bay is dominated by a deep niche-window with concave splayed reveals almost semicircular in section. (FIG. 54.) An independent pediment, like those used above the nave windows, is set above the crown of the niche-window header. It is a small triangular pediment with a concave break at the pinnacle around which the same moldings employed for the raking cornices continue. Located in the spaces to each side of the window are two niches, one superimposed above the other in a manner similar to that on the Cathedral and the Compañía de Jesús. Only the upper two of each pair have pediments, the raking cornices of which are broken to form scrolls just where they would otherwise connect at the pinnacle.

The lateral bays of the retable, it will be remembered, are rather narrow in proportion to the central bay. The columns rest on a rather high podium treated as an unbroken unit without the usual mitered pedestals on column centers. It is divided into a base, a central panel or dado, and a projecting cornice on which the engaged pairs of solomonic columns rest. The column shafts have a slight diminution and the spirals of each wind up in opposite directions from shaft to shaft. Very simple Tuscan capitals complete the columns.

The entablature supported on the columns of the side bays runs across the whole retable, breaking miters on column centers. The architrave is a rather narrow two-fasciae band while the frieze, rather high in proportion, is unadorned except for the coat-of-arms of the Hapsburgs with the double-headed eagle placed on door center. The cornice which completes the first story is of moderate projection and is composed of rather uncomplicated moldings.

The second story begins with a low podium or Roman attic which breaks miters on column centers forming pedestals under each of the four columns. The pairs of superimposed niches and the main niche-window of the central bay rest on this podium too. The column shafts of the second story are exactly like those of the first, except that they are much smaller in diameter and in a proportion corresponding to the lower height of the story as a whole. Ionic capitals, rather than Tuscan, crown the shafts. The entablature is like that of the first story, breaking miters on column centers. Each pair of columns is surmounted by a broken segmental pediment, the ends of the curved raking cornices terminating in small tight scrolls. The third story or *remate* is now entirely missing.

The layout of the niches recalls that of the Cathedral and La Compañía de Jesús. (FIGS. 42, 83.) Each bay is occupied by a pair of niches, one above the other, which fill the entire space between columns. The lower niche in the first story is treated with a segmental pediment, while the one above has a broken triangular pediment. The pilasters which frame the openings of each niche are broken halfway up their length by a horizontal molding. The niche recesses are apsidal in plan. In the upper ones a statue is still *in situ*.

The treatment of the pairs of niches of the second-story lateral bays is a little different, owing perhaps to the smaller height of the space to be filled. The lower niches lack pediments. The entablatures which run between pilasters serve as the floor of the niches directly above. The upper pair of niches terminate with small segmental pediments supported on unadorned pilasters lacking the horizontal moldings seen in the lower story. All four niches of the second story still have statues *in situ*.

Little can be said concerning the original appearance of the twin towers which flank the central retable, because they are in such bad repair. (FIG. 53.) Apparently the belfry of the north tower was once treated with corrugated pilasters like those seen on El Calvario. (FIG. 113.) This type of pilaster is post-1717 in date.

vi. The West Gate

The main gate in the west wall of the convent grounds is of the same style as the central retable of the church façade. (FIG. 52.) But the overall treatment is more ornate, largely because of the *ataurique* decoration which once covered all plane surfaces. The gateway is formed by a section rising above the rest of the precinct wall and is conceived of as a triumphal arch, the actual entry being through a central semicircular-headed opening. The jambs and archivolt are treated with mixtilinear molding profiles to accent the opening which is flanked by piers adorned with niches framed by solomonic columns. Two columns, four in all, flank each side of the niches, and rest on a high podium. The shafts spiral up in opposite directions from each other and are surmounted by Ionic capitals. The niches have independent segmental pedimentals above, for there are no pilasters at all to either side. The niche recesses are squeezed into the narrow space between columns so that the corners of the pediments almost touch the column shafts. The soffits of the half domes of the apsidal-shaped niche recesses are treated with convex flutes. A coat-or-arms done in plaster relief fills the space above the niches.

The four columns in each bay are not in the same plane, the inner two nearest the niches being set back while the outside two are set forward. The entablature which runs across the whole gate breaks miters only over the projecting columns. The proportions and composition of the entablature are like those of the church retable-façade, except that here the frieze is covered with geometric and floral patterns in stucco. The cornice which projects very boldly is supported on a dentil course of normal design.

A broken pediment consisting of a pair of scrolls reaches part of the way across the central bay between the two columns nearest the door opening. Visible above the central bay are remains of some scrolls of what probably was once a freestanding pediment surmounting the door opening and which completed the composition as a whole with a vertical accent.

vii. The Cloister

The state of preservation of the convent is very bad, the remains being a mass of tamped-earth walls of uncertain character and not clearly defined, extending for almost one hundred meters to the south of the church. The main cloister adjacent to the church is also in a very bad state of repair, though enough remains to reveal the general plan. (FIG. 51-k.) Originally the cloister was two stories high. Largely in ruins now, the stairway in the second long room in the entrance once gave access to the upper floor of which hardly a vestige remains today. (FIG. 51-l.)

The arcading of the main cloister is gone now and all that remains are two piers and part of a third. However, the emplacement of the rest of the piers is quite apparent from the broken bits of masonry still visible in the corridors. Each pier measured 1.65 m. by 1.1 m., while those in the corners were even more massive measuring 1.65 m. square except for a chamfered outer corner. A low parapet wall once ran between piers and was broken only between the pairs of piers on the two main axes. Vestiges of a fountain and a pool of complicated design are still visible in the central open area. This fountain must have been somewhat the same as that in the cloister of La Merced which has been extensively restored, however. (FIG. 62.) The actual water spout was located in the center of the pool in which fish and aquatic plants were to be found, a type of fountain similar to that in the convent of Santo Domingo described by the English friar Thomas Gage.[43]

2. *La Merced: Church and Convent,* ca. *1650–1690, 1767*

(FIGS. 5–7, 67–75.)

Historical Data

THE Mercedarian order was first established in 1538 when Marroquín brought Fray Juan Zambrano and Marcos Dardón with him on his return to Ciudad Vieja from Mexico where he had been consecrated as the first bishop of Guatemala.[1] Besides the convent in Ciudad Vieja, another had been founded by two other monks in Ciudad Real, Chiapas, perhaps a year before, in 1537.[2] The founding date of 1538 was also given by the order itself in 1741, referring to a

[43] Gage, ch. XVIII, pp. 283 ff.

---

[1] Ximénez, I: p. 203.

[2] Juarros, I: pp. 124 ff. Gil González Dávila, *Teatro* I: p. 144, gives the founding dates of the order in Ciudad Vieja as 1538, and as 1537 in Ciudad Real. See also Markman, *San Cristóbal*, pp. 27–29, for data on La Merced in Chiapas, Mexico.

cédula, dated February 26, 1538, authorizing its establishment.[3] Marcos Dardón left almost immediately for Chiapas while Zambrano remained in Guatemala. In a letter to the king dated February 4, 1540, he bemoaned the fact that he had neither candles nor oil for divine services nor money to buy any. He also asked help to finish the conventual building.[4] Judging by the general architectural character of Ciudad Vieja at that time and by the poverty of the sole occupant of the convent, it may be safely assumed that the building referred to could hardly have been much more than an adobe and thatch affair. But by 1542 the order already had six monks in residence and the local inhabitants offered to help build the church and convent.[5]

In 1543, two years after the destruction of Ciudad Vieja, a site was assigned the order in Antigua.[6] Three years later, on December 17, 1546, the Mercedarians informed the *ayuntamiento* that their church had been completed except for the *capilla mayor* which could not be built because of the lack of space on the site assigned. They, therefore, petitioned for a parcel of land to the rear of the building.[7] In 1548 the petition was once again presented, for it had apparently not yet been approved. The new request stated that the church which they planned to extend in length had a thatch roof and that the land required was about thirty feet.[8] In 1561, however, the roof was already described as being of a more permanent character, as an *artesón*.[9] The church was not to remain standing very long before it was seriously damaged, if not completely destroyed, in the earthquake of 1565.[10] It is apparent that during most of the sixteenth century both church and convent were, more likely than not, of nondescript architectural character, all vestiges of which were probably completely eradicated in the subsequent erection of buildings of a more formal and monumental character. No formal convent had been built in 1572. Not until 1583 did the *audiencia* authorize the enlargement of the conventual buildings to the north of the church. These were so seriously damaged in the earthquake of 1586 that they had to be rebuilt.[11]

In the seventeenth century very few notices appear which give more than vague information concerning the church and convent. In a *cabildo* dated June 17, 1687, the *ayuntamiento* granted the order permission to occupy some more land, about four and one-half *varas* or approximately fifteen feet, in back of the church to the east and to close off the street which once ran through there.[12] The church was probably not completely furnished until the last decade of the century to judge by the fact that a retable for the chapel of Santo Cristo was ordered in 1691.[13] Just how much of the original sixteenth-century structure had survived and was incorporated into the seventeenth-century construction is difficult to tell. Fuentes y Guzmán writes that when he was eight years old a controversy broke out in the convent of La Merced at which time the *comendador* Fr. Zapata, who had spent 70,000 *pesos* in building the *capilla mayor* and the new two-story cloister, took refuge in Espíritu Santo.[14] Fuentes was born about 1643,[15] which would make the time he was referring to about 1651, and the completion date of the church and the cloister mid-seventeenth century. To judge by the fact that the small parcel of land to the east of the church needed to enlarge the building was ceded in 1687, it would seem that not until the last decade of that century did the church plan take on its final shape. By the turn of the century, however, the interior furnishings were all installed and the contract to gild the main altar let to Pedro Lorenzo and José Veles, *estofadores y doradores*.[16]

The first notices from the eighteenth century speak of the church and convent as having been ruined in the earthquake of 1717. About six months after the destruction, the monks were reported living in thatch huts.[17] As late as 1719, the church was still in bad condition with only the crossing and the choir in good shape, so that divine services had to be held in a temporary structure of thatch while repairs were being made.[18]

The crown had inquired concerning conditions in Antigua soon after the 1717 destruction. The Mercedarians replied in a letter dated in August, 1720, reporting that the order lived from its *haciendas*, and was not too well off. Curiously enough, nothing is said about the convent and church

[3] *AGG*, A 1.11 (1741) 5025–211, *BAGG* **10**: 2 (1945): pp. 162 ff.

[4] Ximénez, *loc. cit.*

[5] Fuentes y Guzmán, **2**: p. 374; and Remesal, **1**: pp. 218 ff.

[6] Remesal, *loc. cit.*

[7] *Efem*, p. 10.

[8] *BAGG* **8**: 1 (1943): pp. 13 ff.

[9] Pardo, *Guía*, p. 96. In 1551 it was still roofed with thatch. A formal convent had not yet been built either, nor was there one even as late as 1572, see Mencos, *Arquitectura*, ch. III, nn. 119 and 121, quoting *AGI*, Guatemala, 168, "Información de Fr. Marcos de Ardón, en Guatemala, a 17 x 1551."

[10] Fuentes y Guzmán, **2**: p. 440.

[11] Pardo, *op. cit.*, p. 97; and Mencos, *op. cit.*, ch. III, n. 122, quoting *AGI*, Guatemala, 172.

[12] Pardo, *loc. cit.*, p. 97; also *Efem*, p. 105.

[13] *AGG*, A 1.20 (1691) 1189–120 vuelto, also *BAGG* **10**, 2 (1945): p. 158.

[14] Fuentes y Guzmán, **3**: p. 120.

[15] See *ibid.* **1**: p. xix, a biographical note by Juan Gavarrete.

[16] *AGG*, A 1.20 (1704) 739–183 vuelto, also *BAGG* **10**, 2 (1945): p. 160.

[17] *AGG*, A 1.2–5 (1718) 15.766–2207–16.

[18] *Ibid.* A 1.2–5 (1719) 15.776–2207–71, also *BAGG* **8**,1 (1943): p. 126.

which had been so badly damaged in 1717.[19] Ximénez, a contemporary of the cataclysm of 1717, gives some information on this score, saying that the barrel vaults of the cloisters (they are actually semihexagonal vaults) had been hurt and that half the church as far as the choir had been destroyed.[20] He goes on to say that the church had been damaged once before, in the earthquake of February, 1689, and temporarily repaired, and that these patches were exposed by the ruin of 1717. Land had been ceded for extending the length of the church toward the east in 1687, implying, therefore, that the damages referred to by Ximénez as having occurred in 1689 must have been to a building either in construction or one very recently completed. The damage could not have been very extensive, for in 1691 a retable was ordered for one of the chapels which would hardly have taken priority if the structure had still been in need of repair. It may be assumed that the repairs required were slight and certainly completed before the retable was ordered two years later.

Nevertheless, Ximénez does concur with the report from contemporary documentary sources that the choir had remained intact, which if actually were the case, would imply that the later eighteenth-century reconstructions comprised the area to the east of the choir, that is, the nave, *capilla mayor*, and apse. He is silent, however, concerning the crossing which is mentioned in the same document as also being unhurt.[21] The last sixteenth-century structure had been covered with an *artesón*, and though no specific mention is made of how the seventeenth-century building was roofed, it is more likely than not that the same type was employed. Ximénez describes the cloister vaulting incorrectly, it is true, yet makes no mention of vaults in the church, implying that there were none there. Another contemporary author also confirms the information given by the Mercedarians in the letter of 1719, mentioned above, actually quoting its contents and saying that the main body of the church was ruined but that the crossing and choir were in good shape.[22] On the other hand, still another contemporary of the destruction of 1717 claims that the church and convent were so badly damaged that they did not know where to begin the repairs, yet makes no mention of the choir or crossing.[23]

As was already noted in the same letter of 1719, the information is given that repairs had already begun. But these repairs must have been carried on very slowly for they were still in progress in 1730 when, in compliance with a *cédula*, a report was rendered concerning the reconstruction of the church and convent.[24] Ten years later, in another report detailing the history of the order from its beginnings to about 1740, nothing concerning the condition of the church and convent is mentioned.[25] Slowly indeed did the rebuilding progress, for thirty-one years after work had first been started it had still not been completed. On May 26, 1749, the *maestro mayor de obras* of Antigua reported to the *ayuntamiento* that some errors had been made in tearing down the church, and others committed in the new plan because of the inexpertness of the brickmasons. He recommended that a commission be sent to investigate whether the brickmasons as well as those directing the job had the title of *maestro*.[26]

When a violent earthquake struck the city again in 1751 the rebuilding and repairs of the church and convent were probably not yet complete, for an appeal was made to the *ayuntamiento* for help in repairing the new damages.[27] Ten years later the *ayuntamiento* donated two hundred *pesos* for the purpose of installing an image of the Virgin, which had previously been venerated in the atrium, so that the cult might continue in the same manner as before in the new construction.[28] And finally, in 1767 the new church was placed in service.[29]

The rebuilt church and convent had been standing only six years when the whole city of Antigua was laid waste by the earthquakes of September and December, 1773. The information concerning how the buildings in question fared is very confusing. As usual, the civil authorities exaggerated the damage and the ecclesiastical authorities, who did not wish to move to the new site, minimized them. To judge by the repairs known to have been carried out in the middle of the nineteenth century, damages to the church must have been extensive. Brasseur de Bourbourg, writing in 1855, mentions that the church had been recently rebuilt and suggests that it would be a good idea to rebuild the cloister also.[30] The damages to the convent, however, need no contemporary witnesses. To this very day the ruins bear evidence of the destruction. Only a small part of the debris of the convent buildings remains because much of the material was carried off for the church of San Felipe de Jesús lying just north of the city, not far from Jocotenango.[31]

[19] *AGG*, A 1.11 (1720) 16.779–2292.

[20] Ximénez, **3**: p. 355.

[21] See fn. 18 above.

[22] *Isagoge*, p. 405; see also *Efem*, p. 149, and fn. 18 above.

[23] Vásquez, **4**: p. 393.

[24] *Efem*, pp. 168 ff.

[25] *AGG*, A 1.11 (1741) 5025–211, also *BAGG* **10**, 2 (1945): pp. 162 ff.

[26] *Efem*, p. 202.

[27] *ASGH* **24** (1949): p. 377; see also *Efem*, p. 205.

[28] *ASGH* **25** (1951): p. 147.

[29] *Ibid.*, p. 165; also *Efem*, p. 235.

[30] *ASGH* **24** (1949): p. 165.

[31] Díaz, *Romántica ciudad*, p. 46.

Just what damages the church suffered in 1773 are difficult to ascertain because of the mid-nineteenth-century repairs. According to Victor Manuel Díaz, the building lost only its cupola, the towers, and part of the vaulting while the walls were all intact and left standing.[32] This is indeed a puzzling statement, for an examination of the cupola and especially the towers will reveal that they are comparable to the rest of the building in general workmanship, design, and decoration. The situation is further confused by the fact that the building was again damaged by earthquakes in 1917 and 1918, when one of the towers supposedly came down, and these damages were repaired in 1919.[33] This information must be only partially correct, for the towers are absolutely identical. They are of equal height and the pilasters which decorate the outer surfaces are identical. Both, however, had lost their roofs, and were covered with some wood framing and corrugated sheet iron until about 1960, at which time the present reinforced concrete roofs were built.

The statue of the Virgin which once occupied the ledge of the niche-window over the main door was inadvertently damaged in 1955 or 1956 when a workman who was doing some minor maintenance work slipped and clung to the head of the statue which broke off and came to the ground. A new head of obviously different style was replaced shortly afterward. (FIGS. 67, 68.)

Architectural Data

i. The Church Plan (FIG. 62.)

The *lonja* which runs around the west and south sides of the church measures between 33.50 m. and 33.70 m. on the west side, not including the one step still visible, and is uniformly 12.20 m. deep directly in front of the façade. (FIG. 62-A.) The southwest corner has been chamfered off. A step runs around the whole of the *lonja* most of which is almost completely hidden on the south side. It is possible that more steps would be uncovered should the area adjacent to the *lonja* in the atrium be cleared, for it appears that its present level is higher than it was originally. The two merlons on the step are lined up to either side of the door of the west façade. The south side of the *lonja*, measured in a similar manner and including the 12.20 m. of extension on the west side, is 58.00 m. long and uniformly 6.00 m. wide, terminating at the south transept where some recent repairs seem to have been made.

The exposed area of the west façade abutting on the convent entrance, and in part hidden by it, measures 27.60 m. (FIG. 64.) The total width of the façade at ground level including the corner buttresses of the towers, the north one being hidden by the two-story structure of the convent entrance, is 32.00 m. (FIG. 73.) Except for this detail, the façade is absolutely symmetrical in every other respect.

The twin towers which flank the central retable of the façade are about 9.00 m. wide, the north one, as noted above, obscured by the convent entrance. These towers are not independent units projecting beyond the nave, but actually line up with the side aisles. The church interior is divided into three aisles, yet the façade is pierced by a single door giving access to the central nave only. (FIG. 65.)

The overall exterior dimensions of the church are 73.70 m. long, including the small chapel or apse which projects from the *capilla mayor*, by 32.00 m. wide, including the towers. (FIG. 62.) Only at the level above the side aisles are the towers freestanding, extending beyond the width of the clerestory. (FIG. 66.) The plan of the church is cruciform with projecting transepts and a quadrilinear apse.

The interior dimensions of the church attest to the symmetry of the plan despite the fact that the lower part of the north tower is hidden by the convent entrance. The total width from side-aisle wall to side-aisle wall is 24.38 m. of which the nave, measured from pier center to pier center, is 11.32 m. wide and each of the side aisles, from pier center to wall face, is 6.53 m. The nave is six bays long, the first two bays just beyond the door representing the choir. (FIG. 62-B, C.) The following are the longitudinal dimensions of the bays when measured on pier centers; the first choir bay, 8.80 m.; the second choir bay, 8.15 m.; the third, fourth, fifth, and sixth bays each 6.85 m. None of these nave bays are square. The seventh bay or crossing is, however, approximately square in plan. (FIG. 62-D.) Its width, like the rest of the nave, is 11.32 m. on pier centers, and its length is just slightly less, measuring 11.20 m.

Except for the first two which comprise the choir, the third, fourth, fifth, and sixth side-aisle bays are approximately square in plan. They are 6.53 m. wide on pier centers to wall face and in length 6.85 m. each, or about 0.30 m. off square. The two side-aisle choir bays are oblong in plan, measuring 8.15 m. long from pier center to pier center, the first bay being even longer when measured to the interior wall face of the façade. One would expect half piers or pilasters to end the nave arcading; instead, a whole pier abuts on the wall. When the measurement of the first choir bay is taken on pier centers disregarding the wall, the span is the same as that of the second choir bay. The use of a whole pier rather than a pilaster seems to be an improvisation providing a constructional clue to the fact that the façade and interior plan must represent separate construction

[32] Díaz, *Bellas artes*, p. 273.

[33] Díaz, *Romántica ciudad*, p. 48.

operations. The superstructure of the towers rises above the first bay, so that at roof level there are five small domes over each of the side aisles between the rear of the towers and the front of the projecting transepts. (FIGS. 66, 70.)

An exterior side door pierces the third bay of the south wall giving access to the nave from the *lonja*; directly opposite another door opens into a short passageway which leads into a small chapel outside the church plan proper. (FIG. 62-K.) One wall of this passageway sweeps back on an angle while the other describes an arc. This interior chapel door is not of the same design or workmanship as the exterior side door, and was doubtless cut through at a later date to provide access to the chapel which lies outside the church plan proper within one room off the cloister corridor.

The projecting transepts are asymmetrical. (FIG. 62-E.) The south transept walls and buttresses are different in workmanship from the rest of the building indicating that some alterations had been carried out there. The north transept is somewhat shorter in projection than its counterpart on the south side. When measured from the interior angle buttress to the outside corner buttress, the south transept projects 6.00 m. or equal to the width of the *lonja*. Measured from the church wall, however, the total projection of the south transept is 7.20 m., implying thereby, that the corner buttress must be a later addition. This conclusion is borne out by the fact that it does not rise to the total height of the transept corner, nor does a similar counterpart buttress exist on the north transept. It must have been added to strengthen the corner now hiding an older buttress underneath, which it was designed to reinforce. The corner of this exterior reinforcing buttress is hollowed with a square chamfer so that it does not quite line up with the interior angle buttress of the church wall or with the return of the transept wall. It does, however, line up with the edge of the *lonja*. Beyond the chamfer on the line of the *lonja* it projects 1.00 m. Its south face is 3.55 m. long or about the same as on the west face, except for the chamfer. It returns 1.00 m. where it clearly abuts on a small exposed portion of the original corner buttress which it covers. The south wall of the south transept is of the same thickness as its counterpart in the north transept, except for the recess formed by the niche.

The interior of the transepts also shows some variation in size and in arrangement. The interior of the south transept measures 5.00 m. by 6.50 m. Its walls have no recesses for niches. The reveals of the piers to either side of the opening are simple, flat surfaces. The north transept is considerably smaller, measuring 3.50 m. by 5.00 m. A niche 0.70 m. deep forms a recess in the north wall. The west and east walls also have recesses, the west recess being treated as a niche about 0.50 m. deep. Opposite is a similar recess, once probably a niche too, but which is now pierced by a very narrow door slightly less than 1.50 m. wide. The thickness of the wall here is also 1.50 m., equal to its counterpart of the south transept. The door opening, completely unadorned and with nothing to set off the jambs and lintel, was probably cut through after the wall itself was already standing. The simple treatment of this opening is markedly different in character from the reveals of the piers to either side of the opening of the transept giving access from the crossing. These piers are treated with small pilasters of the same style and workmanship as those on the piers in the nave.

The *capilla mayor*, presbytery, is raised two steps above the nave floor. (FIG. 62-F.) Its dimensions when measured from the edge of the top step to the rear wall and from side wall to side wall are 10.50 m. by 7.50 m. The projecting apse beyond the *capilla mayor* is quadrilinear in plan. Its interior dimensions are 8.35 m. wide by 5.30 m. deep. Windows are placed rather high up in the north and south walls.

On a line with the side aisles and to either side of the *capilla mayor* are two rooms of practically equal dimensions. The one to the south measures 6.35 m. wide by 5.70 m. deep. Its opposite, north of the *capilla mayor*, is 6.30 m. by 5.80 m. To either side beyond are still two more rooms which line up with the projecting transepts. The room behind the south transept is somewhat smaller than its counterpart behind the north transept, its interior dimensions being 4.15 m. wide by 7.00 m. deep. Considerable alterations seem to have taken place here, to judge by the fact that the south wall is about 2.00 m. thick while its counterpart in the room on the opposite side is but 1.20 m. A long flight of stairs abuts on the east wall which gives access to a caracole stairway leading to the roof. A curious element, a solid bit of masonry walling 1.70 m. by 1.20 m. and about 2.00 m. high abuts on the south half of the north wall. Just what this may represent is hard to say. It is covered with fresh stucco making it impossible to see how it is bonded, if at all, to the adjacent walling. It may very well represent a pier of a stair landing which once supported a stairway, now gone. The ceiling in this room is on the same level with the transept ceiling. If this low corner pier does represent the landing of a flight of stairs, one may suppose that this room originally had a mezzanine floor, and that the present unusually long flight of stairs represents a later addition necessitated when the mezzanine floor collapsed in one of the many earthquakes.

The small room behind the north transept is quite different from its counterpart behind the south. Its interior dimensions are 4.40 m. by 7.40 m., and thus a little larger. A small caracole stairway rising from floor level occupies the southwest corner where it is contained within a pier 1.10 m. by 3.50 m. The north wall of this room, as has been noted above, is considerably thinner than its counterpart in the room behind the south transept. A door in this wall leads to the convent proper. Another door, probably a later addition, already mentioned above, gives access to the north transept.

ii. The Church Vaulting

The vaults of the central aisle of the two lower choir bays are rather shallow and supported on low flat arches. False ribs in imitation of intersecting vaults decorate the soffits. These vaults might best be described as pendentive "half-watermelon" domes; that is, half ellipsoids cut by four vertical planes. Similar elongated ellipsoid pendentive domes cover the side-aisle bays of the lower and upper choir, false ribs dividing the soffit into eight sections. The vaults of the central aisle of the nave, including those over the two bays above the upper choir, are also ellipsoid domes on pendentives fitted into the oblong plans. The clerestory runs the length of the nave and includes the second bay of the upper choir. The towers occupy the first side-aisle bays. Ranging along either side are five large octagonal windows which light the interior.

The side-aisle nave bays, four in number, are covered with small domes on pendentives. These domes are true hemispheres, more or less, since the side-aisle bays are only approximately square, and each is surmounted by a small lantern. (FIG. 70.) However, over the side-aisle bays of the crossing elongated ellipsoid or "half-watermelon" pendentive domes similar to those of the two choir bays are set so that their crowns are at the same level as those of the central aisle of the nave. In other words, the transepts rise above the level of the side aisles and like the nave have a clerestory, each with a single octagonal window looking out over the side-aisle roofs. (FIG. 70.)

The crossing which is square in plan is roofed with a cupola set on true spherical pendentives, the drum of which is circular in plan and pierced by eight octagonal windows. (FIG. 71.) The exterior appears as a sixteen-sided polygon owing to the fact that the short spaces between the windows are treated as plane surfaces. Each window is set within a frame of pilasters supporting truncated pyramidal pediments, and surmounting each pilaster is a rather naïvely executed, hollow terra-cotta lion. The lions, sixteen in all, sit in an upright doglike manner and are treated with bright yellow glaze for the body and a blue glaze for the mane. They add a colorful accent to the otherwise drab dirty white stucco veneer of the cupola. The scheme is completed by a dome with sixteen false ribs converging on an eight-sided lantern above. The eight small windows in the lantern are separated by small pilasters with corrugated faces and volute profiles, the scrolls of which are at the bottom. Another dome, lower than the cupola of the crossing, without a drum, but with the same lantern and exterior rib treatment, covers the *capilla mayor*. The apse behind is roofed with an even smaller dome of the same sort. It is lower still. The exterior effect is very striking, with the two domes stepped out and rising to the cupola over the crossing.

The rooms to either side of the *capilla mayor*, lined up with the side aisles, are roofed with ellipsoid domes each divided into eight sections by an equal number of false ribs. The two rooms beyond these, lined up just behind the transepts, are roofed with simple barrel vaults.

iii. The Church Walls and Piers

The six pairs of freestanding square piers which divide the choir and nave bays are all of exactly the same style and dimensions and are treated with pilasters 1.80 m. in width at the base. (FIG. 5.) The pilasters which face the nave pro-

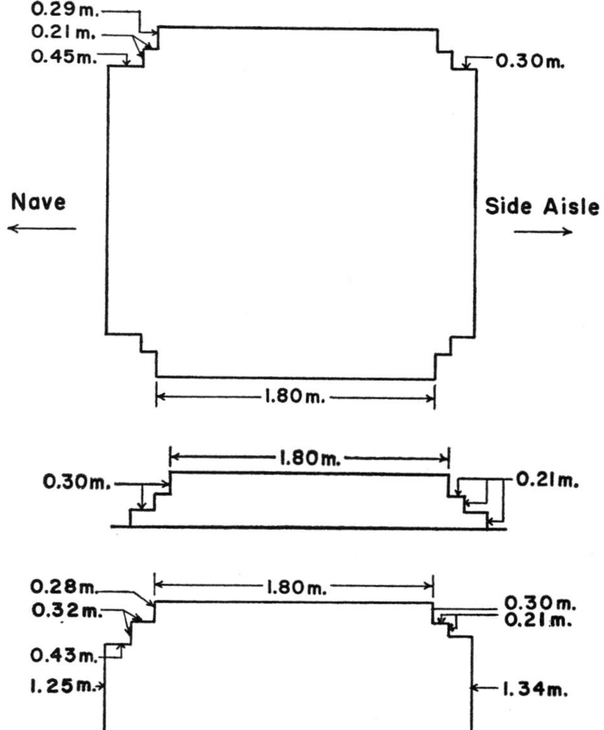

FIG. 5. La Merced. Church piers: top, arcade piers; center, side aisle pilasters; bottom, piers engaged to façade wall.

ject half as much again as those of the other three faces, namely, 0.45 m. as against 0.30 m. The pilasters engaged to the walls in the side aisles are of the same dimensions as the latter, that is, 1.80 m. by 0.30 m.

The first pair of piers, backed to the rear of the façade wall, are most unusual. They are treated exactly like the rest of the nave piers, except that the pilasters are cut off as they merge with the wall. The pilasters facing the nave are but 1.25 m. long, or 0.55 m. narrower than the rest of the pilasters. In every other respect they are alike, projecting 0.45 m. on the nave face and 0.30 m. on the side-aisle face.

This is a curious arrangement, for one would naturally expect these first two piers of the nave arcade to have been treated as simple pilasters much like those engaged to the side-aisle walls. A possible explanation might be that the nave arcading and the plan of the church interior were arrived at independently of the façade wall. In laying out the bays the builders had to adjust the scheme by using these partial piers since they could not be integrated into the façade wall already standing. Ordinarily the façade would be laid out and constructed so that the first pair of arcade supports would be integrated with it. The documentary and literary evidence points to the fact that in the earthquake of 1717 the choir was left standing and the crossing also remained intact while the nave alone was destroyed. Sometime near 1749 some errors had been made in dismantling the ruined church, though exactly which parts were not specified in the report. The layout of the church interior and the design of the nave piers indicate that they represent a unified single-construction program. But the fact that the first two piers are backed up to the façade leads to the conclusion that this wall dates from before the earthquake of 1717 and was utilized in the new construction.

The homogeneity of the interior plan of the church is further borne out by the manner in which the pilasters are fitted into the corners of the side aisles of the first choir bays. These too are cut off where they engage to the façade wall, and have the same plan and projection as the rest of the side-aisle pilasters. Another clue to the uniformity of the interior layout is observed in the pair of piers which divide the crossing from the *capilla mayor* and the corner pilasters in the far corners on which the partition walls of the *capilla mayor* and the side rooms abut. All are of the same design as those of the nave arcade.

The exterior walls of the church building are, however, not uniform in construction. They vary in thickness, a fact which may be accounted for as the result of utilizing older walls in the eighteenth-century reconstruction. The south wall is 2.50 m. thick at ground level and is uniform throughout, except for the part which encloses the projecting transept where it is slightly less, or 2.20 m. As has already been noted, a superimposed reinforcing buttress partially hides the original southwest corner buttress of the transept. The north transept is smaller than the south, probably because its dimensions had to be adjusted to fit into the existing convent buildings.

The same conclusion is reached after an examination of the north wall which abuts on the convent. It varies from 3.70 m. to as much as 5.10 m. thick in places. Here, however, there is evidence of much postcolonial reconstruction and alteration. At ground level the total thickness of the wall which separates the north side-aisle and the small chapel off the third bay is 5.10 m. (FIG. 62-K.) As has already been noted above, the door is not of the same design or size as the door opposite leading to the *lonja*. The area occupied by the chapel itself was originally not connected with the church, but was part of the convent. A bricked-up door in front of which a small retable is now located once gave access to the cloister. The same thickness is noted in the party wall between the room in the convent with a niche located just below the north transept and the last bay of the north side aisle. But elsewhere the north wall is 3.90 m. thick.

An inspection of the top of the north wall where it projects above the side-aisle roof shows that it is composed of two independent walls, one belonging to the church and another to the convent. (FIG. 70.) The buttresses on the exterior of the side aisles which exist on the south wall are not matched here. Apparently the former party wall shared by the convent and the original church wall was utilized to reinforce the side-aisle vaulting, explaining why the present wall in places is as thick as it is, 5.10 m. or almost three meters more than the south wall opposite. All the rooms off the cloister abutting on the church have been renovated in recent years obscuring all former constructional details. (FIG. 62.) On the plan, the outline of the buttresses is indicated to show that theoretically the church could have been laid out absolutely symmetrically had the party wall been taken down, but that this was not done because of the expense.

The short corridor with the concave curving wall which is 5.10 m. thick at this point, may be explained as part of a no-longer-existing caracole stairway. When the chapel was built in the nineteenth century, the opening in the wall was utilized as a corridor to connect it to the church.

The twin towers line up with the side aisles and are perfectly symmetrical in ground plan indicating that the façade wall was left standing after 1717 as demonstrated in the historical data. (FIG. 64.) Though La Merced is a three-aisle church, there is but a single door on the main front which opens on the nave only. Normally, smaller doors to either side would lead to the side aisles. Furthermore, the north tower is absolutely symmetrical with the south despite the fact that its lower story is partially hidden by the convent entrance. The upper two stories of the towers, however, are not contemporary with the rest of the façade, and date from the eighteenth-century reconstruction as the decoration would seem to imply. (FIGS. 66, 72.) The cornice of the convent entrance is not at the same level as that of the first story of the towers. (FIG. 73.) A low parapet wall was added and leveled with the Roman attic just above the first-story cornice of the tower, in order to integrate both structures. It would seem, then, that the convent entrance is posterior to the original church façade built in the mid-seventeenth century. The structure of the façade wall and the returns of the bottom story of the tower walls on the south and north sides show that both the central retable and the towers are part of a single constructional unit. The exterior scheme of the façade gives the impression of being composed of independent elements, a central retable and two towers. The bottom stories are treated as simple massive walls each pierced by two windows, while the second and third stories of the towers are successively set back. But when the walls are examined along the top of the side aisles, it is immediately apparent that the upper two stories of the towers are also integrated into the masonry of the clerestory. The only safe conclusion possible is that the church façade represents a single unified design from the seventeenth century which was utilized when the new interior of the church was built after the middle of the eighteenth century. The upper stories of the towers with their nonarchitectonic pilasters were added at this time since they are bonded to the clerestory walling. This explains the curious arrangement of the first pair of piers of the nave arcade backed up against the façade wall. It seems, therefore, that the interior of the church is the result of a single building operation. The unified three-aisle plan and the vaulting point to this fact. The retable-façade of the seventeenth-century church was retained, but the upper two stories of the towers were probably added at the time the nave vaulting with its clerestory was built.

The repairs in the nineteenth century must have been more in the nature of patching rather than of reconstruction, as for example the reinforcing buttress of the south transept, the room in back of the south transept with the caracole stair, the small portion of the convent now inhabited by the parish priest to the west of the cloister, and the alterations and repairs of the rooms off the south corridor of the cloister. The little chapel off the third bay was probably built in the nineteenth century, and the narrow unadorned door in the north transept was very likely cut through at that time.

iv. The Church Façade (FIGS. 64–69.)

The layout of the church façade with its massive twin towers is typical of the seventeenth century. But the seventeenth-century layout of the façade is puzzling in the light of the extravagant stucco ornament which decorates the central retable and the upper two stories of the towers. An interesting feature is the fact that the façade does not reflect the interior three-aisle arrangement, the retable being lined up with the central nave. The interior piers which take up the nave arcading do not coincide with the exterior arrangement, and are placed so that they overlap the outside columns of the retable and the slight indentation which marks off the tower. It would seem that the foregoing is another indication that the whole façade was built independently of the church interior.

a. The Twin Towers. The bottom story of the south tower is treated on the façade side as a large, smoothly stuccoed panel with a projecting stone corner buttress and two octagonal windows. (FIG. 66.) The lower window is smaller but it is set within a round-headed shallow recess. Only the dado of the corner buttress is of stone, and a veneer at that, very likely of eighteenth-century origin when stone veneers began to be employed in Antigua. The buttress is treated with three separate fasciae, each successively recessed. The first story of the tower is unified by a cornice with a broad geison above which a short parapet or Roman attic is set with a rather elaborate merlon in the corner.

The north tower is exactly like the south, except that the space which would have been occupied by the buttress is covered by the convent entrance. (FIG. 73.) The cornice, as has been noted above, is slightly lower than that which tops the convent entrance, the difference in level having been adjusted by the parapet wall. The elaborate corner merlon is absent here because this portion of the tower is built into the second story of the convent.

The upper stories of the towers seem structurally independent of the central retable, being set back slightly so that the retable side bays overlap them in part. (FIG. 64.) The second story, approximately square in plan, is set back considerably from the corner of the bastionlike first story. The third story, octagonal in plan, is set even farther back. It must have originally been surmounted by a domical or pyramidal roof with a finial.[34] The third story cornices of the

[34] The roofs of the towers were rebuilt about 1960 with reinforced concrete. Before then, makeshift roofs of sheets of corrugated galvanized iron supported on wood rafters kept out the weather.

two towers are almost level with the highest point of the central finial of the central retable.

Each of the three exterior sides of the second story is pierced by two round-headed openings and decorated with very fanciful pilasters, a pair on each side of the exposed corners and another pair between the windows on the exterior north and south faces. (FIGS. 2, 66, 72, 73.) On the west side, the main façade, that is, and on the east side facing over the side aisles, only one pilaster is set between the windows. And since the inside corner is hidden behind the central retable, none exist there.

The pilasters are noteworthy of comment. A number of superimposed horizontal moldings run around the base of the second story forming a sort of stylobate or podium broken by miters on pilaster centers and serving as the sills of the arched openings. The pilaster shafts are composed of a number of scrolls and are quite nonarchitectonic in conception. The fluid and malleable character of the plaster, the material from which the pilasters are modeled, is evident in the design. Nor is there any pretense of imitating structural forms. An architrave of horizontal superimposed moldings breaks miters over pilaster centers and is separated by a flat, unadorned band or frieze from an equally simple cornice. Merlons are set on the outside of the cornice in the space where the chamfered sides of the octagonal third story are set back.

The third story serves as a belfry and rises above the upper choir and the nave vaulting so that at this level all eight sides are exposed. A single large round-headed window, twice the size of the windows of the second story below, pierces each of the four main sides. The diagonal faces of this belfry, octagonal in plan, are treated with corrugated nonarchitectonic pilasters. (FIGS. 2, 72, 73.) Atlantids flank the window openings on the other two exterior sides. (FIGS. 66, 73.) The sides of the belfries facing the side aisles and nave roof lack these figures. They are worked in plaster and stand on corbelled pedestals fully clothed with rather curious, short, kiltlike garments, not unlike the figures decorating the pendentives of the Cathedral cupola. (FIG. 49.)

Another pair of pilasters consisting mainly of high bases with small grill-like shafts are set in the space between the corrugated angle pilasters on each of the two chamfered or diagonal sides of the belfry. (FIGS. 2, 72, 73.) Though fanciful in design, they are, nevertheless, less ornate than those of the second story. The mixtilinear element of the bottom member was added about 1960.

A rather plain entablature breaks miters over pilasters completing the order of the third story of the towers, immediately above which is a low parapet wall with a merlon set over each angle of the octagonal plan. Most of these merlons are now missing.

b. The Portada or Central Retable. The total effect of the central portion of the façade, the *portada*, is that of a retable or a screen wall set in front of the twin towers which covers them in part. (FIGS. 64, 65, 67–69.) It is laid out in three vertical bays by three horizontal stories, the third being the finial or *remate* and confined to the central bay only. The applied orders are relatively uncomplicated retaining a frank architectonic quality. But in contrast to this simplicity of architectural elements, the surface decoration is very ornate, so much so that the whole of the *portada* seems covered with lace. This lacelike quality of the surface treatment is further enhanced by the painting of details which stand out against the contrasting background color. Since the church is in use today and has been maintained and repaired since the mid-nineteenth century, it is impossible to tell what the original color scheme was like.

The central bay of the retable is the widest of the three. The first story is completely taken up with a large round-headed opening or recess, a deep niche, into which the door is set. The door has a horizontal header, thus leaving a semicircular tympanum or lunette between the door lintel and the soffit of the recess arch. The reveals of the door recess are treated with a dado which is connected with the dado of the lateral bays of the retable. The rest of the reveal is treated with a repeated design of a four leaf rosette worked in rather high plaster relief. Two rows of four large rosettes, eight in all, run up the length of the wall. Smaller six-petal rosettes are fitted into the spaces formed by the cross pattern of the larger. An interesting variation is to be noted in the bottom pair of the large four-petal rosettes. These have only three petals, the two horizontal and the top vertical one, for the bottom vertical petal could not be fitted into the space. The same design of large four-petal and smaller six-petal rosettes continues around on the soffit of the arch above, and there too the bottom pairs of rosettes just above the springing have only three petals.

The decoration of the wall surface of the lunette over the door consists of an oval medallion with a convex face from the center of which a series of foliated scrolls in imitation of leaves or vines undulate to fill the space. The spandrels to either side of the extrados of the recess arch are filled with small curly-haired cherub heads set at an oblique angle with large wings spreading to the springing below and crown above. The entablature which runs across the whole of the retable is set immediately above the door.

The second story of the central bay is entirely occupied

by a large niche-window with splayed reveals and a trumpet-shaped header, the *abocinado* type. (FIGS. 67, 68.) On the deep ledge formed just above the cornice of the first-story entablature, a figure of the Virgin is placed on a rather elaborate pedestal. The reveals of the niche window are treated more or less like those of the door below. Above an unadorned dado the space is treated with a design, the principal element of which is a fluted vase shape. Another smaller inverted vase is placed on top like a cover from which foliated scrolls in imitation of vines with leaves emerge flowing over the rest of the surface. Lotus shapes and both eight-petal and star-shaped rosettes are symmetrically distributed in the pattern.

The soffit of the niche is splayed. In profile it is a straight line and not a cyma or cavetto as is normal in *abocinado* or trumpet-shaped niches. Its surface is striated with deep semicircular flutes separated by fillets which in turn are incised with deep though narrow rabbets or channels. The lower half of the flutes, those immediately adjacent to the window, are filled with convex fascicles. The window opening at the back of the niche has a semicircular header concentric with the larger outer arch and lights the upper choir. The archivolt of the window arch is a simple molding ending in small tight scrolls on the imposts. The window opening occupies the total height of the niche, but rests on a short piece of screen wall fitted between the window jambs just above the dado. This screen is also decorated with *ataurique*.

The pairs of engaged Tuscan columns of the first-story lateral bays rest on a low podium and support an architrave which break miters on column centers. (FIG. 69.) In the space between the columns, niches with statues are located. The podium, mitered on column centers, is divided into two pedestals and a recess between. The three faces of the pedestals and the surface of the recess are treated as panels framed with moldings. A cornice, the dentils of which are wider than high, terminates the podium. The column bases consist of a very thin plinth and two narrow, half-round convex profile moldings separated by a half-round concave profile molding of equal size, that is to say two astragals separated by a scotia, all of equal size. The column shafts have no diminution and seem somewhat stumpy. The Tuscan capitals are composed of two flat moldings separated by narrow astragals and surmounted by exceedingly thin two-molding abaci.

Worked in plaster, a delicate vine ornament with pendant fruit and leaves spirals up the total length of the shafts. The fruit seems to imitate bunches of grapes, but may also be identified as pine cones. In outline, each bunch is more or less a pointed oval within which appears a regular pattern of rows of little globules. The Tuscan capitals are also covered for the most part with *ataurique*, except for the necking and the small astragal, immediately above, which are plain.

The entablature which breaks miters on column centers and runs across the three bays of the retable is also very heavily decorated with molded plaster or stucco ornament. (FIGS. 67, 68.) The very narrow architrave is decorated with a broad, curly scroll-leaf repeated over its total length. A flat undecorated fillet separates the architrave from the frieze which is more than three times as high as the architrave and is laden with a repeated ornament based on a lozenge from which scroll leaves emerge to either side. This pattern is repeated on each of the projecting mitered portions over column centers, as well as in the intercolumniations. Two such designs are fitted in the central bay and are separated by a blazon which breaks across the entablature and the archivolt.

A very small dentil molding, a cyma reversa with an egg-and-dart design, and a cyma recta separate the frieze and cornice. The latter is comprised of two main moldings: a flat band or geison decorated with a guilloche pattern; and, immediately above, a very widely-projecting cyma recta with a flat fillet of almost equal height.

The niches are framed by ornately decorated pilasters. (FIG. 69.) Hollow indented chamfers join the short raking cornices of the pediments. The wall above the pediments is filled with a lozenge design, the four corners of which continue as loops from which scroll leaves emerge to fill the entire space. All plane surfaces on the pilasters, moldings, tympanum, and pedestals are equally covered with plaster ornament. Heavily draped male figures in the round fill the recesses of the niches which are treated like small apses, semicircular in plan and roofed with quarter spheres. The interior surface behind the figures is likewise decorated with plaster ornament.

The same decorative scheme is employed in the second-story lateral bays. (FIG. 67.) The columns and the patterns of the stucco ornament, however, are more elaborate and ornate than below. The order rests on a proportionately shorter podium or Roman attic with pedestals on column centers. The front faces of the pedestals are covered with a floral design consisting of a central rosette with eight broad leaves radiating from it. The leaves are arranged with the apex toward the central rosette. The four leaves on the diagonals terminate with small fleur-de-lys which fill the corners. The recessed spaces of the attic between pedestals are filled with an oblong medallion with rounded corners from which scroll leaves emanate to fill the rest of the surface.

Surmounting the second-story columns is a type of composite capital consisting of a pair of tight little volutes connected by a perfectly straight horizontal band at the top. The space between the volutes is filled with a single row of elongated acanthus leaves. The abacus is made up of two molding profiles, the top one a cyma reversa with a leaf pattern. The bases are exactly like that of the first story.

The column shafts are very ornate and, like those of the first story, do not have any diminution. The first third of the shaft immediately above the base is decorated with two nude dancing figures flanking a central element in the form of an oval medallion from which foliated scrolls or vines undulate to fill the rest of the space. The upper two-thirds of the shaft is worked with narrow spiral fluting. The spirals on each pair of columns wind away from each other from the bottom up so that the diagonal lines converge toward the center of the bay. The same spiraling effect is observed in the vine ornament on the column shafts of the lower story.

The entablature which breaks miters on column centers is of the same proportions as that of the first story, though smaller in size. The very narrow architrave is divided in two, the lower part decorated with a rectilinear guilloche and the upper with a leaf pattern. A narrow cavetto molding devoid of surface decoration crowns the architrave separating it from the frieze which is at least twice as high as the architrave. The surface of the frieze is covered with a continuous scroll-leaf pattern broken only by four-leaf rosettes on the projecting portions where the frieze breaks miters. The cornice is rather wide in projection and is built up of three main moldings: first, a small dentil course below; second, a geison decorated with a pattern of four-leaf rosettes and darts; and third, an undecorated massive cyma recta with a broad fillet on top.

The third story or *remate*, confined to the central bay, is dominated by a large niche flanked by undulating, mixtilinear scroll half-pediments to either side. (FIG. 65.) A small niche with an ogee header is located in each pediment. Completing the design above is a pediment with swagged raking cornices and a crown set on the apex. The *remate* as a whole, somewhat like a roof comb, is set back slightly behind the small pedestals of the low podium or Roman attic on which it is set. The pedestals break miters on column centers and act as the finials of the applied orders of the retable-façade. These pedestals were once surmounted by merlons of which only vestiges now remain. It is noteworthy that the third story is bare and devoid of any surface ornament.

c. The Stone Cross. The cross which stands in the atrium, the open area in front of the church, fits into the overall scheme or design of the façade. (FIG. 73.) Whether or not it is the original and contemporary with the façade, whether it was placed there in the nineteenth century, or whether one was ever there in the first place cannot be known for certain. On the scotia of the base of the cross proper the date 1765 is inscribed. On the shaft a heart-shaped blazon with a smaller shield divided in three with a slim trefoil above is incised. Directly underneath the blazon the date 1688 appears. Both dates, 1688 and 1765, are significant in the history of the construction of the church, but there is no way of knowing whether the cross as a whole or only part belongs to La Merced.

The cross is set on a rather high, square pedestal with chamfered corners. The lower of the two courses is of a much lighter stone and not nearly as weathered as the upper course which is treated as a molding. It would seem that the lower course of less weathered or newer stone postdates the rest of the cross. Immediately above is a three-member base consisting of a wide sweeping scotia surmounted by a narrow torus, and finally another scotia the profile of which is more like a shallow flute and whose surface is striated with horizontal bands. The date 1765 is inscribed on the lower scotia. Immediately above is a large globe from which the shaft of the cross emerges. This globe is laid in mortar on the base and not cut from the same block of stone. The globe is of an even darker stone and much more weathered than the base. The shaft is made of three separate cylinders. The cross member and the projection of the shaft are all of one piece. The shaft has the inscription 1688. It would seem that the material of the present cross represents parts from three different monuments.

v. The Convent (FIG. 62.)

Except for part of the main cloister, the entrance vestibules, and a few rooms contiguous to the church nothing of the convent remains today. The exact site occupied by the convent buildings cannot be ascertained because the surrounding area is now completely built up. A great deal of damage was also incurred when much of the material was removed in postcolonial times.

Access to the convent is gained from a special entrance consisting of a vestibule and a porch with a triple archway facing the *lonja*. (FIGS. 62-J, 73, 74.) The center opening, with a special decorative treatment, is the doorway through which one passes into the vestibule. A door in the east wall of the vestibule leads to another room from which another turn to the left or north leads into the main cloister. (FIGS. 62-H, 63, 75.) This entrance unit is two stories high, as was the cloister originally, and abuts on the north corner of the church façade. The entrance building is more or less contemporary with the church façade, both dating from the second half of the seventeenth century but constructed in separate building operations, the church probably slightly earlier. However, both the convent entrance and the church façade were renovated when the church interior was rebuilt during the third quarter of the eighteenth century.

There are a number of rooms along the south side of the cloister on a line back of the vestibule. (FIG. 62-K.) These are the only parts of the convent still intact, but have gone through considerable transformations due to rebuilding and repairs in postcolonial times. One room has been converted to a chapel and is connected with the north side-aisle of the church.

The convent could also be entered from the church through the door in the north wall of the last room north of the *capilla mayor*. Still another door cuts through the west wall of the north transept giving access to the same room through which one enters the convent. (FIG. 62-E.) The convent room immediately adjacent is rather large and is divided a little off-center by pilasters which support a half-circle arch. Each side of the room is roofed with a pendentive dome, *bóveda vaída*, with false ribs. The dome on the northernmost side, in the slightly larger of the two rooms, has ribs which cross diagonally on the soffit connecting to a circle on the crown. The smaller half of the room, that adjacent to the church, has a dome decorated with more ornate false ribs of plaster. The soffit is divided into six parts by six ribs describing mixtilinear curves. Four emerge from the corners and join a circle at the crown, while two others bisect the vault from east to west. In both vaults the ribs are treated as free decorative forms in plaster.

To the west and immediately adjacent to this divided room is a small square room with a pendentive dome overhead. Beyond it is still another room, rectangular in plan and with a niche in the south wall. A door opens to a narrow corridor which is covered with a barrel vault leading directly to the cloister. The latter two rooms, the rectangular one with the niche and the barrel-vaulted corridor, are not in their original state, having been subjected to repairs and alterations in recent years.

The larger part of the convent has disappeared. The cloister, though in a bad state of repair, is still the best-preserved part. (FIGS. 62-H, 63.) Twenty-eight piers once surrounded the open central area, of which only fourteen are still *in situ*. An equal number of piers, now all gone, supported the second-story arcade, the emplacements of which are still visible on the roof of those parts of the lower corridor still standing. The overall dimensions of the cloister, exclusive of the rooms which still exist on the south side and those which once ranged around the other three sides, are 40.10 m. by 41.40 m. Each side once had eight piers, counting corner piers twice. The central area, almost totally filled by a fountain and pool, measures 29.50 m. by 31.10 m. The corridors are about 5.00 m. wide.

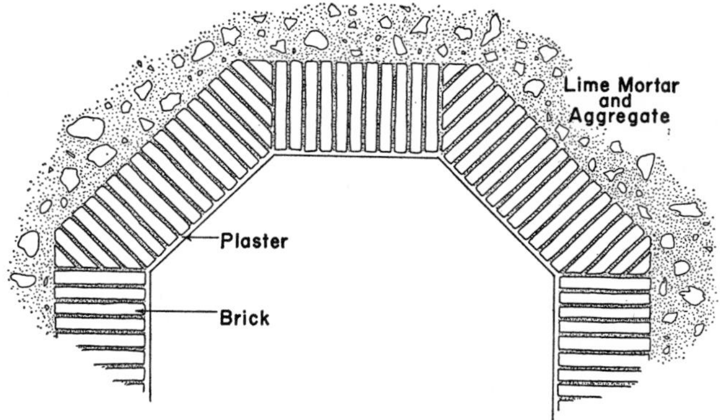

FIG. 6. La Merced. Vaulting, cloister corridors.

The ceilings of the corridors are extraordinary in that they are formed by continuous semihexagonal- or isosceles trapezoid-shaped vaults which are mitered where the corridors meet at right angles in the corners. (FIGS. 6, 75.) Brasseur de Bourbourg (see Historical Data above) assumed that these vaults represent an example of the afterlife of preconquest Maya building methods. In fact, he goes on to say that even the *portada* of the church, which he noted was wider than high, also imitated the Indian manner of construction.

But an examination of the extant masonry work of the vaults leads to a quite different conclusion. Rather than recalling a preconquest form, the vaulting is indeed more imitative of Spanish carpentry than Indian masonry. The semihexagonal shape of the corridor ceilings is more likely the result of a desire on the part of the builders to imitate the *mudéjar* inverted bathtub type of wood-paneled ceiling, *artesonado en forma de artesa*, rather than a Mayan corbeled vault. Also, in keeping with the *mudéjar* character of the vault, are the octagonal piers.

In order to achieve the effect of a wood *artesonado* in masonry materials, a wood centering in the shape of a truncated pyramid was doubtless first built; that is, a timber framework composed of sloping rafters and horizontal collar or tie beams. Planking was then nailed on the upper side of this wood framing, exactly as would be done in constructing an *artesonado*, but which in this case served as an unbroken solid surface to support the subsequent masonry work.

With the centering thus completed, the brickmason then laid a single course of bricks as a veneer to serve as the soffit of the completed vault and at the same time provide the form or mold for the wet mortar and aggregate which were thrown into the void formed between the "haunches" of the vault and the wall supported on the arcade piers. The break in the ceiling where the no-longer-existing east corridor once joined the south exposing the brickwork, clearly shows that the large, thin square bricks of the veneer were *not* laid as one would expect in constructing a corbel vault, that is, with each succeeding course cantilevered out a bit beyond the one below. Instead, the bricks are laid here at right angles to the plane of the sloping sides of the centering rather than level or parallel to the ground. Over the horizontal portion of the ceiling, corresponding to the narrow surface at the apex supported by the collar beams in a wood *artesonado*, the bricks are set vertically on end, that is, again at right angles to the plane of the centering. Once the brick veneer and the lime mortar with its aggregate of boulders, broken brick, and rubble had set up, the whole became a monolith in much the same manner as occurs in modern poured-concrete construction work. The soffit of the ceiling was then finished with various coats of stucco and a final one of plaster smoothly troweled and painted.

The arcades which face the open central area consist of short, squat *mudéjar*-style octagonal piers supporting segmental arches, the soffits of which are chamfered to follow the same profile as the three inside faces of the piers. (FIGS. 7, 63, 75.) The total height of the piers is only 2.10 m. while the plinths, also octagonal, are 1.50 m. by 1.50 m. if squared in plan.

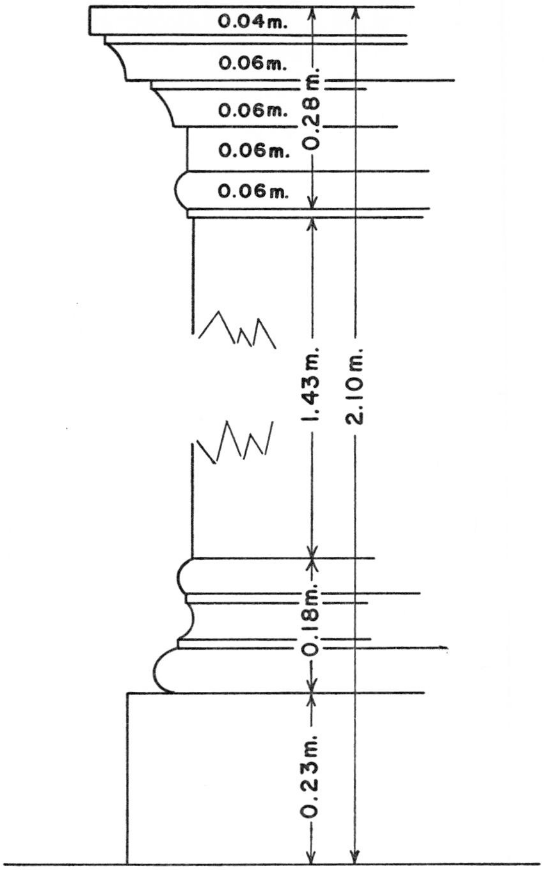

The fountain has been restored in recent times, but is probably close in appearance to what it might have been like originally. (FIG. 62-I.) The pool occupies almost the total open area of the cloister.

The *portada* of the convent entrance is worthy of special note. (FIG. 74.) The wall is now pierced by three arched openings on the first story. As late as the 1920's, the two side openings were filled with walling and pierced by small rectangular windows which were subsequently removed at some undetermined time.[35] The central arched opening, which seems to have escaped alteration, is decorated with solomonic columns covered with very delicate *ataurique* ornament. Each column is set on a high pedestal, the surface of which is treated as a panel. The twisted column shafts are very finely tapered and treated with a continuous spiraling vine of broad and very finely worked leaves. Undecorated Tuscan capitals support the entablature above. The thin architrave is a series of superimposed undecorated moldings, and the frieze a narrow band with curious plaster decoration, part of which seems, on first glance, to be imitative of Arabic script. The elements of this design are seemingly fortuitous in form, nor are any repeated. The total effect is that of a continuous scroll pattern between the undulations of which linear elements like writing are placed.

FIG. 7. La Merced. Cloister columns.

Above the frieze is a very narrow dentil course, and above that the cornice. The geison is decorated with two entwined strands of plaster forming open loops. The raking cornice above is set so low that there is almost no tympanum in the pediment at all. Each side of the raking cornice ends in a volute and thus frames the bottom of a niche in which a statue is placed.

The archivolt and spandrels of the arch over the central opening are decorated with a floral pattern and winged cherub heads in the corners. On column centers above the entablature merlons are set. The niche above the broken scroll pediment of the door is very elaborately framed with a series of vertical moldings, each decorated with a different pattern in plaster. These support a segmental pediment with a massive projecting cornice on which very small merlons are set to either side and a third on the crown.

[35] See a photograph in Díaz, *op. cit.*, p. 47.

The niche opening proper is semicircular in plan and roofed with a half dome like an apse. The soffit is fluted in the same manner as the niche-window of the church *portada*. In the space to either side of the niche are half-pediments worked in low relief and comprised of very elaborate mixtilinear moldings consisting of two scrolls meeting halfway up the raking portion. These scrolls enclose an area which is completely covered with plaster ornament consisting of a continuous pattern of lozenges with loops at the outside corners. In the remaining space thin strands of plaster twist and turn imitating a vine without leaves.

3. *Santa Teresa: Church and Convent, 1687 and after 1738*

(FIGS. 76–78.)

Historical Data

THE movement to found a convent for nuns of the Carmelite order in Antigua began in the mid-seventeenth century. But it was only after the special efforts of a secular clergyman, the presbyter Bernardo de Obando who had donated a site and some houses and also obtained the agreement of the *ayuntamiento*, that the convent was finally established in 1677.[1] Some property had also been turned over previously, in 1675, for the purpose of founding this convent.[2] Various sums of money, some of it donated by a couple in Peru, were also received that same year.[3] Official permission in the form of a royal *cédula* to found the convent had also been emitted in 1675 confirming the authorization already granted by the *ayuntamiento* through the intercession of Obando.[4]

Three nuns arrived in Guatemala from Peru about two years later in 1677[5] and were temporarily housed in the convent of Santa Catalina until their own was ready later in the same year. The church had not yet been built, so some temporary place of worship within the convent was prepared for them.[6] A formal church was begun five years later in 1683 and completed in 1687 when it was formally dedicated.[7] A vote of thanks from the *ayuntamiento* was extended to the Capitán Don José de Aguilar y Revolledo who had borne most of the cost.[8]

These buildings were damaged in the earthquake of 1717.[9] Despite the assertion of Ximénez that only the convent entrance was hurt, and not seriously at that, the nuns were terrorized and afterwards lived in thatch huts in the convent garden.[10] According to Juarros, however, who was writing almost a century later, the convent had been damaged by earthquakes, though he does not specify when, and was partially rebuilt sometime around 1738 by the bishop Pedro Pardo de Figueroa who was in office from 1737 to 1751.[11] This fact is corroborated, except for the exact date, in a document from 1740 in which it is stated that at that time the nuns were still living in temporary quarters in the convent garden.[12] In 1751 the church was once again damaged in the earthquake which rocked the city on February 4.[13] Almost twenty years later in 1770 both church and convent were still in bad condition, and the *ayuntamiento* granted permission to collect alms to be used for repairing the building.[14] The great catastrophic earthquake of 1773 caused a total destruction of the church, the ruins of which are still visible today. The description given by González Bustillo that the *artesonado* had fallen must be taken with a grain of salt, for the ruined temple was actually roofed with vaults.[15] That the pinnacle of the *portada* had fallen and that the walls were out of plumb, are the only facts which jibe with his description. The convent which dates *ca.* 1738 or later is in a far better state of repair and was in use as a prison until the 1950's. The upper floor is largely gone, but the lower story with its cloister and cells for nuns was readily converted to a prison with but small alterations necessary.

Architectural Data

THE church façade does not altogether conform to the usual layout of the retable-façade, particularly in the design of the first-story central bay. (FIG. 76.) Normally, the whole

[1] *AGG*, A 1.2.2 (1674) 11774–1780; *Efem*, p. 84; *AGG*, A 1.20 (1678) 1480–28, 1480–35 and 1480–40 vuelto; also *AGG*, A 1.20 (1679) 683–191. For the latter four documents of 1678 and 1679, see *BAGG* **10**, 3 (1945): pp. 229 ff., 232 ff., 234 ff., and 235 ff. See also Mencos, *Arquitectura*, ch. v, who cites a number of documents referring to the founding of Santa Teresa, *AGI*, Guatemala, 16, dated in 1657.

[2] *Efem*, pp. 84 and 85.

[3] *AGG*, A 1.20 (1675) 9093–600; A 1.20 (1675) 1322–803; see *BAGG* **10**, 3 (1945): pp. 243 ff. and 245 ff.

[4] *Efem*, p. 85; *AGG* A 1.23 (1675) 10975–1520–220: also *BAGG* **10**, 3 (1945): p. 247; Larreinaga, *Prontuario*, p. 206; see also *AGI* (1688), *Aud. Guat.*, *Ramo eclesiástico*, Est. 65–Caj. 1–Leg. 30/1.

[5] Molina, *Memorias*, p. 132; *Efem*, p. 87.

[6] Ximénez, **2**: p. 380; Vásquez, **2**: p. 370; Juarros, **1**: p. 134; *Efem* pp. 87 and 88; *AGG*, A 1.2–2 (1677) 11775–1781, also *BAGG* **8**, 1 (1943): pp. 32 ff.

[7] Juarros, *loc. cit.*; Vásquez, *loc. cit.*; *Efem*, pp. 97 and 104; Villacorta, *Historia*, p. 236.

[8] *AGG*, A 1.2.2 (1687) 11777–1783, also *BAGG* **10**, 3 (1945): p. 248; *Efem*, p. 104.

[9] González Bustillo, *Razón puntual*; also *Extracto*; see also *Ciudad mártir*, pp. 77 and 130.

[10] Ximénez, **3**: pp. 356 and 391.

[11] Juarros, **1**: pp. 134 and 208.

[12] *AGG*, A 1.18 (1740) 5022–221, also *BAGG* **1**, 2 (1936): pp. 124 ff., and **10**: 3 (1945): pp. 249 ff.

[13] *AGG*, A 1.10.3 (1751) 31346–4049.

[14] *ASGH* **36** (1951): p. 169.

[15] *Razón particular*, also *Ciudad mártir*, p. 103.

of the first-story central bay with its large half-circle door is set in a plane noticeably behind the applied orders and surmounted by a low pediment, whereas, in the case of the church of Santa Teresa, it actually projects beyond the plane of the façade wall. In effect it is as if the central bay with the door were a triumphal arch almost independent of the rest of the façade wall. A similar trait is to be seen on the façade of San Pedro. (FIG. 33.)

Another interesting feature is the absolute symmetry of the two towers which are now reduced to corner buttresses flanking the central retable. The design of the applied orders adheres to classical prototypes and are comparable to those employed in Los Remedios, San Sebastián, the Cathedral, and San Agustín. (FIGS. 26, 29, 42, 81.) However, the focal point where the emphasis of the design of the retable is concentrated is the first-story central bay with its quasi "triumphal arch." The second story is not as organically integrated with the first story, as is normally the case in the usual retable-façade, because of the arrangement of the niches and the noticeably larger columns on the first story in comparison to the abnormally shorter ones employed above. The first-story columns are about one and one-half times the height of the second-story columns, that is, a proportion of 1½:1. The more common proportion is nearer 1¼:1. As is the case in the Cathedral, the same type of low podium is employed. The taller columns of the first story thus allow space for two superimposed niches in each of the side bays.

The proportions of the side bays are very much like those of the Cathedral too, namely, very tall in relation to width. Framing the side bays are classical Tuscan columns, the shafts, and other members as well, devoid of surface ornament. The superimposed niches are shallow and rectangular in plan with square headers, another distinctive feature; for ordinarily, niches are apsidal in plan with curvilinear headers. The lower of the two niches is topped by a broken pediment with raking cornices ending in volutes. The upper one has a normal triangular pediment. The only ornate features in the execution of the niches are the small framing pilasters to either side. The lower pair are treated somewhat like solomonic columns, while the upper pair are corrugated and topped by small brackets. Each of the corrugations is square at the sides giving the appearance as if small sections had been cut from a strip of wooden molding and each piece set one on top of the other to form the pilaster. This same type of niche pilaster is seen on the little church of Santa Isabel dated in mid-eighteenth century (FIG. 179.)

The entablature breaks miters on column centers and runs across the façade. But where the entablature crosses the central bay, the miters reverse so that the central portion projects even farther out than those sections on column centers in the side bays, thus producing the effect of a triumphal arch. This feature is duplicated on San Pedro.

The entablature has a three-fasciae architrave, each fascia projecting slightly more than the one below, and a triglyph and metope frieze. The triglyphs, however, are exactly as wide as the metopes and have four flutes rather than the classical three. Six conical guttae are arranged below each triglyph and under the fillet which runs across the whole frieze. The cornice above is quite simple consisting of a large corona with some smaller crown moldings. But under the projecting corona, small dentils are evenly spaced.

The treatment of the central bay is comparable to that of San Pedro. (FIG. 33.) In many respects it recalls a somewhat similar treatment of doors in some churches of sixteenth- and seventeenth-century date in Peru and Mexico.[16] The Mexican and Peruvian examples are not part of retable-façades, and the comparison is valid only with regard to the arrangement of the door with flanking pairs of columns and the pediment over the central bay. The door design of Santa Teresa is fitted into the usual retable scheme of Antigua.

The door itself is set within a deep recess or niche. Both have semicircular headers but are not concentric. The jambs of the niche are much higher than those of the door, so that a large space is reserved between the intrados of the niche arch and the extrados of the door arch. The same arrangement is also seen in the doors of the Cathedral, San Francisco, and the Compañía de Jesús. (FIGS. 43, 53, 83.) A horizontal molding stretches between the two arches at the level of the springing of the niche arch and just above the crown of the door arch to outline a tympanum shaped like a lunette in the center of which a diminutive niche with pilasters and pediment is set. To either side of the niche the wall surface of the tympanum is decorated with a single large rosette or flower. The jambs of the door opening are very ruinous, but originally they were splayed and made up of a number of molding profiles. The spandrels above are covered with an open interlaced pattern in stucco.

The niche or recess in which the door proper is located is

[16] For example, the *portada* of the church in Tecali, Puebla, Mexico, Angulo, *Historia del arte* **1**: p. 370, fig. 496, and the *portada* of the church at Paucarcolla, Peru, Angulo, *op. cit.*, p. 647, figs. 751 and 772; and Marco Dorta, *Iglesias renacentistas*, p. 712, fig. 2. Because of the striking similarities with San Pedro, in which the youthful José de Porras (II) had an important part in building, it is not unreasonable to identify him with the construction of Santa Teresa too. The use of two superimposed niches in the side bays, a feature appearing on the Cathedral which was built under his direct supervision, is another clue to his hand in the design of Santa Teresa. See ch. VI, and ch. XII, no. 4, above. Angulo, *Historia del arte* **2**: p. 58, says that Santa Teresa has been attributed to him.

squeezed between the two inside columns of the side bays leaving but a narrow flat and undecorated strip by way of jambs. Outlining the spandrels is a protruding molding thus forming a panel in which a single flower or rosette is arbitrarily rearranged to fill the more or less triangular spandrel area. The leaves are not of natural or uniform size, but vary to fill the space as they radiate from the central boss.

Centered in the entablature with its triglyph and metope frieze is a large blazon set directly over the door and actually breaking across the architrave. The whole of the central bay, as remarked above, projects slightly. The entablature over this section supports a low triangular pediment of classical proportions with a very small tympanum in which an oblong, undecorated blazon with upcurving striations emerging from either side is worked in stucco. The lower central bay thus has the appearance of an independent element set in front of the façade.

The second story, however, follows the more normal retable arrangement so common in Antigua, but appears to be set back more than is usual because of the projecting central bay with its triumphal-arch effect. Actually the applied orders are set in the same plane as those below. A low podium or Roman attic runs across the entire second story portion of the retable and is broken only by the pediment of the "triumphal arch" below. The applied orders here are Ionic and quite Vitruvian in appearance. The bases are of the classical type as are the capitals above, the volutes of which are rather tight little scrolls. The entablature above breaks miters on column centers and is also of classical Ionic inspiration in that the frieze is a continuous band covered with a decorative floral pattern in stucco relief. The second-story central bay as a whole appears excessively set back because of the projecting pediment of the triumphal arch below. Between columns in each of the side bays there is a normal round-headed apsidal niche surmounted by a broken pediment supported on pilasters ending in brackets and decorated with rosettes. The space between the pinnacle of the broken pediments of the niches and the second-story architrave is occupied by a large oval blazon with ribbons and festoons worked in stucco which fills the upper part of each side bay.

Dominating the second-story central bay is a large square-headed window with splayed reveals, the outside header of which is formed by a segmental and rather flat arch. The soffit is treated with flutes giving it a seashell effect. The spandrels to either side of the archivolts are filled with a floral pattern, that is, a rosette of similar design as that of the door spandrels below in the first story.

Of the finial, the third story, not a trace remains. Nor is there enough material left so that its outline or general shape might be reconstructed.

Framing the central retable to either side are the towers, now reduced to massive buttresses which protrude beyond the width of the nave. The height of these towers is equal to that of the second-story entablature of the central retable, not including a finial of some sort which once surmounted each but is now gone. But even with the finial in place, the towers were considerably lower than the central part of the façade since the central third-story bay, the *remate*, once added to the height of the façade. The use of absolutely symmetrical buttresses, really vestigial towers, to either side of the central retable is also employed on Nuestra Señora de los Dolores del Cerro, Santa Clara, and Santa Rosa. (FIGS. 100, 140, 171.) Likewise, in churches such as Los Remedios and Las Capuchinas (FIGS. 25, 146) framing buttresses of this type are employed, though their effect in these instances is altered by the presence of side chapels which project and make for an overall asymmetrical façade design.

The interior of the church is in total ruins, and rubble from the ruined ellipsoidal vaults of the type common in the eighteenth century still clutters the interior. The remains of spherical pendentives indicate that the crossing was once roofed with a cupola or dome. The engaged piers or pilasters are remarkably like those of the Cathedral in design except that they are left plain. (FIG. 77.) It is interesting that, as in the Cathedral, the entablature is confined only to the space over each pilaster and does not continue around the walls. (FIG. 47.) The plan of the church is the single-nave type.

The convent probably dates for the most part from about 1738 and later, the most interesting feature being the central cloister which seems to be almost a replica of that of Las Capuchinas. (FIG. 78.) It is two stories high and surrounded on four sides with arcading. The short, squat Tuscan columns of enormous girth which support the arches are very close both in design and proportions to those of Las Capuchinas. (FIG. 152.) The lower-story columns, however, unlike those of Las Capuchinas, have hardly any diminution at all and, as a result, are not as graceful in appearance. The columns of the second story are of the same proportions but are smaller in size and have somewhat more diminution, thus adhering somewhat more closely to those of Las Capuchinas.

The arches connecting the columns vary slightly from those of Las Capuchinas which are constructed on the basis of the secant of a circle, that is, they are little less than a semicircle. Those of Santa Teresa are considerably flatter and are based on three centers or three different points of origin. Actually they seem to have been done almost freehand and are not as true in outline as those of Las Capuchinas.

4. *San Sebastián: Church, 1692*

(FIGS. 8, 79–82.)

Historical Data

THIS parish church, founded as a hermitage after the earthquake of 1565, was formerly located on the hill of San Felipe in the northern sector of Antigua near Jocotenango. In 1582 a new church building was erected on the present site to which the cult of the hermitage of San Sebastián was then transferred.[1] When the English Dominican Friar Thomas Gage entered the city in 1627 coming on the road from Chimaltenango, this was the first church he saw. He mistakenly wrote that it was the only parish church in town and a "... new built church, standing near a place of dunghills, where were none but mean houses, some thatched and some tyled."[2] Apparently the hermitage had been raised to a parish a few years after its transfer to the new building in the new location.[3]

No descriptions are given of the sixteenth-century structure, though it is known to have suffered some damage in the earthquake of February 18, 1631, the cost of the repairs of which the *ayuntamiento* agreed to pay.[4] Some additions to the original structure were also carried out soon after September of 1668 when the master brickmason, Pedro de Barrientos, signed a contract to build a vault over the *capilla mayor*.[5] It is very likely that a small sixteenth-century structure with a wood-and-tile roof was enlarged by the addition of a vaulted *capilla mayor* at this time.

In 1689 the sixteenth-century structure, by then more than one hundred years old, was completely demolished in the earthquake of that year.[6] It was rebuilt immediately and inaugurated on January 18, 1692.[7] By this time this church served a very populous neighborhood, for in 1700 the parish was reputed to have had more than 8,000 communicants.[8]

The damage which this new church suffered in 1717 is not exactly known, nor is there agreement in the different contemporary accounts. Ximénez insists that the building remained intact while Arana reports the contrary.[9] A document dated June 3, 1718, however, being a report to the crown of the damages to Antigua in general as a result of the earthquake of 1717, mentions this church among others as one specifically in need of repair.[10] In 1720 the *audiencia* complied with a royal *cédula* requesting an estimate of the cost of making the repairs, indicating that the building had indeed been damaged in 1717.[11] In 1738 the building was still in bad condition, and the *ayuntamiento* donated 100 *pesos* toward the construction of the vaults.[12] One may assume that the vaults referred to were either over the *capilla mayor* or in the little chapel, probably the *sagrario*, projecting to the north. At any rate, to judge by these sparse references it would seem that the principal damage to the church was restricted to the roofing and that the façade, still intact today, was not badly hurt in the 1717 earthquake.

Quite the contrary was the case in the great earthquake of 1773. According to González Bustillo, the *artesonado* fell and the walls were also cracked.[13] Parts of the vaults and arches of the chapel on the north side also are reported to have come down. The chapel is still intact today, and so must have been repaired. The church was in use through most of the nineteenth century until 1874 when an earthquake caused the roof to cave in leaving the building in its present ruinous state.[14]

Architectural Data

ALL that remains standing of this church today are the walls including the main façade. All of the roofing is gone except a small dome on pendentives in the little chapel projecting to the north of the main façade and entered from the first bay. The most interesting feature of the church is the retable-façade which is typical of Antigua serving as a datum point for comparison with others since its date is well known. (FIGS. 79, 81.)

The dimensions of the façade as a whole are as follows: total width, 18.20 m., not including the projecting chapel which is set back 1.50 m.; the central *portada* or retable,

[1] Remesal, **2**: pp. 309 ff.; Vásquez, **1**: p. 267 and **4**: pp. 381 ff.; Juarros, **1**: p. 147; *Efem*, pp. 25 and 27.

[2] Gage, ch. XVIII, p. 265.

[3] Juarros, *loc. cit.*

[4] *Efem*, p. 50.

[5] *Ibid.*, p. 79.

[6] Juarros, *loc. cit.*; Ximénez, **3**: pp. 355 ff. If the date which Mencos gives for the inspection of the building by José de Porras, Bernabé Carlos, Andrés de Illescas, and Agustín de Cárcamo is correct, then the church was already in deplorable condition the year before the earthquake, that is, in 1688. This committee of experts all agreed that the church should be rebuilt from the ground up. See his *Arquitectura*, ch. IV, n. 125, citing *AGI*, Guatemala, 180.

[7] Juarros, *loc. cit.*; *Efem*, p. 111.

[8] Ximénez, **3**: p. 168.

[9] *Ibid.*

[10] *Efem*, p. 149.

[11] *Ibid.*, p. 150.

[12] *Ibid.*, p. 184.

[13] *Razón particular*, also *Ciudad mártir*, p. 91.

[14] Díaz, *Romántica ciudad*, p. 43; Pardo, *Guía*, p. 85.

11.80 m.; the towers, 3.20 m. each; the total height to top of the segmental pediment, 13.20 m.; the height to top of cornice of the second story, 11.70 m. It is interesting to note that the proportions of the retable set between the flanking towers is about 1:1, except for the *remate* or finial which adds about one-seventh to the total height. (FIG. 8.)

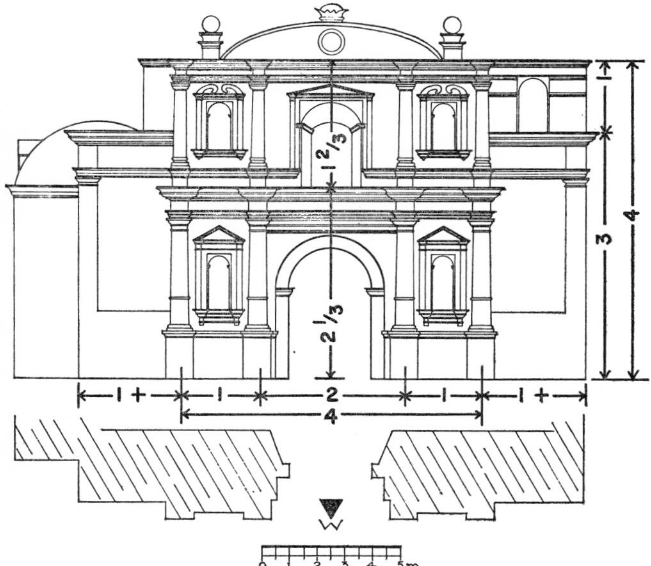

FIG. 8. San Sebastián. Proportions of retable-façade.

The dimensions of the various parts of the retable are: the total width, as already mentioned, 11.80 m.; the height of first story, 7.00 m.; the height of second story, 4.70 m.; the height of the *remate*, that is, the segmental pediment, 1.50 m., making a total height, as already mentioned, of 13.20 m.; the side bays on column centers, 2.80 m.; the total width of each bay at the podium, 3.80 m.; the central bay on column centers, 5.40 m., or approximately equal to the combined width of both side bays; the door opening, 3.00 wide; the door jambs, 3.30 m. high to spring line of the arch; the total height to the crown on the intrados of the arch, 4.75 m. more or less; the niche in which the door is recessed, 3.90 m. wide, and the height to the crown on the intrados of the arch, 5.20 m., more or less. The same dimensions in width are repeated on the second story, except for the window which has splayed reveals and measures 2.20 m. wide on the outside decreasing to about 1.60 m. on the inside.

The overall proportions of the retable are 1:1. In order to facilitate calculations one may divide the dimensions into modules or units of 3.00 m. each since the retable portion of the façade is approximately 12.00 m. by 12.00 m. Thus the proportions of height to width are 4:4. (FIG. 8.) The lower story as a whole is 4:2⅓ and the upper is 4:1⅔ approximately. The central bay of the lower story is almost 2:2⅓ and 2:1⅔ for the upper. The lower side bays are 1:2⅓ and the upper 1:1⅔. The side bays as a whole are 1:4 and the central bay 2:4. In other words the square proportions of the façade are divided into sections in such a manner as to obscure the equal dimensions of height and width. It is, however, noteworthy that the main door repeats the overall square proportion of the façade and is approximately 1:1, or 3.00 m. wide by 3.30 m. to the spring line of the semicircular arch.

The towers to either side are symmetrical in the lower portions only. Both are 3.20 m. wide. Only the south tower rises to the total height of the retable, 11.70 m. The north tower is but 9.00 m. high, or equivalent to the height of the first story of the south tower. The remaining 2.70 m. represents the height of the belfry, absent on the north tower. Instead of a belfry, a short piece of freestanding masonry about 1.00 m. wide frames the retable on the north. The proportions of the towers are then a total of a little more than 1:3 for the first story, or a little more than 1:4 for the total height. Actually the towers are about 20 cm. wider than the side bays of the retable, hence the modular width of a little more than 1.

The style of the façade is quite Vitruvian, *renacentista*, in its simplicity except for the *ataurique* ornament which covers the plane surfaces of the retable. (FIGS. 79, 81.) The column shafts have simple Tuscan capitals resting on equally simple bases. The lower story begins with a high podium so as to reduce the total height and keep the columns in a proportion adhering to classical prototypes. The podium breaks miters on column centers to form pedestals. The entablature, also breaking miters on column centers, is divided into three main members. The architrave is comprised of a series of narrow superimposed moldings, the frieze a flat band with stucco ornament, and the cornice of a number of superimposed moldings with the usual exaggerated projection so common in Antigua. The upper story repeats the same pattern as below, except that here the podium is a low Roman attic commensurate in proportion to the total height of the story and the columns. (FIG. 80.)

The finial or *remate*, the third horizontal division of the usual three-by-three scheme, is exceedingly simple in design consisting of no more than a segmental pediment crowned with a papal crown on the very summit and framed to either side by a merlon. Though it fits the general scheme of the façade, there is no way of knowing whether the segmental pediment is the original *remate* of 1692 or whether it is an eighteenth- or even a nineteenth-century addition. It is more likely than not postcolonial, for it is improbable, though possible, that this bit of freestanding masonry could have resisted all the earthquakes which plagued the city through the eighteenth century and later.

There are only four niches on the whole façade, one in each of the side bays in the first and second stories. (FIG. 82.) Brick and stucco statues are still *in situ* in three of them. The first-story niches are treated with triangular pediments, while those of the second story have broken segmental pediments ending in volutes at the apexes. A deep niche-window with splayed reveals is located on center of the central second-story bay. One must assume that a statue once occupied the floor of this window. (FIG. 80.)

The only remaining roofed part of the church is the small chapel which projects to the north and is set back from the main façade. The thickness of wall at the towers is reduced to 2.50 m. from the 3.00 m. at the retable. The chapel interior shows no such break and is not set back. Its walling is merely reduced in thickness to allow for the break on the exterior and only in that part where its plan projects beyond the façade. In other words, the inside face of the chapel wall is on a line with the inside face of the façade wall occupying part of the space behind the north tower. It is square in plan and roofed with a dome on pendentives which is pierced with dormerlike windows, one looking to the northeast and another to the northwest. The south tower projects partially beyond the main body of the church, so that the façade as a whole is really a gigantic overlapping screen set in front of the building, bearing no direct relationship to the interior layout.

5. *Compañía de Jesús: Church, Convent, and School, 1698*

(FIGS. 83–87.)

Historical Data

AS early as 1561 the first bishop of Guatemala, Francisco Marroquín, petitioned the crown that the Jesuit order be established in Guatemala. But this was denied.[1] In 1580 the *ayuntamiento* submitted a similar request to the *audiencia*, and still no positive action was taken, nor again when it was repeated two years later.[2] It was not until about forty-six years after the first petition had been made that the Jesuits finally came to Antigua in 1607 and established a convent and school.[3] Their first convent building was probably no more than a thatch hut.[4] A permanent church structure was built for the first time about 1626,[5] possibly by one José de Porras, a master mason of the same name as the one who more than fifty years later worked on the Cathedral.[6] He could not have been the same José Porras, for the latter who died in 1703 was but twenty-six years old in 1665.[7] The Porras family name is associated with the building trade from the sixteenth century on.[8]

An entirely new building was built toward the end of the century and inaugurated in December of 1698.[9] Just when work began is not exactly known. The 1626 building, it seems, was replaced by still another structure built soon after 1646, for in that year the Jesuit order received a bequest of 30,000 *pesos*, most of which was employed in the construction of a church and a school. According to P. Manuel Herrera, reporting to the crown in the mid-eighteenth century, the 1646 building had completely disappeared because of earthquakes and weathering.[10] The 1698 church was in all likelihood built from the ground up after the earthquake of 1695.[11]

In 1717 the new church and school suffered some damage,[12] and as usual Ximénez differs with Arana who claimed that the tower had cracked, the upper third of the *portada* had caved in, and that the school was practically uninhabitable.[13] Ximénez disagrees claiming that there was little damage except for some cracks in the vaults and that the school required but few repairs which had already been carried out. Yet in 1738 help is asked of the *ayuntamiento* to reconstruct the school building.[14] It may very well be that the building was patched up after the 1717 earthquake, as reported by Ximénez, but had deteriorated in the following two decades.

The damage wrought on the conventual buildings and the church in the earthquake of 1751 was quite serious for some of the vaulting caved in.[15] The *ayuntamiento* was asked for the sum of 4,000 *pesos* to carry out the repairs on the school.[16] A year later the rector petitioned again for help for the same purpose, but there is no information if this help

[1] *Efem*, p. 16.
[2] *Ibid.*, pp. 25 ff.
[3] *AGG*, A 1.18 (1740) 5031–211, also *BAGG* **1**, 2 (1935): pp. 141 ff. and **10** (1945): pp. 194 ff.
[4] Villacorta, *Historia*, pp. 229 ff.

[5] Vásquez, **4**: p. 363; and Juarros, **1**: p. 126.
[6] *Efem*, p. 48.
[7] *Ibid.*, p. 130.
[8] See ch. VI, "Brief Notices of Architects," above, for biographical data on both men and others bearing the family name of Porras.
[9] *AGG*, A 1.2.2 (1698) 11.779–1785, also *BAGG* **10**, 3 (1945): p. 193; *AGG*, 1.2.9 (1698) 25.348–2840, also *BAGG* **6**, 3 (1940/41): p. 429; also *Efem*, p. 123.
[10] *AGG*, A 1.18 (1740) 5031–211.
[11] Pardo, *Guía*, p. 100.
[12] Vásquez, **4**: p. 393.
[13] Ximénez, **3**: p. 356.
[14] *AGG*, A 1.10.3 (1738) 31.313–4047.
[15] Juarros, **2**: p. 248.
[16] *Efem*, pp. 204 ff.

was ever accorded.[17] The Jesuits continued to occupy their convent, school, and church until the year 1767 when the royal decree arrived from Spain expelling the order from Guatemala.[18] The buildings were abandoned and left in the care of the Dean of the Cathedral, remaining thus until the earthquake of 1773.[19] Within a few months after the departure of the Jesuits, the *audiencia* recommended that the buildings might serve to house the offices of the *Estanco y Administración de Tabaco* which was never put into effect, perhaps because a scant six years later the whole complex of structures was ruined in the earthquake of 1773 and the city of Antigua itself abandoned.[20]

The ruins which occupy a whole square city block have been much mutilated in the course of time, especially in recent years when the public market, formerly held in the Plaza Mayor, was transferred to the Compañía de Jesús continuing there to this very day. Considerable demolition was necessary to make the area safe, with the result that the convent and school buildings have been considerably altered and rebuilt making it well nigh impossible to make out the original plan. The walls of the church are still standing but the little *plazoleta* in front of the façade is cluttered with the wooden stalls and shacks of the market people.

Architectural Data

THE church façade which faces east is not too well preserved and has only a single massive square tower with but one window flanking it to the north still intact. (FIGS. 83, 84.) There may have been a symmetrical one on the south, but this is not certain because this side has been dismantled almost completely. Above the tower window are the remains of the arcaded belfry which originally had two openings on each exterior side but of which only one is still in place. A chapel was probably once located in the lower part of the tower. There is no corresponding belfry to the south, though there may have been originally.

The central retable part of the façade is divided into the usual three vertical by three horizontal divisions. The third story, the finial or *remate*, is completely gone, but the remains of some volutes or scrolls indicate that it originally consisted of a freestanding wall with a central niche flanked by mixtilinear half pediments. The applied orders and the manner in which they are employed as well as the layout of the niches are all very close to the scheme of the Cathedral façade. (FIG. 42.) Pairs of simple Tuscan columns set rather close together mark off the side bays and also frame the central bay. The door is located in the lower story central bay and a niche-window in the upper. Two niches, one directly above the other, are set between each pair of columns in the side bays. Two niches are also set in the same fashion in the space between the niche-window of the second story and the inside columns to either side, thus duplicating the niche and column arrangement of the Cathedral. (FIG. 84.)

The orders are very simple, almost severe, with a minimum of breaks and miters in the entablatures. A simple broken pediment surmounts the door. Of the total of twelve niches, eight have triangular pediments, two have segmental pediments, and another two are surrounded by very boldly projecting horizontal cornice moldings. Some of the painted decoration is still visible on the surface of the façade which was otherwise unadorned and totally lacking in stucco relief. The best-preserved painting is on the soffit of the arch over the door and the jambs which are protected from the weather. (FIG. 85.) The colors are red, green, and yellow against the white background of the plaster. Continuously intertwined floral motives are used on major molding surfaces, while geometric patterns are employed on the smaller plane surfaces. The walls of the façade and the tower were originally covered with a repeated geometric pattern of lozenges two sides of which are concave and the opposite two convex. (FIG. 83.) It is not certain whether the engaged columns were painted or not, but since no remains of paint are visible now it may, therefore, be assumed that they were undecorated.

The interior of the church is in complete ruin except for the outside walls. (FIGS. 86, 87.) Originally it was divided into three aisles, the central nave being the widest and terminating in an enormous arch at the crossing, which is now missing, corresponding to one still in place and located at the entrance to the *capilla mayor*. Rectilinear in plan, the latter projects beyond the transepts like an apse. The transepts are of the same width as the side aisles and do not project beyond them so that at ground level the plan is not cruciform, though it must have been at clerestory height.

Nothing of the piers or columns which supported the vaulting above the nave and side aisles remains. But vestiges of the pendentives at the springing of the arch at the entrance to the *capilla mayor* indicate that the crossing was once roofed with a cupola of some sort. Remains of pendentives on some of the side walls indicate that like the Cathedral the church was roofed with domes over each side-aisle bay, and probably over each central-nave bay, too. Windows are placed in each of the lunettes formed by the lateral arches of the one exterior side wall, the south on the street.

17 *AGG*, A 1.11 (1752) 2082–97.
18 *Efem*, p. 236.
19 Pardo, *Guía*, p. 100; Díaz, *Romántica ciudad*, p. 38.
20 *Efem*, p. 236.

The north side wall abuts on the convent and school building. As revealed by the bits of pendentives, the vaulting of the central nave as well as that over the transepts must have been the same in height and both somewhat higher than the vaults over the side aisles, allowing the conclusion that the nave was lighted through a clerestory. (FIG. 87.)

Massive pilasters engaged to piers line the side walls and mark off each bay. These piers project into the church space and appear on the outside as equally massive buttresses, while the walls between are considerably thinner. Free-standing piers comparable in style and dimensions no doubt made up the nave arcading. The structure may, therefore, be conceived of as a series of piers on which the roof vaulting was supported, the intervening space between each being filled with screen walling. The pilasters are very simple in design and are surmounted with an angular capital over each. The entablature, especially that in the transepts, projects very markedly and has a dentil course under the cornice. The individual dentils are of gigantic size and more like brackets set at intervals quite some distance from each other, so that two could fit in the space between.

6. *Fountains and Public Water Supply*

FROM the very day of the establishment of the city in the sixteenth century the public water supply system was the concern of the municipal authorities. Yet, not until the seventeenth century was water actually piped to private houses as a normal procedure. Before that time public fountains were the sole source of supply for all the inhabitants. Even today, as in the colonial times, in the outlying neighborhoods the humbler houses still do not have water on tap inside, and it has to be carried from the nearest fountain. All the plazas had such public fountains, as they still do, and in some cases these were equipped with large individual troughs around the basin where laundry could be done.

In 1580 title to the then sole source of water, located near the hermitage of San Juan Gascón, was vested in the *ayuntamiento* and the *Ramo de Aguas* was organized. But not until 1618 was the price of water actually fixed. In the seventeenth century the supply of water had to be augmented at various times since the city had grown considerably. The municipal authorities, therefore, had to seek new sources in order to fill the needs of the growing population.[1]

The sources of water in Antigua are so abundant that to this very day it is customary to allow the public and private water fountains (*pilas*) to run continuously. In colonial times water was not paid for by the quantity consumed, but rather by the *paja*, that is, by a standardized diameter of the pipe supplying the house regardless of the amount used. The *paja* was bought in full at the time of installation. At the end of the eighteenth century the standard *paja* measured half an inch.[2] At the time Antigua was destroyed in 1773 the municipality owned a total of 1,873 *pajas*, each worth 100 *pesos*.[3] Since some of the ecclesiastical establishments had more than one *paja* of water and private houses a fraction of one, it is difficult to ascertain the number of houses which had an interior water supply, although Juarros says that the water was so abundant in Antigua that it was rare that even a humble house did not have three or four *pilas*, water repositories, or fountains.[4] This must be an exaggeration and probably refers only to those houses in the center of town, and not those of the surrounding lower-class neighborhoods. It is certain, however, that the city had many public fountains in the streets and plazas where those inhabitants without water piped to their houses were supplied.

Most of these fountains were frequently destroyed and had to be rebuilt as a result of the many earthquakes. Those not damaged in 1773 fell into disrepair after the destruction of the aqueducts and water mains and the abandonment of the city. The fountain in the Plaza Mayor, though largely restored, is exactly in the same spot where it stood during most of the colonial period. Other fountains such as that in the Alameda del Calvario, La Fuente de las Delicias, the fountain near La Merced, and the fountain of the Dominicans have survived and are discussed below.

i. Pila de la Alameda del Calvario, also known as Pila del Campo, 1679 (FIGS. 88–90.)

Historical Data

The Alameda was completed in 1679 and part of its layout included the fountain (FIG. 88), water for which was supplied from the spring of Santa Ana 2,500 *varas* distant. The estimated cost of the mains to bring the water this distance was given as 3,000 *pesos*. Another thousand for the construction of the fountain[5] was paid for by the *ayuntamiento* to which D. José de Aguilar y Rebolledo, *alcalde ordinario* in 1679, added the sum of 500 *pesos* privately.[6] The water supplied by this fountain was apparently very copious for in 1716 a petition was presented to the *ayuntamiento* re-

[1] Chinchilla Aguilar, *El Ramo de aguas*, p. 19 ff.; *Efem*, pp. 44, 57, 122 ff.; Fuentes y Guzmán, I: p. 134.

[2] *Efem*, pp. 122 ff. The length of the colonial inch, *pulgada*, is not known.

[3] Chinchilla Aguilar, *op. cit.*, p. 21.

[4] Juarros, I: p. 63.

[5] *Efem*, p. 91; *AGG*, A 1.2.2 (1679) 11.776–1782, also *BAGG* **8**, I (1943): p. 36.

[6] Fuentes y Guzmán, I: pp. 134 and 154.

questing that some of it be shared with a nearby neighborhood.[7] In 1731 some repairs were carried out on the fountain itself.[8] It is in an excellent state of preservation and still the major source for the people of the neighborhood where it is located.

Architectural Data

The fountain is constructed of a dark grayish stone and consists of an octagonal pool with a stem and bowl set in the center. The plan of the octagon is based on a square 5.45 to 5.50 m. on each side, the four corners of which were chamfered off to form the eight-sided plan. (FIG. 89.) Those sides on the square measure 2.30 m. while those on the diagonals measure 2.23 m. The total height of the wall of the basin is not visible since the grade level of the roadway has been raised owing to silting as a result of the many floods, both in colonial times and later, caused by the overflowing of the Río Pensativo about a hundred or so yards to the north. The interior depth is 0.90 m. of which only 0.35 m. is visible on the exterior. The step which appears in the plan and the exterior height of the water basin, calculated at 0.70 m., were not actually measured. This would have necessitated digging a pit down to foundation level, permission for which was not forthcoming because the fountain sits in the right-of-way of the street. The step is shown here on the basis of analogy since such a feature is present on the other fountains discussed below. The wall of the basin is 0.40 m. thick, its exterior face scribed with channeling to form a panel on each of the eight sides.

The upright support, 2.30 m. high, rests on a plinth 0.80 m. square and is treated like the stem of a wine glass. The largest dimension in breadth of the different members, all square in section, is 0.56 m. The topmost member of the stem, however, is quite different and is more like an inverted truncated pyramid, flaring from about 0.40 m. at the bottom to about 1.10 m. at the top just under the bowl. On each of the four faces of the inverted pyramid there is carved a full-front head with the face surrounded with what may have been meant to indicate a full wig. The workmanship is very rough in character, and undoubtedly not from the hand of an experienced sculptor.

Immediately above the stem is a wide-flaring shallow bowl 2.40 m. in diameter and 0.40 m. high. (FIG. 90.) The surface of the bowl is carved with eight full-front faces set on center with each side of the pool. Wings flank each head, the tips of which touch those of the adjacent heads to either side. Short pieces of pipe are fixed in the mouths of each by way of a spout.

Set in the center of the bowl is a shorter round stem 0.70 m. high and of quite simple profile. (FIG. 89.) Immediately above is a smaller bowl 0.70 m. in diameter and 0.25 m. high. The surface of the upper bowl is carved with a floral pattern. Four pipes are set at equal intervals around the circumference.

The total height of the fountain from the present grade level is 4.10 m. The original height including the parts hidden by the fill, that is, the step and the lower part of the pool wall, is estimated at 4.60 m. The appearance of the fountain is like a champagne glass or a two-tiered centerpiece for a dining-room table.

ii. Fuente de las Delicias, Seventeenth Century (FIG. 91.)

Historical Data

Just when this fountain was built is not known for certain. It may date from soon after 1620 when the *ayuntamiento* decided to dismantle an older fountain and build a new one of stone. If, indeed, this was the fountain referred to in the document in question, its location was given as being in the middle of the street in front of the main entrance of the convent of La Concepción,[9] then it was probably not built until 1626.[10]

The convent, located in its present site since the late sixteenth century, was completely rebuilt in 1694. An earlier building had existed there in 1620 about the time the fountain was supposedly built. It may very well be that the 1626 fountain was reconstructed again when the convent was rebuilt in 1694. But this is only conjecture.

Architectural Data

The basin which is quatrefoil in plan seems to have been repaired in recent years. (FIG. 91.) It consists of two heavy, superimposed members with half-round profiles, the surfaces of which are covered with a heavy stucco coating. The stem supporting the bowl is composed of members resembling some of those of the Pila de Campo. The bowl, however, is not as wide-flaring as that of the Pila de Campo. But like it, it has a shorter stem and smaller bowl set in the larger, both of which are unadorned.

iii. Fuente de los Dominicos 1618 (?) (FIGS. 92–94.)

Historical Data

This fountain was supposedly located originally in the cloister of the convent of Santo Domingo. It was set up again at some undetermined time at the entrance of the city on the road from Guatemala City where it stood until

[7] *Efem*, p. 144. [8] *Ibid.*, p. 170; and *ASGH* **23** (1948): p. 231.

[9] *Efem*, p. 45.
[10] Díaz, *Romántica ciudad*, p. 24.

shortly after 1951 when it was reassembled in a small plaza in the northeast section of town. Various sources[11] identify this fountain as that from the cloister of Santo Domingo; if these are correct, it was built by the Dominican friar Felix de Mata in 1618.[12]

Architectural Data

The plan of the basin is based on a square from which the semicircular elements like exedra project on each side so that sharp corner angles and semicircular shapes alternate. (Fig. 92.) The basin is approached by a low step of the same plan. The surface of each of the four semicircular projections is decorated with a relief depicting a coat-of-arms of the Dominican order flanked by two dogs sitting on their haunches and with torches in their mouths, according to Gage. (Fig. 93.) He also says they have a globe of the world at their feet.[13] These are now missing, but it is interesting to note that each dog has the inside paw upraised under which a globe could have fitted. There is no indication or markings on the stone surface, however. It may be supposed, therefore, that globes were once attached rather than carved. The rest of the surface is treated with deep sunken bands just above the projecting base and below the projecting cornice to form raised panels which continue around the whole of the fountain in the areas between the coats-of-arms and the dogs.

The stem is a simple unadorned short column shaft with a modified Tuscan capital, the necking band of which is just under the bowl. (Fig. 94.) The bowl itself is fluted on the spherical underside. A flat band above is adorned with eight full-front heads with wings, the tips of which are contiguous with those of the neighbors' to either side. In the mouths of each, a short piece of pipe acts as a spout. The actual rim of the bowl is composed of two simple fillets, the upper one projecting slightly and rounded off on the upper edge. Whether there ever were a smaller stem and bowl above is not known. When last seen none were present. It is not certain also whether the stem and bowl belonged originally to this fountain or not.

iv. Fountain near the Convent of La Merced, Seventeenth Century (Figs. 95–98.)

Historical Data

No literary or documentary data have come to light by means of which this fountain may be identified. It is located in the open area on the south side of La Merced. Stylistically it too may be placed in the seventeenth century, and may be the one formerly known as the "Fuente de los Pescados" which once had a sculptured group of Neptune and his horses done by one Juan Perales which has disappeared.[14]

Architectural Data

The plan of the basin is based on a square with semicircular projections like the Fuente de los Domínicos. (Fig. 95.) But here the projection occupies almost all of each side so that the appearance of the whole is that of a quatrefoil with sharp arrises inserted where each semicircular element meets its neighbor. The basin is set on a circular platform, about 0.40 m. high, which is really a step with the tread varying in depth depending on the space left between its edge and the basin wall. Its narrowest portion, 0.40 m., is located at the projecting semicircles.

The basin is somewhat off-square, the four sides measuring 4.22 m., 4.16 m., 4.02 m., and 4.13 m. respectively. The semicircular projections vary in diameter too, ranging from 2.56 m. to 2.74 m. But all project a uniform 1.00 m. The remaining bits of the straight walls which join at the corners are more like arrises and vary from 0.69 m. to 0.76 m. in length. The greatest overall dimension of the pool is about 6.20 m. The diameter of the circular platform or step is approximately 7.00 m. The basin wall is about 1.75 m. high and about 0.40 m. thick.

The entire surface of the basin wall is carved with reliefs. (Figs. 95–98.) On the walls of the projecting semicircular members a large oval area limited by a rope is flanked by two little nude male figures with perukes. In the space to one side and behind the nude figure a thirteen-leaf rosette is carved. On the short section of straight wall at each corner there is a relief of a mermaid who raises her hand to join that of the mermaid on the other side of the corner.

The stem and bowl may not be the original with the fountain. A photograph taken in the early 1930's shows in its stead a high pedestal with a bust of Bartolomé de las Casas of recent origin.[15] This statue is now located nearby on the south side of the church. The bowl and stem now in place may possibly be from some other fountain. The present stem seems too short in proportion to the bowl, and may represent only a portion of its original height. The fact that a statue of Neptune is said to have been located in this fountain further fortifies the conclusion that the present stem and bowl are not from the same ensemble.

The stem and bowl total a height of 1.35 m. and are rather squat. (Fig. 97.) The stem is about 1.00 m. high and

[11] Castellanos, *Relación sintética*, pp. 74 ff.; Pardo, *Guía*, p. 89; Gage, ch. XIX, p. 339, and ch. XVIII, p. 284.

[12] Ximénez, **2**: p. 224.

[13] Gage, ch. XIX, p. 339.

[14] Pardo, *Guía*, p. 27.

[15] Diáz, *Bellas artes*, p. 428.

is comprised of two principal parts. The lower is 0.70 m. high, square in section, and consists of three members: two tori about 0.70 m. wide on each side and separated by a deeply recessed scotia. The upper part of the stem, also about 0.70 m. square, is quite massive and about 0.30 m. high. On each of its four sides there is a relief of a catlike face carved with deep incisions running laterally from the nose across the cheeks.

The bowl above is 1.60 m. in diameter and about 0.35 m. high. The underside is carved with elongated leaves with heavy raised rims. On the upper portion of the bowl the principal carved decoration consists of four peruked heads with pipe spouts in their mouths. Between the heads there are reliefs of blazons and also some quatrefoil rosettes in the remaining empty spaces.

7. *Espíritu Santo: Hermitage, after 1702*

(FIG. 99.)

Historical Data

THE neighborhood where this little hermitage is located was one of the poorer of Antigua in the seventeenth and eighteenth centuries. The inhabitants were for the most part Indians and a few *ladinos* who were given to drunkenness and were repeatedly embroiled with the law usually on the complaint of their wives whom they kept in rags.[1] Some sort of church already existed there in the seventeenth century which must have been of the simplest sort despite Fuentes y Guzmán's statement that it was vaulted and had an excellent *portada*, though he admits that the furnishings of the sacristy and altars were of the poorest.[2] The existence of the church, which may very well have been no more than a hut in the seventeenth century, is further attested to by Fuentes y Guzmán's story that one of the contending parties in a controversy in the convent of La Merced left that convent and took refuge in Espíritu Santo.[3]

References to this church in the colonial literature are scant indeed. It is mentioned in passing by Vásquez as one of the hermitages existing in Antigua in 1695.[4] Juarros, who is usually replete with information, gives no details except to say that this hermitage along with those of San Jerónimo, San Antón, and Santa Ana was founded in the eighteenth century,[5] a statement at variance with the information given by Vásquez and Fuentes y Guzmán. A specific reference to the church building is recorded in a petition, dated in 1702, on the part of the Indians of the neighborhood who asked that a fourth part of their *tributo*, taxes, be refunded in order to build their church.[6] This may very well have been the first formal structure planned for the hermitage, and explains why Juarros gives the eighteenth century as the founding date. According to González Bustillo, the church of Espíritu Santo was completely destroyed in 1717. This would imply that he was referring to a building built after 1702.[7] If this be true, then a new church was built after 1717 for which the parishioners requested help in repairing after the earthquake of 1751.[8] Though no specific mention has been found as to the fate of the church in the earthquake of 1773, the present ruins of the building are evidence enough that the destruction was total.

1 Fuentes y Guzmán, **1**: pp. 389 ff.
2 *Ibid.*
3 *Ibid.* **3**: p. 170.
4 Vásquez, **4**: p. 385.
5 Juarros, **1**: p. 152.

Architectural Data

NOTHING remains of this church except one small portion of the façade which faces west. (FIG. 99.) Of the interior nothing is visible above ground, and the area once occupied by the building, now private property, is tilled as part of the contiguous cornfields. A very small portion of the central bay and one lateral bay still remain to the height of the first story, enough to reconstruct the first-story layout. The retable-façade proper was divided into three vertical bays. The central one was at least two stories high, above which there was most likely a third story in the form of a *remate*. The second story of the central bay was treated with a steeply pitched triangular pediment in the tympanum of which the remains of an octagonal window is still to be seen. The *remate* above may have been a simple, arcaded belfry, though this is conjectural.

The door itself, now filled with adobe bricks, has a semicircular header. The deep niche in which it is recessed has a rather flat segmental arch set on jambs considerably taller than those of the door, so that a section of wall space is reserved between the soffit of the niche arch and the crown of the door arch. A small niche with ornate framing is set in this space and is flanked with a figure in relief to either side. In the space above the crown of the segmental arch of the door niche and below the entablature of the applied order, three small niches are placed. The central one is square in elevation with a corbeled floor which projects and on which a statue is placed. The other two remaining niches are rather odd-shaped with squat, flat trefoil arch headers.

6 *AGG*, A 1.10.3 (1702) 31.278–4047.
7 González Bustillo, *Extracto*.
8 *ASGH* **34** (1949): p. 377; also *Efem*, p. 205.

The second story of the central bay, as has been stated above, consists of a triangular pediment with an octagonal window occupying almost all the area of the tympanum. Small figures are placed in the corner angles between the horizontal and raking cornices. The rest of this bay is difficult to reconstruct, for all that remains is part of one engaged column set on center with the inside one of the pair of the first story below. Apparently the central bay was flanked by two columns which in turn supported the finial or *remate* above.

The lateral bays are only one story high. Part of the second story of the one extant bay remains and indicates that the upper portions of the lateral bays once consisted of mixtilinear half-pediments which rose from behind low pedestals at the extremes to join the columns flanking the central bay. The applied order in the first story consists of a pair of columns. Small pedestals surmounted by merlons line up on center above the column. But only the outside one exists now.

The columns of the first story rest on a high podium which breaks miters to form pedestals on column centers. The columns are rather stubby and have Tuscan capitals with a twisted-rope pattern with open loops on the necking. The shafts are adorned with a molding which spirals around the shaft three times. The spirals run in opposite diagonals on the two columns, pitching down toward each other and to the niche which occupies most of the lower portion of the area between. The entablature above breaks miters on column centers. The niche between the columns has a flat segmental pediment, while in the space above and just under the entablature an eight-leafed floral pattern is worked in low stucco relief. A headless statue is still *in situ* in the one remaining niche.

The façade is not framed by any wall area to either side, lacking even vestigial towers or buttresses, but stands independently like a freestanding retable. Only the central bay has three horizontal divisions; that is, two stories and a finial. The lateral bays have but two divisions, a single story and a finial. The change in design is logical in view of the small size of the building, for to have employed the usual three-by-three scheme in so small a façade would have resulted in making the scale of the orders almost miniatures. Furthermore, the three-by-three scheme would have been completely impractical in view of the fact that the height of the nave roof was about at the same level as that of the raking cornice of the pediment in the central bay, which in turn would have required a proportionately Tom Thumb-sized door in keeping with the small dimensions of the façade as a whole.

8. *Nuestra Señora de los Dolores del Cerro: Hermitage, 1710*

(FIGS. 100, 101.)

Historical Data

THERE were three hermitages under the advocation Nuestra Señora de los Dolores in Antigua which were distinguished one from another by appending the particular name of the neighborhood where each was located. One was Nuestra Señora de los Dolores "del Manchén," another "del Llano," and the third "del Cerro," all in the northeast sector of the city.[1] Of the first two hardly a trace remains. The latter, Nuestra Señora de los Dolores del Cerro, is the only one of the three of which some remnants still exist. The original colonial street-plan of the neighborhood served by this small church has been obliterated and is now a suburban area with no houses at all. In fact, the church is located in a cornfield. Corn is actually planted inside the nave. All that remains standing today are parts of the main façade, the *capilla mayor*, and bits of the long walls.

The earliest reference to this hermitage is in a *cabildo* dated July 3, 1703, in which Marcos Francisco Ortiz and Manuel Calvo de Alegría speaking for the neighbors of the *barrio* of Santa Inés ask license to build a chapel dedicated to Nuestra Señora de los Dolores.[2] A small chapel was built about this time which served the community for about seven years until a new and enlarged structure took its place in 1710, paid for by the *alférez* Juan de Estrada.[3] The only other reference to this building dates from 1733 and is concerned with a rogative procession which had been organized in the church, indicating that it had no doubt weathered the earthquake of 1717.[4] We have no information that it had been damaged then, or in the earthquake of 1751. The ruins which exist today are testimony enough to what this building suffered in the earthquakes of 1773 and later.

Architectural Data

THE plan of building is very simple indeed, a single-nave church of four bays. The *capilla mayor* is at the north end from which a smaller chapel protrudes. The façade, facing south, consists of a retable without flanking towers. (FIG. 100.) The layout of the façade is the usual three-by-three divisions, the third story being the finial or *remate*. All of the *remate*, except for some parts of volutes, and the entire west

[1] Juarros, I: pp. 152 ff.

[2] *AGG*, A 1.2.2 (1703) 11.780–1786, also *BAGG* 8, 1 (1943): p. 108; *Efem*, p. 130.

[3] Juarros, *loc. cit.*

[4] *Efem*, p. 175.

side bay of the *portada* have disappeared. All that is left standing of the façade are the first two stories of the central bay and the east side bay. The façade is devoid of any surface decoration.

The one lateral bay still extant consists of a podium with pedestals which project and on which the columns rest. Each is treated with an upright volute like a bracket. The columns which frame the niche, an elongated octagonal in outline, have plain shafts with diminution, slight entasis, and Tuscan capitals above. The same type of columns appear on the second story but they are smaller. They too frame a similar niche of elongated octagonal outline. Remains of statues built of brick and plaster are still visible in each niche.

The entablature of the first-story order is very simple. The only decorative feature present is applied to the frieze in the form of pairs of vertical rows of lozenges spaced at equal intervals across the whole façade. The cornice runs unbroken across the façade except over the head of the niche where it describes a very small semicircle. The entablature of the second story is simple too, but here the frieze is treated with triglyphs and metopes.

The dominating feature of the central bay, which projects slightly, is the door which occupies the total width. The door opening itself is semicircular at the head and is set within a deep recess with a similar semicircular header on jambs which are longer than those of the door. The springing of the half-circle niche arch is higher than that of the door resulting in a semicircular space or tympanum between the intrados of the niche arch and the extrados of the door arch, a feature noted on the Cathedral, San Francisco, and elsewhere. (FIGS. 42, 53.) The tympanum is occupied by a small, shallow niche with pairs of scrolls filling the space to either side.

The third story or *remate* is completely missing except for the remains of scrolls over the one existing lateral bay. On the analogy of other existing structures in Antigua one may safely assume that it consisted primarily of an elevated free-standing wall with a central niche from which mixtilinear half-pediments undulated down to the side bays.

Nothing remains of the roofing over the nave. It is interesting to note that the four bays are each separated by a massive pier between which the walls are considerably thinner. This would imply that the piers probably supported transverse arches which crossed the nave and also lateral arches engaged in the side walls between piers, and that a domical system of vaulting was employed. But the foregoing is only a supposition, for the roofing over the nave could just as easily be conceived of as having consisted of wood and tile.

The *capilla mayor* is unique. (FIG. 101.) On the exterior it is square in plan. On the interior the rear corners are filled so that beyond the arch which divides the nave from the *capilla mayor* there are five walls, making a partial octagon in plan. The vault above, however, is a complete octagon. Squinches fill the angles above the diagonal fill in the rear corners. The *capilla mayor* is divided into two parts, the square area adjacent to the nave, and the rear half with the filled or chamfered corners. A smaller chapel, an isosceles trapezoid in plan like the rear half of the *capilla mayor*, protrudes on the north-south axis projecting from the rear wall of the *capilla mayor*.

The arch which divides the nave from the *capilla mayor* is half a hexagon or an isosceles trapezoid in outline. A similar arch divides the square portion of the *capilla mayor* from the smaller trapezoid part behind. Four flat squinches, inverted triangles in elevation with plane surfaces, fill the four corners or haunches between four isosceles trapezoidal arches and thus provide an octagonal plan from which an eight-sided flat dome rises. The soffits of the eight main sections of the dome are each divided into four coffers separated by moldings. Each coffer is a plane surface, thus giving the dome the appearance of a many-sided polygon. Nor are there any spherical surfaces at all in the bay as a whole, for the squinches, the walls, and the dome above are all treated with plane surfaces. Piercing the crown is an oculus surmounted by a lantern with four windows framed by scroll brackets.

Despite the outward polygonal effect, the construction of this dome was carried out by means of curvilinear arches. The trapezoidal arch which faces the nave has lost some of its archivolt revealing a typical flat brick curvilinear relieving arch beneath the surface. The dome was built with flat arches, and the spherical soffit was hidden by the coffers worked in plaster. The interior polygonal effect is only skin deep, and not structural, for the exterior of the dome is spherical.

9. *Palacio Arzobispal, 1711*

(FIGS. 40, 41, 102–104.)

Historical Data

VERY little of the eighteenth-century construction remains of what was the former palace of the archbishop of all Central America. Occupying half of the east side of the Plaza Mayor and immediately adjacent to the Cathedral, this building was altered and occupied in the nineteenth century. (FIGS. 40, 41.) Originally it had been two stories high, but at present it is a one-story structure in which the local post office and some other government offices are housed. Small retail shops are located in the rear half on the street leading from the Plaza Mayor. Parts are also used as private dwellings.

The history of the building goes back to the sixteenth century, long before the Cathedral had been raised to metropolitan rank and its principal prelate from bishop to archbishop in 1743.[1] The present structure was actually built about thirty-two years before that event as a residence for the bishop. The original location of the bishop's palace in the sixteenth century was diagonally opposite the Cathedral on the southeast corner of the Plaza Mayor on a site now occupied by the Capitanía. But the Bishop Francisco Marroquín started a new palace adjacent to the Cathedral sometime before he died in 1563.[2] Just what this new sixteenth-century palace looked like may never be known, for toward the end of the seventeenth century it was rebuilt and extended all the way to the corner fronting not only on the Plaza Mayor but also on the street running in an easterly direction from the plaza as it does today. This alteration or completely new construction was carried out by the Bishop Fr. Andrés de las Navas y Quevedo who occupied the episcopal seat from 1683 to 1702.[3] Fuentes y Guzmán, writing about 1690, must have known the older palace as well as the newer one built by de las Navas y Quevedo, but unfortunately he does not distinguish between the two. He simply states that the episcopal palace adjoins the Cathedral to which it is connected by a passageway, that its main doors face the Plaza Mayor, and that the *lonja* of the Cathedral continues in front of it too.[4]

Further work was done on the building during the office of the Bishop Mauro Larreategui y Colón who in 1711, the year he died, completed the construction begun by his predecessor.[5] Six years later the structure suffered some undisclosed damage in the earthquake of 1717. Ximénez refutes Arana's statement that the episcopal palace was a total ruin and the bishop was afraid to go back.[6] This is the last reference to the building except one from 1760 relating to an order requiring that subterranean conduits be built to carry sewage away from the Palace which had heretofore emptied above ground into the Plaza Mayor.[7] In this respect the Palacio Arzobispal was probably not as comfortable a building to live in as the convent of Las Capuchinas built about 1726 and which was endowed with many amenities including an adequate plumbing system. At any rate, it is quite possible that the Palacio remained more or less unchanged throughout the eighteenth century and that it was not so badly damaged in 1717.

[1] Juarros, 1: p. 168. [2] Pardo, *Guía*, p. 72.

[3] Juarros, 1: p. 204; Pardo, *loc. cit.*; Díaz, *Romántica ciudad*, p. 27; and Villacorta, *Historia*, p. 224. [4] Fuentes y Guzmán, 1: p. 139.

[5] Juarros, 1: p. 205; Villacorta, *loc. cit.*; and *Efem*, p. 138.

[6] Ximénez, **3**: p. 354.

[7] *Efem*, p. 217; and *ASGH* **25** (1951): p. 146.

Architectural Data

THE complete original layout of the building may never be known for only the lower part of the original façade facing the plaza exists today. (FIG. 40.) It was once two stories high as is proven by the existence of the two-story cloister in the interior and a sketch of the Cathedral façade dated 1784.[8] (FIGS. 41, 102.) Only the lower arcade of the cloister and a few bits of the upper story still exist. The walls which enclosed the other side of the corridors are still for the most part two stories in height.

Another very interesting feature is the very heavy cornice which runs around the exterior of the building on both the plaza and street sides. Three very unusual doors with stone jambs and lintels face the plaza and are from the original construction. (FIG. 104.)

The layout of the cloister comprises an open square area surrounded by four arcaded corridors. The only access to the cloister today is from the side street and through a wide vestibule, *zaguán*, the floor of which is slightly inclined like a ramp up from the street to the higher level of the cloister. Whether there once existed an entrance from the Plaza Mayor it is not possible to tell because of the modern alterations of the building.

The columns of the cloister arcade are most unlike those of the cloisters of San Francisco, Escuela de Cristo, La Merced, Las Capuchinas, the Ayuntamiento, and the Capitanía, all of which are uniformly of squat and massive proportions, at most two and one-half diameters high. (FIG. 102.) These are slender by comparison, about four diameters high. The shafts are set on low, square plinths and are surmounted by simple and unadorned Tuscan capitals consisting of an astragal, a necking band, a small echinus, and a two-member abacus. (FIG. 103.) The arches which spring from the abaci are described on three centers, that is, secants of a large circle for the crowns and two smaller circles for the haunches. The extradoi of the arches are left completely unmarked by any moldings. An entablature set but a short distance above the crowns of the arches runs around the whole of the cloister. Its profiles are uncomplicated. The architrave consists of a series of narrow moldings of but slight projection and the frieze is a simple flat band. The cornice above alone provides the only pronounced projection. It is made up of a flat band with slightly stepped-out superimposed fillets,

[8] *AGG*, A 1.10.3 (1784) 15.091–2123.

then a large flat geison above which another band of fillets supports a smaller geison, and finally above that completing the ensemble, a cyma reversa with a narrow fillet on top. A square pier with half-column responds stands in the corners of the cloister and takes up the thrust of each arcade.

Of the second-story arcading only one corner pier with its responds remains. (FIG. 102.) Judging by the shape of this pier, one may safely conclude that the columns were different from those below, probably octagonal in section like those of La Merced, but of much slimmer proportions.

Remains of painted decoration still preserving some of the red color is still in evidence on the shafts of the columns, on the wall surfaces of the arcade, and walls of the cloister. (FIG. 103.) The patterns were apparently applied with a stencil and painted while the plaster was still damp. The principal patterns are fleur-de-lys as well as individual petals arranged in a zigzag along the borders. Checkerboard patterns with the red and white squares laid out diagonally were also employed on flat walls. In general, the decoration is not unlike that on the façade of the Compañía de Jesús. (FIG. 83.)

The columns were built of flat bricks cut to conform to the circular cross section of the shafts. These were then covered with a rough mortar of lime and sand as well as a finishing coat of pure lime plaster on which the painted decoration was applied. The walls of the building were built in the typical *antigüeño* fashion, namely, string courses of brick separating larger sections of rough stone and boulders laid in thick mortar.

Facing the plaza are three doors with stone jambs and lintels of rather unusual design and execution. (FIG. 104.) In plan the jambs are set at a slight angle in from the wall face so that the lintel overhead is slightly recessed giving the whole a concave appearance. The jambs are treated as pilasters with high bases and shafts. These are worked with two elongated panels or flutes and surmounted by squared Tuscan capitals carved from the same block as the top section of the shaft. The lintels are really straight arches built of a keystone and two blocks to either side, making a total of five voussoirs. The outer face of each lintel is recessed with a flute or panel repeating the motif of the jambs with but a small variation. Running through the center of the panel and cutting through the moldings at either end is a raised band or fillet which divides the recess in two. The last voussoirs to either side, those directly over the jambs, are set in the same plane as the jambs making an obtuse angle with the rest of the lintel thus resulting in a concave effect. The part of the lintel directly over the door opening is straight as is the threshold below.

A heavy projecting cornice runs around the plaza and street sides of the building. Its projection is great enough so that it may have possibly served as a balcony and might have been adorned with an iron rail. Directly underneath the cornice proper an entablature worked in plaster also runs along the wall. It consists of a two-fasciae architrave, a frieze which is unadorned except for some vertical scroll brackets or consoles set at regular intervals directly under each of the very heavy widely spaced dentils of the cornice. Running in the spaces between the consoles and seemingly broken by them, there is a heavy molding made up of a number of superimposed half-round members which project each beyond the one below as they ascend. The dentil course immediately above is laid out so that the dentils are widely spaced at regular intervals about three times the width of each. The cornice or balcony itself projects far beyond the dentil course. It consists of a narrow band of three fasciae each projecting beyond the one immediately below. Above is a large, widely projecting cyma reversa and a fillet. A low parapet wall set back in the same plane as the wall below runs along the top of the present building, and may very likely be modern having been added at some undisclosed time after this building was put in use again in the nineteenth century. This parapet wall may in fact represent part of the original walling of the second story which has completely disappeared.

10. *La Recolección: Church and Convent, 1717*

(FIGS. 105–109.)

Historical Data

THE official title of this monastic establishment was "Colegio de Misioneros de Cristo Crucificado," though it was also referred to as "Convento del Colegio de Cristo de Recolectos de Nuestro Padre San Francisco."[1] The earliest notices of the establishment of this order in Guatemala appear in the last decade of the seventeenth century when Fuentes y Guzmán, as recorded in a *cabildo* dated December 14, 1691, proposed that a letter be sent to the "Comisario del Orden del Señor San Francisco en México de la Nueva España" asking that a mission be established in the Calvario.[2] It is possible, however, that some monks of this order were already present in Guatemala by 1667, for, according to the early eighteenth-century Franciscan chronicler Vásquez, they had established themselves in a house near Santa Ana on an informal conventual basis, since royal license to found

[1] *AGG*, A 1.18 (1740) 5027–211, also *BAGG* **1**, 1 (1936): pp. 139 ff.; and Ximénez, **3**: pp. 159 ff.

[2] *AGG*, A 1.2–2 (1691) 11.778–1784, also *BAGG* **8**, 1 (1943): pp. 62 ff.

a convent had not yet been granted.[3] In another hand and as an addition to the Vásquez manuscript, it is noted that the Recolectos had been definitely in Guatemala by 1683.[4] At any rate, Fuentes' proposal of 1691 was not immediately accepted. The municipal authorities were rather less enthusiastic about establishing another monastic order in Antigua, pleading that times were hard in Guatemala and that there were no means for supporting the missionaries, and further suggesting that the governor, that is, the Capitan General, himself be consulted about this matter. Many letters passed back and forth between the various ecclesiastical and secular authorities for the next three years. The *ayuntamiento* was generally against the founding of any more conventual establishments which had no visible means of support and which would, therefore, have to depend on the local citizens who would have to bear the burden of the expense involved.[5] The movement to found the order did not abate despite the refusal of the *ayuntamiento* to agree, and in 1695 the matter was again brought up, this time by one Fray Melchor López from the convent in Querétaro, who asked for permission to establish a convent in the church of El Calvario.[6] The *ayuntamiento* was still opposed to this move and continued to be even as late as 1699.[7] But to no avail, for in 1700 an *auto* was signed by the crown and presented to the *audiencia* in May of 1701 ordering that the monks not be impeded from founding their convent.[8] The patent from the mother convent in Querétaro, Mexico, arrived in April, 1701, and that same year title to two parcels of land was given the Recolectos on which to build their convent.[9] The date of 1698 when the royal license arrived given by Ximénez is at variance with the documentary evidence.[10]

Within less than two months after the patent arrived from Mexico, the *ayuntamiento* granted the monks permission to build their church and convent in the *barrio* of San Jerónimo where the parcels of land donated to them were located.[11] The cornerstone for the church was laid on the eighth of September of that same year.[12] In the meantime a temporary structure of thatch served as a church while the permanent one was in construction.[13] The *ayuntamiento*, completely defeated in its opposition, reversed its position once the crown intervened and agreed to cooperate with the monks and take part in the procession and ceremony of placing the altar in the temporary church structure on the seventh of June. About two weeks later, on June 20, the *ayuntamiento* assigned a water allotment to the order, and in November further demonstrated its good will by interceding on behalf of one of the monks, pleading that he not be sent to Mexico since he was engaged in planning the conventual buildings and church.[14]

Actual construction probably began in 1703 when an adjacent parcel of land seventy *varas* wide was sold or donated to the order by one Agustina Ramona, probably an Indian woman, in return for masses. The land was specifically for the church building.[15] This date is repeated by a mid-nineteenth century author who states that Diego de Porras, *maestro de arquitectura*, was in charge of the building operation. The supply of bricks for the building works was contracted for with one Miguel Antón, an Indian from El Tejar, and the carpentry contract given to Ignacio de Cárcamo and Tomás de Santa Cruz.[16]

The buildings were inaugurated in 1717, the very year the city was laid waste by an earthquake. The church was damaged to the extent that the bell tower, one bay of the church, and half the cupola came down also destroying the crossing.[17] The cupola was repaired almost immediately,[18] but the *portada* of the convent was in such bad condition that the top story had to be dismantled.[19] Later on, in 1730, a clock built by the curate Juan de Padilla, a teacher of mathematics, was installed in one of the church towers.[20] The establishment was again damaged in the 1751 earthquake, but to what extent is not known.[21] The earthquake of 1773 left both the convent and church in total ruins. Of the church *portada*, only one half-circle arched door opening remains. (FIG. 107.) Parts of the side walls and a single arch, that which separated the nave from the *capilla mayor*, are still in evidence, but none of the vaulting except bits here and there remain. The conventual buildings are in a state of complete ruin, but some reconstruction work has been done and parts are now used as a private dwelling.

[3] Vásquez, **4**: p. 303.
[4] *Ibid.*, p. 365.
[5] See fn. 2, above; *Efem*, p. 115.
[6] *AGG*, A 1.12 (1695) 16.795–2294, also *BAGG* **5**, 2 (1940): pp. 158 ff.; *Efem*, p. 117.
[7] *Efem*, p. 124.
[8] See fn. 6, above.
[9] *Efem*, pp. 126 and 127.
[10] Ximénez, **3**: pp. 159 ff.
[11] *AGG*, A 1.2–2 (1701) 11.780–1786, also *BAGG* **8**, 1 (1943): pp. 62 ff.; *Efem*, p. 127.
[12] *Efem*, p. 128.
[13] Vásquez, **4**: p. 365, the part in another hand.
[14] *AGG*, A 1.2–2 (1701) 11.780–1786, also *BAGG* **8**, 1 (1943): pp. 101 and 102; *Efem*, p. 128.
[15] *AGG*, A 1.20–(1703) 738–24 vuelto.
[16] García Peláez, **3**: p. 27.
[17] Vásquez, **4**: p. 393.
[18] Ximénez, **3**: pp. 159 and 355.
[19] Vásquez, *loc. cit.*
[20] *Efem*, p. 167.
[21] *Ibid.*, pp. 204, 205.

Architectural Data

THE church interior was divided into a central nave and two side aisles, each with a door on the main façade. (FIG. 105.) The transepts do not project beyond the side aisles. The exact dimensions and details of the interior layout are difficult to ascertain because all the nave piers are missing and the church is still filled with debris. (FIG. 106.) The width of the nave, however, was more or less equal to the distance between the two piers which divide the crossing from the *capilla mayor* and on which a single half-circle arch, one of the original four which once supported the cupola, still rests. (FIG. 108.) The crossing was probably not square in plan, a conclusion reached by locating the other two missing piers of crossing on a line with the pilasters still visible in the side aisles. If this was actually the case, one may further conclude that the crossing was roofed with a dome of elliptical plan. The side-aisle bays are approximately square. The first side-aisle bays, in the choir, and the corresponding nave bays are markedly oblong. It may be assumed, therefore, that the vaulting over all the church was of the modified *bóveda vaída* type, that is half-ellipsoid or "half-watermelon"-shaped pendentive domes forced into oblong bays. This conclusion is borne out by the remains of the vaulting in the transepts which are of this type. (FIGS. 108, 109.)

The *capilla mayor* is approximately square in plan and projects beyond the nave in a manner reminiscent of the Compañía de Jesús. (FIGS. 105, 108, 86.) But here to either side there is a narrow, elongated room entered from the transepts as well as from the *capilla mayor*. It is not possible to tell whether the floor of the *capilla mayor* was raised above the nave level because of the debris which still fills the whole area. A long corridorlike room runs across the entire width of the church in back of the *capilla mayor* and the rooms to either side.

The plan of the convent was only partly obliterated by the ruin of 1773, and to some extent again by recent reconstructions and alterations carried out in order to accommodate the buildings for private use. The outstanding feature of the convent is that it has two cloisters and possibly a third. The plan published by Villacorta is an acceptable reconstruction though no details exist on the basis of which the exact dimensions of piers and the width of the corridors which surrounded the open areas of the cloisters might be determined.[22] The cloisters were originally two stories high to judge by the remains of four staircases.

Some of the architectural details such as the pilasters and the entablatures of the church interior and the one façade door still *in situ* are interesting. (FIG. 107.) The capitals are simple unadorned Tuscan type, except that the necking is pulvinated rather than a plane surface. The door header is a simple half-circle arch, the surface of the archivolt of which is broken by a series of narrow superimposed projecting moldings.

The pilasters engaged to the side walls on the interior of the church are quite similar to those in El Carmen and Santa Clara of later date. (FIGS. 108, 119, 142.) Each pilaster rests on a rather high two-fasciae plinth surmounted by a squared Attic base. The shafts with deeply inset flutes are almost duplicated in the two buildings mentioned above. The flute itself is divided by a sharp arris which rises to the outer plane of the pilaster face. The capital is the squared Tuscan type, but with a pulvinated necking.

The entablature which once ran around the whole interior of the church gives a restless staccato effect and has its counterpart in the later church of La Concepción. (FIG. 109.) The architrave is composed of six narrow superimposed moldings with concave and convex profiles, the total effect being that of horizontal lines or depressions and projections.

The frieze itself is also a series of concave and convex moldings. But here the horizontal continuous line is interrupted by volute brackets or consoles billowing at regular intervals and supporting the dentils of the cornice. The dentils are also treated as smaller consoles made up of two courses of corbels looking like a cyma reversa with a small scroll below where it meets the frieze bracket. The spaces between the main dentils are taken up by smaller dentils treated with vertical striations which give a rippling effect interrupted by the presence of another still smaller bracket of rather low projection whose profile is also a cyma reversa, the top and bottom ends of which terminate in tight little volutes. The frieze and cornice brackets may be considered as two parts of a larger single unit supporting the widely projecting and plain geison which is topped by an even more projecting but narrower superimposed series of crown moldings. The total effect of the entablature is an utter cacophony of restless surfaces.

[22] This plan was drawn shortly before 1942, at which time it is quite possible that the debris from the fallen cloisters had not yet been cleared, and the emplacement of the piers was still visible. The church plan appears on the same drawing. The general arrangement of the nave piers may be accepted as shown, but all dimensions are, of course, purely conjectural since the church is still piled high with debris more than three meters deep in many places. See figs. 105, 106, 107.

XIV THE THIRD PERIOD: 1717–1751

THE use of nonarchitectonic pilasters on retable-façades is the most remarkable innovation of the third period. (FIG. 2.) The corrugated and candelabra types are the most common. In general, the *antigüeño* nonarchitectonic pilasters, which with some reservation may be considered *estípites*, are comparable to those employed on Hispanic retables. They are designed and executed in a manner somewhat akin to that of the cabinetmakers' craft displaying the same freedom of conception possible in wood turning. In Antigua, however, these nonarchitectural, invented, or free forms follow an independent development due, to a large extent, to the very nature of the different craft and materials in which these *estípites* were carried out, plaster over brick cores. The earliest invented or "free-form" pilaster, the corrugated type, is seen on the gatehouse of El Calvario. It is composed of superimposed squared Attic bases. A variation, looking more like superimposed pots, shaped somewhat like squat amphoras appears in a single example on La Candelaria. Another unique form, employed on the façade of Santa Cruz, is the multiple colonnette incised with a single, deep concave flute. Another type, the pilaster shaft composed of multiple strips or ribbons, appears for the first time on San José. Candelabrum- or baluster-shaped pilaster shafts are employed on Santa Clara and Escuela de Cristo. But the most commonly employed type of all is that first seen on El Calvario, namely, the corrugated pilaster.

In this period the exuberant tone of the retable-façade is not restricted to the invented forms of the applied architectural orders, this same spirit also pervades the design of door and niche openings where there is a tendency to use mixtilinear headers in lieu of the simple semicircular arch so common in the preceding two periods. Likewise, a variety of broken entablatures, including pulvinated friezes, appears in this period. Pediments also become quite ornate and tend to deviate from simple geometric forms in favor of an unrestrained combination of both rectilinear and curvilinear outlines. Ogee-shaped arches are employed for the doors of La Candelaria and El Carmen. The complicated and involved profiles of the soffit of door headers are carried down to the jambs. These are no longer simply rectilinear in plan with the face and reveal of the jamb meeting at right angles. Instead, the splayed jamb is made up of a series of moldings which undulate back to the door opening.

But the simple door recess with a semicircular header is not altogether abandoned. It is still preferred for those churches where the rest of the façade gains its exuberant effect from other elements such as *ataurique* decoration and nonarchitectonic pilasters.

Some variations are also to be noted in the arrangement of the retable-façade; for example, on El Carmen where, in lieu of niches with statues between columns in the side bays, an extra pair of columns is set between and in the front of the main pair of the bay. This makes possible a greater freedom in the mitered treatment of the entablatures and pediments above. In addition, the surface of the columns and other areas of the façade are covered with floral ornaments in plaster. On the churches of La Candelaria and Santa Cruz this type of workmanship reaches a quality unequaled in any other period.

The façades of the church of El Calvario (which stands directly behind the gatehouse), Escuela de Cristo, and San José el Viejo do not adhere strictly to the usual three-by-three scheme of the retable-façade. That of El Calvario has two diminutive towers which are lower than the central portion of the façade. The façade design, however, is dominated by the door set in a large trumpet-shaped niche, the type known as *abocinado*. The soffit is treated with elongated petals, or it may be conceived of as fluted like a seashell. This is the only example of this type of door in all Antigua. *Abocinado* headers are more frequently employed for niche-windows. The proportions and the layout of the façade of San José el Viejo mark a departure from the usual three-by-three scheme, for the side bays do not have second stories. Furthermore, the towers are designed to give the façade the appearance of having double side bays, thus stressing the low horizontal proportions of the façade as a whole. The façade of the Escuela de Cristo, much remodeled in postcolonial times, has side bays three stories high rather than the usual two-story arrangement. In general, however, the three-by-three scheme is still adhered to especially in those buildings of older date which were remodeled in this period, as for example Santa Cruz. The conventual churches of Santa Clara and La Concepción are oriented so that one long side faces the street. In the case of the latter, the façade of the choir end is rather plain; but not that of Santa Clara which, though lacking a door, is still treated as a retable-façade.

The interior treatment of churches is quite ornate in this period. The bracket type of frieze first employed on the interior of San Francisco and later in La Recolección is further developed to produce an even stronger staccato effect in the case of La Concepción. And in El Carmen, a rippling cornice is added to the bracket frieze to achieve an even greater display of motion, both staccato and undulating. A complicated type of pilaster fluting is introduced in the gatehouse of El Calvario. The same fluting is employed on the terminal piers of the arcades of the Ayuntamiento, on the pilaster shafts of the church of Santa Clara, and elsewhere in the succeeding period. (FIG. 10.)

There are some definite improvements in construction technique in this period when, as in the cases of El Carmen, San José, Santa Clara, and others, all built for the first time in this period, the walls are reduced to minimal importance structurally. These serve rather as screens between the system of massive piers which emerge as buttresses on the exterior and appear as pilasters on the interior. The half-ellipsoid pendentive dome is the typical type of roofing in the period, even being added to older structures which had previously been roofed with wood and tile.

Stone veneer is also employed for the first time to cover walls of brick. The stone exteriors of Santa Clara, Escuela de Cristo, and the Ayuntamiento are only skin deep. In the case of the Ayuntamiento, the veneer is made to appear like regular coursed ashlar masonry by means of false joints in white mortar. These have no direct relation to the joints between the random-sized slabs of stone of which the veneer is constructed.

A change in proportions of arcade columns is noted in this period, possibly a conscious attempt to make the buildings somewhat more earthquake-resistant. But this is not certain since massive piers were employed in the cloisters of La Merced and San Francisco in the preceding period, and may represent an aesthetic preference equally as well as a structural one. The relatively slender columns of the cloister of the Palacio Arzobispal are abandoned in favor of those of rather squat proportions, almost 1:2, as for example, the Tuscan columns in the cloister of Las Capuchinas, on the arcaded façade of the Ayuntamiento, and the cloister of Santa Teresa. The simple Tuscan column is not the favorite, for others of more complicated shape appear, as for example, those in the cloister of Escuela de Cristo. (FIG. 11.)

In this period the construction of true pendentive domes over square bays is attempted for the first time in the cloister of the convent of Las Capuchinas. Here some improvisation was necessary because the geometric principles involved were not fully understood by the builders. However, a few years later the arcaded façade of the Ayuntamiento and the cloister arcades of the Escuela de Cristo were constructed true to geometric form.

The ebullient imaginative creativity of the period is perhaps best characterized in the last monument, the church of Santa Rosa de Lima, built about 1750. The façade of this church is, without doubt, the most ornate in all Antigua and on which motifs derived or even copied directly from Santa Cruz, San Francisco, the Cathedral, Compañía de Jesús, and the Pila de Campo are employed.

The entablature with its pulvinated frieze and atlantids is a direct imitation of that seen on Santa Cruz. The abaci of the pilaster capitals are exactly like those used on the pot-form pilasters of La Candelaria. The treatment of the second-story entablature with its broken pediments is directly derived from that of Santa Cruz. The niches of this same story have ogee headers like those on doors of La Candelaria and El Carmen. The finial with its little niche is treated with the same type of multiple-strip or ribbon pilasters as used on San José. The central retable is flanked by very narrow pilasters like piers and lacks the twin towers, an arrangement exactly like that seen on the façade of Santa Clara. The side bays are framed with pilasters which are but a slight variation of the corrugated type first seen on El Calvario. The niche treatment to either side of the window of the second-story central bay is like that employed on so many churches of the second period, especially the Cathed al. All in all, the various decorative details of this church seem to recapitulate the ebullient style of the third period.

1. *El Calvario: Church and Gatehouse, 1720*

(FIGS. 9, 10, 110–115.)

THE gatehouse with the church behind stands at the end of the *Alameda del Calvario*, a long tunnel-like avenue with poplar trees lining each side. (FIG. 88.) This avenue is sometimes referred to either as *La Calle de los Pasos* or *Via Crucis*. The church and gatehouse were all part of a scheme for the Stations of the Cross during the Easter processions. The first station was located at the convent of San Francisco and the last inside the courtyard of El Calvario just beyond the gatehouse. During most of the colonial period there was a small permanent chapel at each station. They have all disappeared in the course of time, the ones existing today having been built in the 1940's.

Historical Data

THE site for the hermitage of El Calvario was donated to the Third Order of San Francisco in 1618 at which time a stone cross was erected there in token of possession.[1] (FIG. 110.) No formal structure was built until about thirty-seven years later, in 1655. This building stood until it was totally ruined in the earthquake of 1717. Through the beneficence of Francisco Rodríguez de Rivas, president of the Real Audiencia, a new building was constructed almost immediately and was inaugurated February 11, 1720.[2]

Fuentes y Guzmán describes the seventeenth-century gate building and church, giving an idea of their architectural character. One entered the area of the Calvario through a structure formed by two vaults directly into a garden where three chapels were located. To the west inside there was a *tránsito*, probably meaning the last station of the *Via Crucis*, consisting of a domical vault supported on four columns inside of which was an image of the crucifixion. This domed structure was attached to the church which lay to the south of the entrance gate. The church was oriented north-south and roofed with an *artesón* except for the *capilla mayor* which was covered with a barrel vault. To the west of the presbytery was a sacristy.[3]

According to Vásquez, construction of the gatehouse began in 1619 when the trenches for the foundation of the stone cross were dug by the monks themselves since they did not have the means to hire day laborers. In that same year the four piers of the domed structure mentioned by Fuentes y Guzmán were built. The dome itself was not built right away, but some rough timber beams were laid across the span between the piers and the whole was covered with thatch in order to protect the sacred crucifix. The dome must also have been built by 1655 when the Calvario was completed.[4]

The entrance structure is described in greater detail by Vásquez who says the lower story of the façade consisted of two equal-sized doors with semicircular arches springing from a single central pier. (FIG. 9.) The second story repeated the form of the lower, but in the center there was a *tribuna* with some sculptures (probably a niche with a projecting pedestal) located over the central pier and flanked by two half-circle windows on center with the doors below. There was also a third story, a finial, above which were some merlons. The whole building was plastered with lime mortar.[5]

FIG. 9. El Calvario, 1619. Gatehouse as described by Vásquez.

Enough information is derived from these contemporary accounts to visualize a symmetrical scheme in which the façade wall is pierced by four openings and surmounted with a finial in the form of a pediment with merlons at the sides. The pediment may have been either triangular or segmental, hardly mixitilinear at so early a date. The lower two openings were no doubt the vaults, probably the barrel vaults referred to by Fuentes y Guzmán. The upper story with smaller arched openings and the niche with sculpture was a screen wall and possibly served as an arcaded belfry, *espadaña*, with bells.

Only in 1647 was the decision finally reached to build a formal church in the Calvario. The members of the Third Order of San Francisco personally engaged in manual labor on the project, cutting wood and making adobes, thereby setting an example for many noble gentlemen of the city who, thereupon, also carted lumber, adobes, and roof tiles and even handled shovels and dug trenches for the foundations. But not much was actually done until 1652 when

[1] *AGG*, A 1.20 (1668) 1480–14 vuelto, also *BAGG* **10**, 3 (1945): p. 202.

[2] Juarros, **1**: p. 149, 190; *AGG*, A 1.2–5 (1719) 15.776–2207–71, also *BAGG* **8**, 1 (1943): p. 126; and Ximénez, **3**: p. 357.

[3] Fuentes y Guzmán, **1**: pp. 152 ff.

[4] Vásquez, **4**: pp. 422 ff.

[5] *Ibid.*, p. 425.

construction in earnest began, and the work was finally completed in 1655.[6]

The 1717 earthquake laid waste the whole of the Calvario destroying both the church and the gatehouse.[7] Both buildings standing today, therefore, date from after 1717. To judge by the materials employed in the seventeenth-century construction, mainly wood and adobe, the older structures must have been far simpler than those existing today. Work on the new buildings did not begin all at once and was not yet finished in 1719.[8] Exactly a year later the building was completed and opened for services, the cost of the reconstruction borne by the president of the *audiencia*.[9] Just who was responsible for the design of the new gatehouse and church is not known. However, one Diego de Medina, *maestro de obras*, requested a recommendation from the *ayuntamiento* attesting to the fact that he had in part directed the construction of the hermitage of El Calvario.[10] There is no indication, however, that he drew the plans, nor is it known who did, or whether any were drawn at all.

The gatehouse probably suffered very little damage in the earthquake of 1773, for it does not show any indication of alteration or repairs, except painting and minor touching up of the plaster. It may be safely concluded that it remains today in the original form as it was when inaugurated in 1720. This is not so in the case of the church which was badly thrown off plumb resulting in many breaks in the walling so that the whole was on the verge of collapsing. The little domed structure, the one described by Fuentes y Guzmán as a *tránsito* with a crucifix, did not escape damages either, for part of the dome was dislodged in the later tremors of December, 1773. Inside the church the damage was reported to have been more widespread.[11]

Architectural Data

i. Stone Cross

There is no way of determining whether the cross erected in 1619 is the one which now stands in the unpaved area in front of the gatehouse lined up with the center door, and rests on a stone foundation visible at ground level. (FIG. 110.) The vertical member of the cross is set on a large ball which in turn rests on a circular base made up of two molding profiles. It is topped by a small rectangle inscribed with the letters *INRI*.

ii. Gatehouse (FIG. 111.)

The building site slopes from north to south. The *lonja* which runs across the front of the façade is one step up. Another step up is the actual threshold of the door from which one descends four steps to the interior floor level. At the south side of the building, a final step down leads to the courtyard area. One goes up about 0.40 m., then down 0.80 m. to the interior, and finally down another 0.20 m. on leaving the building, or a total fall of about 0.60 m. The riser of the *lonja* is hidden to about half its actual height by an accumulation of fill in the open area in front of the gatehouse.

The building measures 16.10 m. wide by 7.45 m. deep. The *lonja* in front, extending slightly beyond the ends of the building, is 16.50 m. wide by 1.47 m. deep measured to the face of the pedestals of the pilasters. The plan is a simple rectangle with massive lateral walls to east and west. Attached to the west wall is a small projection like a tower with a circular stair giving access to the roof. The main façade is pierced by three openings in which wooden doors are hung. The opposite south wall is treated as an open arcade of three semicircular arches. (FIGS. 110, 112.) These are set on massive piers carried up the extra height necessary because of the slope. The arches are at the same level as those of the doors in front. Transverse semicircular arches spring from these piers to the inside face of the front wall thus dividing the interior into three bays. (FIG. 114.) These arches provide the support for the wood framing of the flat ceiling and roof above. The fluting of the pilaster shafts on the inside faces of the piers from which the transverse arches spring is noteworthy, for it sets a pattern followed later in the eighteenth century. (FIG. 10.) On the inside of the north wall each fillet is accented by a small astragal, so that in section the concave flute has a smaller convex counterpart emerging from the fillet, and best described with a drawing. The jambs of the piers of the south side have even more elaborate fluting which is set below the plane of the face of the pier. It consists of a flat fillet, a small astragal, another fillet, then a convex profile ending in an arris from which a concave depression goes back to join a two-stepped fillet, then an astragal, and the same scheme repeated in reverse order.

The walls are not of uniform thickness. (FIG. 111.) The west wall is 1.50 m. thick north of the stair tower, but reduces to about 1.30 m. behind. The east wall is 1.60 m. for a

[6] *Ibid.*, p. 429. The church had been in use forty years at the time he was writing in 1695.

[7] A later hand adds a note to Vásquez' manuscript, "Adición, Capítulo Noveno, Libro 5, Tratado 5," **4**: pp. 429 ff., saying that the ruin of the church was total except for the vault of the presbytery and that over the sacred image of Christ, meaning perhaps Fuentes y Guzmán's *tránsito*. But both were still very badly damaged.

[8] *Isagoge*, p. 405; *Efem*, p. 149; *AGG*, A 1.2–5 (1719) 15.776–2207–71, also *BAGG* **8**, 1 (1943): p. 126.

[9] *AGG*, A 1.10.3 (1720) 31.297–4047; *Efem*, p. 150; Juarros, *loc. cit.*

[10] *Efem*, *loc. cit.*

[11] González Bustillo, *Razón particular*.

FIG. 10. Fluting. Top, El Calvario, south piers. Center, El Calvario, pilasters on north wall. Bottom, Escuela de Cristo, nave pilasters.

short distance and then reduces by means of an offset to 1.50 m. The caracole stair is 1.25 m. in diameter. Half the stair cuts into the west wall and the other half is located in the projecting stair tower. It projects 1.40 m. from the face of the west wall and is 3.45 m. long.

The height of the façade to the top of the belfries is approximately 8.00 m. It is divided into two stories and three bays with pilasters framing the door and belfry openings. Each of the three doors is set within a deep niche with a semicircular header concentric with that of the door headers. (FIG. 113.) The width of the doors varies from one bay to another, the center one being 2.32 m. and the widest of the three, while the east and west doors are each approximately 2.18 m. wide. A total of four pilasters, each 4.00 m. high, support an architrave made up of a number of superimposed moldings of varying profiles, but of very small projection.

Immediately above the architrave in each bay there is a trapezoidal pediment, the height of which is equal to the height of the niche arches from the springing to the crown, or about 1.10 m. A heavy swagged molding, actually the cornice of the order, sweeps up diagonally to join a shorter horizontal section of molding to outline the pediment. The cornice molding which marks off the top of the first story continues around to the rear of the building and is the only member above the arches of the open arcade on the south side. (FIG. 112.)

A total of six pilasters is employed on the second story. Two frame each opening and are surmounted by segmental pediments with merlons set on pilaster centers. Mixtilinear half-pediments with scroll profiles sweep down from the sides of each belfry. (FIG. 113.) The width of each belfry, without the flanking mixtilinear half-pediments, is about 2.20 m., or just about the width of the door openings below. The half pediments fill the space between the freestanding belfries giving the whole of the second story a crenelated effect.

The proportions of height to width of the façade are rather squat, as are those of each individual bay. The width of the building is 16.10 m., while its total height to the top of the belfry pediments is 8.00 m., or a proportion of 2:1. The width of the individual bays of the first story measured on pilaster centers is about 4.70 m. The height of each pilaster is 4.00 m., and that of the architrave above 0.50 m., making a total height of 4.50 m. or just a little less than the width of the bay, or a proportion of about 1:1. Each lower-story bay thus forms a square framed by two pilasters and an architrave. In a like manner the belfries define smaller squares aligning with the door openings below. Added height and width are gained in the belfries by the addition of the segmental pediments above and the mixtilinear half-pediments to either side of each.

The four large pilasters of the first story, the corrugated type employed on other later eighteenth-century buildings in Antigua, are noteworthy. (FIG. 2.) The low pedestal on which the shaft rests is 0.70 m. wide projecting 0.35 m. from the wall. Its total height including a many-membered crown molding is less than its width, or 0.60 m. The base of the shaft rests on a plinth which recedes in a hollow chamfer to join a small torus, rectangular in plan like the plinth.

The pilaster shaft consists of a number of superimposed elements each of which in profile looks like a squat open-mouthed pot, quadrilinear in section. The first pot form above the base is shorter than the four superimposed pot forms above, each shaped like the body section of an amphora. Separating each of these pot forms is a small torus and an even smaller scotia, the two members separated by a narrow fillet. Fillets are also used beneath the torus and above the scotia. Surmounting the shaft, the capital, an elaborate variation of the Tuscan type, rests directly on a torus and scotia with fillets.

iii. The Church Building (FIG. 115.)

The façade of the church has been subjected to much alteration, probably even as late as the 1940's when the little chapels for the last two Stations of the Cross were rebuilt. These abut on the façade framing the entrance to either side. The most interesting feature is the door which is set in an apselike recess, the jambs of which have concave splayed reveals. The same concave profile continues on the soffit above. The door itself, set back in the recess, has simple, unadorned jambs with a plain molding to mark the impost of the half-circle header arch. The exterior arch of the recess springs from the same height as marked by the molding which continues around the reveals from the tops of the door jambs. The soffit of the recess is trumpet-shaped, *abocinado*, and decorated with elongated leaves, the rounded edges of which project slightly beyond the face of the archivolt of the exterior arch giving the whole a fascicular or shell-like effect.

A small belfry with a niche crowned with a triangular pediment completes the door bay. It rests on the small horizontal section of the cornice molding of the swagged pediment which rises from the entablature below.

The interior of the church seems to have been completely renovated in modern times and is roofed with tile supported on a simple timber framework.

2. *El Carmen: Church, 1728*

(FIGS. 116–121.)

Historical Data

THE church of El Carmen, located not very far from the Plaza Mayor, was completed and inaugurated in 1728.[1] The conventual building had probably been completed a few years prior to that date, for in 1725 the curate of the church had requested a water allotment of the *ayuntamiento*.[2] This was actually the third church building of the same advocation on the site. The first, dating from about 1638, must have been a very humble building, for the time which elapsed between the granting of the license to build and the inauguration of the structure was only about two months.[3] A second building, probably the first formal structure, was built in 1686.[4] No mention is made anywhere which might throw some light on the architectural character of this second seventeenth-century building, though it too could not have been of very monumental character.[5] It suffered some damage in the earthquake of 1717,[6] but was probably soon repaired.

When the nuns of Las Capuchinas arrived in Guatemala in 1726 they were housed in El Carmen,[7] probably in the conventual buildings, standing since 1725, which adjoined the church to the north.[8] A larger and more sumptuous church building, the present one which supplanted the 1686 structure, was in construction while they were there.[9] No vestiges of either the 1638 or the 1686 churches exist today, nor were any of the older remains incorporated in the structure of the existing building. It may, therefore, be concluded that a completely new church was planned and executed between the years 1724 and 1728, the ruins of which are still standing today.

Architectural Data

THE building is at present surrounded by private houses on all sides except for the main façade, so that it is not possible to tell the exact extent of the area which the conventual buildings had once occupied. (FIGS. 117, 118.) They were located on the north and connected through a large door, now in ruins, in the second bay of the church. (FIG. 116.)

The overall exterior dimensions of the church are 13.10 m. by approximately 52.00 m., making a proportion of width to length of about 1:4. Included in the width are the two corner buttresses, each 1.00 m., so that the actual building behind the façade wall is but 11.10 m. wide. The thickness of the back wall and the buttresses there could only be estimated, since it was impossible to observe and measure these features because of the modern constructions abutting on them. The interior length from the back of the façade wall to the inside face of the rear wall is 45.30 m. The façade wall is about 4.55 m. thick at ground level including the projecting podia. The remaining walls are about 1.50 m. thick throughout the nave except where the niches reduce the thickness. The walls of the front half of the presbytery, like those of the choir, are also 1.50 m. thick and have no niches.

The interior width at the choir is 8.45 m. while the width of the nave from niche wall to niche wall varies slightly be-

[1] Juarros, **1**: p. 150.

[2] *AGG*, A 1.10.3 (1725) 16544–2280; *Efem*, pp. 157 ff.

[3] *Efem*, pp. 53 ff.

[4] Juarros, *loc. cit.*

[5] Vásquez, **4**: p. 384, says the church was in ruins in his day, about 1695. At any rate, it was in use in the early eighteenth century and is mentioned in the minutes of some city council meetings. See *AGG*, A 1.2.2 (1701) 11.780–1786, also *BAGG* **8**, 1 (1943): p. 93; and A 1.2.2 (1704) 11.780–1786, also *BAGG op. cit.*, p. 109.

[6] Ximénez, **3**: p. 351.

[7] *Efem*, p. 161.

[8] *AGG*, A 1.23 (1727) 1526–210, also *BAGG* **10**, 3 (1945): p. 257. See fig. 117 for the Muybridge photograph of *ca.* 1876 with the ruins of the convent still there.

[9] Juarros, *loc. cit.*

tween 9.90 m. and 10.00 m., the niches being 0.76 m. deep. The pilasters of the three nave bays are set 7.62 m. on center. Those of the crossing are 10.00 m. on center, making that bay more or less square in plan as described within the area bounded by the transverse arches and the side walls. The choir bay is quite short, about 3.90 m. from the rear face of the façade wall to the inside return of the first pilaster. The presbytery or *capilla mayor* is divided into two parts. The first is about 5.20 m. between pilasters or slightly less than the nave bays in length. The last pair of pilasters, that is, the two on the west side of the crossing, are narrower in width than the rest. A step 0.50 m. wide leads up to the rear part of the presbytery which is 2.16 m. deep, making a total for this bay of 7.86 m., or approximately the length of the nave bays. The smaller and higher part of the presbytery is reduced in width to 6.75 m. The walls thus line up with the face of the pilasters of the rest of the church.

The pilasters of the nave bays are 3.10 m. wide at the base and project 0.80 m. from the wall. The last pair, the smaller ones mentioned above between the crossing and presbytery, are only 1.85 m. wide. The half pilasters built into the corners where the presbytery is divided are only 0.70 m. wide. The same is true in the case of the corner pilasters of the choir.

The system of construction is really one where seven pairs of piers, about 3.10 m. to 3.15 m. square at ground level, support the transverse arches, the span of which is approximately 6.75 m. The piers emerge as pilasters on the church interior and as buttresses on the exterior.

The walling material throughout the building is rubble stone with occasional leveling courses of brick laid in a thick lime mortar. The walls of the niches between piers are rather thin, only 0.80 m., so that many holes have been punched out by vandals in recent years. The piers are built of the same material as the walls. A finish coat of stucco covered the rough walls.

The original brick floor of the church is still in evidence. Beneath the crossing a stairway once gave access to the subterranean chambers below. In addition to the door in the north wall of the first nave bay, there are two small doors in the presbytery, one on each side. These once gave access to the conventual buildings now obliterated by modern construction. (Fig. 117.)

The roofing of the church consisted of a series of vaults supported on the transverse arches. Over the crossing, to judge by the fact that it is square, one may suppose a cupola probably supported on pendentives once existed. The remains of a short barrel vault, the crown of which is missing, are still to be seen over the smaller rear portion of the presbytery. The rest of the bays, including the choir which once had a mezzanine or upper choir as witnessed by the remains of the springing of what was once a rather flat arch, were all covered with ellipsoidal "half-watermelon" domes forced into the rectangular bays.

The crowns of the transverse arches were higher than those of the half-circle lateral arches embedded in the side walls. (Fig. 119.) These run along the side walls springing from the pilasters forming lunettes in which windows are cut through. The spans of the lateral arches are approximately 5.00 m. and about 2.50 m. high at the crowns. They thus fall about a full meter below the crowns of the transverse arches. Still visible in the space between the haunches of the arches are remains of corbels built out and over the crowns of the lower lateral arches which once reached the height of the crowns of the larger transverse arches. False ribbing in brick and plaster like so many sinuously twisting ribbons once decorated the soffits of the ellipsoidal vaults.

The façade is unique in Antigua. (Fig. 118.) Niches, which are usually located in the lateral bays of the normal three-bay façades, are absent here. Instead, three pairs of freestanding columns occupy the lateral bays of both the first and second stories. The layout of the façade into three vertical and three horizontal divisions, however, still obtains here, but the use of freestanding columns in lieu of niches is a unique feature.

The third story or finial is missing, not enough remaining to venture a conjectural reconstruction of its appearance. But to judge by the rather ornate pediments which surmount each pair of columns, it must have had a mixtilinear profile and provided the setting for a niche with sculpture over the central axis of the façade. (Fig. 120.)

The central bay is occupied by the door in the first story, a large octagonal window in the second, and presumably a niche in the third which is now missing. The three pairs of columns in each of the first-story lateral bays rest on a high podium, the central portion of which with its corresponding pair of columns projects and makes a very bold miter. This mitered effect characterizes the whole of the façade for, in plan, the breaks continue in the second story too. The surface of each section of the podium is treated as a panel and decorated in stucco with a fanciful design combining volutes and palmettes. Each section is also treated with a base molding at the bottom and a crown molding at the top.

The structural part of the façade wall is about 2.30 m. thick at the top of the foundation. The podium projects about 1.15 m. and the central section another 1.15 m. making a total projection of about 2.30 m. beyond the structural wall behind, or a total of approximately 4.50 m. for the façade wall. The height of the façade to the cornice of the second story is about 12.30 m. The width, as already stated above, is 13.10 m., making the main portion of the façade almost square. The finial or third story, it may be assumed, was at most 3.00 m., and probably not less than 2.50 m. high at the very pinnacle on the central axis.

The orders of the columns employed are also uncommon in Antigua. The shafts of all the columns of the second story still retain the overall floral decoration in stucco, while those of the first-story columns are now bare. They were once adorned in the same manner like those above, as attested to by a small fragment still adhering just under the capital of the south pair of the south bay. The order of the first story is Tuscan with some garlands covering the echinuses. That of the second story is Ionic. The architraves are rather thin and consist of a series of superimposed moldings. The friezes are considerably higher and adorned with an overall pattern of spiraled acanthus leaves flanking a palmette, each group separated by a rosette. Above the friezes a very complicated series of superimposed moldings make up the cornices.

Pediments of different shapes surmount each of the three pairs of columns on the first story, and vary again on the second. (FIGS. 118, 120.) The central projecting pair of the first story have a segmental pediment, while in the second story the corresponding pediment is of the same type but more elaborate, for the crown is indented by a concave chamfer or hollow. The pediments over each of the outside pairs of columns on the first story are treated as one-half of a broken pediment, the raking cornices of which spiral back in volutes before reaching the apex. The inside pairs of columns of the first story, those to each side of the door in the central bay, are treated as half-segmental pediments, forming a continuous undulating line with the mixtilinear door header which is set back about 1.15 m. (FIG. 121.)

The corresponding inside and outside pairs of columns on the second story also have half pediments, but here they are triangular. The one raking cornice of each lines up to an apex hidden behind the segmental pediment with hollow crowns of the central pair of projecting columns. In other words, the design on the second story is that of two pairs of columns supporting a triangular pediment in front of which and on the center a third pair of columns with the fanciful segmental pediment projects hiding the apex of the main triangular pediment behind.

The treatment of the central bay follows the more usual arrangement of having a door in the first story, a window in the second, and probably a niche in the third. (FIGS. 120, 121.) The door treatment is one of the most ornate in all Antigua. It is set in a nichelike recess 2.90 m. wide but which splays back to the inside of the façade wall to make an opening there of about 3.55 m. The reveals of the recess jambs and the header arch above are most intricate. The header arch of the recess above consists of two cyma reversa profiles separated by a short horizontal member. Another way to describe the arch would be to say that each leg is an ogee in outline, meeting at a short horizontal member rather than a point. The soffit of this arch is treated as a series of three major molding profiles which continue up from the reveals. The header of the door opening itself must have been a half-circle, for part of a small crescent-shaped area still exists just under the soffit of the recess arch.

The entablature which surmounts the columns of the lateral bays continues across the central bay and above the door recess arch, the spandrels of which are overladen with floral patterns in stucco. (FIG. 120.) In the space just above the entablature a pediment is formed by widely projecting moldings, the outline of which consists of concave raking cornices which join a short horizontal member. This member acts as an emplacement for a statue still *in situ*, behind which an octagonal-shaped niche-window pierces the façade wall to light the upper choir. The four main sides of the octagonal window opening are straight, while the four diagonal sides are slightly concave. The window has splayed convex reveals making the opening considerably smaller in back than in front. The entablatures of the lateral bays of the second story continue across the central bay. The segmental pediment above is the same type as that over the central pair of columns of the second-story lateral bays and is indented at the crown by a hollow chamfer.

The interior order is just as ornate and complicated in detail as that of the façade. (FIG. 119.) The pilasters project boldly, about 0.80 m., from the side walls and seem almost freestanding, for the space between them is occupied by niches an additional 0.76 m. deep. (FIG. 116.) Each pilaster rests on a very high pedestal. For the most part, the bases and lower parts of the pilasters have been destroyed by vandals so that it is impossible to tell what the profiles of the bases were like. The broad surface of the outermost face of the pilaster shafts is adorned with a single, deep, square flute in the center of which an astragal runs between two fillets raised from the inside corners.

The entablature which surmounts the pilasters is exceedingly ornate and complicated because of the number and variety of members employed in its composition. (FIG. 119.) The architrave is rather narrow and is like a projecting crown molding made of a large number of superimposed moldings, each of a different height and profile. Breaks in the entablature as a whole follow the same plan as the miters of the pilasters below. The frieze is punctuated with two types of brackets alternating around the entire building. One, rectangular in plan, is corbeled out in succeeding fasciae which project out as they go up. Between these rectangular corbeled brackets are shallower semicircular brackets. These, likewise, are corbeled up to the cornice above.

The cornice is made up of two principal members, a dentil course and a projecting many-membered molding above. The dentils are not separate and distinct from each other, but blend one into another and are corrugated in plan. The crown molding above undulates in still larger, opener curves than the corrugated dentil course below. The result is that the overall appearance of the entablature is quite rippling and kinetic in effect. The nonstructural fluid character of the entablature is what the designer apparently sought to emphasize, and was able to achieve because of the freedom of execution possible in plaster. A further decorative and nonarchitectonic note is the decoration of the soffit of the dentil course. Just in front of each semicircular bracket on the underside of the dentil molding, the space is occupied by a pair of extended wings, a pattern which is repeated around the entire church.

3. *La Candelaria: Church, Late Seventeenth and Early Eighteenth Centuries*

(FIGS. 122–125.)

Historical Data

THE hermitage of La Candelaria dates from the time of Bishop Marroquín in the sixteenth century, and is known to have been administered by a prior from the convent of Santo Domingo nearby.[1] The inhabitants of the *barrio* consisted of about 213 Indian families for whom services were still held in the Pipil language as late as 1690. Some Spaniards, mulattoes, and Negroes also lived in the neighborhood, but they attended services in the parish church of San Sebastián nearby.[2] In 1754 La Candelaria was raised to a parish after the secularization of the religious doctrine.[3]

Ximénez himself, the Dominican chronicler, spent some time as curate there. But he does not give much information concerning the character of the structure, other than to say in one place that it was not damaged much in 1717 except for one chapel which had also been badly hurt in the earthquake of 1689 and subsequently repaired.[4] Two years after the earthquake of 1751 a petition was made for license to beg alms to be used in the reconstruction of the church. This would imply that the building had been damaged. But no indication is given as to the nature of the work planned.[5]

Though not mentioned specifically by any of the eyewitnesses of the earthquake of 1773, widespread destruction of the church must have occurred at that time judging by the state of the ruins today. This part of the town, the northeast section containing the *barrios* of La Candelaria and Santo Domingo, was the worst hit and leveled to the ground.[6]

Díaz, depending on sources which he does not cite, says the seventeenth-century structure was converted to a pile of rubble after the earthquake of 1717, but that thanks to Ximénez, who was curate there from 1718 to 1721, the church was rebuilt with additions of offices, a sacristy, corridors, a fountain, and a garden.[7] Díaz believes it was Ximénez who had the façade embellished with the stucco decoration still partly in evidence today. This information is at variance with that given by the supposed author of the reconstruction; for, as noted above, Ximénez himself says the damage to the church was slight except for that in one chapel. Within the text of his manuscript Ximénez often gives dates as to when he is actually writing. In Book IV, Chapter 11, he says he is writing in the year 1717, and in the same book, Chapter 70, he says the year then is 1720. It would seem that he was engaged in writing his chronicle during the time he was curate of La Candelaria and supposedly engaged in the reconstruction of the church, yet he does not say anything about this.

The known facts may be collated as follows: a church existed in the seventeenth century as indicated by Fuentes y Guzmán, Vásquez, and Ximénez. This building suffered some damage in 1717. One chapel was completely destroyed and rebuilt soon after. And finally, in 1753, more repairs were needed, probably as a result of the earthquake of 1751.

Architectural Data

ALL that remains today are the lower parts of the façade, a bit of the walling of the north side, and a small portion of the second story of the north bay of the façade. (FIG. 122.) The façade is divided in two main parts: the façade of the church proper and, abutting on it to the south, a narrower structure pierced by a wide door flanked by two fanciful pilasters of an even more fanciful order. (FIG. 124.) This ex-

[1] Juarros, **1**: p. 148.

[2] Fuentes y Guzmán, **1**: pp. 400 ff.

[3] Juarros, *loc. cit.*; Vásquez, **4**: p. 385, mentions it as one of the hermitages of Antigua in 1695. In 1754 the crown ordered that those parishes under the administration of the monastic orders be placed in the hands of the secular clergy.

[4] Ximénez, **3**: p. 356.

[5] *AGG*, A 1.10.3 (1753) 31.351–4049.

[6] Cadena, *Breve descripción*; González Bustillo, *Razón particular*.

[7] Díaz, *Romántica ciudad*, pp. 44 ff.; see also Pardo, *Guía*, pp. 52 ff.

tension may possibly be the chapel to which Ximénez refers as having been completely destroyed in 1717 and rebuilt in his time. Stylistic differences exist in the treatment of the orders which adorn the church and chapel respectively.

The church façade was divided into three bays, the central one the wider and containing the door. (FIG. 122.) The lateral bays are narrower and have niches in each story. The exact nature of the door treatment cannot be known because a gaping hole exists where it once was located. Enough of the reveals remain to indicate that the door itself stood back in a recess, but the header treatment is missing. Orders with solomonic columns frame each lateral bay in both stories. Those of the lower story rise from podia with projecting cornice moldings, while those above rest on a low Roman attic with small pedestals on column centers. (FIG. 125.) The bases of the shafts, resting on square plinths, are composed of two almost equal-sized tori with a guilloche pattern, looking more like a honeycomb, and are separated by a deep, concave scotia. The shaft, overladen with a leaf pattern, twists up to an equally ornate Tuscan capital supporting a three-part entablature. The architrave is composed of a few shallow, superimposed moldings. The frieze is decorated with floral and interlaced scroll patterns in stucco. And above, a widely projecting cornice breaks miters over the column centers.

The order of the second story, much shorter than that of the lower, follows the same scheme. (FIG. 122.) Each pedestal is set over the projecting miters of the cornice below. The central and south bays of this story are gone. Some sort of niche-window with splayed reveals probably served as the emplacement for a statue.

The niches, one to either side of the door, have semicircular header arches. (FIG. 125.) In plan each is semicircular and covered with a quarter sphere like an apse. The niches are framed with simple unadorned pilasters supporting broken pediments with spirals at the end of the raking cornices in the lower story and modified triangular pediments in the upper. Most of the flat wall surfaces, the capitals, and the entablature are all heavily covered with stucco decoration in the form of leaves, rosettes, geometric designs, connecting vines, and other motifs. The whole façade has an appearance as if covered with lace. The workmanship is very delicate.

The smaller unit to the south is just as ornate in surface treatment, but different in style. (FIGS. 123, 124.) The door with a half-circle header is set in a deep niche spanned by an ogee arch. The general outline of this arch is a cyma reversa. It has low haunches, the tops of which are marked by a step or break, from which the upper parts surge up in opposite directions to join in a point over the very center. The extrados of the archivolt of this mixtilinear arch, decorated with little rosettes set at intervals, does not come to a point as does the intrados. Instead, the legs of the arch meet at a short horizontal strip at the crown. The architrave moldings above miter out slightly over the flat crown.

The applied order which frames the door is rather involved. (FIG. 2.) Beginning with pedestals at ground level, a column rises to either side. The shafts consist of a number of superimposed forms shaped like little open-mouthed pots. Each pot form is separated from the one above and the one below by a molding decorated with a scalloped edge. The capitals above are even more ornate, the forms having completely lost their pristine architectonic quality. What would normally be the echinus is now a heavy mass of foliage whose general outline is rather like a stubby pulvin. The abacus above has diagonally chamfered corners. The remaining space between the diagonal chamfers is cut into by two hollow or concave chamfers. It would seem that, beginning with a square abacus, the mason first cut the corners off and then gouged out the rest of each side with two concave cuttings.

The entablature above is likewise extremely complicated. The architrave projects only slightly and consists of a number of superimposed moldings which are mitered out to form corbels supporting tiny nude male figures over the spandrels of the ogee arch of the door below. Set directly below the corbels or projecting portions of the architrave, are the heads and wings of two robed stucco figures with outstretched arms. These form the principal decoration of the spandrels. The cornice does not flow in a level horizontal line. Instead it surges up and then down, and under the floor of the niche above, to form a pediment. The niche with its corbeled projecting floor invades the apex and fills most of the tympanum.

The documentary and literary data do not clarify the date of the buildings. The church façade and the attached "chapel," however, are of different design and were possibly built at different times. On the basis of the stylistic affinities of its stucco decoration to that of the convent entrance of La Merced, dated 1650–1690, of San Sebastián, dated *ca.* 1690, and of the Cathedral, dated 1680, the church façade probably originally dates from the late seventeenth century. Moreover, since the decoration of the *portada* of the extension or "chapel," is indeed more extravagant, this would tend to place this structure later than the church, possibly after 1753 when some construction work was done. This conclusion is bolstered by a comparison of the ogee door arch with similar arches employed in the corridors of the

Universidad and the Seminario Tridentino, both from the third quarter of the eighteenth century. An almost similar door opening was used in the church of El Carmen built *ca.* 1728. It might be safe to conclude that the "chapel" *portada* dates from the second quarter of the eighteenth century at the earliest, or from the third quarter at the latest.

4. *La Concepción: Church and Convent, 1694 and* ca. *1729*

(Figs. 126–130.)

Historical Data

A BEQUEST of 2,000 *pesos* in the will of the first bishop of Guatemala, Francisco Marroquín, in 1563 provided the funds with which the founding of this convent for nuns was made possible.[1] However, it was not until 1577 that license to establish the order was granted, and a year later in 1578 when the first nuns arrived in Guatemala[2] who were housed in some temporary quarters in a private house located on the present site of the convent. In 1579 an allotment of water for the use of the convent was requested of the *ayuntamiento*. As late as 1585 no permanent structure had yet been built to judge by the fact that the nuns asked the *ayuntamiento* to be allowed to use the building of the Hospital of Santiago as a convent and to convert their present convent building to a hospital.[3] It was not until sometime between 1623 and 1641 that a proper structure including a church was built.[4]

By the end of the century the convent building, as apart from the church, must have been extensively repaired and possibly rebuilt as indicated by an inscription just over the lintel of the main *portada* saying that it was completed the twenty-third of February of 1694. (Fig. 128.) In the earthquake of 1717 both church and convent were badly damaged, though Ximénez, as usual, minimizes matters.[5] His statement cannot be accepted as accurate, for even as late as 1723 the nuns were dispersed in various parts of the city having no place of their own to live as yet.[6]

[1] Fuentes y Guzmán, **3**: pp. 130 ff.; *Efem*, p. 17.

[2] Fuentes y Guzmán, **3**: p. 134; Vásquez, **4**: pp. 366 ff.; Juarros, **1**: pp. 132 ff.; *Efem*, p. 23; and Villacorta, *Historia*, p. 125.

[3] *Efem*, pp. 24 and 28.

[4] Juarros, **1**: p. 201; see also Villacorta, *op. cit.*, p. 218; and Díaz, *Romántica ciudad*, p. 26. In 1607 the convent was damaged in an earthquake. See Mencos, *Arquitectura*, ch. III, n. 134, also append. IV, transcribing *AGI*, Guatemala, 176, "Información sobre el convento de monjas de Guatemala, en Guatemala a 23 x 1607."

[5] Ximénez, **3**: p. 356.

[6] *AGG*, A 1.2.4 (1723) 16.192–2245, also *BAGG* **10**, 3 (1945): p. 210; and *AGG*, A 1.18 (1740) 5022–211, also *BAGG* **1**, 2 (1936): pp. 131 ff.

By 1729 the nuns were already living together in their own house again implying that the church and convent had been rebuilt in the meantime. It must have been a vast structure in order to have housed 103 nuns, 140 pupils, 700 servants, and 12 *beatas*. The convent is also said to have had twenty-two fountains with running water in the various patios.[7] In the earthquake of 1751 the church was rather badly damaged, the sacristy completely destroyed, and part of the cupola and vaulting of the choir down.[8] In 1773 the church and convent again lay in ruins, as it does to this day.[9] Most of the surrounding area has since been built up with private houses so that the exact extension of the conventual buildings is almost impossible to ascertain at present.

Architectural Data

NOTHING remains of the convent building except the main gate, which bears the following inscription: "Esta portada ce acabo en 23 de Feb° de 1694." (Figs. 91, 126.) The podia, the door jambs, and the lintel are built of stone. The pediment and upper portions of the wall to either side of the door opening are of brick covered with stucco. Framing the door and rising from the stone podia to either side is a single short pilaster with a deep chamfer or flute giving the shaft the appearance of two separate ribbonlike moldings or vertical fillets. These are joined at the top by a very crudely executed Ionic capital. The volutes are tight little scrolls appended to either side of the top of the pilaster. These are the sole members of the capital except for a little flat horizontal fillet which is really part of the architrave. Immediately above this fillet or architrave there is a very narrow band, presumably a frieze, in which the inscription mentioned above is cut. The order terminates with a simple cornice of slight projection.

The finial or *remate* is the only ornate element in the design of the entrance. (Fig. 128.) Set on center over each of the two pilasters is a small pedestal topped with a pyramidal merlon. A pediment picked out with three niches occupies the space between. Two of the niches are set directly on the horizontal cornice and on center over the reveal of each door jamb, while the third is located a little above directly over the center of the door. The pediment wall itself has a mixtilinear outline with scrolls at either extreme connecting at the merlons.

Small statues are still *in situ* in each of the niches. Some reliefs are worked in the spaces between. Directly under the higher central niche a small nude atlantid in relief supports

[7] *Efem*, p. 167; and Pardo, *Guía*, pp. 125 ff.

[8] *Efem*, p. 205.

[9] González Bustillo, *Razón particular*; Juarros, **1**: p. 132.

the projecting pedestal. Flanking this figure is an emblem of the radiant sun to the left and a crescent moon to the right. The space between the left niche and the merlon is filled with a coat-of-arms. And in the corresponding space to the right, a mounted figure, sword in hand, charges straight forward, presumably representing St. James (Santiago) on the coat-of-arms of the city, Santiago de los Caballeros de Guatemala. The quality of execution of the figures, both in the round and in relief, is obviously not from the hand of a practiced artist, but rather from that of a craftsman, probably a plasterer or mason rather than a sculptor.

The church is in complete ruins and altogether unroofed. (FIG. 129.) The west side which would normally be the main façade, since the building is oriented east-west, was once actually enclosed within the convent proper. The main door is on the long side facing north on the street. The other two sides of the church abutted on the convent buildings. Though the west side within the convent grounds has no entrance at all, the choir is located there as is normal in *antigüeño* churches. The two freestanding exterior walls are undecorated except for rather massive square buttresses which diminish in size just above the springing of the interior transverse arches. At clerestory level the north wall is pierced by octagonal windows. (FIG. 127.) Directly under the cornice at the top of the north wall, a molding with half rosettes is worked in plaster. The narrower west side once had a belfry and probably a mixtilinear pediment above, all of which are now missing. The rest of the west façade is treated without any special ornament, except that the fenestration is different from that of the north side. Three massive piers rise to take up the thrust of the vaulting of the mezzanine floor separating the upper and lower choirs inside. Pairs of rectangular windows with segmental arch headers light the upper and lower choirs. A similar window is also located on the north return in the upper choir. The choir is still filled with the debris of the fallen vaulting. (FIG. 129.)

The church is single nave in plan and laid out in oblong-shaped bays roofed with half-ellipsoid pendentive domes. An entrance to the convent proper is located in the south wall opposite the street door. The *capilla mayor* occupies the last bay of the east end, while the *sagrario*, covered with a rather flat vault with false ribbing, lies to the south projecting into the convent area. The vault has been cracked outlining the crown and may fall in the next earthquake.

The most interesting feature of the church interior is the applied order. The lower parts of the massive pilasters, backed by the buttress on the exterior, are now largely missing. The entablature, however, is in a good state of preservation. (FIG. 130.) Its outstanding feature is a heavy cornice with a high corona supported on very ornate brackets which cut through the frieze. These brackets are made up of two main members: first, supporting the corona is a horizontal projecting bracket with a wide flaring profile and a pendant inverted pyramid at the end; and second, below and joining the first, is a console or half volute set upright on end with the spiral at the top and a bulging convex curve below. In addition to these complicated brackets, the corona is supported by a dentil course. The dentils fill the space between brackets. They are semicircular in plan and are further gouged out by deep, hollow chamfers. The dentils rest on semicircular corbels which invade the frieze below. Rather than an alternation of rectilinear projections and recesses, the heavy dentil molding gives the whole entablature a staccato or corrugated effect. In the center of each dentil another smaller bracket is set which adds another beat to the already involved staccato rhythm of the entablature.

The construction of the church walls is rather shoddy, not all parts of the walls being laid up in regular courses of brick and mortar. There are portions which seem to have been built in a method usually employed for tamped earth. Whole and broken brick, as an aggregate to the lime mortar, were thrown into forms. Stones were added in some places, especially in the pendentives of the vaults. It would seem that in the last construction, that is, of about 1729 or so, debris from the 1717 ruin had been employed as a measure of economy.

5. *Santa Cruz: Church, 1662 and 1731*

(FIGS. 131–139.)

Historical Data

THIS small church has been sometimes confused with another one of almost the same name, Santa Cruz del Milagro, of which hardly a trace remains except for some ruined walling. A further cause of the confusion among postcolonial writers is that the latter is located but a few hundred yards away from Santa Cruz on the opposite side of the Pensativo.

The *pueblo* of Santa Cruz is listed among those under the ecclesiastical jurisdiction of the order of Santo Domingo in 1617. It is unknown whether any church stood there, though it is most probable that some crude and rudimentary type of structure already existed since Santa Cruz was known to be one of the oldest hermitages in Antigua.[1] Later in the century a formal structure including a special chapel for a miraculous image of Nuestra Señora was undertaken by one

[1] Remesal, **2**: p. 610; Juarros, **1**: p. 148.

Fr. Diego de Rivera, perhaps not too long before 1662, the year he died.[2] A contract dated February 4, 1662, entered into by one Blas Marín,[3] describes the church as being more or less at the midpoint toward completion.[4]

During the seventeenth century the population of this neighborhood had apparently grown, to judge by the fact that the water allotment for the use of the inhabitants had been increased in 1656.[5] By the end of the century, *ca.* 1690, the number of families is given as thirty-seven, all of whom were Cakchiquel Indians who spoke Spanish. The church is described as being vaulted, of good architecture and with bells in its tower. The existence of a house for the priest is also mentioned. The location of the church is described as backed up against one of the many hills which surround the city.[6]

The above information is confusing since the church is spoken of as being vaulted. From the contract of 1662 entered into with a carpenter to complete the job, one is led to believe that it was roofed with wood and tile. The vault, however, may have referred to the *capilla mayor*, for the nave was indeed covered by a wood and tile roof. (FIG. 134.) In fine, one may assume that by the end of the seventeenth century a church already existed with an *artesón* and at least one vaulted dependency, probably the *capilla mayor.*

Just what transpired with regard to this building in the earthquake of 1717 is not exactly known, though it is hardly likely that it escaped injury. This conclusion is corroborated by the fact that a church was in construction in 1727 when one Fr. José Vásquez, a Dominican, asked license of the *ayuntamiento* to divert some water from the Río Pensativo, just in front of the church, to be used in the construction work on the hermitage. About eight months later in 1728 this request was granted, so that one may assume that actual work on the new building began at the latest that year.[7]

Three years later, on October 13, 1731, the new church was formally inaugurated and within three days opened for services.[8] The same person who had asked for permission to divert water from the Pensativo for the construction of the church, now asked license to construct a fountain in the area which was granted at the end of the month.[9]

2 Ximénez, **2**, p. 334.

3 An Europeanized Indian from the neighborhood of San Francisco and a master carpenter. "... indio ladino, vecino y natural del Barrio de San Francisco, maestro del oficio de carpintero ...," *Efem*, p. 71.

4 *Ibid.*; Pardo, *Guía*, p. 180; Díaz, *Romántica ciudad*, p. 68.

5 *Efem*, p. 66.

6 Fuentes y Guzmán, **1**: pp. 403 ff.

7 *Efem*, pp. 164, 165, and 168.

8 *Ibid.*, p. 171; *ASGH* **23** (1948): p. 232; Juarros, **1**: p. 148; Villacorta, *Historia*, p 334; Pardo, *loc. cit.*; Díaz, *loc. cit.*

9 *Efem*, p. 171.

In 1746 the *ayuntamiento* allotted a sum of money for some work to be done on the building.[10] Some minor repairs were carried out after the earthquake of 1751.[11] But it was not to be standing long, for in 1773 the church was more or less totally destroyed.[12] The north and south walls, except for the top portion, were left more or less intact and were still standing in 1957. No mention is made of the cupola, which had not been damaged in 1773, but which is completely gone now. It was still intact during the first part of the twentieth century, appearing in a photograph taken some time before 1924,[13] and seems to have been of the same type as the church of La Merced with pronounced decorative exterior ribs converging to a lantern.

Summing up the literary and documentary data, it appears that the church of Santa Cruz went through two major building periods, that is, an original seventeenth-century church was rebuilt in the second quarter of the eighteenth century. The problem of identifying which parts of the earlier building were utilized in the later is clarified by an examination of the architectural remains.[14]

Some restoration work, well intentioned but disastrous, nevertheless, was carried out on this church in 1956 when some of the walls were repaired and the vault of the choir rebuilt with reinforced concrete. One of the outstanding features of this church is its very excellent and fine *ataurique* decoration, but this has been spoiled by the recent restorations which hardly match, let alone equal, the quality of workmanship of the original colonial material. Excavations were also carried out in the atrium and in the area to the north of the building. The overall effect of this work resulted in raising the building above ground level thereby changing the original appearance of its setting.

Architectural Data

THE church is located about forty meters east of the Pensativo. The excavations carried out in the surrounding area, now occupied by a coffee plantation, have revealed vestiges of private dwellings immediately to the north of the church, and must be accounted as the remains of the *barrio* or *pueblo*

10 *Ibid.*, p. 197, *ASGH* **24** (1949): p. 369.

11 *ASGH* **24** (1949): p. 378.

12 González Bustillo, *Razón particular*, also *Ciudad mártir*, p. 94. The destruction is described, "La Iglesia o Hermita de Santa Cruz, de sólida fábrica, y demás que ordinario primor en su arquitectura, cuarteada desde los primeros temblores, según dice el ingeniero; y con los de 13 y 14 de diciembre, expresa el escribano, que cayeron algunas paredes, que miran al mediodía, se rompieron notablemente las de septentrión y se destrozó el tejado del artesón y enteramente la habitación del cura y sacristán."

13 Mislabeled Cruz del Milagro, Elliot, *Central America*, facing p. 52.

14 See Markman, "Santa Cruz; etc.," *JSAH* **15** (1956): pp. 12 ff.

mentioned in the colonial literature. The building itself is backed up against the hill so that the site for the *capilla mayor* had to be excavated from it. The open space between the church façade and the river, the atrium, probably also served as the village square. (FIG. 132.) Its dimensions are approximately 40.00 m. deep from east to west and 29.00 m. wide from north to south. The church is on the east side, the main façade facing west. The high embankment or dike located on the opposite or east side, and which protects the atrium and church from the constant flooding of the Pensativo, was first built some time after 1742.[15] It has been added to from time to time in order to deepen the river bed, so that by now the top of the embankment is about six or seven meters above the level of the atrium, access to which is gained by a staircase built at some unknown date. A stone cross of colonial origin, now broken, stands in about the middle of the atrium.

The plan of the church begins with a low-stepped platform, the *lonja*, approximately 15.00 m. wide by 6.55 m. deep. (FIG. 131.) Three steps, the treads of which are each 0.40 m. wide, bound the west and south sides making the width of the *lonja* 17.40 m. at the lowest step. The paving is now gone. The steps on the south side are in ruins. The north side of the *lonja* is bounded with a wall about 3.00 m. high which abuts on the north tower. This wall probably had a return to the north to enclose the area in front of the curate's house where the fountain, built soon after the dedication of the church, is still to be seen. The wall rests directly on the steps on the north side of the *lonja*, hiding them from view. This indicates that it was built after the *lonja* had been completed. In fact, it is chamfered on the outside corner where it abuts on the tower, again disclosing that it was an afterthought and not part of the original construction of either the *lonja* or façade. The wall has two merlons and an ornamental coping along the top.

The exterior dimensions of the church proper are as follows: length—42.35 m.; width through the nave—10.35 m.; total width of the façade including the projecting chapel—19.50 m.; width of central portion, the retable-façade—10.45 m.; width of the towers—each 2.30 m., thus making the façade proper 15.05 m. wide; the projecting chapel—4.40 m.

The church plan is the single-nave type but is not laid out with the usual clearly marked bays, attesting, therefore, to the fact that the nave portion of the church probably represents the original seventeenth-century structure to which the choir, crossing, and *capilla mayor* were added in the eighteenth century. (FIGS. 138, 139, 134.) The interior width of the church is uniformly 7.80 m. throughout, except in the *capilla mayor* where it is 7.90 m. (FIG. 131.) The longitudinal dimensions are as follows: choir, from the rear of the façade wall to the pier which separates the choir from the nave—5.00 m.; the total length of the nave, between the choir pier and that of the crossing—15.40 m. divided into four unequal bays only one of which has niches; the length of the *capilla mayor* between piers—7.90 m., making that bay practically square; the length of the *capilla mayor* from the pier that divides it from the crossing to the inside of the rear wall—5.40 m. The *sagrario* or chapel which abuts on the façade projecting 4.40 m. to the south is set back from the tower 1.60 m. Its remaining exterior dimensions are 7.85 m. on the south side and 6.70 m. on the return or east side where it abuts on the nave wall. The interior dimensions of this dependency, which shares a party wall with the church, are 5.30 m. wide by 6.25 m. long.

The walling is not uniform in construction throughout the church, an understandable condition in view of the fact that the building represents two main building periods. The nave walls have undergone considerable repairs and modifications, old niches having been filled up and new ones cut. The walls are composed for the most part of large sections of rubble and uncut stone laid in a thick lime mortar and divided by stringer courses of brick in the common *antigüeño* manner. In the case of the sacristy and some parts of the nave, the stringer courses are absent, and the stones are larger and rougher than those of other church walls. The thickness of the walls is not uniform throughout the church. Variations exist ranging from as thin as 0.60 m. for the rear wall of the *capilla mayor* where a niche within a niche enlarges the recess, to a thickness of 1.20 m. in the nave and 1.25 m. in the choir. The variation is even greater in the sacristy where the south wall is 1.40 m. and the east and west walls but 0.80 m. thick.

Another curious feature, again revealing that an older building was utilized in the eighteenth-century construction, is the fact that the buttresses are not arranged symmetrically on the exterior. (FIG. 131.) Nor are they uniform in character structurally. And, furthermore, they bear no structural relation to the interior vaulting. The four buttresses on the south side of the church behind the sacristy were really conceived of as wall reinforcements, one actually lining up almost on the center of a niche. The other three buttresses do not back up the interior piers which support the vaulting of the *capilla mayor* and crossing. On the north side of the nave, the four buttresses there do not line up with those on the south side; one in particular, the third one

[15] *Efem*, p. 193.

beyond the tower, is of different dimensions in section and does not rise to the total height of the wall. Only two of the north-wall buttresses line up with the interior vaulting, that at the choir pier and the last one at the corner of the *capilla mayor*. There are no buttresses at all to back up the piers of the crossing on the north side.

The twin towers, though presenting uniform exterior elevations, are quite different on the interior. (FIG. 132.) The north tower has a caracole stair which gives access to the upper choir, while the south tower is a solid mass of masonry except for a small recess or cubicle which opens on the sacristy. A crack runs from the header of the cubicle where the sacristy abuts on the tower and continues up into the dome. The west and east sacristy walls are not bonded to the church wall, as revealed by the crack, indicating that this dependency was built in a separate operation either before or after the choir portion of the church was constructed. This fact is further attested to by the difference in the three exterior walls of the sacristy and the party wall which it shares with the church.

The fenestration is not uniform throughout the church. For example, in the sacristy there is an octagonal window in the east wall, and a circular window in the west wall. The octagonal window facing west is centered on the exterior in accordance with the dimensions of that portion of the sacristy which projects beyond the south tower. The east window, however, does not line up with the west, but is centered in accordance with the interior dimensions. The nave was once lighted by long rectangular windows set high up in the walls. Since most of the upper portions of the walling are missing, the exact layout of all the windows is uncertain. The few that remain are not of uniform size or design, apparently having been altered or added at different times. The fenestration of the crossing and *capilla mayor*, on the other hand, is uniform in design. Large octagonal windows are located in each bay high up in both the north and south walls of the crossing and *capilla mayor*. A similar one, making five in all, located in the rear wall of the *capilla mayor*, had been bricked up probably before 1773.

The interior pilasters of the choir, crossing, and *capilla mayor* are really massive piers arranged so that they are bonded and integrated with the vaults above, but are unrelated to the nave which was roofed with wood and tile. The piers are uniformly 1.70 m. wide at the base, but vary in projection from the wall. The two piers which divide the choir and nave project a total of 0.80 m., those of the crossing and *capilla mayor* are more massive and project a total of about 1.15 m. The four nave pilasters which are set to either side of the niches are purely decorative, being very shallow and projecting but about 0.15 m. from the wall. The other two nave pilasters are fluted and project a little more, about 0.35 m., from the wall.

The roofing of the church combines masonry vaults and wood *artesonados*, a not uncommon occurrence in Antigua. The pitched roof of wood and tile once extended from the rear of the façade over the upper choir and nave to the crossing. (FIG. 134.) One may safely assume that the original seventeenth-century church was also roofed with wood as the contract entered into with a master carpenter for its completion in 1662 would imply.

A low, flat, ellipsoidal, "half-watermelon" pendentive dome provides the mezzanine floor which separates the lower choir from the upper choir. (FIG. 138.) The soffit of the vault is crisscrossed with false ribs built of brick and stucco. The thin square brick was laid with one corner projecting from the vault surface to provide the foundation on which the molding profiles of the ribs were shaped. In each corner a figure of a *putto* worked in stucco is set at the springing of the ribs. The sacristy is roofed with a flat ellipsoidal pendentive dome, but is pierced by an oculus in the crown. Here too the soffit is crisscrossed with false ribs of similar design and construction as those of the choir vault.

The crossing was once roofed with a cupola consisting of a dome on a drum supported on true spherical pendentives. (FIG. 139.) This was possible because the crossing is square in plan. The supporting arches and the pendentives are still more or less intact. Little corbeled pedestals are worked at the springing of the pendentives on which statues were probably once located.

The *capilla mayor* is roofed by a flat ellipsoidal "half-watermelon" pendentive dome still in good condition. The elevation of this dome is considerably lower than that of the original height of the cupola over the crossing. The effect must have been similar to that observable today in the arrangement of the cupola and vaults of La Merced.

Very little of the priest's house, contiguous to the rear half of the north wall, still exists today. (FIG. 131.) This house must have been in existence before the crossing and *capilla mayor* were added, for a buttress was built on top of the house wall more or less on center with the pier which separates crossing and nave. In other words, the existing house wall was utilized to reinforce the lower parts of the church wall. An unadorned low door, 1.30 m. wide, was cut through the niche in the north wall of the *capilla mayor* to connect directly with the house. Another door in the north nave wall, located just behind the choir, gives access to the area in front of the priest's house where the fountain is located.

The façade, unlike the church structure, represents a unified design and a single construction operation. (FIG. 132.) The sacristy which abuts on the south tower and shares the party wall with the choir must have been built before the façade and may, therefore, be identified as the chapel alluded to by Ximénez as having been built by Fray Diego de Rivera some time before he died in 1662. The sacristy as a whole is set back slightly from the tower. The cubicle is actually a reserved space in the tower wall. In other words, the wall construction of the sacristy probably took place before that of the façade. The façade, which is the common retable framed by twin towers, as a whole is in a very good state of repair except for the belfries of the towers. Even the crowning *remate* or finial of the third-story central bay is complete including the sculpture which fills the niche.

The outstanding feature of the façade is the overall stucco decoration which seems to be hung like fine lacework woven in floral and geometric patterns covering all plane surfaces between the architectonic elements. (FIGS. 135–137.) The applied orders, though conforming to structural logic in general form are, nevertheless, free from all restraint and composed of sculptural nonarchitectonic elements.

The side bays begin with moderately high podia which break miters on column centers to form pedestals. (FIG. 133.) These are treated with two deep flutes of the type seen on the interior pilasters of the gatehouse of El Calvario. (FIG. 10.) The central recessed panel between the pedestals is treated with an octagonal coffer.

The applied orders begin with a narrow plinth, the exterior faces of which are decorated with the *toisón de oro* design which looks like a two-strand necklace on which oblong gems with rounded corners are strung. The same motif decorates a horizontal molding continuing across the recess between the two projecting columns. The column base may best be described as a squat, open-mouthed pot form set on a small torus. (FIG. 136.) The pot form is decorated with modified acanthus leaves. The individual leaves are divided to form pendant loops thus making a continuous pattern which covers the belly of the vase and the underside of the lip.

The column shafts are very complicated. The shaft is made up of three small columnettes, each of which is channeled by a deep concave flute. (FIGS. 2, 137.) Between each columnette a molding, triangular in section, projects, making an abrupt separation between the spherical surfaces of the columnettes. Each columnette is surmounted by an individual simplified Tuscan capital, the echinus of which is decorated with undulating striations. Each projecting triangular molding between columnettes is capped with a small bust-high Atlantid with arms upraised in the act of supporting the abacus common to the different parts of the shaft as a whole. (FIG. 135.) The exterior faces of the abaci are decorated with a variation of the undulating striations on the echinuses below, that is, with angular incisions or diagonal ups and downs to form a band of continuous triangles.

The entablature runs across the whole retable-façade breaking miters on column centers as well as under the triangular broken pediments of the lower side bays. (FIGS. 132, 133, 135.) The architrave is comprised of a number of narrow moldings which corbel out slightly as they approach the frieze. The frieze itself is pulvinated and covered with a very fine pattern which seems to imitate a type of design not uncommon in actual textile lacework. In other words, the pattern looks more like knitting than architectonic decoration. Small nude *putti*, each with a sash across the chest hanging from the right shoulder and with arms upraised, support the cornice above. One is set over each column and one directly over the niche between. Three *putti* are arranged over the central bay, one on center and the other two nearer the extremities of the bay.

In contrast to the highly ornate frieze and architrave, the cornice consists of a single narrow geison surmounted by a cavetto and fillet. The raking cornices of the indented pediments over the side bays repeat the same profiles with some small variations. The rather small recessed space which comprises the tympanum is covered with the same type of lace decoration as the frieze.

The niches of the first-story side bays are crowded into the narrow space between the columns. (FIG. 133.) The decorative treatment of the actual recesses proper with their statues is nonarchitectonic. The niche opening is framed by a band of interconnected circles and bars in stucco. An entablature and a pediment surmount the niche without the benefit of supporting pilasters, nor are they directly connected with the vertical bands of ornament which could have served as quasi pilasters. The remaining space above the pediments is filled by two small figures, one on each of the raking cornices, with one hand extending and touching a staff on which a sunburst or rosette with an inset small cross rests.

The niche recess proper is the usual apsidal type. Beneath the floor level of the niche, a pedestal fills the space between column bases. It is made up of a number of complex molding profiles, the principal one of which is pulvinated and marked with deep striations accentuating the pulvinated profile. The statue rests on a small projecting platform square in plan but set on the diagonal.

The second-story side bays begin with a low Roman attic, the surface of which is completely covered with a stucco design of lacelike appearance. (FIG. 132.) The columns, the same type as those of the first story, are much shorter and thinner than those below in conformity with the shorter height of that story. The entablature is a little different here, for it does not break miters over column centers and, as a result, the whole bay and the wall behind with the recessed niche is set back under the overhanging entablature. The entablature is, however, of the same type as below including the pulvinated frieze with small nude male figures set at intervals. The decoration of the frieze is the same as that of the large molding of the pedestal of the first-story niches, that is, striations which accent the pulvinated profile. The second-story niches are a little more ornate than those below. The unattached pediments are broken with the raking cornices ending in scrolls. Above the main horizontal entablature in each side bay there is an unbroken triangular pediment. Its small tympanum is filled with a small figure.

The lower-story central bay is almost totally occupied by the round-headed door recessed in a large niche with a concentric header. The jambs of the door and the arch overhead are treated with some of the most beautifully executed *ataurique* decoration in all Antigua. (FIGS. 135–137.) The jambs are laid out like pilasters with base, shaft, and capital. The base is high and plain while the shaft is covered with a design based on a lozenge repeated and connected by means of foliage. The capital is simple except that the necking is decorated with deeply incised connected triangles. The jamb pattern is repeated on the archivolt of the arch over the door. The soffit of the door arch is decorated with the *toisón de oro* and interlaced circles.

The spandrels of the niche or recess in which the door is set are decorated with a pattern of intertwined foliage and a small robed figure holding an oval medallion over the haunch of the arch. The *toisón de oro* marks the archivolt of this niche arch. The entablature over the central bay is a continuation of that of the side bays. A low, broken pediment, the raking cornices of which end in scrolls, cuts into the attic of the second story and reaches the same height as the indented pediments of the side bays.

A square niche-window with splayed reveals dominates the second-story central bay. The opening itself is framed with the *toisón de oro*, the necklace of the Hapsburg Order of the Golden Fleece. To either side of the window are pilasters of the corrugated type. In the narrow space between these pilasters and the inside columns of the second-story side bays is a diminutive niche.

The third story, or *remate*, of the retable-façade is mainly confined to the central bay. It consists of a freestanding wall which juts considerably above the original roof line. (FIG. 134.) The central feature is an oblong niche with a square chamfer breaking the line of and jutting above the horizontal header. (FIG. 132.) A relief representing the Crucifixion is still *in situ*, the vertical member of the cross actually going up into the chamfered space above the lintel. The surface to either side of the opening is totally covered with stucco lacelike designs. However, two pilasters of the corrugated type support an entablature on which a segmental pediment serves as the finial of the story as a whole.

Mixtilinear undulating half-pediments composed of scrolls carry the eye down to the side bays and join a rather high merlon set on bay center. The merlon consists of a pulvinated base and an upper pear-shaped member, neither of which is left plain but are both decorated with stucco ornament. The merlon is connected by another scroll which dips and rises to a smaller merlon of the same type on the outside of the bay.

The towers which flank the central retable are symmetrical except that the north one has a caracole stair on the interior which gives access to the upper choir. (FIGS. 132, 133.) Each of the exterior faces of the tower is decorated with very shallow pilasterlike vertical members treated with four flutes separated by fillets with complicated molding profiles. Each pilaster is surmounted by a small human head. The entablature is set on a level below that of the first story of the central retable. The second stories of the towers, now in a bad state of repair, are belfries with two arched openings on each side. These were once surmounted with either domical or pyramidal roofs.

6. *Santa Clara: Church and Convent, 1734*

(FIGS. 140–144.)

Historical Data

LICENSE to found this nunnery under monastic regulations of the Franciscan order was issued by the crown in 1693 or 1695.[1] Funds for the construction of the convent buildings and an endowment, as well, to provide an income for the maintenance of the newly established order were bequeathed by the Maestro de Campo José Hurtado de Arria in 1698 and by María Ventura de Arrivillaga, a widow who had left her fortune shortly before for the same purpose.[2] Thus all was made ready beforehand for the establishment of this

[1] *Efem*, p. 114; Vásquez, **4**: p. 370; Larreinaga, *Prontuario*, p. 207; see also Berlin, *Fundación de Santa Clara*, for Mexican documentary sources.

new order of nuns, the first members of which were brought to Guatemala from Mexico.

Five nuns and one legate, the founders of the new house in Antigua, left their convent in Puebla, Mexico, in October of 1699 and arrived in the valley of Guatemala by the end of November.[3] A very simple church had been built in the meantime, and some houses adjacent to it were purchased and utilized as a temporary convent which was occupied by the newly arrived nuns in January of 1700. They remained in these provisional quarters until 1703 when construction began on a new church and convent which were completed in two years.[4] The number of nuns in residence must have increased in the short intervening time, for the *ayuntamiento*, after due petition, granted an increase in the water allotment for the use of the order.[5]

The new and formal convent building consisted of three main enclosing walls, the fourth was presumably shared with the church. It was two stories high and had the usual cloisters, interior sacristy, refectory, kitchen, work rooms, infirmary and other rooms, and sufficient space to house forty-six nuns.[6] The church apparently had a roof of wood and tile, to judge by the descriptions given after its total destruction in the earthquake of 1717.[7] The convent was left unserviceable then, so that the nuns had to seek temporary quarters in the town of Comalapa,[8] where they remained until the roof of the convent was repaired, probably not long after 1720 when Ximénez was writing his account.[9]

A new church and convent were undertaken a few years later, apparently in 1723, for in June of that year permission was solicited from the *ayuntamiento* to place building materials in the little plaza to the west of the church and to extract earth for the mortar to be used in the construction.[10] Money for the new construction was donated by the president of the *audiencia*, Antonio de Echevers Suvisa (or Subiza)[11] who died in 1733 before the church and convent were finished.[12] That year the conventual authorities informed the king, in reply to an earlier inquiry, that the order had lost more than half of its income in the earthquake of 1717 owing to the destruction of some of its holdings in real property, and, therefore, lacked funds to continue work on the church and convent.[13] Yet, both the church and convent were completed, or at least were formally inaugurated and dedicated, on August 11, 1734.[14]

Of the original church and convent from before the earthquake of 1717 not a trace remains. The present ruined structures represent a single unified plan which had been built between 1723 and 1734. But to judge by the information given in a letter dated April 24, 1733, some changes had been carried out in answer to the complaints of the conventual authorities who claimed that certain grave errors had been committed in the construction of the church.[15] The state of the building today seems to indicate that the errors alluded to, mainly with regard to the disposition of certain dependencies for ritual purposes in the church, had been subsequently corrected prior to the date of dedication in 1734.

In the earthquake of 1773 both the church and convent were left in total ruins.[16] The area to the south of the convent, just in front of the main façade of the church, was built up with private houses in the nineteenth century, some of which utilized the convent structure leaving but a very narrow space in front of the façade through which access to the cloister is gained. Apparently not all of the vaulting of the church had been destroyed in 1773, for a century later in the earthquake of 1874 the cupola came down.[17]

Architectural Data

EXCEPT for the two-story arcade of the main cloister, little of the convent remains today. (FIGS. 143, 144.) Not until recent years, in the 1940's, was the cloister cleared of debris and rubbish, and landscaped. Bougainvillea was planted

[2] *Efem*, p. 123; Juarros, **1**: pp. 135 ff.; also Vásquez, *loc. cit.*; Pardo, *Guía*, p. 132; *AGG*, A 1.20 (1692) 696–23, also *BAGG* **10**, 3 (1945): pp. 253 ff.

[3] *Efem*, p. 124; *AGG*, A 1.2–2 (1699) 11776–1782, also *BAGG* **8**, 1 (1943): pp. 92 ff.; Vásquez, *loc. cit.*; *ASGH* **12** (1935/36): pp. 328 ff.; Villacorta, *Historia*, p. 236.

[4] Vásquez, *loc. cit.*; also Juarros, *loc. cit.*; *Efem*, p. 125; Pardo, *op. cit.*, p. 133.

[5] *AGG*, A 1.2–6 (1703) 25573–2848.

[6] Vásquez, *loc. cit.*

[7] *ASGH* **17** (1941/42): p. 156; Ximénez, **3**: p. 356.

[8] *ASGH* **17** (1941/42): pp. 233 ff.; Juarros, *loc. cit.*; González Bustillo, *Razón puntual*, also *Ciudad mártir*, pp. 77 ff.

[9] Ximénez, *loc. cit.*

[10] *Efem*, p. 154.

[11] Juarros, *loc. cit.*; Pardo, *loc. cit.*; Díaz, *Romántica ciudad*, p. 72.

[12] Juarros, **1**: p. 190.

[13] *AGG*, A 1.23 (1733) 1526–310, also *BAGG* **10**, 3 (1945): pp. 255 ff.

[14] Juarros, **1**: pp. 135 ff.; Pardo, *loc. cit.*; Villacorta, *loc. cit.*; *ASGH*, **24** (1949): p. 196; *Efem*, p. 176.

[15] Berlin, *op. cit.*, pp. 52 ff., and fn. 8, a letter found by Fr. Lázaro Lamadrid in the Museo Nacional de México.

A number of documents relative to the founding of Santa Clara have been discovered by Mencos, *Arquitectura*, ch. v, n. 26 referring to *AGI*, Guatemala, 180, and append. XXVII, transcribing *AGI*, Guatemala, 229, "Reconocimiento del Convento de Santa Clara de Santiago practicado con asistencia de los artífices Diego de Porres, Maestro Mayor de Arquitectura, y Antonio Gálvez, Maestro de Carpintería y Baluartes, en Guatemala a 22 v 1734."

[16] Juarros, *loc. cit.*; Pardo, *loc. cit.*; González Bustillo, *Razón particular*, also *Ciudad mártir*, p. 104.

[17] Pardo, *op. cit.*, p. 134; Díaz, *op. cit.*, p. 73.

which now climbs over parts of the arcade. Walks were laid and a fountain built in the center. The present overall appearance, though very attractive indeed, is hardly what the cloister looked like in the eighteenth century.

The cloister is square in plan with ten piers on each side, counting corners twice, supporting nine half-circle arches. The same scheme is repeated in the second story. The piers, square in section and very simple in design, are of the same squat and massive proportions as those of Las Capuchinas, the Ayuntamiento, and the Capitanía.[18] (FIGS. 152, 156, 201.) In other words, they are the supposed earthquake-resistant type developed in eighteenth-century Antigua. The corner piers are treated with responds to line up with each arcade. (FIG. 143.)

The outside face of each pier is treated with a rather slender pilaster with a Tuscan capital. The inside faces of the piers also have Tuscan capitals which return and are intersected by the pilaster of the outside face. Simple unadorned arches spring from the pier capitals. The same pattern is repeated on the second story.

The entablatures are extremely simple. The architrave is a plain, narrow, flat continuous band surmounted by a projecting taenia. The frieze projects slightly and is an unadorned flat and continuous band. It is surmounted with a series of superimposed moldings each projecting beyond the other to form a corbel on which the cornice rests. The cornice consists of three members, the principal one being a large cyma recta. In the second story the architrave is of even narrower width. The frieze projects slightly, and the cornice consists mainly of a cyma recta. The floor between stories and the roof too were once supported on wooden beams.

The church is located along the west side of the cloister. Its east wall actually abuts on the cloister. (FIG. 142.) The plan is of the single-nave type with the *capilla mayor* on the north and the choir at the south end. The nature of the walling at the south end of the church indicates that the main entrance had probably once been located there, but that, after the complaints alluded to in the letter of 1733, it was bricked up and pierced by the octagonal window still there today. The main entrance to the church, as is the case in Santa Catalina and La Concepción, is through two doors which open directly from the nave on the long west side facing the plaza. (FIG. 141.)

The walling of the church is of the common type so widely employed in Antigua, namely, brick and rough stone in lime mortar. But here the exterior is finished with a cut-stone veneer rather than stucco. There are no exterior buttresses at all, not even for decorative purposes. The builders probably considered the interior engaged piers massive enough to support the interior vaulting and the walls between as screens.

The vaulting is of the usual type seen elsewhere, that is, ellipsoidal, pendentive "half-watermelon" domes over rectangular bays. The soffits are treated with false ribs to give the appearance of intersecting vaults. Large octagonal windows are placed in the clerestory formed by the lunettes under the lateral arches engaged to the upper parts of the wall. (FIG. 142.) None of the nave vaulting remains intact except for that over the choir, yet the walls are complete to their original height.

The interior order recalls that of San Francisco (FIG. 56), though it does not have as ornate an entablature. The piers are treated with pilasters which support a very simple entablature from which the transverse and lateral arches of the vaulting spring. The shafts of the pilasters are each treated with a deep single flute or inset panel as seen in San Francisco. But the deep inset is not left plain here. A keel molding rises to an arris in the center of the flute, dividing it in half. (FIG. 142.)

The mezzanine floor between the lower and upper choir is still *in situ*, as are the stairs leading to the upper choir. A caracole stair is still well enough preserved to be serviceable and give access from the upper choir to the roof over the choir bay.

The most interesting feature of this church is the decorative treatment of the south façade and of the two doors located on the west side of the church facing the plaza. (FIGS. 140, 141.) The principal motif of the decoration is based on a candelabrum or ornamental baluster-shape rather than any normal architectural pilaster form. This type of pilaster was also used on the church façade and the convent entrance of the Escuela de Cristo (FIGS. 163, 164) which, in a general way, is employed in a manner similar to the *estípite* but which is quite different in shape and construction.[19] (FIG. 2.)

The façade layout is the usual three-by-three retable type, except that here all relation to structural logic and architectonic form is completely absent. In appearance it is like a retable such as might be built by a cabinetmaker. In fact, the bricked-up door might well serve as a niche for an altar. It is very difficult to appreciate the layout because of the modern constructions which crowd the space in front of the façade. A large buttress comparable to that which appears on the outside corner of the church of Las Capuchinas has been

[18] See chs. III, no. 7, and VIII, no. 3, above.

[19] See ch. X, no. 1, iii, e, above.

obscured by a wall extending from the nearby house and through which a door now gives access to the cloister. But the corner buttress actually emerges from the top of the modern wall and is almost exactly like that on the outside corner of Las Capuchinas even with regard to the materials employed. (FIG. 146.) The other side of the façade abuts directly on the cloister wall so that the retable is set off on one side only by a buttress.

The lower side bays begin with a rather high podium divided into a central panel, cornice, and base. (FIG. 140.) Above, pairs of candelabrum pilasters, four in all, frame a niche and support a very ornate entablature which also continues across the central bay. The individual pilasters of each pair are not set in the same plane. The inside ones, or those immediately adjacent to the niche, are set back slightly. The design of the baluster or candelabrum shaft is very ornate and involved. The basic pattern of the shafts is the scroll or the *f*-hole such as used on violins and cellos. Two pairs form the general outline, separated by a smaller octagonal element with swagged or concave diagonals. The scrolls swing to their widest dimension toward the center, so that the upper pair is narrower at the top and the lower pair narrower at the bottom. The space between the scroll outline is filled with a leaf emerging from a double calyx with a pair of spirals, upright between the upper pair of scrolls and pendant between the lower. A square rosette fills the space in the octagonal element which unites the pairs of scrolls.

A specially designed capital surmounts this ornate candelabrum shaft. It is of rather slender proportions with deeply concave sides which end in tiny spirals. A leaf and rosette fill the space between the two concave sides. An unadorned abacus completes the capital. The base on which the candelabrum or baluster shaft rests is reduced to a plain-surfaced, rather high plinth.

In contrast to the ornate pilasters, the entablature is very simple and reserved. The architrave is composed of two fasciae, the upper one projecting slightly. The frieze is a very narrow unadorned band with a strongly projecting series of narrow moldings above. The cornice is like that used in the cloister, consisting mainly of a high cyma recta.

The second story of the retable façade is more ornate than the first though the "architectural order" employed is the same. The ornate character results from the overall stucco decoration which covers all plane surfaces including the low Roman attic on which the order rests. Directly over the bricked-up door in the central bay, a large octagonal window lights the upper choir. This window is so large that the second-story entablature which runs across the whole retable breaks over the upper part of the window and follows its outline for the three upper sides of the octagon. The remaining five lower sides of the window are left plain.

The third division, the *remate*, is incomplete at present, the remains consisting of a bit of wall with scrolls undulating down to either side to about the middle of the side bays. Four similar candelabrum pilasters frame a small central niche and provide a high central accent to the whole retable. A final element which rose even higher must be imagined as carrying the total height of the retable up to a pinnacle of some sort above the actual finial wall.

Statues of brick and stucco bonded to the façade wall still remain in place in all the five niches of the façade. The first-story side-bay niches are surmounted by triangular pediments. The recesses, however, where the statues are placed are surmounted by an arch, each leg of which is a cyma reversa in profile with a short horizontal piece joining the two. The upper niche repeats the same treatment except that it is surmounted by a broken scroll pediment.

The pair of doors in the long west wall facing the plaza are treated with pilasters similar in design to those of the retable-façade, but are more complicated in detail. (FIG. 141.) The door opening itself is quite simple with a semicircular header and an unadorned archivolt. In contrast to the plain arched opening, the bay is framed with a pair of candelabrum pilasters which support niches. An architrave, which follows a horizontal line only over the door opening, rises over each pilaster and niche. In the sunken part, that is, in the space formed over the door proper, another niche with a statue is placed.

The pilasters do not rest on pedestals, but rather on bases corbeled out from the wall consisting of a very heavy torus molding which stands out in high relief. The pilaster shafts are like those of the retable-façade, except that each is doubled. That is, a broad pilaster of about the same outline with an overall foliage pattern is set back from and underlies the narrower and higher projecting candelabrum pilaster. The elongated concave-sided capital is absent here, for the shaft is surmounted directly by a multiple molding abacus.

An entablature extends only for the width of the pilaster. It consists of a two-fasciae architrave, a heavy, pulvinated frieze almost like the torus molding base below, and then a cornice which projects very boldly. Directly above, a statue, still *in situ*, stands in a niche. The niche is of the type seen on the retable-façade with a cyma reversa header.

The main entablature which unites the bay as a whole is not very high, and is made up of the normal three members. The architrave is a series of narrow buildings, the principal one with a cavetto profile. The frieze is pulvinated, or more like a torus molding, while the cornice consists of four main moldings which duplicate the general profile, but in larger dimensions of the architrave below.

The space directly over the door is occupied by a niche with the same sort of header as those to either side. It is framed by a pair of small, simple, but slightly projecting plain, flat pilasters. This central niche is actually higher than the lateral ones. Broken scrolls connect its diminutive cornice to the cornice of the main entablature. A large octagonal window is set directly above and on center with this niche and the door below. The finial on the niche cornice actually breaks into the window outline. The whole *portada* bay projects forward beyond the plane of the wall and the window.

7. *Las Capuchinas: Church and Convent, 1736*

(FIGS. 145–154.)

Historical Data

THE movement to establish the Capuchin order for nuns in Guatemala began in 1720 when the principal convent in Madrid petitioned the crown for permission to found a house in Antigua. The members of the *ayuntamiento* were less than enthusiastic about having another religious community established in the city. In a report to the king they claimed that the country was impoverished for lack of commerce and coinage and, therefore, could not support another religious order.[1] Approval was nevertheless given in May of 1725. A small group of nuns arrived in Guatemala in 1726 as founders, temporarily occupying some private houses until a formal convent and church could be built for them.[2]

Work on the permanent convent and church began soon after 1726 and was well under way by 1731,[3] for a year earlier the *ayuntamiento*, obeying a royal *cédula*, had informed the king of the progress made in the construction.[4] By 1731 work on the circular cell block for the novices must have already begun, for the abbess petitioned the *ayuntamiento* for two *reales* of water specifically for this section of the convent.[5]

The convent and church were finally consecrated in 1736.[6] By 1740 there were twenty-eight nuns in residence in Las Capuchinas, although they had been limited to twenty-five in the original *cédula* authorizing the establishment of the order in Guatemala.[7] Within fifteen years the buildings suffered some damage in the earthquake of 1751, and the nuns petitioned the *ayuntamiento* for help in making the necessary repairs.[8] In 1770 again they reported that the church and convent were in bad condition and asked the *ayuntamiento* for a license to collect alms to make the necessary repairs.[9] Finally, in the earthquake of 1773 both the church and convent were left in ruin.[10]

The buildings of Las Capuchinas are especially important since they were constructed in one short building period from the foundations up and, except for what must have been some minor repairs in 1751 and 1770, were not altered subsequently as were so many of the important civil and religious buildings of Antigua constructed before 1717.

Architectural Data

THE total length from south to north is approximately 101.15 m. divided as follows: the church—13.15 m.; the cloister—59.00 m.; the circular cell block for novices—29.00. (FIGS. 145, 146.) The depth of each varies so that an irregular open area lies behind the buildings to the west. The site on which the convent and church stand was once larger in size than it is today. The exterior east wall beginning where it abuts on the entrance hall of the convent runs about 87.00 m. north. There it terminates at a modern wall running east-west for about 67.00 m. The grounds which originally extended farther to the west and north are occupied at present by a small coffee farm and some private houses.

i. The Church (FIG. 145-*a*.)

The exterior dimensions of the church are 11.80 m. wide by 40.50 m. long. The north wall is not included in the measurement of the width for it serves as a party wall with the cloister. The interior of the church is single nave in plan divided into five bays. An extra room or chapel lies behind

[1] *AGG*, A 1.11 (1720) 16.779–2292, also *BAGG* **8**, 1 (1943): p. 127.

[2] *AGG*, A 1.23 (1727) 1526–210, also *BAGG* **10** (1945): pp. 256 ff.; *Efem*, pp. 159 ff.; Larreinaga, *Prontuario*, p. 206. See also Mencos, *Arquitectura*, ch. VI, n. 157, quoting *AGI*, Guatemala, 368, "Licencia del Arzobispo de Toledo, D. Diego de Astorga y Céspedes, en Madrid a 21 II 1725" and "Consulta del Consejo de Indias, s.l.s.a., acordada a 10 IV 1725."

[3] *AGG*, A 1.9 (1731) 1380–54.

[4] *Efem*, pp. 168 ff.

[5] *Ibid.*, p. 171.

[6] Juarros, **I**: p. 207, also p. 136; *Efem*, p. 180. A document in *AGI* corroborates Juarros' information, see fn. 8 below.

[7] *AGG*, A 1.18 (1740) 5022–211, also *BAGG* **I**, 2 (1936): pp. 122 ff., **10** (1945): pp. 260 ff.

[8] *Efem*, pp. 205 and 206. The cost of building the convent had been met by the bishop Gómez de Parada who was in office from 1728 to 1751. See Mencos, *op. cit.*, ch. VI, n. 167, quoting *AGI*, Guatemala, 361, "Testimonio sobre el convento de Capuchinas de Guatemala, en Guatemala, año de 1758," and Juarros, **I**: p. 207.

[9] *Efem*, p. 239.

[10] González Bustillo, *Razón particular*.

the last bay or *capilla mayor*. It measures 9.75 m. wide by 5.95 m. deep and is covered with a segmental barrel vault. The length of the nave including the five bays from the back of the façade wall to the rear cross wall is 37.70 m. The interior width of all five nave bays is 8.95 m. The first bay just beyond the door of the main façade is the lower choir and is devoid of niches. The flat vault which supports the upper choir floor is still in place. The next three bays have niches 0.85 m. deep on either side except for the third where a side door pierces the south wall. The total width between recessed niche walls across the nave is 10.75 m. The fourth bay is the crossing and the fifth corresponds to the *capilla mayor*. There are some low platforms still *in situ* abutting on the rear cross wall which served as the emplacement for the retables.

A large arched opening pierces the north wall of this last bay at floor level and opens into an adjacent rectangular chapel in the convent where the nuns attended services. The chapel behind the *capilla mayor* has a niche in the short north wall and a small window with splayed reveals in the south. Two doors, one to either side of the main retable in the *capilla mayor*, give access to this room.

The bays are divided by fluted stone pilasters above the capitals of which half-circle transverse arches 8.95 m. in diameter of the same material span the nave. (FIGS. 147–149.) The connecting lateral arches, also half circle, which, though springing from the same height as the transverse arches, cover a shorter span and are approximately 5.00 m. in diameter. The fourth bay or crossing, however, is approximately as long as it is wide so that the crowns of the lateral arches are at the same level as the transverse. (FIG. 147.) A cupola or dome, now missing except for part of the spherical pendentives, covered the crossing.

The roofing of the other four bays presented a very difficult problem because the crowns of the lateral arches are considerably lower than those over the nave. Ellipsoidal, "half-watermelon"-shaped pendentive domes, *bóvedas vaídas*, with false ribs were forced into the rectangular bays. (FIGS. 147, 148.) This was accomplished by building the pendentives as high as possible up the haunches of the arches and then continuing the brickwork up from the crowns of the lower lateral arches to the height of the crowns of the higher transverse arches. An opening, more or less circular in plan, was thus formed in which a shallow saucerlike crown of the dome was built. The vault as a whole may then be conceived of as half an ellipsoid cut through by four vertical planes. The soffits of these domes are also treated with false ribs of brick and plaster to give the appearance of intersecting vaults. It is interesting that in one or two of the bays the vaulting has cracked just where the spherical portions above the crowns of the lateral arches meet the crowns of the transverse arches making a well-defined circle outlining the saucerlike crown of the vault. (FIG. 148.)

Though there were no impediments to roofing the chapel behind the *capilla mayor* with a simple barrel vault since there are no intersections, a segmental or flat arch was used for the tunnel vault. Actually the masons corbeled the wall out at the height of the springing thus reducing the width to be spanned. A very flat vault was built into the intervening space. The soffit is plastered so that the profile of the vault gives the appearance of having been constructed on three centers.

The *lonja* in front of the church measures 13.15 m. by 8.80 m. and continues around the south side ending abruptly on a line with the back wall of the *capilla mayor*. (FIG. 146.) It also serves as a sidewalk on that side of the street.

Between the buttresses which are employed on the street side of the church, except in the bay occupied by the side door, a masonry fill rises to a height of 3.10 m., the top of which is sharply pitched to join the wall about 1.00 m. higher still. A higher fill abuts on the chapel behind the *capilla mayor* and is 1.45 m. wide or the same width as the side *lonja* which ends here.

The walling material consists of brick and rubble laid in a lime mortar covered with a stone veneer in lieu of stucco, but only on those sides of the building facing the street. The party wall abutting on the convent to the north, including even the top part exposed above the height of the adjoining cloister, is simply stuccoed as are the convent buildings.

The façade is very simple in design and is noteworthy for the absence of superfluous nonarchitectonic surface ornament. Two plain, asymmetrical, buttresslike towers frame the central retable portion of the façade. The south buttress is 2.50 m. in plan on both exterior faces. The north buttress which disappears into the wall of the convent vestibule is about 1.90 m. at the foundation level though only about 0.20 m. is exposed. This north buttress emerges from the one-story convent vestibule wall and rises to the total height of the façade, the exposed north corner of which is chamfered reducing its width to about 1.00 m. on the façade side.

The central part of the façade, the retable, is divided into three vertical bays and three horizontal stories. The central bay is the widest. The side bays are considerably narrower and are but two stories high. Two engaged columns frame niches in each story making a total of five niches including that in the third-story central bay which is treated like a finial. Very simple engaged Tuscan columns with unfluted shafts decorate the façade. These rest on unadorned individual pedestals in the first story rather than on podia as is so common elsewhere. The shafts rise from a simple Attic base with two tori separated by a scotia. The entablature as a whole breaks miters over column centers lining up with the pedestals below. The architrave consists of two fasciae and a two-member crown molding. The frieze is a plain unadorned band surmounted with a widely projecting cornice.

The niches, semicircular in plan, are not framed, nor do they have any special decorative treatment. A small corbel provides a larger space at the bottom of each niche for the emplacement of statues, now missing. The same austere treatment is repeated in the lateral bays of the second story. The columns here are slightly shorter, though of the same proportions. They stand on a low Roman attic which breaks miters over column centers to form short pedestals.

A large door with a semicircular header occupies the central bay of the first story, and, except for the normal masonry treatment of the voussoirs, the archivolt is not given any special attention. The reveals of the door jambs are quite complicated in section, the profiles of which continue on the intrados of the arch. The entablature of the side bays continues across the central bay. The only special feature of the second story of this bay is a large window with splayed reveals and a flat header arch.

The third story, really a finial or *remate*, is a free-standing wall above the roof line and consists, for the most part, of a niche set on door center. Its shape is exactly like that of the other four niches, and it too is flanked by Tuscan columns of commensurate proportions. Very simple mixtilinear half pediments flank the niche and carry the eye to either side down to the lower parapet wall which joins the buttresses at either extreme of the façade. The pediments and parapets are accented with projecting moldings terminating in spirals.

ii. The Convent (FIG. 145-*b*.)

The main cloister is entered from the street through a vestibule projecting from the main body of the convent and abutting on the church and the *lonja*. Access to the church is also provided through the vestibule. Originally the main cloister was two stories high, but only the lower exists today. (FIGS. 150–153.) The rooms which range around the arcaded corridors are about 5.00 m. in width, but vary in length. The room in the southeast corner served as the chapel where the nuns could attend services, the party wall of which is shared with the church. It is pierced by a wide opening in which a grill may have been originally located. The fixed iron grill there now may be modern. The east and west walls of this chapel have niches in which altars were probably once placed. In the rooms directly on the east-west axis of the cloister remains of stairs still lead up to the second story. The west stair is in a better state of repair. Underneath, another narrow flight leads down from the cloister corridor to a basement below. The east stair has been altered considerably.

Passageways from the east and west arcaded corridors lead to two small courts on the north side separated by the kitchen. The north wall which extends across the two courts and the kitchen, terminating on the exterior of the structure, is not squared up with the east and west walls. The west court is 12.75 m. long on the north-south axis and the east court on the other side is 12.55 m. long. The area to the north of these two kitchen courts is a jumble of adobe and tamped-earth walls in a state of almost complete ruin. Beyond this ruined area is the circular block of cells for the novices, treated separately below.

The main cloister including the four corridors is 27.80 m. on each side with but very slight variations. The corridor width is 4.45 m. more or less from the wall to the edge of the step. The central unroofed area is 18.70 m. square measured on the east-west and north-south axes. Ranged around the central open area are twenty short, squat Tuscan columns. The four in the corners are treated as half-column responds attached to square piers at the end of each of the four arcades. (FIG. 150.) Each of the twenty-four bays thus marked off by the columns is roofed with a small pendentive dome. When measured on column centers and to the face of the wall behind, the bays are approximately 3.90 m. square at the floor level. (FIG. 151.) The distance from column center to column center varies from 3.85 m. to 4.00 m. At the height of the springing of the arches these dimensions change, and the square plan at floor level is altered because the added thickness of the arcade wall reduces the opening (from the face of the back wall to the inside face of the arcade wall) about 0.34 m. The width of the bay at the springing thus is approximately 3.56 m. instead of from 3.85 m. to 4.00 m. at ground level.

The actual span, at the springing, of the arcade columns is 3.00 m. The arches, however, are not complete semicircles. (FIG. 151-*a* and *c*.) The transverse arches which span the corridor vary from 3.56 m. to 3.50 m. and are also segmental. The crowns of both the arcade and transverse arches are of equal height, about 3.55 m. from the floor. The total height of the arcade to the top of the cornice is 4.55 m. The pendentive domes, *bóvedas vaídas*, were forced into the off-square openings and were laid out on four separate centers. (FIG. 151-*a*.) Each bay of the cloister was no doubt conceived of as a cube, for the height to the cornice, 4.55 m., is exactly the width of each bay on the arcade side measured on column centers, and the same dimension from the step to the face of the rear wall. (FIG. 151-*a* and *c*.)

The columns, the plinths of which are 1.30 m. square, are set 3.90 m. on centers. Subtracting half of the plinth width, a dimension of approximately 3.90 m. to the wall face behind results. At floor level, therefore, each bay thus measures 3.90 m. square on centers. But the *maestro de obras* did not take into account that, once he set the arches of the arcade in place, half the girth of the arcade wall above went beyond the column center thus reducing the span of the corridor to the wall behind from approximately 3.90 m. at ground level to about 3.56 m. at the level of the springing of the arches. Approximately half a column diameter had been added at the springline of the corridor arches.

The miscalculation was probably observed too late, perhaps at the time the transverse corridor arches were being constructed. In order to level up the crowns of all four arches in each bay, segmental rather than half-circle arches were employed. (FIG. 151-*a*.) But in filling the space between the haunches to form the pendentives, an elliptical rather than a circular opening resulted. The domes are then not true pendentive domes, *bóvedas vaídas*, that is, hemispheres cut through by four vertical planes meeting at the corners of the square bay. Instead, they are rather flat half-ellipsoids. The rough brick work was then covered up with a coat of plaster and finished to give a spherical appearance to the whole.

A more logical solution might have been achieved on the level of the springing either by having added 0.65 m. to the arcade wall (it is actually 1.70 m.) in order to square the bay, or else by constructing corbeled arches 0.325 m. on both the rear wall and the interior face of the arcade wall. (FIG. 151-*b*.) Had such a solution been chosen, it would then have been necessary to reduce the width of the soffits of the transverse corridor arches also. Though logical geometrically, a solution of this sort would have been rather cumbersome and probably also have weakened the structure. The solution resorted to, though theoretically illogical, has had the virtue of stability having withstood two earthquakes in the eighteenth century, and an unknown number in the nineteenth and twentieth centuries, and is still standing for the most part today.

The proportions of the columns and the arcade elevation are worthy of note. (FIGS. 151-*c*, 152, 153.) The total height of each column is approximately 2.54 m. divided as follows: the base, 1.30 m. square on the plinth, 0.57 m. high; the shaft—1.48 m.; and the capital—0.41 m. The lower diameter of the shaft is 1.01 m. and the upper, 0.85 m. The total height of the story to the top of the cornice, as has already been stated, is 4.55 m. Using the square plinth as a module, the intercolumniation is exactly 2. Each bay, inclusive of two whole columns, is then 4 plinths wide. The total height of the elevation is then 3½ plinths. The total height of the column itself is slightly less than 2 plinths. The space between the spandrels of the arches above is equal to the upper diameter of the column, or about ¾ of a plinth. The height to the crown of the arches is 2¼ plinths or slightly more than an intercolumniation. The elevation of each bay is then almost a square. The points of origin of the segmental arches were found to be located at the intersection of the diagonals of a square formed by the axes of the columns, the floor line below, and the top of the lowest member of the cornice above. (FIG. 151-*c*.)

iii. Novices' Quarters (FIGS. 145-*d*, 154.)

This dependency is a two-story circular structure, the upper floor of which appears on the plan.[11] The lower story is really a basement partly below grade. It is entered through one of the cells and also from a room in the part of the convent now in almost complete ruin. A very massive central circular pier about 3.00 m. in diameter supports a barrel vault about 4.50 m. wide which runs around the circular plan of the structure. The vaulting springs from the central pier and the perimeter wall. The latter also serves as the foundation, or bearing partition, for the inside wall of the cells on the upper story. Another concentric, exterior wall continues up and also serves as the outer wall of the cells above. On the basement level, the space between the two concentric, circular walls is occupied by a sewage trench where a continuous flow of water once ran. This trench lies directly below the water closets with which each cell on the second story was equipped. Adjacent to this circular base-

[11] Circular cloisters are indeed rare. That of Las Capuchinas is probably without precedent in the New World or in Spain. One such cloister is described by Philibert de l'Orme in 1626. It was located in Montmartre, Paris, and is quite different in many respects from the circular cell building of Las Capuchinas. In the Parisian example a portico fronts the cells, and the whole structure is roofed with an elongated timber-framed dome. It is most unlikely that there is any direct connection between the French cloister and that of Las Capuchinas of more than a century later. See Markman, *Las Capuchinas*, p. 33, fn. 16.

Mesa and Gisbert, *El edificio circular*, are of the opinion that the circular structure was probably a public bathing establishment of seventeenth- or early eighteenth-century date which was incorporated with the convent proper when the latter was built. This is an interesting suggestion but, unfortunately, neither in the contemporary literature nor in the archival documents has any evidence come to light to indicate specifically that public baths ever existed in Antigua Guatemala. In the vicinity of Antigua there still exist some hot mineral springs known and used continuously since before the Conquest which the authors feel may have inspired the construction of steam baths in town.

ment room are a number of bathrooms with large masonry bathtubs.

The upper story where the cells are located is approached by a sloping passageway from the east kitchen court. The ramp turns slightly west and has two sets of steps in order to gain added height without increasing the slope of the ramp. The total diameter of the cell block or tower, as it is sometimes called, including the open central circular area and the cells as well, is approximately 24.00 m. The diameter of the open central court is 12.00 m. exactly.

The eighteen cells, arranged around the perimeter of the structure, are not of equal size, nor do their side walls radiate from the center point of the open court. (FIG. 154.) Each cell is entered through a door approximately 0.85 m. wide. A window in each cell approximately 1.20 m. wide pierces the exterior perimeter wall. Just inside the narrow entrance of each cell is a water closet recessed in a niche. An opening in the seat lines up with the sewage trench in the basement below. A circular opening in the vaulted roof above serves as a vent. Besides the water closet, there are two recesses of different sizes in the entry in which a wooden clothes press and a cupboard respectively were once located. The vaulted roof over the entry supports a terrace above. The wall separating the entry from the cell proper rises about 3.00 m. above this terrace floor and once supported one end of the wood rafters of a shed roof over the cells proper. The other ends of the rafters rested on the outside circumference wall.

In addition to the eighteen cells and the entrance passageway, there is also an air shaft which occupies almost as much space as a cell and which formerly served as a vent for the sewage trench in the basement. In one of the cells in the southwest section a winding stair leads down to the ruined area of the convent and to the basement.

On the east side of the exterior of this circular dependency the remains of a wide stair once led to the upper parts of the convent adjacent to the kitchen courts. At ground level around the exterior of the building are a number of small recesses just large enough for a person to stand. These once had wooden doors, as the remains of some of the hinges prove. To what use these small recesses were put it is hard to say, perhaps for solitary prayer or perhaps punishment.

8. *Ayuntamiento, 1743*

(FIGS. 155–160.)

Historical Data

THE present structure dates from 1743, but was preceded by a far less monumental building. The building which housed the municipal government was sometime also referred to as the Casas Consistoriales or Casa de Cabildo. In 1629 the *ayuntamiento*, city council, contracted the master carpenter Damián Rodríguez to erect the pillars of the new corridor of the "Casas de Cabildo."[1] Some further improvements were carried out three years later, in July, 1632, at which time the city council paid the sum of 3,000 *maravedis* for repairing the "Sala de Cabildos" which had been damaged in the earthquake of February 18, 1631.[2] Again in 1668 it was necessary to rebuild the roof of the building and also to repair the tile roof of the corridor which faced the plaza, no doubt the one built by Damián Rodríguez in 1629. These repairs were completed by July 9, 1669.[3] In 1673 the appearance of the building was probably enhanced with the addition of iron railings in the corridor at the cost of 6,000 *maravedis*.[4] In the earthquake of February, 1689, the building was slightly damaged and the next month a committee was appointed to see about the repairs necessary.[5] These were carried out the following year in October, 1690, at which time chains were put in the corridor across the front of the building to keep unauthorized people out.[6] Despite these repairs, the general condition and utility of the building must not have been commensurate with the aspirations of the *ayuntamiento*, for five years later, in 1695, the *audiencia* granted a license to reform the whole structure.[7] Work was still in progress in February of 1696 at which time 400 *pesos* were still required to complete the job.[8]

This was, then, the seventeenth-century patchwork Ayuntamiento that was standing when the earthquake of 1717 laid the city of Antigua waste. The building was very badly damaged, but there were no funds available to repair it immediately.[9] Within a year or two this situation had altered for the better, and the building was repaired and put in use again.[10] Thus a makeshift building which had been extensively repaired in the previous century continued in use until about 1740 when the construction of a brand new building from the ground up was undertaken, as may be deduced from a contract in which one Eugenio Alberto de Bocanegra agreed to deliver twenty-four wagon loads of stone daily at a *real y cuartillo* per load, and an equal number of loads of brick at twelve *reales y medio* each.[11] Within three

[1] *Efem*, p. 49.
[2] *Ibid.*, p. 50.
[3] *Ibid.*, pp. 78, 79.
[4] *Ibid.*, p. 83.
[5] *AGG*, A 1.2–2 (1689) 11.778–1784.
[6] *Efem*, p. 109.
[7] *AGG*, A 1.2–2 (1695) 11.779–1785.
[8] *Ibid.*, A 1.2–2 (1696) 952–39.
[9] *BAGG* **8**, 1 (1943): p. 123.
[10] *Ibid.*, p. 126.
[11] *ASGH* **24** (1949): p. 360.

years the work was finished for the most part and the first *cabildo* was celebrated in the new building on November 19, 1743.[12] A brief account of the construction work as well as a description of the plan is recorded in this same *cabildo* where it is noted that in 1739 the decision to build a new municipal building was reached owing to the influence of Sr. Dn. Pedro de Ortiz de Letona, *alcalde ordinario*, and that Sr. Dn. Juan Batres supervised the building operations. There is no indication as to who the architect was, but it may be supposed that Batres had much to do with carrying out the plan since he represented the city council during the construction.[13] It may be that Manuel José Ramírez, *maestro mayor de obras* may have had something to do with the technical aspect of the construction, but this is not certain. His name appears in connection with the building just four years after it was occupied when he advised the *ayuntamiento* that some of the vaults had been damaged because of dampness and that the bell tower was in danger of falling for the same reason.[14]

The building occupies only half of the north side of the main plaza. In the eighteenth century, and as is still the case today, a number of private houses occupied the other half. In the *cabildo* of December 7, 1761, it was proposed that the *ayuntamiento* buy these adjacent houses in order to continue the arcade across the whole north side of the plaza. These were finally acquired in 1762 for 20,000 *pesos*.[15] A remaining bit of real estate on that side of the square, a two-story house, was purchased in 1763.[16] But nothing was done right away until August of 1766 when Francisco de Estrada, *maestro mayor de arquitectura*, presented some plans and estimates for the project to the *ayuntamiento* which were approved. Manuel Batres was put in charge of the work as the representative of the city council; he may have been related to Juan Batres, the superintendent in charge of the 1740 construction.[17]

[12] *Efem*, p. 195.

[13] Batres Jáuregui, *América Central* **2**: p. 526, assumes that Juan Batres was the architect. But this is not specifically stated in any of the contemporary literature or documents though he did, however, turn the building over to the *ayuntamiento* in October, 1743; *BAGG* **8**, 1 (1943): p. 131.

[14] *Efem*, p. 199. Angulo, *Hist. del arte* **3**: p. 22, believes that Diego de Porres (sometimes given as Porras), who supposedly worked on the Capitanía and who was *maestro mayor de obras*, may possibly have had something to do with the design of the building. In 1740, when work on the Ayuntamiento began, he was already an old man who have been active during the earthquake of 1717 and earlier still in the work on the convent of La Recolección, see García Peláez, **3**: p. 27. He died and was replaced as *maestro mayor de obras* by Juan de Díos Aristondo in 1741, two years before the building was completed; *Efem*, p. 191.

[15] *ASGH* **25** (1951): pp. 148 ff.

[16] *Efem*, p. 224.

[17] *ASGH op. cit.*, p. 160.

But the project was never begun, let alone carried out, for the tenants residing in the acquired property were strongly opposed to leaving. The matter was brought before the *audiencia* and litigation actually continued until the earthquake of 1773.[18] Even had the city not been destroyed and the civil and ecclesiastical capital not moved, the proposed addition might never have been built, at least not in the eighteenth century. In March of 1773 a decision had been rendered supposedly in favor of the *ayuntamiento*, but with the proviso that construction begin only after the expiration of the tenants' leases. These leases had been made for *dos vidas*, for two lifetimes, in 1762 and 1772, and the first tenants had not yet died.[19]

The building suffered relatively little damage in the earthquake of 1773, for even according to González Bustillo, who tends to dramatize the destruction, only one of the inside vaults had fallen and the bell tower had cracked.[20] But when the building was abandoned it probably fell into a bad state of repair. It was restored and put into service again in the mid-nineteenth century through the efforts of the *corregidor* J. María Palomo y Montúfar.[21] (FIG. 21.)

Architectural Data

THE building is two stories high, the most interesting side, the arcaded façade, facing the plaza to the south (FIGS. 155, 156.) The west wall abuts on some one-story private structures in front of which there extends a corridor supported on wooden columns and roofed with tile. Though of modern construction, it is probably not too far different in appearance from the one Damián Rodríguez built in 1639. The north side abuts on private houses while the east wall, covered with a stone veneer faces a side street. Except for the cornice continuing around from the main façade and the five windows in the upper story and the two doors at ground level, this side is unadorned. The appearance of the veneer from the distance is that of large square blocks of coursed ashlar masonry. But on closer inspection each large block is seen to be made up of smaller stones set with thin joints in a dark mortar. The appearance of large stones is achieved by means of interspersing joints of light mortar every two or three courses of the smaller blocks. All walls throughout the building including the veneered east wall are 1.25 m. thick.

[18] *Efem*, pp. 161, 172.

[19] *Ibid.*, p. 244.

[20] *Razón particular*; also *AGG*, A 1.18.6 (1774) 38.306–4502.

[21] Brasseur de Bourbourg, in a letter dated July 25, 1855, *ASGH* **24** (1949): pp. 164 ff. The Muybridge photo (fig. 21) mentioned, taken in 1876, shows the building to be exactly as it is today except for some minor details due to pointing up of the masonry and exterior painting.

The overall ground plan of the building is approximately square. The width of the corridor is 38.17 m. on the top step of the stylobate measured from the outer corner of the piers at each end. A stair with four risers leads up to the stylobate, the bottom riser continuing as part of the sidewalk on the east side. A small merlon is set in the angle where the sidewalk turns the corner to the side street. The ground level falls to the west so that the lowest riser is greater at that end, and so must be considered as a leveling course of the three-stepped platform which is 0.50 m. high.

The columns of the first story are set 3.65 m. more or less on center. (FIGS. 157, 160.) Each plinth is exactly 1.00 m. square and the actual opening between the plinths is 2.65 m. The corridor is 4.05 m. wide from the edge of the stylobate to the wall in back, the resulting open area between the plinths and the wall thus is 3.05 m.

At the east end of the corridor four risers, lower than those on the main façade, lead up to the stylobate from the sidewalk. A quarter-column respond is engaged at this point to the rear wall to balance the half-column engaged to the pier opposite. At the west end of the corridor there are no steps, but a slightly inclined ramp of modern construction leads down to the wooden corridor fronting the adjacent buildings. The lack of the steps here no doubt points to the fact that the project of continuing the arcading of the Ayuntamiento was already considered when it was first built in 1740–1743. This conclusion is further borne out by the absence of a corresponding engaged quarter-column on the wall at the west end.

The corridor is divided into ten bays, each approximately square in plan. (FIG. 159.) Each is covered with a small pendentive dome, *bóveda vaída*, which is geometrically accurate, more or less. The span from the column centers to the wall face behind is 3.65 m., and the same between centers of the columns, thus forming the square bays. The arcade arch, the two transverse arches, and the face of the wall behind mark the four vertical planes set at right angles to each other which cut through a hemisphere whose diameter is the diagonal of the square thus formed.

The height of the first-story columns is 2.30 m. more or less. (FIG. 160.) The diameters of the half-circle arches are 0.65 m. less than the distance from column center to column center, that is, 3.00 m. The crowns of the arcade arches are about 1.50 m. high, the same dimensions being true for the transverse corridor arches, both being complete semicircles. The total height from the stylobate to the intrados of the crowns is 3.75 m. more or less. The total height of the order to the top of the cornice which divides the two stories and forms the stylobate of the second is 4.65 m. Each bay is thus a square in elevation, 4.65 m. high and the same in width when measured to include the columns to each side.

The second-story arcade repeats the same scheme as below. (FIG. 157.) The columns are spaced on the same centers and each bay is likewise roofed with a *bóveda vaída*. The second story is 4.75 m. high, approximately 0.10 m. more than the lower. The total height of the building including the low parapet wall above and the steps below is about 10.40 m., making a proportion of about 1:4, height to width of the façade as a whole.

But though the second story follows the same scheme as below, there are some variations in the details making it appear somewhat lighter and not as massive as the lower. The reason for this effect is revealed in the dimensions of the columns which are 2.20 m. high or about 0.10 m. less. The shafts too are smaller in diameter and have three flutes less than those below. The bases and capitals, however, are of about the same dimensions as those below. The semicircular arches which spring from the abaci are slightly wider than those below, a little more than 3.00 m. in diameter. The crowns thus are slightly higher than those of the first story. But the soffits are still the same height from the floor and preserve the same dimensions for the bay. This commensurability is gained by the shorter height of the columns. The result is that the proportion of void to solid is greater in the second story than in the first giving the whole a lighter appearance.

The columns are especially noteworthy. (FIGS. 160, 155.) The first-story columns are about 2.30 m. high divided as follows: the plinth—1.00 m. square and 0.27 m. high; the base consisting of three members, a torus, a scotia, and a smaller torus—0.31 m. high; the shaft—1.38 m. high, with sixteen concave flutes coming to fillets; the lower diameter of the shaft—0.73 m., and an upper 0.68 m.; the height of the Tuscan capital from the astragal and the necking to the top of the abacus—0.38 m. The proportions are, therefore, very squat if either the plinth or lower torus is used as a module, less than 1:2½, and not much more slender if the bottom diameter of the shaft is used, about 1:3.

The second-story columns are lighter though more than 0.10 m. shorter, that is about 2.20 m. (FIG. 157.) The heights of the capitals, plinths, and bases are about the same as in the lower story. The shaft, however, is about 1.25 m. high. The upper and lower diameters are correspondingly reduced so that only thirteen flutes with fillets, rather than sixteen as below, fit on the perimeter. The fluting and fillets in the lower-story columns are 0.07 m. wide. There is a slight variation above. The flutes also measure 0.07 m., but the fillets are slightly broader measuring 0.09 m.

The arcades on both stories end in massive piers, 1.55 m. by 1.45 m. in plan, with the responds in the form of engaged half columns. (FIG. 158.) Another respond is attached to the corridor side of the pier. Eight flutes are cut on the main façade faces of the piers and seven on the side surfaces. The fillets have an added astragal which projects to counter the concavity of the fluting, a device noted in El Calvario. (FIG. 10.) The responds engaged to the inside faces of the piers are somewhat less than half columns in diameter. The soffits of the arches of the arcades are convex and scored with the same fluting as the shafts. The cornices which cap each story are about 0.50 m. high and are made up of a series of superimposed moldings projecting markedly.

A low parapet wall about 0.50 m. high runs the length of the façade above the top cornice. (FIGS. 155, 156.) Merlons consisting of a square base and ball are set on the column centers. None exist at the corners today, though they may have been there originally. The clock tower on the roof replaced the original one. The present one is probably of nineteenth-century origin and does not fit the façade at all.

The interior of the building has been considerably reformed since it was put in use in the nineteenth century, but much of the walling and the vaulting is still intact. The plan reveals a central patio with the rooms arranged around it. Various stairways give access to the upper stories and the roof as well.

9. *Escuela de Cristo: Church and Convent, after 1720 and after 1740*

(FIGS. 10, 11, 161–166.)

Historical Data

OFFICIAL sanction to found this order, the full name of which was Convento de la Congregación de San Felipe Neri y Templo de la Escuela de Cristo, was first granted in a papal bull dated May 25, 1683, confirmed by the Consejo de Indias on October 10, 1697, and transmitted to the *Real Audiencia* on March 16, 1699.[1] The convent had been founded much earlier *de facto* when Bernardino de Obando established the church and convent in 1661.[2] This is further borne out by the fact that funeral services for Pedro Betancur, the founder of the Hospital and the Convent of Belén, were held in the church of the Escuela de Cristo in April of 1667.[3] Further evidence that the establishment existed before 1683 is given by Vásquez who relates that a sanctuary had been erected about fifty years before the time he was writing, that is, about 1645, called the Escuela de Cristo which served as an oratory for the Congregación de San Felipe Neri and which was located on the site where the convent of San Francisco had first been built. This earlier structure of San Francisco had been converted to a hermitage under the advocation of La Vera Cruz and served as a chapel for the Indians of the neighborhood.[4] This small hermitage had deteriorated with the passage of time, and may possibly have been the same one used by the Escuela de Cristo until its destruction in the earthquake of 1717.

A new church and convent were begun in 1720 with the funds donated by the president of the *audiencia*, Francisco Rodríguez de Rivas, who held office between 1716 and 1724. Only the church was finally completed in 1730.[5] As late as 1740 the old convent was still in such bad condition that it could not be inhabited. The new one begun twenty years before must have been completed not long after.[6]

In 1751 the church was damaged somewhat in the earthquake[7] and was not yet repaired by 1753.[8] In the earthquake of 1773 the new convent, which must date from about the middle of the century, was almost completely destroyed. The church, however, is one of the few which has remained standing. It has doubtless been changed by subsequent repairs. In 1957 the cloister of the convent was in the process of being restored. The plaster which had fallen from vaults, walls, and piers was being applied anew. A new brick floor was laid in the corridors. Some restoration work of the same sort had already been done on the church in the 1940's at which time the south belfry of the church façade was repaired, and in 1957 the north belfry was restored.

Architectural Data

THE church plan is similar to that of San Francisco a block away, single nave with salient transepts. The nave is three bays long, each with an ellipsoid, pendentive dome over the rectangular plan. The transepts and the projecting presbytery are also rectangular in plan and roofed with similar

[1] *AGG*, A 1.18 (1741) 5029–211, also *BAGG* **1**, 2 (1936): pp. 136 ff.

[2] Fuentes y Guzmán, **2**: p. 78.

[3] *Efem*, p. 76.

[4] Vásquez, **4**: p. 384.

[5] *AGG*, A 1.11 (1720) 16.779–2292, also *BAGG* **8**, 3 (1943): p. 129; Ximénez, **3**: p. 420; Juarros, **1**: p. 190; also García Peláez, **2**: p. 159. See also Mencos, *Arquitectura*, ch. v, and append. XXVII, transcribing *AGI*, Guatemala, 309, "Reconocimiento de la Escuela de Cristo y tasación efectuado por el alarife Diego de Porres, en Guatemala a 11 VI 1720," and ch. v, n. 96, quoting *AGI*, Guatemala, 369, "Información de la Congregación de San Felipe Neri y Escuela de Cristo de Guatemala, en Guatemala 18 XI 1730."

[6] *AGG*, A 1.18 (1741) 5029–211, also *BAGG* **1**, 2 (1936): pp. 136 ff.

[7] *AGG*, A 1.10.3 (1751) 18.807–2448.

[8] *Ibid.*, A 1.10.3 (1753) 18.808–2448.

domes. (FIGS. 161, 165.) The crossing, however, is square in plan and roofed with a dome on pendentives and is surmounted by a lantern through which light is admitted. The interior of the church has been subjected to renovation in the course of time, but the eighteenth-century layout is still in evidence; as for example, the pilaster fluting which is almost a replica of that employed in the church of Santa Clara. The single deep flute has an inset molding which simulates the general outline of a keel molding composed of a quarter round, a fillet, a scotia, another fillet, and then on the very top an astragal or half round. (FIG. 10.)

The walling is not unlike that of the Ayuntamiento, Santa Clara, and Las Capuchinas, stone veneer over a brick-and-rubble wall. (FIGS. 161, 156, 141, 146.) The veneer is generally made up of the large rectangular slabs of stone with joints pointed up with a white mortar which may have been renovated in years past. Only on the west or principal façade is the stone laid in regular courses; on the side and rear walls the blocks tend to be smaller and rather square and are generally uncoursed.

The main façade which is oriented to the west is preceded by a *lonja* rising about 1.50 m. at the highest point above the sidewalk in front. It is cut through by stairs directly in front of the church door. (FIG. 161.) A low parapet wall girdles the *lonja*. Merlons topped with spheres flank the staircase, and a small freestanding column is located in the corner where the *lonja* returns on the street running east-west.

The façade layout is remarkably simple, even more so than that of Las Capuchinas. The central retable is flanked by two towers surmounted by belfries with a single arched opening on each of the four sides. Both belfries have been extensively restored. The scheme of the retable is like that of other *antigüeño* churches in that it has three vertical and three horizontal divisions. But in this case the finial is not confined to the central bay and continues across all three bays. A low balustrade above runs between the two belfries.

In each of the lateral bays on all three stories there is a large, rectangular inset panel rather than the customary niche. (FIG. 163.) It is no more than an unadorned, shallow depression in the masonry wall where statuary could not have been set. The central bay, scarcely wider than the lateral bays, is pierced by a large door with a semicircular header set in a deep recess with a concentric header. A small but deep niche is located in the second-story central bay where a stone cross with the symbols of the order is placed. The third-story central bay has a large rectangular window with complicated profiles at the corners, so that its outline is, more properly, octagonal.

The applied orders which adorn the façade and which flank the recessed panels are nonarchitectonic. The pilasters of the first and third story are shaped like candelabra or balusters. The architrave above is made up of a number of fasciae. Above that a pulvinated frieze is surmounted in turn by a very boldly projecting cornice. The entablature breaks miters over each of the pilasters.

The second-story order is shorter and quite simple. (FIG. 161.) The shafts of the pilasters are flat bands accented with four deep concave flutes and surmounted with an uncomplicated Tuscan capital. The architrave is completely eliminated in this story except for small sections over each pilaster. The frieze is an unadorned flat surface, while the cornice above is of smaller projection than that of the first story. This cornice also extends across the width of the towers to either side.

The pilasters of the third story are the same candelabrum type as those of the first story, though not as complicated in detail. The architrave is also completely eliminated here. Instead, the abacus moldings of the Tuscan capitals of the pilasters continue across the entire retable. The cornice continues across the towers as well. The balusters of the balustrade which connects the two belfries are also a simplified version of the candelabrum pilasters. Still another version is employed on the convent entrance. (FIG. 164.)

A dominating feature of the exterior design is the cupola over the crossing which has false exterior ribs and is surmounted by a lantern. (FIG. 165.) Lacking a drum, the windows are cut through the lower part of the haunches. At intervals, especially where walls make returns at corners, merlons with pear-shaped finials are set and provide an accent to the exterior design. Many of these merlons may actually be modern; they are of the same type seen on the Capitanía, most of which were restored in the latter part of the nineteenth century.

Little remains of the convent except the two-story cloister. The dimensions and proportions of the arcades are very close to those of the Ayuntamiento. (FIGS. 162, 166, 160.) The columns are set 3.56 m. on center. The total height of each story from the floor of the corridor to the top of the cornice is 4.48 m. The second story is exactly the same height and repeats the same layout as below. The height of each column is 2.01 m. The height to the crown of the semicircular arches is 3.30 m. at the archivolt and 3.20 m. to the soffit. The span at the springing is 2.67 m. at the archivolt or extrados and 2.45 m. at the rounded soffit or intrados. The height of the crown from the point of origin on the spring line is about 1.30 m. at the extrados and about 1.20 m. at the rounded soffit, or just half the diameter. The cornice is made up of two major members, the total height of which is 1.00 m. with a projection 0.38 m.

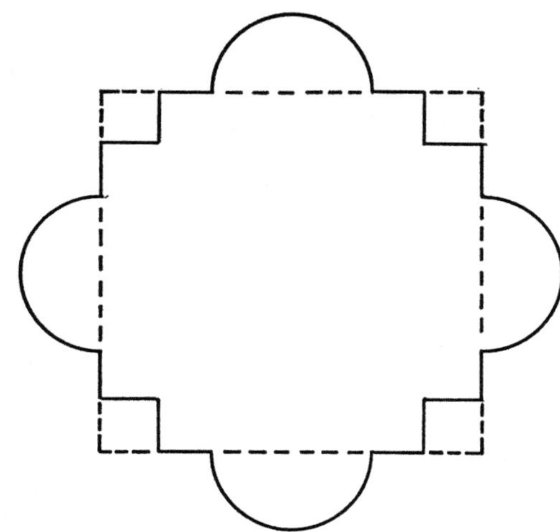

FIG. 11. Escuela de Cristo. Cross section of cloister column shafts.

The columns are very interesting, especially in section, for the same outline is frequently used for window openings in eighteenth-century *antigüeño* buildings. (FIG. 11.) The plinth which is 0.17 m. high and 1.25 m. square has rounded corners. The base is a simple three-member Attic type with two tori separated by a scotia. The shaft is quite complicated in section. Beginning with a square, each corner is cut back about 0.05 m. from each face to make a square chamfer. In appearance the effect is as if to each side of a square pier a broad, flat pilaster had been attached. On the flat surface between chamfers a half-round column is added, the profile of which continues up as a rounded soffit on the intrados of the arcade arches. (FIGS. 162, 166.) The same profiles are used for the transverse arches of the corridor. On the exterior toward the open central court, the half column emerges above the capitals and continues up to the cornice at the top of the story thus framing each arched opening of the arcade. (FIG. 165.)

10. *San José el Viejo: Church, 1740 and 1761*

(FIGS. 167–170.)

Historical Data

A MOVEMENT to found this hermitage was begun approximately in 1740.[1] In 1742 the *ayuntamiento* reported to the crown that alms had been collected and construction on the building started. Approval in the form of a license was, therefore, requested to legalize the work already done.[2] The authorities in Spain took a dim view of this procedure, and in June of 1744 ordered the hermitage closed and the building torn down. The municipal authorities pointed out the need for a place of worship in the populous neighborhood where San José is located, and urged the crown to allow divine services to be continued.[3] The order to close the church had actually arrived in Guatemala in January of 1745, almost three years elapsing between the date when the original petition for the license had been made and the arrival of its denial, or five years after construction had begun. Apparently work on the building did stop when the order arrived, but the completed sections were not torn down, and the church remained standing in an incomplete state. In the meantime, during the earthquake of 1751, the partially built structure was hurt, but there is no record of what damages exactly were occasioned other than that a straw hut had to be built temporarily in the atrium.[4]

Whether religious services had continued in the incompleted building after it had been ordered demolished is difficult to ascertain. But apparently by 1759 the suit for reinstituting religious services was brought once again before the civil authorities, the *audiencia*.[5] This action must have been successful, for building operations began again and were completed by 1761, including some additions to the original structure. The church was finally inaugurated February 20, 1762.[6] The finishing touches were carried out the year following in 1763.[7]

In the earthquake of 1773, to judge by the present remains, the building did not suffer much, only some of the vaulting having come down. It was sound enough to serve as a temporary convent for the Carmelite nuns of Santa Teresa.[8] Just how the building fared during most of the nineteenth century is difficult to say, but as late as the 1930's a tannery was still installed in the structure which was removed only when the church was declared a national monument soon after.[9]

Architectural Data

THE plan is the single-nave type divided into six bays (FIG. 167): the first just behind the façade is the choir; the next three comprise the nave; the fifth is the crossing; and the sixth bay, raised 0.10 m. above the rest of the church floor, is the presbytery or *capilla mayor* from which a small rec-

[1] Juarros, **1**: p. 154.

[2] *Efem*, p. 193.

[3] *Ibid.*, p. 196; also *ASGH* **24** (1949): p. 368.

[4] *Efem*, p. 205; also *ASGH* **24** (1949): p. 377.

[5] Pardo, *Guía*, p. 146.

[6] Juarros, *loc. cit.*; also *Efem*, p. 220; *ASGH* **25** (1951): p. 148.

[7] *Efem*, p. 225; *ASGH* **25** (1951): p. 154; see also González Bustillo, *Razón particular*, also *Ciudad mártir*, p 93, who says the building was dedicated thirteen years before its destruction in 1773.

[8] Díaz, *Romántica ciudad*, p. 77.

[9] Pardo, *loc. cit.*, p. 146.

tangular room projects. The exterior dimensions are as follows: the width of the façade—17.20 m. including the towers; the width at the nave—11.50 m., the towers actually project 2.60 m. to each side; the total length—38.00 m., not including the projecting room beyond the *capilla mayor*. The interior dimensions are: length—33.90 m.; width—7.00 m. from pilaster base to pilaster base, 8.50 m. from wall to wall, and 9.80 m. from niche wall to niche wall; the small room off the *capilla mayor*—1.95 m. wide by 2.75 m. long.

On centers the length of the bays measures as follows: choir—approximately 4.90 m. from façade wall to pier center; nave bays—approximately 5.30 m. each; crossing—8.50 m.; *capilla mayor*—4.30 m. pier center to rear wall. Except for the square crossing, all bays are oblong, measuring 8.50 m. wide by the lengths indicated immediately above.

The third bay has a door on the north side. Otherwise all bays have niches, except for the choir and *capilla mayor*, all uniformly 0.65 m. deep. The bases of the pilasters on ground level actually project and overlap the niche openings so that on the plan, figure 167, they appear in part behind the piers. In fact, the shafts are narrower than the bases shown so that the niche openings fit between pilaster shafts.

The façade wall is more like a stage backdrop or screen. (FIG. 168.) The body, *cuerpo*, of the church is conceived of as a series of pairs of piers, treated as pilasters on the interior and as buttresses on the exterior, which support the roof vaulting. The walls between the piers are no more than screens 1.50 m. thick which, after the depth of the niches is discounted, remain but 0.85 m. in thickness. It is interesting that the rear wall of the church which has no niches is also but 0.85 m. thick. The exterior buttresses are 1.70 m. wide and project from the wall approximately 1.10 m. at ground level. They rise to about the height of the haunches of the interior transverse arches just above the sill of the octagonal clerestory windows. From the top of each buttress a waterspout projects. Smaller buttresses, really pilasters, rise from the main buttresses to the top of the wall where they are surmounted by merlons. These smaller buttresses are decorative and enclose the drains which run from the roof to the waterspouts. The space between buttresses, except in the bay with the side door, is partially filled with masonry to a height of 1.30 m. in front and 2.00 m. behind. (FIG. 167.) The depth of the fill is 1.03 m. set slightly back from the buttress face. But the top surface, which is pitched dropping 0.70 m. lower from back to front, is 1.41 m. long.

The nave wall at clerestory level above the niches is slightly less in thickness than the rest of the walling, or about 1.25 m. Brick is used for arches, miters, and angles, but otherwise the walling material is for the most part roughly dressed, uncoursed stones laid at random with smaller stones and mortar in the interstices. All wall and vault surfaces are covered with a heavy coat of stucco, the same material being also used for decorative architectural details both on the interior and the façade.

The vaulting of the church is of the type common in eighteenth-century Antigua, namely ellipsoidal, pendentive domes over oblong bays. The transverse arches which span the church from pier to pier are semicircular while the lateral arches, engaged to the side walls forming lunettes pierced by octagonal windows at clerestory level, are really only skin deep and consist of archivolts built up by corbeling and plaster. These lateral arches are semicircular too, but their crowns are considerably lower than those of the transverse arches. In effect, each bay is really covered with an odd-shaped "half-watermelon" cut through by four vertical planes set at right angles, as is common in pendentive domes, *bóvedas vaídas*. Each of the vault soffits is decorated with false ribs and supposedly supports a very flat, small, circular saucer vault at the crown which is marked out by a circle of moldings of the same profile as the false ribs. The ribs actually do not conform to structural logic, joining the circle like a four-spoked wheel.

The crossing is square in plan and once was roofed with a dome or a cupola supported on spherical pendentives, parts of which are still *in situ*. The upper and lower choir are separated by a mezzanine floor supported on a very flat, ellipsoidal, pendentive dome. Most of the remaining vaults were still in place in 1963, but were very badly cracked threatening to come down in the next earthquake.

The façade is unique in Antigua as much for its ornate character and changes in the proportions as for its lack of strict adherence to the three-by-three divisions normally employed in the retable-façade. (FIGS. 168–170.) Only the central bay has three horizontal divisions, two stories and a *remate* or finial. The side bays have but two horizontal divisions or, strictly speaking, a single story and a finial above. The second story with niches is absent from the side bays. A version of this same scheme was also employed on the little church of Espíritu Santo. (FIG. 99.) Pairs of ornate engaged half-columns frame the side bays and another pair is repeated in the second-story central bay. (FIGS. 2, 169.)

The façade as a whole which is lower than usual, is also quite wide in proportion to its height owing to the addition of the even shorter twin towers. (FIG. 168.) They are actually lower than the pinnacle over the central bay. In most façades the horizontal cornices which cross the central retable normally end at the towers, but here they continue across the towers, the lower stories of which are treated as additional side bays. The tower belfries are actually lower than the horizontal cornice of the second story of the central retable. A very low-ribbed domical roof surmounts each tower. Though the lower portions of the towers are integrated into the design of the retable by the horizontal cornice of the first story, the belfries, however, are independent of the second story. The space above each of the side bays of the retable proper is filled with a mixtilinear half pediment set immediately in front of a blind wall which is surmounted by a truncated pediment with concave raking cornices and topped by a merlon. The upper choir runs the total width of the retable, so that the blind wall with its pediment actually encloses this space.

Thus the first story of the retable as a whole may properly be thought of as having five bays: the central one with the main door; the two lateral inside bays with niches; and the extra outside bays of the towers which are set back slightly. The surface treatment of the three exterior sides of the towers is that of recessed panels separated by narrow pilasters supporting a corbel table of mixtilinear arches directly under the entablature common to the whole façade. (FIG. 170.) The pilasters on the façade side are fluted, while those on the other sides are not. The outline of the corbeled mixtilinear arches recalls the door header of the church of El Carmen (FIG. 121); namely, an ogee profile consisting of a cyma recta and a cyma reversa combined, the two legs of the arch meeting in a short horizontal strip rather than a point. The rest of the members of the entablature on the towers are the same as on the rest of the façade.

Immediately above the entablature is the small belfry with a segmental arched opening on the main façade side and semicircular arches on the other two sides. (FIGS. 168, 170.) Fluted pilasters decorate the corners and support a low attic. The domical roof, as has been already noted, is hardly visible from close by and is surmounted by a merlon.

The side bays of the retable sit on high podia with pedestals breaking miters on column centers. The applied order is very complicated. (FIG. 169.) A plinth supports the base which is composed of the usual three moldings, two tori and a scotia. The tori are of rather small diameter and the scotia even smaller.

The shafts are composed of five little, plain, striplike pilasters each of which is divided into two by a horizontal molding. Each strip is set on a small, separate base consisting of a torus with a small astragal above and below. The total effect may be described as a shaft with very narrow and deep flutes separated by broad flat fillets interrupted by torus and astragal moldings at the midpoint of their total height. The scheme of breaks established in the shaft is also employed to alter the appearance of the Tuscan capital. Above each pilaster there is a bit of the necking, each section of which is flat rather than convex. The echinus follows the same scheme. A common abacus unites the various parts below. Its surface is broken by having chamfered corners as well as a pair of hollow chamfers between as seen on La Candelaria and Santa Rosa. (FIGS. 124, 173.) Describing the abacus from another point of view, the corners of a square abacus were cut off to form diagonal chamfers, and the space between them hollowed out with two concave depressions meeting in a sharp arris in the middle.

The decorative treatment of the side-bay niches is also very ornate. (FIG. 169.) A pair of corrugated pilasters and a small truncated pediment with swagged raking cornices frame the actual niche opening. The pilasters are composed of a series of superimposed, squared pot-forms, or elongated, squared Attic bases with high scotiae, giving the whole a corrugated effect. These niche pilasters are of the same type seen on the gatehouse of El Calvario. (FIGS. 2, 113.) Directly above each pilaster and flanking the pediment is a small fluted merlon shaped like a teardrop or a pear with a smaller globe at the pinnacle. The niche itself is set on a high pedestal made up of molding profiles that follow the pattern set by the pilasters to either side. The arch which covers the niche is exactly of the same mixtilinear profile used in the corbel table on the towers, the type seen in the main door of the church of El Carmen and in the niches of the south façade of the church of Santa Clara. (FIGS. 121, 140.)

The entablature, as has already been stated, runs across the entire façade, including the towers, breaking miters on column centers. (FIG. 168.) The architrave is very narrow but broken up by means of four deep concave channels or semicircular horizontal flutes separated by projecting convex fillets of the same profile in reverse. The pulvinated frieze is broken at regular intervals by sets of five vertical astragal moldings set closely together and is remotely inspired by the classical triglyph and metope frieze. The cornice consists of three main elements, the lowest a series of very narrow moldings which corbel out to support a very thin corona, and above that a wide flaring cyma.

The central bay of the first story is entirely occupied by a large round-headed door which has apparently undergone considerable remodeling in postcolonial times. The second story begins with a low attic which breaks miters on column centers and continues across the total façade including the towers. The attic pedestals are treated with some vertical grooves like triglyphs. Triglyphs are also set below the pilasters which frame the niche-window of the central bay, and are repeated again on center and on the corners of the towers. The second-story side bays have already been described as mixtilinear half-pediments with a blind wall behind. Only the central bay, properly speaking, has a complete second story and a third-story finial above.

The large niche-window with splayed reveals and header dominates the second story. Its general outline may best be described as an elongated octagon. The jambs to either side are quite long and the sill below is broken by slight inclines in the corners, while the header above is treated as a very shallow arch of complicated, mixtilinear scalloped profile. The window itself is framed with pilasters and a pediment which repeat the same scheme as those which frame the niches in the side bays below. The central bay as a whole, however, is framed by engaged half-columns like those of the first-story side bays.

Niches are fitted in the space to either side of the window. These are framed with diminutive pilasters which are a variation on the engaged half-columns of the main order of the façade, that is, the shafts are made up of vertical strips. The niches are surmounted by half-round pediments, a change from the truncated, triangular swagged pediments of the niches in the side bays below, and by merlons on pilaster centers.

The finial of the third story over the central bay is rather low. It is a freestanding bit of wall of undulating outline in the center of which a small niche is set directly on center with the window and door. Merlons located on center with the columns below frame the pediment. A third merlon once surmounted the wall and served as the very pinnacle of the whole façade.

11. *Santa Rosa de Lima: Church,* ca. *1750*

(FIGS. 171–175.)

Historical Data

THIS church was part of a nunnery for white women known as Beaterio de Santa Rosa de Lima. It was sometimes referred to as Beaterio de Gentes Blancas to distinguish it from the nearby Beaterio del Rosario for Indian women, also known as Beaterio de Indias.[1] The actual founding date of the organization is not known for certain. But it was probably late in the sixteenth century, perhaps around 1580 when the site was purchased.[2] Nothing is known of the first convent which the nuns, *beatas,* occupied or of the church where they held services.

Some simple sort of informal convent building must have served the community for about one hundred years prior to 1677 when it was either built anew or altered and improved, since glass costing about 1,000 *pesos,* specifically for the church, was ordered from Spain.[3] More building activity of an unknown sort continued on in the early eighteenth century to judge from the fact that the curate of the church of Yajalón (a town now in the state of Chiapas, Mexico) had collected money sometime before 1708 destined for the construction of the church of Santa Rosa.[4] But not until the beginning of the eighteenth century was a formal conventual house built for the first time.[5] The church may have been completed about the same time. It was not in use for very long, for it is mentioned, among others, as having been damaged in the earthquake of 1717.[6]

In 1732 the order petitioned the *ayuntamiento* for a larger allotment of water, implying that the number of nuns, *beatas,* inhabiting the convent, *beaterio,* had increased.[7] About seventeen years later in 1749, the *audiencia* granted the order license to collect alms for the purpose of paying for the rebuilding of their church.[8] It would seem then that the early eighteenth-century church had been either destroyed in 1717 or had become inadequate for the needs of the order. The present ruined structure, therefore, must date from about the middle of the eighteenth century. In 1766 the order formally converted from a *beaterio* to a *convento,* that is, from a lay order to a regular cloistered order.[9] González Bustillo,[10] who tends to exaggerate the ruin of 1773 since he was of the party in favor of moving the capital, merely mentions Santa Rosa among those public and religious buildings that were damaged but gives no specific details. It is quite possible that the church and convent were not badly hurt in 1773 but were ruined later in the nineteenth century after Juarros' time, continuing in use even after the city of Antigua was abandoned.[11]

Architectural Data

THE site where this church and convent stand is now privately owned and quite overgrown with brush and trees. (FIGS. 171–173.) Hardly more than part of the façade is visible. The interior of the church has not been cleared of the rubble from the fallen vaults still piled up to the height of

[1] Juarros, **1**: p. 137; Pardo, *Guía,* pp. 140 ff.; and Díaz, *Romántica ciudad,* p. 52.

[2] Juarros, **1**: p. 138.

[3] Ximénez, **2**: p. 381.

[4] *Ibid.* **3**: p. 247.

[5] Juarros, *loc. cit.*

[6] González Bustillo, *Extracto,* also *Ciudad mártir,* p. 130.

[7] *AGG,* A 1.2 (1732) 15808–2212.

[8] *Efem,* p. 203.

[9] Juarros, *loc. cit.*

[10] *Razón particular,* also *Ciudad mártir,* p. 104.

[11] Juarros, *loc. cit.,* writing in the early nineteenth century, speaks of the church and convent as though they were still in use in his day.

the capitals of the pilasters. The nave is overgrown with grass and weeds. (FIG. 174.) Of the convent building not a single trace remains.

The church has many architectural details directly derived from Santa Cruz, San José el Viejo, the Cathedral, El Calvario, the Compañía de Jesús, La Candelaria, El Carmen, Espíritu Santo, and the fountain of La Alameda del Calvario. It appears that the builder of this church was well acquainted with the buildings mentioned and utilized motifs from all for his decorative scheme with the result that the façade of Santa Rosa is probably the most ornate in all Antigua.

The side bays begin with a rather high podium divided into a central depressed panel with pedestals to either side which break miters and project on pilaster centers. (FIG. 171.) The treatment of the podium is more or less like that of Santa Cruz with regard to the deep flutes which accent the pedestals. The central panel is treated with a coffer design like Santa Cruz, but here it is square rather than octagonal.

Above the podium a pair of very massive pilasters, almost as deep as they are wide, frame a niche and project so boldly from the plane of the façade wall that they look almost freestanding. (FIG. 173.) They are the same corrugated type seen on the gatehouse of El Calvario and the stem of the fountain of La Alameda del Calvario. Not a direct imitation, but a variation. (FIGS. 113, 89.) The base begins with a very narrow undecorated plinth on which a high full torus molding rests and is separated from the plinth by a deeply recessed but very narrow fillet. Above the torus is a molding which looks like a Tuscan capital.

The shaft proper is made up of three broad pot-forms reminiscent of elongated squared Attic bases or amphorae. It is as if three pots are set one upon the other to form the shaft proper. Each one is composed of a number of moldings: first a very narrow plinth; then a very narrow scotia with a thin fillet above; third, a large torus with a slightly splayed fillet; fourth, a very high scotia, taller in height than the torus; fifth, a crown molding composed of a fillet below, a cyma reversa, and a slightly broader fillet above. This same scheme is repeated three times to form the shaft. The crowning member is absent on the topmost unit of the shaft and its place taken by an even more complicated Ionic capital.

The horizontal moldings on the sides of the capital where the echinus would normally be sweep up from the corners on the front to form a truncated swagged pediment. The diminutive tympanum is filled with a full-front head treated like a rosette with petals and a pair of wings. Immediately above, a pair of volutes are pendant in the spaces left reserved by the swagged raking cornices of the pedimental form below. The bolsters or side elevations of the volutes are decorated with deep incisions. The abacus which completes the capital is composed of the same moldings as the crowning members of the pot forms below, namely, a fillet with a cyma reversa and a second, slightly broader fillet above.

The entablature seems to have been directly copied from that of Santa Cruz. (FIG. 135.) The architrave is made up of a number of narrow moldings, each of which projects slightly beyond the one below. The pulvinated frieze above is decorated with little male figures as is the case of Santa Cruz. Here they are robed on the first story. On the second story they are nude and exact copies of those of Santa Cruz with upraised arms holding the cornice like Atlantids. (FIGS. 171, 172.) The frieze surface is left plain, in contrast to the lacelike *ataurique* decoration on Santa Cruz. The cornices are made up of moldings similar to those employed on Santa Cruz, and are also of extremely wide projection.

The niche treatment is rather plain, the openings occupying almost the total width of the space between the pilasters. (FIG. 173.) Two superimposed Attic bases form the pedestal on which the statues stand. Narrow undecorated moldings are set to either side and return as lintels across the top of the openings. The niches are surmounted by narrow entablatures with pediments, the raking cornices of which meet in a hollow chamfer, that is, a small semicircle cuts into the tympanum.

The second-story side bays repeat the scheme as below, except for a variation in the capitals of the main corrugated pilasters and in the niche treatment. (FIG. 172.) A low podium or Roman attic runs across the whole of the retable-façade breaking miters on pilaster centers. It is treated with molding profiles, the dominant features of which are two scotiae separated by a torus crowned with a molding similar to those used on the pot forms of the pilaster shafts.

The pilaster capitals are quite different in the second-story side bays. The first member, resting directly on the topmost squared pot, is an inverted trapezoid or wedge shape decorated with a full-front head treated like a rosette. Immediately above is an abacus of the type seen on La Candelaria and San José el Viejo (FIGS. 123, 169), namely, the corners of the abacus have been cut off or chamfered, and the remaining portion of the sides gouged out with two hollow chamfers resulting in a scalloped effect.

The entablature of the second story is almost an exact imitation of that of Santa Cruz even to the nude little Atlantids with arms upraised and sashes across one shoulder. Over each of the side bays, broken pediments forming miters on pilaster centers return to the façade wall behind where the raking cornices meet in a point. These are the same type seen on the first story of Santa Cruz. The niche proper, set between the pot-form pilasters, is crowned with a normal triangular pediment. The niche, semicircular in plan, has an ogee-header arch like that seen in La Candelaria, El Carmen, Santa Clara, and San José. (FIGS. 121, 123, 140, 169.)

The central bay of the retable-façade again repeats motifs known from other building both of seventeenth- and eighteenth-century date, but also includes some innovations not seen elsewhere. (FIG. 171.) The door opening occupies the total width of the first story and is set within a niche with a concentric semicircular header. The sides or exterior vertical faces of the niche are very narrow with the result that the jambs of the actual door opening, though set back, dominate the scheme. The main or larger niche itself is set back but a few centimeters from the plane of the wall so that the narrow molding which runs from the ground to the springing of the arch continues past the spandrels and returns across the bay just under the main entablature to form a frame or modified *alfiz*. Fully robed female winged figures in stucco relief fill the spandrels. The jambs of the actual door opening are fluted and a repeated lozenge decorates the archivolt of the semicircular arch.

Dominating the second-story central bay is a deep niche-window with a half-circle header and splayed reveals decorated with vertical moldings which continue on the soffit producing a striated effect. (FIG. 172.) The Roman attic which runs across the entire width of the façade terminates abruptly at the window opening where it returns for a short distance on the window reveals and then billows out to be transformed into a semicircular pedestal on which a statue of the Virgin and Child still remains *in situ*.

The arrangement of niches to either side of the window, two on each side one directly above the other, is reminiscent of the scheme seen on the façades of San Francisco, the Compañía de Jesús, and the Cathedral. (FIGS. 54, 83, 42.) These diminutive niches are almost like framed pictures with their heavy projecting moldings and little statues resting on tiny corbeled pedestals.

Only the central bay has a third story. (FIG. 172.) As in the case of Santa Cruz, it consists of a freestanding wall with a niche in the center and undulating mixtilinear half-pediments to either side. The niche itself has a trefoil header arch such as seen in Espíritu Santo. The niche is framed by pairs of little pilasters which, except for a small variation, seem to be replicas of those employed on the façade of the church of San José el Viejo. (FIG. 169.) Here, however, the little pilasters are flat rather than half round. The shafts are channeled with two deep flutes divided in the middle by horizontal moldings. The outside pilasters of each pair are set slightly back. An entablature connects all four pilasters. Immediately above, a truncated pediment, with swagged raking cornices on which a foliated crown is placed, forms the topmost pinnacle of the façade. Decorating the half-pediments to either side are robed figures worked in stucco relief similar to those in the spandrels of the main door. Parts of rather fanciful foliated merlons are still visible over the side bays and once formed part of the undulating broken outline of the retable finial.

The interior of the church, as has already been noted, is filled with the debris of the fallen vaulting to this very day. (FIGS. 174, 175.) The pilasters are buried to the height of the capitals. The lateral and transverse arches spring from small sections of entablature set over each pier. The entablature, as in the Cathedral and Santa Teresa, does not run along the side walls to connect piers. (FIGS. 45, 77.) The frieze is of the pulvinated type treated with a triglyph form set at regular intervals. Pulvinated friezes with exactly the same treatment appear on the façade of San José el Viejo and in the chapel of the Seminario Tridentino of later date. (FIGS. 168, 186.)

The pendentives of the ellipsoidal, "half-watermelon" vaults in each bay are treated with some very fine stucco decoration. (FIGS. 174, 175.) A large oval medallion filled with anthemion motifs and foliage is framed by a border which follows the shape of the pendentive to form a triangle, the sides of which indent and project alternately like a meander until coming to a point at the very springing. The soffits of the transverse arches were treated with extremely ornate forms. The central portion is channeled with two broad flutes coming to fillets with a half-round molding in the center such as is also seen in El Calvario, the Ayuntamiento, and elsewhere. (FIG. 10.) Framing the fluting is a triple-meander pattern, the indented portion of which is left hollow so that when viewed from the side, the edge of each arch is sawtoothed. The rest of the decorative details are buried in the debris.

XV THE FOURTH PERIOD: 1751–1773

THE duration of this period is but about twenty years. Beginning with an earthquake of relatively minor destructive consequences in 1751, it ends with the cataclysm of 1773 which brought about the final ruin and abandonment of Antigua. The only buildings of any importance undertaken during this twenty-year period were the Capitanía, the Seminario Tridentino, and the University. Church construction was limited to such small and unimportant structures as the hermitages of Santa Ana, Santa Isabel, and Santísima Trinidad. The only convent undertaken was that of the Beaterio de Indias which was largely a reconstruction of an older structure of tamped earth. The school of San Jerónimo, though somewhat pretentious in plan, was also partly built of such a humble material as tamped earth. But it was never completed having been expropriated by the civil authorities. The church of Jocotenango, dating in great part from the eighteenth century, represents a reconstruction and an enlargement of an older seventeenth-century structure. It is quite possible that some reconstruction work on this church was carried out in the nineteenth century.

The extravagant decorative style introduced in the preceding period continues, and is seen on two buildings of lesser importance, the churches of Santa Ana, where the corrugated El Calvario-type of pilaster is employed, and Santa Isabel where a new pilaster form, the lyre shape, is invented. Ogee arches, some even more ornate and complicated than before, are employed for the cloister arcades of the Seminario Tridentino and the University. The plaster ornament which decorates the cloister arcade of the University and the oratory of the Seminario Tridentino is of a high quality.

The arcaded façade of the Capitanía which balances the composition of the Plaza Mayor with that of the Ayuntamiento on the other side, was largely destroyed in 1773. More than half of the present structure was reconstructed in the latter part of the nineteenth century. Only the lower story had true pendentive domes over each bay. The upper story was roofed with wood and tile.

The last major construction undertaken before the final destruction of the city was the Beaterio de las Indias of which scarcely a physical trace remains. The actual construction drawings, however, still exist in the Archivo de Indias in Seville. The church was oriented like that of the other nunneries, but was built of materials no longer common in eighteenth-century Antigua, a combination of tamped earth, stone, and brick.

1. *Santa Ana: Church, Mid-eighteenth Century*

(FIGS. 176, 177.)

Historical Data

JUARROS, writing at the beginning of the nineteenth century, mentions this little church in passing saying that it and some others were but little chapels in the various *barrios*, the history of which offers nothing notable.[1] The founding of the church may date, however, from the seventeenth century and possibly even from the sixteenth.[2]

Writing *ca.* 1690, Fuentes y Guzmán mentions this neighborhood or village on the outskirts of Antigua saying that it was inhabited by forty Europeanized Cacchiquel Indian families, and that there was a beautiful little church there.[3] This information is in part corroborated by a document dated 1689 in the *Archivo Arzobispal de Guatemala*, quoted in full by the editor of the Vásquez manuscript, in which it is stated that the church was under the ecclesiastical administration of the Franciscan convent of San Juan del Obispo, was inhabited by Europeanized Indians, that services were held in Spanish, and that there were sixty-seven confessants.[4] The number of inhabitants differs somewhat from the account of Fuentes y Guzmán who gives the number of families as opposed to individuals, confessants, not including little children. In either case the figures indicate a rather small

[1] Juarros, **1**: p. 152.
[2] Pardo, *Guía*, p. 158.
[3] Fuentes y Guzmán, **1**: p. 403.
[4] Vásquez, **4**: p. 38.

population and that some sort of church building must have existed at the end of the seventeenth century.

Though not mentioned in any of the accounts of the earthquake of 1717, this church must have been damaged and, consequently, have fallen into disrepair. A document dated 1739 mentions a proposal to cede a fourth of the income from the tax contributions, *tributo*, of the local inhabitants for the purpose of reconstructing the building.[5] And in 1743 a physical on-the-site inspection of the building was ordered for the purpose of determining its exact condition.[6]

The present building is in a good state of repair and in use today. One may assume, in view of the above documentary evidence, that the present church was built anew or reconstructed some time after 1743, but exactly when is not certain. Nor is the possibility of a nineteenth-century date to be excluded. The fact that it is in use today might well be taken to indicate that it was repaired in the mid-nineteenth century, or even later, when so many other buildings in Antigua were put into service again. There is also the possibility that this little church is not the same one referred to in the seventeenth- and eighteenth-century documents and contemporary literature cited above. Díaz, writing in the 1920's, speaks of the ruins of the hermitage of Santa Ana, and in the next sentence mentions that the town has a small church. In other words, if he is correct, we are dealing with two separate and distinct buildings.[7] The present building, however, is the only one there now and there are no remains of a destroyed hermitage to be seen. The building was no doubt remodeled and repaired during the last 185 years or so since the final destruction of Antigua in 1773, and the façade patched up and painted even in recent years.

Architectural Data

THE design of the façade is that of a central retable framed by two lower towers set back slightly and projecting beyond the width of the nave. (FIG. 176.) Like San José, the second-story central bay has little niches to either side of the central window. The first-story entablature of the central retable continues across the face of the towers as in San José. (FIG. 168.)

The applied orders are quite free of architectonic restraint and consist primarily of corrugated pilasters, which are a variation of those employed on the gatehouse of El Calvario and the façade of Santa Rosa. (FIGS. 177, 2.) The side bay begins with a high podium which breaks miters on pilaster centers forming narrow pedestals with striated faces like triglyphs between which the dado is treated like a panel with a pulvinated face. The pilasters are placed on rather thin bases, square in plan, and treated with a small torus profile on each exposed face. The scotia above is very narrow and flat, rather than concave. It is bounded by an astragal below and a two-fasciae crown molding. Above the base, a plinthlike member surmounted by a shallow scotia provides the transition to the shaft proper.

The shaft is square in section and consists of four elements or groups of moldings which are a variation on the base. Each element is made up from bottom to top as follows: first a narrow torus with a small fillet above and below; second, a very high slightly concave molding, like a scotia, which is crowned by a series of very narrow but flat moldings projecting one above the other almost like a capital; third, a smaller plinthlike piece with plane faces; and fourth, a short slightly concave molding like a scotia on which the next series rests repeating the same scheme four times for the total height of the shaft. The last one, however, also serves as the capital of the shaft.

The entablature breaks miters on pilaster centers and runs across the entire façade uniting the main retable with the side towers. (FIG. 176.) It has a pulvinated undecorated frieze. Exactly the same scheme is repeated in the second-story side bays. The whole second story rests on a low attic or podium. The entablature here does not carry across to the towers. They are quite lower and are not integrated with the retable as in the first story.

A niche is located between each pair of pilasters in both first- and second-story side bays. (FIG. 177.) The first-story niches rest on half-round pedestals made up of a number of superimposed moldings filling the space between pilasters very much like those on San José. Each niche is framed by pilasters broken halfway up the shafts by a horizontal molding reminiscent of those on the main engaged columns of the façade of San José. (FIG. 169.) The shafts here have an added decorative touch in that they are fluted. Small triangular, truncated swagged pediments are fitted between the little niche pilasters. The niche opening itself is semicircular in plan and crowned with an ogee mixtilinear arch of the same outline as the door niche in the central bay.

The second-story niches are quite different in that their general outline is oblong but they are framed with pilasters like those below. (FIG. 176.) Small segmental pediments are fitted between the pilasters. The niche itself is smaller and rests on a pedestal supported on an element which looks like a two-branched candlestick.

[5] *AGG*, A 1.10.3 (1739) 31318–4047.
[6] *Ibid.*, A 1.10.3 (1743) 31322–4048.
[7] Díaz, *Romántica ciudad*, p. 82.

The first-story central bay is occupied by the large door set in a deep recess. The door proper has jambs and a half-circle arch of stone. The niche header, however, is mixtilinear and of the type seen in the cloisters of the Seminario Tridentino and the University. (FIGS. 184, 193.) The outline of the opening is as follows: the haunches consist of a cavetto; then there is a break by means of an interior angle; and finally a cyma reversa. The niche header may be described as an ogee broken by a step or square chamfer.

The second-story central bay is dominated by an octagonal window which lights the upper choir. (FIG. 176.) To either side is a small niche set on a very high corbeled pedestal and crowned with a triangular pediment. The third-story or *remate* seems to be of recent construction. It does not seem to fit the scheme of the retable. A small niche is centered on the highest portion over the central bay.

A further ornate and nonarchitectonic note is added at the corners of the main retable wall which are set back from the applied orders. The surface at the edge of the retable wall is treated with the same corrugations as the pilasters, thus weakening the structural line where the retable juts out from the projecting towers.

The towers which frame the central retable project beyond the width of the nave. The nave proper is contained within the limits of the retable. The first story of the towers, united with the retable by means of the cornice which runs across the whole of the top of the first story of the façade, is left rather plain except for a pair of rather elongated and thin pilasters. These are set away from the corners, where they would normally be to provide a structural accent, so that a narrow bit of blank wall is left at the corners. These pilasters, set on pedestals, are incised with a single, deep flute of the type seen on the inside piers of El Calvario.

The second stories of the towers contain the belfries, each with a single arched opening surmounted by a small dome. The north tower has a door at ground level which gives access to a caracole stair leading to the belfry. Over the door there is an octagonal niche.

Parts of the façade are decorated with stucco ornament some of which is reminiscent of the work seen in the University. (FIG. 177.) Small three-leaf-cloverlike patterns decorate the large scotiae of the main pilasters of the retable. Small medallions of similar material decorate the spandrels of the arch of the main door. In general, the stucco decoration is not of very good quality and may very likely be recent. All the niches, including that of the finial, have statues built of brick and stucco which may or may not be of colonial origin.

2. *Santa Isabel: Church, Mid-eighteenth Century*

(FIGS. 178–179.)

Historical Data

IN 1673 the neighborhood of Santa Isabel was under the ecclesiastical jurisdiction of the convent of San Francisco in Antigua.[1] By 1690 there were 210 Indian parishioners living there, thus implying that some sort of church must have already existed.[2] Another bit of information concerning Santa Isabel is mentioned in a *cabildo* dated April 16, 1735, in which the *ayuntamiento* was notified of an order prohibiting sixteen families from moving away from Santa Isabel to a location near the slaughterhouse where they wished to found a community of their own.[3] Apparently the chief occupation of the inhabitants was that of soapmakers,[4] possibly explaining why they would naturally want to live near the source of supply of the fat by-products of the slaughterhouse needed in their trade.

The neighborhood must have been no more than a collection of thatch huts, while the church itself, as witnessed by the ruins still extant today, was of the simplest type. In none of the reports made by González Bustillo after the earthquake of 1773, nor in his recapitulation of the earthquakes which had plagued the city in all its history, is the church of Santa Isabel mentioned. Nor does Arana who reported the ruin of 1717 mention it either. Juarros merely lists it giving its ecclesiastical affiliation in his day, under the parish of Nuestra Señora de los Remedios, but he has nothing more to say about it.[5]

No information is forthcoming as to the building date of the present church. A structure of sorts must have existed by the last quarter of the seventeenth century which, in all probability, was either damaged or destroyed in the earthquake of 1717, a *terminus post quem* for the present building. On the basis of the nonarchitectonic pilasters employed on the façade, it may be possible to place this building after 1750.

Architectural Data

ALL that remains of this little church is the main façade, parts of the long walls, and portions of what may be presumed to have been the priest's house or a chapel abutting on the north side set back slightly from the façade. (FIG.

[1] *AGG*, A 3.2 (1673) 15207–825, also *BAGG* **10** (1945): p. 128.

[2] Fuentes y Guzmán, **1**: p. 379; and a document dated in 1689, quoted in full by the editor of the works of Vásquez, **4**: p. 38, *Archivo Arzobispal de Guatemala*, A 4.5–2.

[3] *Efem*, p. 178.

[4] Díaz, *Romántica ciudad*, p. 81; Pardo, *Guía*, p. 159.

[5] Juarros, **1**: p. 82.

178.) The open area in front of the church probably served as the neighborhood plaza as well as the atrium of the church in a manner similar to that seen in Santa Cruz, San Cristóbal el Bajo, and Santa Ana. Dominating the space is a large stone cross in a remarkably good state of repair.[6]

The façade is a modest retable type lacking towers. In fact, the design of the architectural decorative details of the façade betrays unmistakable influence from carpentry as in church retables for altars. A unique type of pilaster is employed on the façade which is nonarchitectural to an extreme.

The exposed walling of the church is very crude when compared to some of the more important buildings of Antigua. Rather large, undressed stones are employed for the walls. Brick is used only at openings and as a backing for the decorative features finished in stucco.

The central bay is not very much wider than the side bays, nor is the second story much shorter than the first. The overall proportion of the first two stories is almost a square. The height is not quite a meter more than the width. The third story, an *espadaña* with three arched openings, is a high freestanding wall which adds to the vertical effect. But its height is only slightly less than that of the second story.

The first-story side bays are set on rather low podia with paneled pulvinated dadoes framed by flat moldings. (Fig. 179.) The pilasters are absolutely nonarchitectonic in form. Each is set on a squared base consisting of a large shallow scotia surmounted by two narrow tori and an abacus or cyma recta with a fillet. The pilaster shaft is composed of six superimposed lyre forms, that is, six pairs of right and left S-curves with the larger sweep or curve at the bottom. (Fig. 2.) Inside the lyre shape thus formed foliage is appended to the curves and partially fills the interior space. The shaft is surmounted by a square Tuscan capital which corbels out far beyond the width of the shaft to form a very wide flaring and massive support for the entablature.

The entablature breaks miters on pilaster centers. It is relatively simple consisting of a two-fasciae architrave, a frieze with rosettes at intervals and small heads set on pilaster centers, and finally a widely projecting high cornice. The entablature runs across the total width of the façade. (Fig. 178.) It is now partly missing in the central bay.

A niche of the usual apsidal shape is set between pilasters in each bay. It is framed by moldinglike pilasters which do not rise from the podium below, but begin on a small corbel. (Fig. 179.) The shafts of these narrow pilasters are made up of alternating tori and scotiae to give a corrugated effect. The sides end abruptly making a sharp edge with the return face. It is as if a carpenter had first cut a molding into short lengths and then nailed them one on top of another to achieve the effect of a pilaster. The pilasters are surmounted by small Tuscan capitals. The entablature above breaks miters on pilaster centers and supports a pediment with sharply rising raking cornices which do not join in a point but rather in a shallow concave chamfer. The tympanum is decorated with a single winged-head. The statues are still *in situ* and stand on little corbeled pedestals within the niche proper.

The second story lacks the podium or Roman attic so common in retable-façades of Antigua. (Fig. 178.) The columns emerge directly from the massive widely projecting cornice of the first-story entablature. The same type of pilaster of the first story is repeated here. A similar corbeled base supports a shaft built up of right and left S-curves or lyre forms, but with a slight variation. The lyres do not sweep out as much as below. The foliage used to fill the interior space consists of a vertical stem to which oval pointed leaves are attached. The capitals set on the shafts are the same as below, as is the whole entablature too. The niche treatment is also the same except that the pediments are the broken scroll type.

The central bay lacks certain features common to the other retable-façade churches of Antigua and is of more simplified construction. For example, the round-headed door opening is not set within a niche with a concentric header; instead there is but a single opening in which the doors were hung, the door jambs actually abutting on the side bays. The reason for the lack of the door niche may be that the central bay is not sufficiently wide to allow for one. The spandrels of the door arch are undecorated, the only adornment being a little scalloped pattern on the archivolt of the arch.

The second-story central bay is dominated by a large octagonal niche-window now in ruins. That part of the first-story architrave on which the window once rested is missing. The window is framed in the same manner as the niches of the second story including the molding-strip pilasters surmounted with a broken scroll pediment. A further decorative note is added by means of winged-head rosettes on some of the sections of projecting torus molding of the pilaster shafts. Winged heads are also placed in the spandrels formed by the upper diagonal sides of the octagonal window, and a shell ornament is placed just above the outside corners of the pediment.

The third story is a freestanding wall, an *espadaña*, with

[6] The cross was no longer there in August of 1963.

three arched openings where the bells were once hung. A small, round half-column is engaged to each of the two central piers dividing the openings. Pilasters of the corrugated type are set at the extremes of the wall where half-pediments carry the eye down to the side bays. The half-pediments which sweep down to either side are partially undulating and partially rectilinear. Beginning with a diminutive spiral in the outside corner of the half-pediment, a short stretch of raking cornice emerges and is connected in turn to a short vertical piece which indents before billowing out into a large open spiral. Directly in the space under the spiral, a mermaid is worked in stucco relief. A high conical merlon is set at either extreme of the *espadaña.* A round medallion, set directly over the central arch, terminates the *espadaña.* It is flanked by undulating curves which sweep up ending in horns to each side.

3. *San Jerónimo: Church and School Late Seventeenth Century, 1759 and after 1765*

(FIGS. 180–182.)

Historical Data

THE neighborhood where this building once stood has not changed much in the last 250 years. In 1963 the streets were not paved and were lined with modest houses of nondescript architectural character. In the late seventeenth century, according to Fuentes y Guzmán, it was one of the slum areas of the city and inhabited mainly by Indians, and some *ladinos,* all of whom were given to violence and drunkenness. The church which served the neighborhood, under the ecclesiastical administration of La Merced, was most humble.[1] (FIG. 181.) Under the advocation of San Jerónimo, it had probably been established earlier in the century, for it is mentioned in a document dated in 1676 recording the purchase of some houses by the sculptor Mateo Zúñiga located in front.[2] The existence of a church of this name, a hermitage, is also attested to by Juarros, who informs us that the school which La Merced had built without prior license was located right next to it.[3]

In 1739 the order of La Merced petitioned the *audiencia* for permission to found a school on land already acquired to the rear of the church of San Jerónimo. They gave as reasons that, since the Dominicans had the school of Santo Tomás and the Franciscans the school of San Buenaventura, it was, therefore, fitting and proper for the Mercedarians also to have a school.[4] Work was begun on the school building at some indeterminate date after this request had been made in 1739, and was completed by 1759. But apparently the school had been established without previous license from the crown, and so was ordered closed in 1761. At that time there were eight monks and five students in residence.[5] The school had not fared well even before the order of closure, for, with the secularization of the religious doctrine in 1757, the church of San Jerónimo was taken away from La Merced and put under the ecclesiastical jurisdiction of the parish of San Sebastián, and the door between the church and the school bricked up.[6] The school was ordered demolished in 1763. This was not carried out since it was decided to turn the building over to the civil authorities and to convert it into a customshouse and a military barracks.[7]

The monks and students were not removed from the building until 1765, which was then altered and utilized as a customshouse and tax office. Some alterations were also carried out in the structure in order to accommodate a company of dragoons and their horses. Some further changes were projected in 1772 which never materialized because of the destruction of the city in 1773. But we are fortunate in having the actual plans, drawn by Diez de Navarro in 1767 and 1772, both for the original conversion of the building into a customshouse and barracks as well as for the alterations proposed at a later date.[8] (FIG. 180.)

Architectural Data

VERY little remains of the school building except for a short piece of wall on the north side with the remains of a single door which may be tentatively identified on Diez de Navarro's plan of 1767 as "2" and on that of 1772-No. 1, as "B," that is, the main door on the street running east-west toward La Recolección. (FIG. 182.) It once gave access to an anteroom through which one passed to a large cloister.[9] The building was two stories high with a large central cloister square in plan and surrounded by covered corridors sup-

[1] Fuentes y Guzmán, **1**: p. 389.

[2] *Efem*, p. 86.

[3] Juarros, **1**: p. 125.

[4] *BAGG* **8**, 4 (1943): pp. 402 ff., referring to *AGG*, leg. 161–exp. 14.

[5] Pardo, *Guía*, p. 49.

[6] *ASGH* **25** (1951): pp. 153 ff.

[7] *Ibid.*; *Efem*, p. 225; see also *AGG*, A 1.11 (1763) 2113–98; and Lamadrid, *ASGH* **18** (1942/43): pp. 279 ff., "Los Estudios Franciscanos en La Antigua Guatemala"; see also Angulo, *Planos* **4**: p. 422, who gives the date ordering the demolition of the school as October 13, 1764.

[8] Angulo, *op. cit.*, pls. 163 and 164, **2**: pp. 71 ff. and 76 ff., **4**: pp. 422 ff.; also Torres Lanzas, *Planos*, nos. 69 and 192 referring to *AI*: 100–3–20(1) and 101–4–12(1) respectively; see also *Efem*, p. 225.

[9] Angulo, *op. cit.*, pl. 164, **2**: pp. 76 ff., for the plan of 1772.

ported on seven piers on each side, counting corner piers twice. In 1767 there were some small houses abutting to the east of the cloister and facing the street running north-south which Diez recommended be bought in order to carry out the alterations necessary for erecting stables for some 200 horses in the area. It is interesting to note that these houses were still the property of the order of La Merced. Apparently they had not been included in the original expropriation and must, therefore, have not been part of the original school plan. This would imply that the area for the quarters for the dragoons represented an addition to rather than an alteration of the existing school building.

The original structure had been built in great part with walls of tamped earth, parts of which Diez de Navarro recommended be torn down and others be reinforced with piers and courses of brick, "hechándole rafas y berdugados de ladrillo." He also recommended that a storm sewer, *thaugia*, be constructed in the large patio so that during the rainy season the rainwater might be carried out to the drainage ditch in the street. The building must have been quite damp, for in the proposed alterations of 1772 he suggested that the floor level of various dependencies be raised so that the humidity would not seep in.[10] The level of the street running east-west was one *vara* (approximately 0.875 m.) lower than that with the drainage ditch running north-south, and probably the cause of the dampness of the interior of the building.

The main entrance on the north side, that is the door to the cloister, was walled up, probably when the building was remodeled in 1765. The threshold is considerably below the present street level which has been raised since the building was first erected. The lower part of the door jambs were equally buried until about 1957 when the accumulation of debris was cleared away.

The design of the door is comparable to that on the gatehouse of El Calvario. (FIGS. 182, 110.) The actual opening has a semicircular header with broad archivolt which continues down to the jambs below. The springing of the arch is accented by imposts treated as pilaster capitals.

The door is framed by a pair of corrugated pilasters with shafts built of squared pot-forms or squared Attic bases, that is, with alternating scotiae and tori set on pedestals recalling those of El Calvario and Santa Rosa. (FIGS. 2, 113, 173.) A simple entablature runs across the space over the door connecting the two pilasters. Set directly over each pilaster center is a rather ornate merlon built up of squared lower members and spherical upper members. The horizontal line of the cornice sweeps up over the door to a short horizontal strip forming, thereby, a rather squat swagged pediment. Directly above there is a niche surmounted by a small steep triangular pediment with pyramidal merlons to either side.

The little church which dates from the latter part of the seventeenth century is adjacent to the east wall of the convent building. (FIG. 181.) It faced an open area, formerly the atrium still discernable though now overgrown with vegetation. It was set back from the street, now obliterated, which ran parallel in a north-south direction to that known in colonial times and still known today as the Alameda de Santa Lucía. The church is approached now from the street which runs in an east-west direction along the north side of the convent toward La Recolección. (FIG. 15—No. 29.)

The building is in an extremely ruinous state of preservation, only the main façade wall still being more or less intact. Remnants of the side walls still exist; the rear wall is missing, but there are distinct traces of its having abutted on the convent wall. The walling material of the main façade is brick coated with a heavy layer of stucco crudely finished. It is but one story high surmounted by a finial, *remate*, and is divided into three bays by unpretentious flat pilasters set on equally plain pedestals. The effect is more like that of shallow buttresses than that of an applied architectural order.

In each side bay there is a shallow round-headed niche executed in very crude fashion which is filled with a statue worked in stucco on a brick core bonded to the façade wall. Another small niche of even more rustic character is set directly over the door in the central bay. The door occupies almost the total width of the central bay and is spanned by a wooden lintel. The extremely humble and impoverished character of the building is further emphasized by the use of the wood lintel implying that the mason chose this cheap, simple method of spanning the opening rather than the more complicated expensive system of constructing an arch. Though the type of the wood could not be immediately identified since it still was covered with remains of stucco and paint, it may in all probability be either *zapote* or *chichicaste*, varieties of tropical hardwood well known since preconquest times for their extremely durable and weather-resistant qualities, and commonly employed to suspend bells in church belfries.[11]

[10] See his plan of 1772, no. 2, Angulo, *loc. cit.*, pl. 164, **2**: p. 78.

[11] Wood of this type is often employed for statuary of church façades, as for example in the still largely Cacchiquel Indian town of Santa María de Jesús a few kilometers away on the slope of the Volcán de Agua. The wooden figures, surprisingly enough, have required less attention in the last two hundred years owing to weathering than the masonry wall they decorate.

The finial or *remate* presents a rather sorry aspect revealing the unsuccessful attempt of an unskilled craftsman to imitate the undulating outline such as employed on the finials of the richer and more monumental churches of contemporary Antigua. A square window, also spanned by a wood lintel, pierces the central portion of the *remate* directly above the door. The pinnacle above is surmounted with the remains of a crown worked in stucco.

The only decorative note, a rather frail attempt at exuberance, is confined to the four small urnlike forms set at the bottom of the finial on pilaster centers. The two outside ones are missing. The interior surface of each is decorated with a motif which looks like a conventionalized rendering of a crucifix. The urn itself rests on a hemispherical member decorated with two very crudely executed leaves.

The side walls were originally built of adobe, in all probability, explaining the fact that only traces of their former presence are discernable. The roof may be conceived of as having been of wood and tile. But in view of the crudity and poverty of construction of the main façade, and, in keeping with the slum character of the neighborhood and the low economic and moral state of its inhabitants, it is not outside the realm of possibility that the roofing material may very well have been thatch when the church was first built.

4. *Seminario Tridentino*, ca. *1758*

(FIGS. 183–187.)

Historical Data

THIS seminary was first founded at the end of the sixteenth century, *ca.* 1597, but where it was located is not certain. However, by the eighteenth century it occupied its present site.[1] Financial support for the seminary was derived from the income from some parishes over which there was some controversy early in the eighteenth century.[2] Again in the middle of this same century some question arose concerning payments to the seminary from curates in various parishes.[3] It was just about this time, in 1758, that construction of the new university building began, located next door in the same block, the two actually sharing a party wall. It was deemed feasible, therefore, to construct a new seminary building too.[4] The building is more or less intact today and seems to have escaped major injury in the 1773 earthquake.[5]

[1] Juarros, **1**: p. 121.
[2] *AGG*, A 1.11 (1710) 2043–94 and A 1.11 (1713) 2049–95; also Ximénez, **3**: pp. 252 ff.
[3] *AGG*, A 1.11 (1752) 2087–97.
[4] *Efem*, pp. 206 and 213.
[5] Cadena, *Breve descripción*; González Bustillo, *Razón particular*, also *Ciudad mártir*, pp. 29 and 110.

Architectural Data

THE building, now in private hands, is divided in half by a partition wall running right through the middle of the main cloister and serves as two separate houses. The principal door is still intact. (FIG. 183.) Another one has been opened near the corner toward the Plaza Mayor to give access to that half of the building.

The door treatment reveals many details which are exact replicas of elements on the façade of the church of San José el Viejo. (FIGS. 168, 169.) The door proper has a stone lintel and jambs, the latter treated as simple pilasters. Framing the door are pairs of half columns set on high podia, the shafts of which are exactly like those used on the façade of San José being made up of small, flat vertical bands. Even the capitals, a modified Tuscan type, are exactly like those of the church including the same type of abaci diagonally chamfered at the corners and scalloped with hollow chamfers between. The columns are set very close together leaving no space for niches. Fitted into the narrow space, instead, are corrugated pilasters made up of superimposed square pot-forms or Attic bases, a type first seen on El Calvario. (FIG. 2.) The entablature miters out over each pair of columns. The frieze is pulvinated and has an arrangement of open spaces alternating with triglyphlike patterns consisting of five narrow vertical bands separated by four deep incisions.

Exactly on door center there rises a roof comb or finial with a niche in which a statue is still *in situ*. To either side of the niche are pairs of small pilasters which are a variation on the engaged columns below and exactly like those used on the third-story finial of the church of Santa Rosa. (FIG. 172.) The individual pilasters of each pair are not located in the same plane; rather, the outside ones are set back slightly behind the inside ones exactly as in Santa Rosa.

A broken triangular pediment is located above the niche. Between the ends of the raking cornices in the tympanum a large square blazon is fitted. Topping the whole composition is a large crown with the remains of an orb inside. Attached to either side of the finial are volute half-pediments. Large ornate merlons are set on center over the pairs of engaged columns to either side of the door below to complete the composition.

A balustrade, decorated at regular intervals with a lyre-form motif in stucco relief, runs around two exterior street sides of the building. The ornament is made up of two opposed S-curves to outline a lyre shape.

The cloister has been subjected to a great many repairs, for the building was converted into two private houses. (FIG. 184.) The supports of the arcading are simple round

Tuscan columns of the type seen in Las Capuchinas, but not nearly as massive or as tapered. They are more like the lower-story columns of the cloister of Santa Teresa. (FIG. 78.) The arches which connect columns are of the ogee type seen in various other eighteenth-century *antigüeño* buildings such as La Candelaria and El Carmen, but are closest in design to those of the façade of Santa Ana. (FIGS. 123, 121, 176.) Each leg of the arch is in profile like a cyma reversa with a square chamfer where the curve changes direction. The soffits of the arches are incised with three whole flutes in the center portion and half flutes to either side. A flat surface lines the very edges of the soffit. The archivolt is accented with a small molding of slight projection. The roof over the arcade is of wood and tile. Decorating the walls at intervals, especially over doors of the rooms behind and in the spandrels of the arcade, are papal coats-of-arms worked in low stucco relief.

The rooms off the cloister have been modified in post-colonial times. There is one dependency in the building, the chapel or oratory, which has remained more or less intact except for some repairs to the floor. It is a large room roofed with a barrel vault which rises above the rest of the one-story structure. (FIGS. 185, 187.) At the east end is a low gallery which supports the upper choir on a flat arch and a half-ellipsoid vault. The barrel vault is pierced through the haunches by windows, the reveals of which are decorated with *ataurique* decoration. Likewise, the soffit of the main barrel vault over the room proper is treated with false ribs of plaster worked into an overall sinuous pattern. The ribs, more like moldings, undulate from springing to springing coming together and separating to form more or less oval spaces at regular intervals in many of which other designs such as blazons are worked in very low relief.

The pilasters and entablatures are very ornate. (FIG. 186.) A narrower pilaster projects from the wider one directly contiguous to the wall. This is not uncommon in Antigua, where as many as three pilasters, so to speak, are worked one in front of the other making three steps when viewed in section. But it is only the outermost surface, that is, the furthest away from the wall and the narrowest which ordinarily is fluted. But here in the chapel of the Seminario Tridentino all of the exposed surfaces of the pilasters are fluted. The fluting is of the type seen in the gatehouse of El Calvario (FIG. 10), but rendered even more ornate by the addition of a Doric-leaf pattern running the length of the convex surfaces of each composite flute.

The entablature is exactly like that seen on the door of the building including a pulvinated frieze with the triglyph design of the type also seen on San José el Viejo and Santa Rosa. In this case, a leaf pattern fills the spaces between triglyphs.

Another interesting feature of the building is the corner window facing the Plaza Mayor, a type common to many private houses in Antigua but which may have been added when the building was converted to private use, possibly in the nineteenth century. The window is set on a high stone pedestal in the very corner of the building with two openings, one on each side of the corner. The inside mullions are of stone as are the lintels. The corner mullion, shared by both openings, is a small stone column which supports the wall above. An iron grill of unknown date encases the two windows.

5. *University of San Carlos,* ca. *1763 and Later*

(FIGS. 188–196.)

Historical Data

THE institutional history of the University of Guatemala has been treated from various points of view by colonial, nineteenth-century, and contemporary authors.[1] But in no case does one find an account of the buildings which housed the university. The present, late-eighteenth-century building is still in a good state of repair. It was preceded by another, located in a different part of town on an unidentified site owned by and near the convent of Santo Domingo.[2] Some account of this first building is given in various colonial documents making it possible to form an idea of the physical plant of the university in the last quarter of the seventeenth century.

As an academic institution, the university was founded in 1676.[3] A year later, in 1677, it was decided to install it in the building which had formerly housed the Colegio de Santo Tomás de Aquino and to make the necessary repairs and alterations. These required two years to be completed.[4] In the meantime the faculty had already been selected in 1678.[5] After the arrival of a royal *cédula* in 1680 authorizing what the university council had agreed upon with regard to the building and other matters as well, classes began with seventy students on January 6, 1681.[6] Apparently the building

[1] Juarros, **1**: pp. 116 ff.; Castañeda Paganini, *Historia*; Salazar, *Historia intelectual*; Lanning, *Eighteenth Century*, and *The University*.

[2] *AGG*, A 1.3 (1760) 13769–2003.

[3] *ASGH* **13** (1936/37): pp. 84 ff.; *AGG*, A 1.3.1 (1676) 12235–1882; Molina, *Memorias*, p. 142; *Efem*, p. 85.

[4] *Efem*, pp. 86, 87, 90; also Fuentes y Guzmán, **3**: p. 240.

[5] Molina, *loc. cit.*

[6] *AGG*, A 1.23 (1680) 10.076–1521–221, also *BAGG* **9**, 2 (1944): p. 117; Molina, *op. cit.*, pp. 149 ff.; *AGG*, A 1.3 (1680) 12245–1885, also *BAGG* **9**, 2 (1944): pp. 119 ff.; Ximénez **2**: p. 414; *Efem*, p. 93.

had been built largely of tamped-earth or adobe, the raw materials for which were taken right from the patio which subsequently required fill. The chapel and other dependencies were either not yet completed or not entirely satisfactory, to judge by some criticism and dissatisfaction voiced later that same year.[7] Classes had been running but a short time when one professor registered a very strong complaint that something be done about class attendance, since some of the students were always cutting classes, coming late, or not coming at all, especially those studying theology.[8] The constitutions or bylaws of the university were finally published in Madrid in 1686.[9]

The chapel which had caused some controversy, especially with regard to the cost of construction,[10] still lacked a retable in 1683 at which time bids were asked to build one.[11] The famous *ensamblador* Agustín Núñez was low bidder for 1,600 *pesos*. He agreed to complete the work by June of 1684 according to the plans and designs. In the specifications the retable was described as six *varas* wide by five *varas* high and was to include various sculptures. The work was to be delivered plain, *en blanco*, ready for gilding and finishing. Apparently the price was deemed too high and new bids were asked offering a maximum of 1,100 *pesos*. Four times the offer was repeated in the year 1683, but there were no takers until one Cristóbal de Melo offered to do the job for 900 *pesos*. Whereupon, Núñez asked to be paid for the designs he had made. Melo then offered to reduce the price by 50 *pesos* and make a design of his own or pay Núñez for his. A new bidding was then opened and Núñez got the job as low bidder, agreeing to do it for 830 *pesos*!

There is no mention of how the university building fared in the earthquake of 1717, but it is hardly likely that it escaped without some damage. At least, in 1735 it was in need of repair and permission was asked to drag the lumber needed for the job through the streets, a request which the *ayuntamiento* denied.[12] In the earthquake of 1751, however, the building was rendered totally useless. Only two rooms were left intact while the rest of the building was threatening to cave in. The main *portada* of stone was destroyed as well as the *artesón* of the assembly room.[13]

In 1758 the Rector of the Seminario Tridentino, Juan González Batres, offered to donate the houses adjacent to the seminary. These were to be altered and rebuilt to house the university whose building was still in a ruinous state. This offer was accepted by the university authorities the following year.[14] Before making the final decision to move to the new site and acquire the houses and land, in 1759 estimates were taken of the cost of the new construction proposed. At the same time the ruined university building and its land were also evaluated by two outside experts, José Ramírez, *maestro de albañilería*, and Manuel de Santa Cruz, *maestro de carpintería*.[15] The site of the seventeenth-century university measured 96½ by 56½ *varas*, or approximately 264′ by 154′, and was evaluated at six *reales* the *vara* or a total of 4,306 *pesos* and one *real*. The building was described by the two experts as being in a ruinous state, but the masonry and carpentry parts were calculated to be worth a total of 14,546 *pesos* and one *real*. The walls were described as being of *pisón*, that is, tamped or stabilized earth. They estimated the cost of repairs to make the ruined building serviceable again at 11,856 *pesos*.

An argument in favor of moving to the new site adjacent to the Seminario Tridentino was that, since most university students also attended the seminary, they would thus avoid the evils of the street in going to and from classes from one school to the other. Another good reason given for moving was that the enrollment had fallen off because the university was located at some distance from the center of the town. A decision was reached at last in 1763 and license to move given with the condition that the royal treasury not be obliged to pay any of the costs since these would be defrayed from the receipts of the sale of the property they were leaving, and the cost of the new construction would be met with the money donated by Batres, rector of the seminary. It was also ordered that the royal arms be engraved above the main entrance of the university and that no door be opened in the party wall between the university and the seminary.

The houses existing on the site were not torn down it would seem, but those parts which could be utilized were incorporated into the new construction. The land and the ruined buildings near the convent of Santo Domingo which had been abandoned were apparently sold a year after the university had moved and its new building inaugurated in

[7] *AGG*, A 1.3–20 (1681) 13145–1956; *Efem*, p. 95.

[8] *AGG*, A 1.3.8 (1681) 12445–1899.

[9] Juarros, **1**: pp. 116 ff.

[10] See fn. 7 above.

[11] *AGG*, A 1.3.3 (1683) 12388–1896, also *BAGG* **9**, 4 (1944): p. 233.

[12] *Efem*, p. 179. A city ordinance of 1723 had prohibited this practice, *AGG*, A 1.2 (1723) 15806–2212. This custom still prevails in San Cristóbal de las Casas, Chiapas, where rough-sawn planks or timbers are lashed to mules and dragged through the streets raising clouds of dust and tearing up the unpaved surface.

[13] *AGG*, A 1.3 (1763) 1157–45; *ASGH* **17** (1941/42): pp. 376 ff.; *Efem*, p. 204; *AGG*, A 1.3.21 (1751) 13160–1957, and 13161–1957.

[14] *Efem*, pp. 215, 216.

[15] *AGG*, A 1.3 (1763) 1157–45, also *ASGH* **17** (1941/42): pp. 376 ff.; see *AGG*, A 1.3.21 (1764) 13162–1957, also *BAGG* **7** (1941/42): p. 480, and *ASGH* **25** (1951): p. 153.

1763.[16] It would seem, however, from a petition for monetary help sent to the *ayuntamiento* in 1765, that some work still remained to be done.[17]

Just who was responsible for the design of the new university building is not certain, although it is very possible that the two craftsmen, Ramírez the brickmason and Santa Cruz the carpenter, were in part responsible for the general layout in view of the fact that they presented some sketches along with their estimates for the alteration and reconstruction of the houses destined for university use.[18] Since this was a civil building, the military engineer Luis Diez de Navarro, who was in Guatemala by that time, no doubt had some authority in the supervision of the construction and possibly even in the very plan. But this is not certain, though he is mentioned in the official minutes of the university cloister as having supervised the construction.[19]

Within ten years of the installation of the university in the new building the whole city of Antigua was laid waste. Apparently both the university and the Seminario Tridentino next door were largely spared.[20] Two years later there was some agitation to move the university to Guatemala City, but this did not materialize at this time.[21] Some work must have been carried out in the next few years, for in 1782 the royal authorities registered a complaint to the effect that the papal coat-of-arms had been placed on the building alternating with the royal arms.[22] Actually they alternate in the cloister, while on the exterior only the papal coat-of-arms is to be seen at present. (FIGS. 190, 195, 196.) It would seem that the cloister and possibly the exterior walls too were repaired and the coats-of-arms placed there long after the building had been completed.

In 1788, fifteen years after the great earthquake of 1773, the roof of the library was in need of repair.[23] In 1790 the building was judged to be in bad condition generally, and help was asked to make the necessary repairs which are spoken of as if comprising the construction of a new building.[24] Apparently the building was indeed completely renovated or rebuilt, for an account of the money raised for this work is rendered as late as 1809, and a final accounting of all the money spent on the work is given in 1815.[25]

The present main door, *portada*, is actually postcolonial dating from 1832 when the building was turned into a public school. The large room which had probably served as the university chapel was converted into the municipal theatre of the city of Antigua and utilized as such for many years right into the twentieth century.[26] A colonial museum is now installed in the building. As recently as 1951 the floor of the arcaded corridors surrounding the cloister was renewed. It was interesting to watch the workman shape each floor tile by hand cutting the clover leaf pattern from the large flat thin colonial-type bricks still manufactured today and setting them one by one in a rough mortar made of *arena amarilla* and lime. Both the materials and the workmanship were no different from those employed in the eighteenth century, and the new flooring is now indistinguishable from that laid during the colonial period.

Architectural Data

THIS building has been in continuous use more or less since it was built and, as a result, has been subject to many repairs and alterations even as late as the 1950's. (FIG. 188.) The exterior dimensions of the building, fronting on two streets and abutting on the Seminario Tridentino to the west and the Colegio de Indias to the south, are approximately 52.00 m. by 45.00 m. The interior cloister is not square, measuring 32.07 m. by 29.95 m. The dimensions of the open central space are 23.62 m. by 21.73 m. The north-south corridors vary from 3.72 m. to 3.78 m. wide and the east-west from 4.21 m. to 4.23 m. The north-south corridors thus average about 3.75 m. wide and the east-west about 4.22 m. or approximately 0.50 m. wider. This discrepancy in the square made it impossible to set the columns at equally spaced intervals around all four sides. The columns line up on the same centers only on the sides opposite to each other; 3.48 m. on center in the north-south corridors and about 3.25 m. in the east-west. The off-square plan of the cloister as a whole is rectified in part by the fact that the east-west corridors are slightly wider than the other two.

The present octagonal fountain in the center of the open court is approximately 4.00 m. by 4.00 m., and the water basin 3.15 m. by 3.15 m. (FIG. 193.) It is doubtful, however, whether it is the original one of the eighteenth century, or whether a fountain existed in the building in the first place.

The width of the rooms which surround the cloister av-

[16] *AGG*, A 1.3.21 (1764) 13162–1957; *Efem*, p. 225.

[17] *Ibid.*, p. 228; *ASGH* **25** (1951): p. 157.

[18] *AGG*, A 1.3 (1763) 1157–45; also *ASGH* **17** (1941/42): pp. 376 ff.

[19] Pardo, *Guía*, p. 63.

[20] Cadena, *Breve descripción*; González Bustillo, *Razón particular* also *Ciudad mártir*, pp. 29, 110; *AGG*, A 1.3.21 (1773) 13163–1957.

[21] *AGG* A 1.3.25 (1775) 13253–1961.

[22] *Ibid.*, A 1.3 (1782) 1163–45, and A 1.3.4 (1782 and later) 4410–49.

[23] *Ibid.*, A 1.3.21 (1788) 13164–1957.

[24] *Ibid.*, A 1.3.21 (1790) 13165–1957, also 13166–1957.

[25] *Ibid.*, A 1.3.21 (1809) 13168–1957, and A 1.3.21 (1815) 13170–1957.

[26] Pardo, *Guía*, p. 64; Díaz, *Romántica ciudad*, p. 62.

erage about 5.50 m. to 5.60 m. in width, the variation being due to the addition of plaster coatings to the walls during various reconstructions and repairs. Door openings in general average about 1.60 m. wide. The chapel, on the west side of the court, is 9.80 m. wide by 36.50 m. long and is used today as the main exhibition hall of the colonial museum housed in the building.

The main entrance to the building on the north side is of postcolonial construction and bears the following inscription over the door:

ACADEM PUBLICAS ESEN^{zas} EXISTE A MERCED DEL S. D. MARIANO GALVEZ
EXITADA POR LA MUNICIPALIDAD DE LA A. 1832.

The door gives access to a wide vestibule two steps up from street level and from which still another two steps rise to the cloister. (FIG. 193.) The long narrow corridor directly opposite on the other side of the court through which one enters a small open area to the rear of the building seems to have been constructed in postcolonial times.

The roofing over the rooms and the large chapel as well is of wood and tile. The corridors are roofed with very flat continuous segmental barrel vaults, the soffits of which are decorated with a band at the crown running the whole length of the vault like a ridge beam. (FIG. 192.) Transverse bands of equal width are set at the intervals where the crowns of the arcade arches are centered. A pent roof, a right-angle triangle in section, of timber and tile protects the exterior of the corridor vaults. (FIG. 189.)

There is a great variation in the thickness of the walling, the exterior one being an average of about 1.15 m. thick. The north and south walls in the large chapel on the west side of the court are almost 2.00 m. thick and probably represent part of the original walling of the house which had been altered for the university *ca.* 1763. Some of the cross partition walls in the rooms surrounding the cloister are only 0.65 m. thick and have been reset in recent years so that the original long dimensions of the rooms have been changed. The width of these rooms has of course remained constant.

The walling materials are brick and stucco. The exterior northeast corner is treated with stone veneer. A large upright post, inset into the wall, is flanked by large, square stones set in courses breaking joints which are filled with thick, white mortar. (FIG. 190.) This bit of stonework forms a sort of high pedestal for a fanciful stucco decoration which is a variation on the papal coats-of-arms decorating the exterior of the building just under the cornice. Some stonework in the form of a rusticated pilaster is set on the exact outside corner angle on each side of which, that is, on each face of the building, a half papal shield with surrounding festoons is placed. Instead of the papal emblems, the interior motif is now a large grotesque face or Medusa head cut in half. (FIG. 191.)

At regular intervals, but not in relation to the present interior arrangement of the rooms, are large octagonal windows placed high on the wall just under an entablature which runs like a coping or crowning element around the entire exterior of the building. Supporting the entablature are short brackets or consoles in low relief arranged at regular intervals between windows so that each is flanked by a pair. These same brackets are used in a similar manner in the corridors of the cloister to frame door and window openings.

Each of these brackets begins with an inverted triangle with a pendant modified palmette filling the area. Immediately above is a triglyph with four flutes flanked by opposed S-curve scrolls. Immediately above is a capital with all the parts normally found in the Tuscan type. The necking, however, is pulvinated and is picked out with deep, closely set, diagonal incisions which tilt toward the center, meeting in a single pointed oval. Immediately above are an echinus and abacus made up of a number of moldings, one projecting beyond the other, and in appearance somewhat like a crown molding. These brackets support an entablature which breaks miters on bracket centers and in which the architrave and frieze are very narrow in relation to the projecting cornice which is the dominating member.

A low parapet wall is set back from the cornice to the same plane as the wall below. It also breaks miters, but of very shallow projection, on bracket centers forming, thereby, pedestals on which pyramidal merlons are set. The outer face of the parapet is decorated with exactly the same motif as that on the parapet wall of the cloister arcade; namely, vertical half-round moldings shaped as if they had been turned on a lathe and looking like slim stems of candlesticks alternately resting on the bottom of the parapet or hanging from the top. (FIG. 189.)

The cloister arcade is perhaps the most ornate in all Antigua. (FIGS. 192–196.) The soft curves of the arches are unlike the contemporary tortured baroque forms of Mexico or Spain.[27] The *ataurique* ornament here represents a culminating and final example of the use of this type of decoration in Antigua. Though not as fine in quality as that of La Candelaria or Santa Cruz, it is, nevertheless, more fanciful and involved.

27 According to Toussaint, *Arte mudéjar*, p. 56.

Surrounding the four sides of the open court are twenty-eight piers, not quite square in section, of a proportion common to those employed in other post-1717 *antigüeño* buildings. (FIGS. 188, 189.) The corner piers are not doubled as normally occurs in other cloisters but are, in fact, slightly smaller since they lack the pilasters common to the rest of the arcade piers. (FIG. 194.) Measured on the base these corner piers are 1.05 m. square. They are fluted on the inside faces toward each arcade in exactly the same manner as the inside faces of the intervening arcade piers. In other words, the corner piers lack the courtside faces with the pilasters, and are treated with fluted side elevations only. Measured at the base, all the other piers are approximately 1.05 m. by 1.15 m. The additional 0.10 m. correspond to the dimensions of the narrow pilasters engaged to the pier facing the court. The inside faces of the piers have four deep flutes which are exactly the same in section as those seen on the inside pilasters of the gatehouse of El Calvario. (FIG. 10.) The three projecting fillets which would normally have flat faces have been rounded off and the center fillet face is reduced somewhat by the addition of small square chamfers at the origin of the rounded part.

The shafts of the piers are also not square in section, and are 0.80 m. by 0.87 m. The pilaster adds an extra 0.05 m. making the pier 0.92 m. in section through the pilaster. The total height of the pier from the base, including the Tuscan capital, is 2.45 m., making a proportion of approximately 1:2⅓. The total height of the order including the entablature which runs above the crowns of the arches of the arcade is 5.05 m. Using the base of the pier as a module, but bearing in mind that the entablature does not rest directly on the piers and that the spandrels of the arches intervene, the proportions are slightly less than 1:5.

The piers, as has already been mentioned, are not exactly lined up or evenly spaced on all four sides. But, in general, the intercolumniations average about 3.25 m. on centers. The spaces left void are thus about 2.25 m., leaving a proportion of solid to void of a little more than 1:2 and the spacing of pier centers about 1:3.

The arches which connect the piers are of the stepped ogee type, but with rounded crowns. (FIGS. 189, 194, 195.) The mixtilinear shape begins with a rather erect cavetto at the springing. This joins a square chamfer above which a very flat cyma recta or open S-curve makes up the haunch. On top of the cyma recta there is another square chamfer, from which the segmental crown of the arch emerges. The height of the opening to the sofft of the crown of the arch is 3.45 m. The space from the bottom of the entablature to the extrados of the crown is 0.30 m. more, making a total height of 3.75 m. from the floor to the entablature. The total height of the entablature is 1.30 m., and the parapet wall above, an added 0.60 m. The merlons are very elongated and rise 0.90 m. above the parapet.

The little pilasters engaged to the court-side face of each pier rest on high plinths and a base consisting of a scotia and torus. The pilaster shaft is plain and 0.30 m. wide. Instead of sharing the same Tuscan capital as the pier, it has its own capital in the form of a triglyph striated by four square flutes. The triglyph cuts through the main pier capital and serves as the base for a short pilaster which divides the spandrels of the contiguous arches. This spandrel pilaster is the type seen on El Calvario, namely the corrugated type, and is made up of three squared, superimposed Attic bases or squat pot-forms plus a capital which is a variation on the Tuscan. (FIG. 2.) To either side of these diminutive spandrel pilasters are four scrolls arranged one above the other in connected pairs to form two cursive capital E-letters facing away from the pilaster. A very ornate repeated egg-and-dart pattern stretches across the top of the bay just under the entablature and connects with the E-scrolls adjacent to the pilasters. Each egg is foliated and rather pointed and is flanked by little scrolls. The darts have little horizontal members near the bottom giving them the appearance of pendant crosses.

The main entablature which runs around the arcade on all four sides is perhaps the most ornate in all Antigua. (FIGS. 189, 195, 196.) The architrave, made up of many small moldings divided into three main bands, breaks miters on pilaster centers. A triglyph and metope design fills the frieze. There are five triglyphs and four metopes in each intercolumniation. The triglyphs, if such they can be called, have four flutes. The metopes are more or less square and are filled with opposed S-spirals or double volutes. An egg-and-dart molding of more normal classical appearance runs directly above the frieze.

The cornice is very high and projects far beyond the base of the piers below and thus acts to throw the rainwater well into the open court away from the corridors. Immediately above the cornice and set back from it is the low parapet wall with the type of candlestick decoration mentioned in connection with the parapet on the exterior walls of the building. The cloister parapet breaks miters on pier centers and thus forms pedestals for the rather elongated merlons.

The openings or intercolumniations located directly on the two axes of the court are treated with special decorative elements. Over the central intercolumniations on the north-south axis, blazons with the royal coat-of-arms are set and rise above the parapet wall abutting on either side. (FIGS.

193, 196.) The cornice of the entablature below changes to a short segmental arched section directly under the coat-of-arms. Two pilasters also frame the blazon, the shafts of which consist of the *toisón de oro*, the chain of the Order of the Golden Fleece. Filling the space between these pilasters and the blazon itself are the same E-scrolls seen in the spandrels of the arches of the arcade below. A truncated pediment with swagged raking cornices and with merlons set on top and to either side completes the composition. Papal coats-of-arms are set over the central intercolumniation on the east-west axis, the decorative composition and details of which are exactly the same as the latter, except that the central portion is filled with the papal emblem. (FIG. 195.)

6. *Capitanía.* *Palacio de los Capitanes Generales, 1769*

(FIGS. 197–206.)

Historical Data

THIS building originally housed the *audiencia*, the seat of the colonial government of the whole of Central America. The mint, Casa de Moneda, where coinage for Central America was produced, was located in the same building. The remains of the building which exist today have been subjected to much reconstruction in the nineteenth century and, except for the western half of the arcade facing the plaza and some of the exterior walls, most of the structure is actually modern.

The *audiencia* was first established in 1543 in Gracias a Dios, Honduras. It was moved to Antigua in 1550 and installed in some houses which Lic. Cerrato had bought the year before.[1] It remained in Antigua only until 1563 when it was again moved, this time to Panama, and the province of Guatemala was put under the jurisdiction of Nueva España.[2] Then in 1566 it was moved back to Antigua permanently where it began to function by 1570.[3]

Just what sort of structure the offices of the *audiencia* were housed in during the sixteenth century is impossible to tell. There is also no indication as to where the houses Cerrato had originally bought in 1549 were located. However, by the seventeenth century a building for the *audiencia* was already in existence which was repaired and rebuilt at various times between 1656 and 1682. It occupied the same site as the present Capitanía, a whole square block on the south side of the Plaza Mayor.[4]

The earliest reference in the eighteenth century to the building gives an account of some money spent for reconstruction work just before 1711.[5] An undated plan, possibly from mid-century, shows a building with a two-story wooden colonnade. The same general layout, as appears on later plans, including the Casa de Moneda, is indicated. Some plans submitted to Spain later by Diez de Navarro give a detailed account of the Capitanía and the Casa de Moneda, and show that the earlier structure was built mainly of adobe and rammed earth with stone.[6] Along with all the others of Antigua, the buildings suffered damages in the various earthquakes, especially in 1717. Ximénez, who tends to minimize the destruction of 1717, since he was on the side opposed to moving the capital, disagrees with Arana who said some rooms and walls were destroyed. Ximénez claims that only some roof tiles and one partition wall of *tierra muerta* (adobe, or rammed earth) had fallen.[7]

What repairs were needed for the damage of 1717 is hard to tell, but at various times during the next forty years until the earthquake of 1751 a number of documents give accounts of the money spent to repair and reconstruct the building.[8] The building was again left in an exceedingly bad state of repair in the 1751 earthquake.[9] The part housing the chancellery had to be demolished because funds were lacking for a wholesale reconstruction. Repairs were carried out only on the "Palacio Chico" and the cracks in the walls of the rest of the building were pointed up. These repairs were makeshift, for in 1754 the building was once again in a very bad condition. In 1755 Diez de Navarro reported what work had already been done and what remained to be completed, indicating this on a plan submitted to Spain.[10] (FIGS. 198, 199.) He recommended that as much of the remains of the ground floor walls as possible should be utilized in the new work, and that these be reinforced with some stone

---

[1] Schaeffer, *Indice* **2**: no. 2636; also *Efem*, p. 12.

[2] *Efem*, p. 17.

[3] *Ibid.*, pp. 18 ff.

[4] See Mencos, *Arquitectura*, ch. VI, and append. XXIX, transcribing *AGI*, Guatemala, 28, "Certificación de los gastos y reparos hechos en el palacio de los Capitanes Generales desde el año 1682." Fuentes y Guzmán, **1**: p. 139, says it contained ". . . los tribunales Reales, del Acuerdo de Justicia, Audiencia, oficio de Provincia, Real Capilla, sala del despacho del Real sello, escribanía de Cámara y sala de Armas . . . ," treasurer's office, and others.

[5] *AGG*, A 1.10.1 (1711) 14902–2101. For the undated plan see *Cartografía de ultramar* **1**: p. 39, no. 11, and **2**: pl. 11.

[6] See Angulo, *Planos* **2**: pp. 56 ff., and **4**: pp. 416 ff., pls. 157–161 and 165, 166, who gives an account of the history of the Capitanía.

[7] Ximénez, **3**: p. 354.

[8] *AGG*, A 1.10.1 (1736) 14903–2101; A 1.10.1 (1736) 14904–2101; A 1.10.1 (1740) 14905–2101; A 1.10.1 (1746) 14906–2101. See also Mencos, *op. cit.*, append. XXXI, transcribing *AGI*, Guatemala, 314, "Declaración de Diego de Porres, Maestro Mayor de Arquitectura, sobre la Real Casa de Moneda de Santiago, en Guatemala a 18 de agosto de 1738." See also Angulo, *op. cit.* **4**: pp. 425 ff.

[9] *Efem*, p. 204.

[10] Angulo, *loc. cit.* **2**: pp. 56 ff.; **4**: pp. 416 ff., pls. 157, 158.

buttresses and courses of brick in order to serve as load-bearing walls for the second story. He also proposed to use the old timbers of the ground floor for scaffolding and thus entail a minimum cost of 55,245 *pesos*, "... quedando con la maior seguridad para la resistencia de los temblores de que esta ciudad es molestada." The façade facing the north side, formerly a two-story wooden portico, is now shown as a stone arcade.

In 1760 an order was emitted that an inspection of the building be made to determine its condition and that an account be given of the work done to date.[11] Diez de Navarro carried out this order aided by Francisco Javier Gálvez[12] who, only three years before, had been accredited by the city council as *maestro mayor de arquitectura civil y de carpintería*.[13] After making the inspection, Diez de Navarro again drew up some plans reporting the work done and the work in progress and/or remaining to be done. These plans were sent to Spain.[14] In 1761 more funds were allocated for the work needed for completion[15] and Gálvez was put in charge of the job,[16] but no doubt under the supervision of Diez de Navarro.

When the new president of the *audiencia* arrived in 1761 and saw the building still partly in construction and partly in ruins, and on the opposite side of the plaza the bright new *ayuntamiento* building, he suggested that the city council provide the funds needed to build a similar two-story arcade with brick vaulting for the Capitanía, promising that one day the royal treasury would reimburse the city.[17] Finally, in 1763, a royal *cédula* authorized the expenditure of 65,183 *pesos* on the work.[18]

The new plans of 1760 were very much the same as those submitted in 1755. Apparently very little had been done in the previous five years for Diez notes on the plan that economy will only be possible if the work is completed without delay and the older parts utilized before they are completely destroyed. If not, he warns, the cost will be double. Work proceeded very rapidly in 1768 bringing the building to completion in the following year.[19] A final plan of the Capitanía was drawn by Diez de Navarro in 1769 and sent to Spain. (FIG. 200.) On it some minor changes in the interior arrangement appear. On the plan he indicates that both the repairs and the new construction are on the point of completion. The arcade in front he describes as "2. Portales con pilastras y arquería de cantería," exactly as he had shown on the 1755 plan.[20] He adds a further note on the plan that the work of completion was not carried out under his supervision since he was away from the city while this was being done.

Thus the building stood until the destruction of Antigua in 1773. That the Capitanía was left almost a total ruin, is attested to by contemporary sources[21] and from photographs taken a century later which show half of the two-story arcaded north façade in ruins.[22] (FIG. 201.) But some of the destruction to the building was willful when it was dismantled for the purpose of obtaining second-hand materials to be used in the construction of the public buildings of the new capital twenty-five miles away in La Nueva Guatemala, the present Guatemala City.[23] Materials taken from other public buildings were stored in the Capitanía, their excessive weight causing some sinking of the building.[24] When Muybridge visited Antigua *ca.* 1876 he took a photograph of the façade in which only the west half is shown standing and the whole of the eastern half missing. A photograph taken from a position opposite Muybridge's, not dated but which may be placed in the late nineteenth century, shows the same ruined character of the façade.[25] Various repairs were carried out on the building in the nineteenth century, the principal one between 1889 and 1890 when the eastern half of the façade was rebuilt and parts of the rest of the structure were put in condition to serve as governmental offices again.[26] (FIGS. 202, 205, 206.)

Though he had a major part in the construction, it is difficult to assert that Diez de Navarro was the sole designer of this structure. He is known to have arrived in Guatemala around 1741 or perhaps a year later,[27] just about the time

[11] *AGG*, A 1.10.1 (1760) 1421–64.
[12] *Efem*, p. 217.
[13] *Ibid.*, p. 214.
[14] Angulo, *op. cit.*, pls. 159, 160.
[15] *AGG*, A 1.10.1 (1761) 1422–64.
[16] *ASGH* **25** (1951): p. 148; see also *Efem*, p. 219.
[17] Angulo, *op. cit.* **4**: p. 418.
[18] *AGG*, A 1.10.1 (1769) 184–8 and A 1.10.1 (1769) 1423–64.
[19] *Ibid.*, A 1.10.1 (1768) 31.220–4044, A 1.10.1 (1768) 31.222–4044, and A 1.10.1 (1768) 31.223–4044. For Navarro's comments see Angulo, *loc. cit.*, pls. 159, 160, **2**: p. 63.

[20] Angulo, *op. cit.*, pl. 161; also Mencos, *op. cit.*, append. XXX, transcribing *AGI*, Guatemala, 657, "Regulación e informe del palacio de Capitanes Generales de Santiago, ejecutado por el ingeniero Luis Diez de Navarro, en Guatemala a 22 de abril de 1760 y 14 de febrero de 1769 respectivamente."
[21] González Bustillo, *Razón particular*.
[22] Muybridge, *Pacific Coast*, negative no. 4264. Also fn. 25 below.
[23] *AGG*, A 1.10.2 (1774) 1642–66 and A 1.10 (1777) 4524–69; *Efem*, p. 192.
[24] *AGG*, A 1.10.3 (1777) 4571–76.
[25] *ASGH* **18** (1942/43). Neither the name of the photographer nor the date is given.
[26] Díaz, *Romántica ciudad*, p. 34. The columns of the lower story are uniform throughout the entire length of the arcade thereby implying that the original ones were employed in the reconstruction.
[27] Berlin, *ASGH* **22** (1947): pp. 89 ff. See also Angulo, *op. cit.* **4**: p. 416, fn. 1, for some bibliography on Diez de Navarro's career in Central America. See ch. VI, above, for biographical data.

the Ayuntamiento was being completed, the façade of which served as a model for that of the Capitanía. The various plans he submitted to Spain were really more in the nature of progress reports rather than working drawings. He was, however, in charge of the work except for the final period during which the construction was completed. The two-story arcade was begun after 1761 with Gálvez in direct charge of the job. The prototype of the arcade is located directly across the square. Since it was specifically intended to reproduce it, the Ayuntamiento façade no doubt served as a ready reference while that of the Capitanía was going up. The Casa de Moneda which formed an integral part of the structure was actually built by Diego de Porras before Diez de Navarro had arrived in Guatemala.[28] This independent structure, though integrated with the Capitanía, was completed by 1738.[29] (FIG. 197.) This dependency figured in each of the plans which Diez de Navarro submitted with his reports and estimates of the construction work carried on during his time. To judge by his disclaimer of having had anything to do with the completion of the Capitanía, it would seem that the general scheme, especially of the arcading, was agreed on from the first, a fact reinforced by Diez de Navarro's plan of 1755. Only in 1761, when money was borrowed from the *ayuntamiento* and Gálvez put in charge of the job, was the arcade built. The design of the building was largely improvised by Diez de Navarro who, for reasons of economy, recommended reinforcing and using the older parts still standing after the earthquake of 1751. In a sense the whole building, including the arcade, was the result of a series of improvisations based on the use of parts of the older structure and a reproduction of the façade of the Ayuntamiento across the square.

Architectural Data

THE floor plan of the building still remaining today is but a fraction of what it was in 1769. It was restored in such a manner that very little of the original character of the interior of the building can be gained from a study of the existing structure. (FIGS. 201, 202.) The building once occupied an entire block. Except for the north façade, half of which has been rebuilt, the sides facing the other streets are of little architectural interest. Diez de Navarro's last plan shows a two-story structure with various rooms and other dependencies arranged around a number of open courts. (FIG. 200.) The building could be entered through the arcade by means of two or three doors which gave access to both the offices of the *audiencia* and the area devoted to the mint. The rear or south wall had no doors, but the east and west walls each had an entrance which gave access to the rear parts of the building and its patios.

The north façade with the two-story arcade is of particular architectural interest. Each story of the arcade consists of twenty-six complete half-circle arches supported on columns of the same design as those of the Ayuntamiento. The arcades end in massive piers with responds. The lower-story columns have almost exactly the same dimensions as those of the lower story of the Ayuntamiento. Likewise, the individual bays with pendentive domes follow the Ayuntamiento scheme. (FIGS. 205, 206.)

A remarkable change occurs, however, in the second story. Though the columns here, as is the case of the Ayuntamiento, are shorter than those below, they are, nevertheless, unlike those of the Ayuntamiento, noticeably greater in diameter defeating the purpose of making the second story lighter and more open. (FIG. 201.) The second story as a whole is about one-fifth shorter than the first story. This is achieved not only by means of the smaller columns, but also because the arches of the upper arcading are somewhat less than semicircular. The point of origin and springline of the arches are actually below the level of the capitals of the shafts from which these second-story arches emerge. The second story is roofed with wood and tile except for the three central bays which have pendentive domes.

The horizontal accents of the façade show a change from the Ayuntamiento where the same cornice is employed at the top of each story. (FIGS. 155–157.) In the case of the Capitanía, a cornice molding of only moderate projection marks the top of the first story. The top of the second is heavier and strongly accented projecting well in front of the plane of the arcade. The total height and projection are well over twice that of the cornice below. A low parapet runs the length of the façade with mixtilinear pear-shaped merlons set on column centers.

The three central bays, that is, the twelfth, thirteenth, and fourteenth intercolumniations, received a special treatment. (FIGS. 203, 204.) The four columns of the three lower-story bays are altered and take the shape of a pier with engaged half-columns to either side, between which a fluted pilaster runs up the outside face. The girth of the supports at these points is about doubled. The columns of the second story above are even heavier still, and are treated in a most unusual fashion—a cluster of four columns, two facing out and two facing the interior of the upper-story corridor. The columns which make up the cluster are of only slightly

[28] Angulo, *op. cit.*, pls. 165, 166. See fig. 197. This plan is reproduced in *ASGH* **17** (1942/43): p. 216.

[29] Juarros, **1**: p. 156; see also Díaz Durán, *Casa Moneda*, pp. 191 ff.; *Efem*, pp. 184, 204; Larreinaga, *Prontuario*, p. 59; and García Peláez, **2**: pp. 143 ff.

smaller diameter than those of the rest of the arcade on that story.

The outline of the horizontal cornice is broken over the three central bays by a riser so as to give the façade a little added height here. The cornice moldings do not continue across the three bays in a straight line but drop down on center over the two inside columns framing the middle intercolumniation. Set directly over this intercolumniation, the thirteenth one of the façade, is a finial with the coat-of-arms of Carlos III bearing the date 1763. Each side of the circular coat-of-arms is flanked with small columns. To either side, mixtilinear half-pediments decorated with rampant lions complete the ensemble. Blazons with coats-of-arms are also located over each of the two fifth bays, counting in from both corners. (FIGS. 201, 202.) The oval-shaped east one, which was no doubt put back in place when that half of the façade was reconstructed in 1890, bears the names of the Captain General Alonso Fernández de Heredia and of Luis Diez de Navarro, and also bears the date 1764. The coat-of-arms on the west side, already there in 1876, is that of the Republic of Guatemala which probably replaced an earlier Spanish one sometime after the independence in 1821.

7. *Santísima Trinidad: Chapel, Late Eighteenth Century*

(FIG. 207.)

Historical Data

THE building, now in almost complete ruin, is located in the northwestern sector of Antigua. Almost nothing is known of its history, and even its very name is not certain. A hermitage called Santísima Trinidad was filial to one of the three parishes into which Antigua was divided after 1773.[1] It is uncertain whether or not this was the chapel which was located in the neighborhood of "el Chajón" and which continued in use until 1804 when it was ordered closed.[2] Beyond these few scant references there is no mention of any church of this name in Antigua, making it difficult to ascertain just when it was built.

Architectural Data

ALMOST nothing remains of this small church except parts of the long walls, the rear wall, and a large single arch which divides the nave from the *capilla mayor.* (FIG. 207.) Nothing whatever of the façade remains. The building had been apparently roofed with an *artesón* of which not a trace is visible.

The only features of special interest are the massive arch and the pilasters from which they spring. The design of the pilasters is of the type seen on the façade of the church of Santa Cruz; namely, a cluster of columnettes. However, the shafts of the columnettes here are not fluted but do have a smaller projecting molding triangular in section between each. The capitals are a variation on the Tuscan type and repeat the contours of the shaft below. Above the capital a short section of entablature, also conforming to the contours of the shaft and capital, has a three-fasciae architrave, a pulvinated frieze, and a rather simple cornice. The arch which springs directly from the entablature is semicircular. Its soffit repeats the contours of the pilasters below. In the angle of each spandrel the papal coat-of-arms is worked in stucco relief.

The side walls of the nave and *capilla mayor* are very massive and are pierced by large octagonal windows set high up. A large octagonal niche is set in the center of the rear wall. The lower part is now filled with adobe brick.

[1] Juarros, I: p. 56.

[2] *Ibid.*, p. 153; Pardo, *Guía*, p. 162.

8. *Jocotenango: Church and Capilla de la Virgen, Seventeenth Century and Eighteenth Century*

(FIGS. 208–212.)

Historical Data

JOCOTENANGO lies to the north just outside the city limits, and is a separate municipality in the Departamento de Sacatepéquez, though its street plan merges with that of Antigua. The village was founded soon after the destruction of the first capital at Ciudad Vieja in 1541.[1] The first inhabitants of the place were all Utatleco Indians who numbered 4,070 by the end of the seventeenth century. The majority were either masons, bricklayers, or producers of brick and roof tile.[2] Even in the late eighteenth and early nineteenth centuries, most of the Jocotecos were engaged in the building trades, principally as masons. When Antigua was abandoned soon after 1773, the majority of the families of Jocotenango moved away too and were employed in building the new capital where they established a town of the same name, also to the north of the new city.[3]

The English friar Thomas Gage passed through Jocotenango on his way to Antigua early in the seventeenth century, but did not stay very long, saying he could come back anytime. He remarks, "The Frontispiece of the church of

[1] Díaz, *Romántica ciudad*, p. 40.

[2] Fuentes y Guzmán, I: pp. 391 ff.

[3] Juarros, I: p. 58; Dunn, *Guatimala*, p. 156.

this town is judged one of the best pieces of work thereabouts; the high altar within is also rich and stately, being all daubed with gold."[4] On the basis of this rather general description it may be concluded that a church already existed in Jocotenango early in the seventeenth century.

From another early seventeenth-century source, *ca.* 1610, we learn that this church and a small convent adjacent to it were built by Padre Fray Juan de Morales and Padre Fray Rafael de Luján[5] probably just before January of 1602 when formal possession of the building was given to the Dominican order.[6] The church and convent were given the rank of vicarage and the nearby towns of San Felipe, San Luís de las Carretas, and Pastores placed under its ecclesiastical jurisdiction continuing in this relationship through the seventeenth century, and very likely even through the eighteenth.[7]

The ceremony in which the church and convent were turned over to the Dominicans is described as taking place in the patio and cemetery in front of the church.[8] This area may be tentatively identified as lying just north of the present church and in front of the chapel, Capilla de la Virgen, attached to a wall which projects at right angles from the presbytery of the present church. (FIGS. 211, 212.) In the 1920's vestiges of an old cemetery were still visible here.[9]

From a notice concerning the earthquake of 1717 we learn that half the *portada* fell down on the church bringing it to the ground too.[10] The building referred to must be the chapel, for incontrovertible documentary evidence exists substantiating the fact that the present church was built during the eighteenth century, and actually completed after the abandonment of Antigua. In 1761 the Indians of Jocotenango asked to be relieved of extraordinary contributions since they were then engaged in building their church.[11] Later on, in 1773, a report was rendered concerning the completion of the church,[12] and in 1781 the parish priest asked for monetary help to furnish the convent.[13] The chapel must, therefore, be the church referred to in the seventeenth century.

Cortés y Larraz was there in 1770, yet does not mention the church at all. He does inform us, however, that the population consisted of 747 families or 3,735 individuals whose only vice was drunkenness and that the school was located in the *cabildo* of the town (where it exists today).[14] It is interesting that the population had remained more or less the same for over a century.[15] But by 1790 so many of the population had moved to Guatemala City that the parish priest presented a formal request asking that the town be moved there too.[16]

There is a fountain of colonial date in the open area in front of the church today, though most likely it has been renovated since its construction. Just when it was first built is hard to tell, but a notice from 1733 indicates that a fountain had been completed in that year and placed in service.[17] The water basin is octagonal in plan and is surrounded by an openwork railing made of brick.

[4] Gage, ch. XVII, p. 264.
[5] Remesal, **2**: p. 246.
[6] *AGG*, A 1.20 (1602) 10.171–16 vuelto, also *BAGG* **10**, 2 (1945): p. 95; see also *Efem*, p. 35.
[7] Fuentes y Guzmán, *loc. cit.*
[8] *AGG*, document cited above in fn. 6.
[9] Díaz, *loc. cit.*
[10] Ximénez, **3**: p. 358.
[11] *AGG*, A 1.10.3 (1761) 4679–236, fol. 1.
[12] *Ibid.*, A 1.10.3 (1773) 31.358–4049.
[13] *Ibid.*, A 1.10.3 (1781) 1772–73.
[14] Cortés y Larraz, fol. 16.

Architectural Data

TWO separate buildings are involved here: the late eighteenth-century church proper, and the little chapel which abuts on the wall projecting at right angles from the north side of the church and which probably represents the original seventeenth-century church damaged in 1717 and subsequently repaired. (FIGS. 209, 211.)

i. The Church (FIGS. 208–210.)

The church façade was doubtless remodeled in postcolonial times. The church is approached from a broad open plaza in which a fountain of colonial date is still in use. A low two-stepped *lonja* with four low merlons runs across the width of the façade. The merlons have rather squat pyramidal finials and are probably postcolonial. The façade is laid out like a retable with the three vertical bays and three horizontal stories. The façade looks almost square, for the belfries of the towers rise to the same height as the third story of the central bay. The towers are not entirely freestanding. About half their width overlaps the side bays of the retable.

The lateral bays of the first and second stories are flanked by pairs of rather thin unevenly finished solomonic columns. (FIG. 210.) The spirals, no two of which are alike, wind in opposite directions on each pair. An astragal molding in the hollows between spirals further accentuates the winding effect. A niche is located rather high and is almost contiguous to the entablature above in each of the side bays.

The first-story central bay is pierced by a door with a semicircular header set in a similar recess. The door and recess arches are not concentric, the springing of the recess arch being a bit higher than that of the door. A crescent-shaped area is formed between the extrados of the door arch and the intrados of the recess arch as appears on Los Remedios.

[15] Fuentes y Guzmán, *loc. cit.*
[16] *AGG*, A 1.10.3 (1790) 1729–73.
[17] *Efem*, p. 175

A large octagonal window which lights the upper choir is located in the second story of the central bay. (FIG. 208.) Flanking the window and framing the bay are two pilasters of the corrugated type. The third story of the central bay is really a finial, and is horizontal in outline. It consists of a low freestanding wall with three niches set rather close to one another. Framing the central niche are two pilasters of the same corrugated type which frame the octagonal window of the second story immediately below. At either end of the finial is a scroll half-pediment. A low swagged truncated pediment crowns the wall as a whole, to either side of which are low merlons with steep, pyramidal finials. A merlon of different design rests on the horizontal portion of the pediment.

The exterior side walls of the church are rather curious in that they do not have a horizontal cornice at the top. A freestanding wall juts out at right angles at the rear of the north wall, probably part of a transept bay which was never completed. (FIG. 211.) It is pierced by an octagonal window exactly like those in the other bays of the interior of the church. Furthermore, this wall abuts on the chapel, merging into its south tower and filling the belfry. It seems that this freestanding wall was an afterthought representing a part of the church which was never finished. The long north wall of the church facing the former cemetery is also not uniform. The last bay just in front of the projecting freestanding wall has a large rectangular recess in the lower portion. This probably represents a door which was planned to open into the incompleted transept and which was bricked up later. Had the transept been constructed, part of the façade of the chapel, first built in the early seventeenth century, would have been covered up.

ii. The Capilla de la Virgen (FIGS. 211, 212.)

The façade of the Capilla de la Virgen is more interesting than that of the larger building. It is approached by a *lonja* with four merlons, one at each corner and one on a line to each side of the door. The north corner merlon is missing, so that only three remain. The *lonja* once was higher than the patio in front, the area having been filled and its level raised covering the *lonja* steps. The façade layout consists of a single bay two stories high, including the finial, flanked by twin towers with belfries which rise above the pinnacle of the finial.

The single bay of the first story consists of a round-headed door set in a recess with a concentric header. The door jambs and arch are of stone. The jambs are treated like pilasters and consist of small pedestals supporting unadorned shafts topped by Tuscan capitals which serve as the imposts of the arch. Framing the door are two Tuscan columns set on pedestals supporting an entablature with a pulvinated frieze. The entablature breaks miters over the columns. Immediately above and part of the second story is a low attic which also breaks miters on column centers. (FIG. 212.) Above it and on column centers merlons are placed, the pyramidal pinnacles of which are separated from the bases below by a rather high scotia.

In the center of the second story and as a focus of attention is a niche with a crucifixion scene worked in high relief. The outline of the niche is rectangular; projecting upward from the lintel is a smaller rectangle. Or looking at it from another point of view, the outline of the niche is a long rectangle the top corners of which are indented by square chamfers. A similar niche appears in the finial of Santa Cruz. (FIG. 132.) The niche itself is framed by narrow pilasters with additional bases set halfway up their height as on San José el Viejo. (FIG. 169.) A simplified Tuscan capital surmounts each, and a horizontal cornice runs between them as a modified entablature. A short piece of cornice is set on center above each pilaster. In the space between, a truncated pyramidal pediment with swagged raking cornices fills the area. Completing the composition are scroll half-pediments to each side made up of two volutes, both spirals of which turn toward the niche.

The whole of the finial in which the niche with the crucifix is located has an outline which follows that of the niche itself. The upper horizontal cornice drops down to form square chamfers and continues across the face of the twin towers to either side. Partially filling the voids thus formed in each corner are scroll half-pediments. A segmental pediment surmounts that portion of the cornice. Merlons are set to either side, while a third rests on the crown of the pediment.

The towers are quite simple and are divided into three stories by horizontal moldings. The north belfry has a single arched opening on each of the four sides. The south tower which is merged into the projecting freestanding wall of the church has a belfry with blind openings. A low, ribbed dome roofs each belfry. Merlons are set on the crown of each dome and also in each of the four corners of the belfries.

Abutting the chapel to the north is another small dependency with a door facing the patio. (FIG. 211.) This door is framed by an entablature and oval-shaped pediment supported on corbeled brackets. The top of the pediment is rounded, that is, the raking cornices do not come to a point. On center with the corbeled brackets are merlons of the same type as seen on the rest of the façade.

XVI NONEXISTENT STRUCTURES, STRUCTURES OF NONDESCRIPT CHARACTER, AND STRUCTURES REBUILT IN POSTCOLONIAL TIMES

THERE were a number of buildings of which now not a trace remains. Among these are the monastic establishment of Santo Domingo with its convent and church, the Beaterio de Indias, the hermitage of Santa Cruz del Milagro, the church of Santa Lucía, as well as others of lesser importance. It will suffice to list these buildings, give the references which have been culled from the available contemporary sources, and make some notes on their architectural character.

1. *Beaterio de Indias del Rosario, 1762 and 1771*

(FIGS. 213-214.)

THIS monastic institution for Indian women was founded under the aegis of the order of Santo Domingo probably after the middle of the seventeenth century. (FIGS. 213, 214.) Formal license, however, did not come till much later, in the middle of the eighteenth.[1] By 1735 the quarters which they occupied were in need of repairs, and financial aid was sought from the *ayuntamiento*.[2] Help was again sought to repair some damages after the earthquake of 1751.[3] Finally, in 1762 the convent structure burned to the ground and a new convent was begun almost immediately and probably completed that same year, the cost of rebuilding borne by Domingo López de Urrelo.[4] The church still remained to be built, and was completed in 1771.[5] The new church and convent were destroyed in 1773, so that only some rubble remains on the site today. The convent moved to Guatemala City in 1779.[6]

The plans submitted in 1762 to the authorities in Spain were drawn by Luis Diez de Navarro. The church, which at that time was in a bad state of repair and needed to be reconstructed, was twenty-three *varas* long by six wide. It had been built of a combination of brick and tamped earth which Diez de Navarro proposed to rebuild entirely of masonry and roof with vaults. On his plan the church is shown with one of the long sides to the street and the other three sides integrated into the convent. The convent contained two cloisters, the principal one abutting on the church, in the center of which was located a large square fountain with small exedrae projecting from the four sides. Surrounding the convent grounds was a wall. In the open area in back of the convent, beneath a roof on piers, he made provision for a large open water receptacle with sufficient space for ten people to wash clothes. This type of *pila* or *lavadero* is still quite common in many of the small towns of Guatemala, and even in the poorer neighborhoods of Guatemala City itself.

[1] Juarros, **1**: pp. 137 ff

[2] *Efem.*, p. 178.

[3] *Ibid.*, p. 205; also *ASGH* **24** (1949): p. 378.

[4] *Efem*, p. 223; also *ASGH* **25** (1951): p. 152; Torres Lanzas, *Planos*, no. 187 referring to *AGI*, Est. 100, Caj. 7–Leg. 22 (6); also Angulo, *Planos*, pl. 154, **2**: pp. 48 ff., and **4**: pp. 412 ff.

[5] *Efem*, p. 241; *AGG*, A 1.11 (1772) 2117–98–7, also *BAGG* **10**, 3 (1945): p. 262; *ASGH* **25** (1951): p. 171.

[6] *AGG*, A 1.11 (1774) 2117–98–13, vuelto, 1, also *BAGG* **10**, 3 (1945): p. 263; *AGG*, A 1.11 (1779) 2117–98–9, vuelto, also *BAGG* **10**, 3 (1945): p. 264; Angulo, *op. cit.*, pl. 155, **2**: p. 49, **4**: pp. 412 ff.

2. *Benditas Ánimas: Hermitage, 1702 and Earlier*

THIS little hermitage, which no longer exists, was built early in the seventeenth century[7] and located on the outskirts of town on the road to San Juan Gascón. In 1702 the *ayuntamiento* donated the sum of 3,000 *maravedis* for its reconstruction.[8] Apparently the little church had been completely covered by the water during the floods of May of 1688.[9] Again in 1745 there is a notice that the building was in need of repair, possibly a complete rebuilding.[10]

3. *Capillas de los Pasos, 1691 and 1942*

THESE little chapels were reconstructed from the ground up sometime around 1942.[11] Originally they were part of

[7] Vásquez, **4**: p. 418.

[8] *AGG*, A 1.2.2 (1702) 11780–1786; *Efem*, p. 129.

[9] Ximénez, **1**: p. 231.

[10] *AGG*, A 1.10.3 (1745) 31334–4048 and 31335–4048.

[11] Pardo, *Guía*, p. 166.

the Via Crucis which extends from the street in front of the convent of San Francisco and ends within the grounds of El Calvario where the last station of the cross is located. The stations of the cross were marked out as early as 1619 by means of pictures or other devices. But nothing of an architectural character, either permanent or temporary, was built to accommodate the different stations.[12] In 1689 a more permanent treatment for the stations of the cross and their exact location was decided on.[13] In 1691, at each station a little domed chapel was built and paintings placed inside. Each chapel was fitted with a door which was opened to the public only on Good Friday.[14]

4. *Cruz de Piedra, 1753* (Fig. 215.)

A STONE cross had been erected in the Calle Ancha de los Herreros sometime before the eighteenth century.[15] It was covered with a little domed structure supported on four piers built after 1753.[16] (Fig. 215.) This little chapel with its cross was intact until 1895 when it was ordered demolished because it impeded traffic and the cross installed in some part of La Merced.[17] It may very well be that the base of the cross in front of La Merced today, which bears the inscribed date 1765, is part of the cross which once stood in La Calle Ancha de Herreros. (Fig. 73.)

5. *San Lázaro: Church, 1734*

THIS building is still standing today and serves as a chapel for the municipal cemetery. It has been altered and reconstructed to the extent that very little of its original colonial character remains. The present structure dates from after the earthquake of 1717 when the former one, built at some undetermined date, was completely destroyed.[18] The earlier building too had suffered repairs and alteration after the earthquake of 1681.[19] A new building was undertaken after 1717 and finally completed in 1734.[20]

[12] Vásquez, **4**: pp. 421 ff.

[13] *AGG*, A 1.2.2 (1689) 11778–1784–14, also *BAGG* **8** (1943): p. 58; *Efem*, p. 107 ff.

[14] Vásquez, **4**: p. 432; Ximénez, **3**: p. 244.

[15] Ximénez, **2**: p. 336.

[16] *Efem*, pp. 202, 206, 209; *ASGH* **24** (1949): pp. 374, 378, 381; *AGG*, A 1.10.3 (1745) 16545–2280 with plan; A 1.11 (1753) 2091–98.

[17] Díaz, *Romántica ciudad*, pp. 42 ff. See ch. XIII, no. 2, above.

[18] Juarros, **1**: p. 151; Ximénez, **3**: p. 357.

[19] *Efem*, p. 96.

[20] *Ibid.*, pp. 150, 176; *AGG*, A 1.10.3 (1720) 18803–2448; *ASGH* **24** (1949): p. 196.

6. *Santa Cruz del Milagro: Church, after 1731*

THIS church has been confused with that of Santa Cruz, not only because of the similarity of their names, but also because they are both located in the same neighborhood within a few blocks of each other on opposite sides of the Pensativo River. The ruins of this building are still visible today, but are in such a deplorable state that nothing of the architectural character except the vague outline of the exterior walls can be made out.

This church was already standing early in the eighteenth century when a petition was brought before the *ayuntamiento* that ten *varas* of adjacent land be ceded so that the church might be enlarged.[21] The building was destroyed in 1717 and not rebuilt until after 1731.[22] It was damaged again in 1751,[23] and in 1773 was completely destroyed.[24]

7. *Santa Lucía: Church, Sixteenth Century, 1695, and Eighteenth Century*

THE origins of this church, of which not a trace remains nor its exact location known for certain, go back to the time when the present city of Antigua was being built after the removal of the first capital from Ciudad Vieja. It was located somewhere between the two towns and served for a few years after 1542 as a place to celebrate mass while the new cathedral in Antigua was being built. Originally it had stood in open country, but by the beginning of the eighteenth century the population of Antigua had grown so that the church was on the very edge of town.[25] The next reference to the building exists in a document dated 1691 in which help is requested for the purpose of buying lumber to repair the roof.[26] Four years later, help was again asked to complete the construction of the church.[27] This would imply that a new building had replaced the primitive sixteenth-century structure.

In 1717 the building was completely destroyed and was not yet rebuilt by 1720.[28] It was damaged again in 1751.[29] In 1773 it was totally ruined and not re-established in Gua-

[21] *AGG*, A 1.10.3 (1703) 31279–4047; A 1.2.2 (1703) 11780–1786; *Efem*, p. 130.

[22] Juarros, **1**: p. 154; *Efem*, pp. 164, 165, 166, 171; *AGG*, A 1.10.3 (1727) 18804–2448 and 18805–2448, fol. 2.

[23] *Efem*, p. 205.

[24] Juarros, *loc. cit.*

[25] Ximénez, **1**: p. 235; Juarros, **1**: p. 148; Vásquez, **4**: pp. 383 ff.

[26] *AGG*, A 1.10.3 (1691) 31271–4046.

[27] *Ibid.*, A 1.10.3 (1695) 31275–4046.

[28] *Ibid.*, A 1.2.5 (1719) 2207–71; A 1.10.3 (1720) 31291–4047.

[29] *Efem*, p. 205; *ASGH* **24** (1949): p. 377; *AGG*, A 1.10.3 (1751) 31347–4049.

temala City. Its furnishings and other religious paraphernalia were given to the church of San Sebastián in the new capital.[30]

8. *Santo Domingo: Church and Convent, Seventeenth Century*

THE order of Santo Domingo and that of San Francisco were the two most important in all Central America. Like San Francisco, this order maintained its principal house in Antigua. The site where it once stood is now partly occupied by a school where some of the vestiges of walls of the church and convent may be seen. But not enough remains to be able even to delineate the plan or the extent of the area the buildings once occupied.

The first convent of the Dominicans goes back to the founding days of Ciudad Vieja on land donated to the order by Pedro Alvarado himself.[31] Structures of some sort to house the monks were first built in 1538.[32] The convent and church which existed in 1544 are very vividly described by Remesal who gives us a clear picture of the very rudimentary and nondescript character of the conventual building of even so important an order as that of Santo Domingo in sixteenth-century Guatemala.[33] "El convento estaba probísimo. Una iglesia de cañas tapadas con barro y tejado de heno; el cercado era unos maderos atravesados, las celdas unas chozuelas apartada la una de la otra." Ximénez gives the same description, probably taken from Remesal, almost word for word.[34]

A chapel for the use of Spaniards only, Capilla de Nuestra Señora del Rosario, was founded in 1559 and built anew a few years before the time Remesal was writing, *ca.* 1619.[35] A new church was begun in November of 1665 and finished in July of 1666. It was roofed with a barrel vault, and probably incorporated the chapel of the Rosary possibly built at the beginning of the century.[36] Though the convent was not as well constructed as that of San Francisco, the Dominican church was, however, the best and the most sumptuous in Antigua until the Cathedral and the church of the Compañía de Jesús were built later on.[37] The church had apparently been erected piecemeal, judging by the various contracts for the building of the *capilla mayor*, the main altar, and the retable.[38] Thomas Gage describes the opulent character of the convent and the luxurious life of the monks, saying that the revenue at the time he was there, *ca.* 1633, amounted to 30,000 ducats.[39] The retable for the main altar, inaugurated in 1657 and painted by Pedro de Liendo, alone cost 15,000 *pesos*.[40]

In the earthquake of 1717 the church was left practically roofless. The cupola came down on the four vaults which converged at the crossing.[41] The final destruction came in 1773 when the neighborhood where Santo Domingo was located was one of the worst hit in all the town. Of the church and convent nothing at all remained standing.[42]

To judge by the description given in the various literary sources, the church was roofed with barrel vaults and was probably cruciform in plan, since the cupola is described as coming down on "... los cuatro cañones que hacen crucero. ..."[43] In this respect it was probably very much like the church of Santo Domingo in San Cristóbal de Las Casas, Chiapas, Mexico, which also dates largely from the seventeenth century.[44]

30 *AGG*, A 1.10.3 (1790) 18823–2448 and 18824–2448.
31 González Dávila, *Teatro* **1**: p. 143; Juarros, **1**: pp. 122 ff.; Pineda, *Descripción*, p. 330.
32 Remesal, **1**: p. 72.
33 *Ibid.*, **2**: pp. 50 ff.
34 Ximénez, **1**: p. 385.
35 Remesal, **2**: p. 401.
36 Molina, *Memorias*, p. 117.
37 Vásquez, **4**: p. 362.
38 *AGG*, A 1.20 (1636) 690–53, also (1636) 690–69 and (1648) 694–668, see also *BAGG* **10**, 2 (1945): pp. 99-102.
39 Gage, ch. XVIII, pp. 283 ff.
40 Molina, *op. cit.*, pp. 98 ff.
41 Ximénez, **3**: pp. 354, 375.
42 González Bustillo, *Razón particular*, also *Ciudad mártir*, p. 98.
43 Ximénez, **3**: p. 354.
44 Markman, *San Cristóbal*, pp. 62 ff.

BIBLIOGRAPHY AND ABBREVIATIONS

ONLY those works specifically cited in the text are listed in the bibliography. Those works preceded by an asterisk (*) are contemporary with the colonial period. When followed by the approximate *floruit* of the author, an edition not always contemporary with the author's lifetime was used for the citations.

The form employed for the citations in the notes consists of the author's surname and an easily recognizable word or two from the title. Where works from the colonial period are referred to with great frequency the author's name alone is given, for example: Díaz del Castillo, Fuentes y Guzmán, Gage, García Peláez, Juarros, Remesal, Vásquez, and Ximénez, all of whom deal with Guatemala and Central America and are each represented by a single work in the bibliography. In these cases, bold face Arabic numerals refer to the volume of the edition quoted and light face Arabic numerals to the pages where the citation may be found. In addition to page references, chapter numbers in Roman numerals are given when citing Gage and Fr. Lorenzo de San Nicolas since editions of great antiquity were used.

Abbreviations of periodicals and other works appear in alphabetical order among the other entries of the bibliography.

A catalogue of the documents cited in the text from the Archivo General del Gobierno de Guatemala follows the bibliography. The descriptive statements in Spanish are transcriptions from the card catalogue in the archives. The statements in English are by the author and are of documents for which cards had not yet been written at the various times they were consulted between 1950 and 1963.

*ACOSTA, JOSÉ DE (*fl.* 1539/1600). 1940. *Historia natural y moral de las Indias* (Mexico).

**AGG. Archivo General del Gobierno.* The colonial archives in Guatemala City, C. A. Document numbers have four parts thus: (1) classification, (2) date, (3) *expediente*, (4) *legajo*; for example, A 1.10.3 (1743) 31.322–4048. See "Catalogue of Documents . . ." immediately following this bibliography.

**AGI* or *AI. Archivo de Indias*, Seville, Spain. Document numbers do not follow a standard form and are given as referred to in other works cited. When read directly in the archives, the most recent catalogue number is given.

AGUIRRE MATHEU, JORGE. 1942/43. "Descripción del valle del Panchoy." *ASGH* **18**: pp. 73 ff.

*ALCEDO, ANTONIO DE (*fl.* 1736/1812). 1786/89. *Diccionario geográfico-histórico de las Indias Occidentales ó América* (5 vols., Madrid).

Anales de la Sociedad de Geografía e Historia (*ASGH*). **1**– (1924/).

An. Estudios Amer. Anuario de Estudios Americanos (Madrid) **1**– (1944/).

ANGULO IÑIGUEZ, DIEGO. 1933/40. *Planos de monumentos arquitectónicos de América y Filipinas existentes en el Archivo de Indias. Sevilla* (5 vols., Seville, Laboratorio del Arte, Universidad de Sevilla).

—— 1946/47. "Eighteenth Century Church Fronts in Mexico City." *JSAH* 5: pp. 27 ff.

—— 1947. "El gótico y el renacimiento en las Antillas." *An. Estudios Amer.* **4**: pp. 1–102.

—— 1952. *Bautista Antonelli* (Madrid).

—— 1945, 1950, 1956. *Historia del arte hispanoamericano* (3 vols., Barcelona).

ANNIS, VERLE. 1949. "El plano de una ciudad colonial." *AntHistGuat* **1**, 1: pp. 48–56.

AntHistGuat. Antropología e Historia de Guatemala (Guatemala) **1**– (1949/).

*ARANA, TOMÁS IGNACIO DE (*fl.* early 18th century). 1717. *Relación de los estragos y ruinas, que a padecido la ciudad de Santiago de Guathemala por terremotos, y fuego de sus volcanes en este año de 1717* (A. de Pineda Ybarra, Guatemala [Antigua] 1717).

ARÉVALO, RAFAEL DE. 1856/57. *Colección de documentos antiguos del Archivo de Guatemala* (Guatemala).

—— 1932. *Libro de actas del ayuntamiento de la ciudad de Guatemala* (1524/30) (Guatemala). Published as a pamphlet in the *Diario de Centro América.*

ASGH. See *Anales*, etc.

ATL, DR., and MANUEL TOUSSAINT. 1924/27. *Iglesias de México* (6 vols., Mexico).

BAGG. Boletín del Archivo General del Gobierno (10 vols., Guatemala, 1935/45).

BAILY, JOHN. 1850. *Central America* (London).

BAIRD, JOSEPH A. 1956. "The Ornamental Niche-Pilaster in the Hispanic World." *JSAH* **15**: pp. 5–11.

—— 1959. "Style in Eighteenth-Century Mexico." *Journal of Inter-American Studies* (Gainesville, Florida) **1**: pp. 261 ff.

BANCROFT, HUBERT HOWE. 1875. *The Native Races of the Pacific States* (5 vols., New York).

BANDELIER, AD. F. 1880/81. "Notes on the Bibliography of Yucatan and Central America." *Proceedings of the American Antiquarian Society*, New Series, Worcester, Mass., pp. 82–118.

BANKART, GEORGE P. 1908. *The Art of the Plasterer* (London).

BARÓN CASTRO, RODOLFO. 1942. *La población de El Salvador* (Madrid).

BATES, H. W. 1878. *Central America, West Indies, and South America* (London).

BATRES JÁUREGUI, ANTONIO. 1915, 1920, 1949. *La América Central ante la historia* (3 vols., Guatemala).

BERISTAIN Y SOUZA, JOSÉ MARIANO. 1816. *Biblioteca hispano americana* (Mexico).

BERLIN, HEINRICH. 1942. "El Convento de Tecpatán." *Anales del Instituto de Investigaciones Estéticas* (Mexico) **3**, 9: pp. 5–13.

—— 1947. "El ingeniero Luis Diez de Navarro." *ASGH* **22**: pp. 89 ff.

—— 1950. "Fundacion del convento de Santa Clara en la Antigua." *AntHistGuat* **2**, 1: pp. 43–54.

*BETANZOS Y QUIÑONES, GERÓNIMO (*fl.* late 17th century). See Lemoine Villacaña, below.

BEVAN, BERNARD. 1938. *History of Spanish Architecture* (London).

BLOM, FRANS. 1945. *Desde Salamanca, España hasta Ciudad Real, Chiapas; diario de viaje 1544–1545* (Mexico).

BRASSEUR DE BOURBOURG, CHARLES ETIENNE. 1857/59. *Histoire des nations civilisées du Mexique et de l'Amerique-Centrale, durant les siècles antérieurs á Cristofe Colomb* (4 vols., Paris).

—— 1871. *Bibliothéque México-Guatémalienne* (Paris).

—— 1947. "Antigüedades guatemaltecas (carta escrita en Rabinal el 9 de Julio de 1855)." Reprinted from *Gaceta de Guatemala* **7**, July 20 and 27, 1855, in *ASGH* **22**, 1 and 2: pp. 99–104.

*BRIZGUZ Y BRU, ATHANASIO GENARO (pseudonym for Zaragoza y Ebri, Agustín Bruno). 1738. *Escuela de arquitectura civil en que se contienen las órdenes de arquitectura, la distribución de los planos de los templos, y casas, y el conocimiento de los materiales* (Valencia).

BUSCHIAZZO, MARIO J. 1940. *La arquitectura colonial en Hispano América* (Buenos Aires).

—— 1941. "Indigenous influences on the colonial architecture of Latin America." *Bulletin of the Pan-American Union* **75**, 5: pp. 257–265.

—— 1944. *Estudios de arquitectura colonial hispano-americano* (Buenos Aires).

CABRERA, VICTOR M. 1924. *Guanacaste: Libro comemorativo del centenario de la incorporación del partido de Nicoya a Costa Rica* (San José, Costa Rica).

*CADENA, FELIPE. 1774. *Breve descripción de la noble ciudad de Santiago de los Caballeros de Guatemala, y puntual noticia de su lamentable ruina ocasionada de un violento terremoto el dia veinte y nueve de julio de mil setecientos setenta y tres* (A. Sánchez Cubillas, Mixco, Guatemala).

CALDERÓN QUIJANO, JOSÉ ANTONIO. 1949. "Ingenieros militares en Nueva España." *An. Estudios Amer.* **6**: pp. 1–72.

CALZADA, ANDRÉS. 1933. *Historia de la arquitectura española* (Barcelona).

CARRERA STAMPA. MANUEL. 1945. "Ordenanzas del noblísimo arte de la platería para el Reino de Guatemala. 1776." *ASGH* **20**: pp. 97 ff.

Cartografía de ultramar: Carpeta IV; América Central—Toponímia de los mapas que la integran y Relaciones históricas de ultramar (2 vols., Madrid, 1957). Published by Servicios Geográfico e Histórico del Ejercito.

*CASAS, BARTOLOMÉ DE LAS (*fl.* 16th century). 1646. *Brevísima relación de la destruyción de las Indias* (Barcelona).

CASTAÑEDA PAGANINI, RICARDO. 1947. *Historia de la real y pontificia universidad de San Carlos de Guatemala* (Guatemala).

CASTELLANOS, J. HUMBERTO R. 1941. "Relación sintética del desarrollo del arte en Guatemala." *Boletín de Museos y Bibliotecas* (Guatemala), ep. 2, 1, Julio 1941, pp. 73–92.

CASTRO SEOANE, JOSÉ 1945. "La expansión de la Merced en la América colonial." *ASGH*, **20**: pp. 39–47, reprinted from *Revista de Indias*, **4**, 13 (Madrid), 1943.

*CAXIGA Y RADA, AGUSTÍN DE LA (*fl.* 18th century). 1914. *Breve relación de el lamentable estrago, que padeció esta ciudad de Santiago de Guatemala, con el terremoto de el dia quatro de Marzo, de este año de 1751.* (*Revista Chilena de Historia y Geografía*, Santiago de Chile, año 4, **12**, 16: pp. 154–169). See also Medina, nos. 225, 113.

CHINCHILLA AGUILAR, ERNESTO. 1953. "El ramo de aguas de la ciudad de Guatemala en la epoca colonial." *AntHistGuat* **5**, 2: pp. 19–31.

—— 1961. *El ayuntamiento colonial de la ciudad de Guatemala* (Guatemala).

—— 1963. *Historia del arte en Guatemala, 1524–1962* (Guatemala).

CHUECA GOITÍA, F. 1953. *Arquitectura del siglo XVI* (Madrid).

CHUECA GOITÍA, F., L. TORRES BALBÁS, and J. GONZÁLEZ GONZÁLEZ. 1951. *Planos de ciudades iberoamericanas y filipinas existentes en el Archivo de Indias* (2 vols., Madrid).

La ciudad martir. Published as a serial in *Diario de Centro-América*, (Guatemala, 1923). Contains complete texts of Cadena, various reports of González Bustillo, and others dealing with the earthquakes suffered by Antigua.

CODI. Colección de documentos inéditos relativos al descubrimiento, conquista y organización de las antiguas posesiones españolas de América y Oceanía, sacados de los Archivos del Reino, y muy especialmente del de Indias. Primera Serie, editada por Joaquín Francisco Pacheco, Francisco de Cárdenas, y Luis Torres Mendoza (42 vols., Madrid, 1864/84). Segunda Serie, publicada por la Real Academia de la Historia (25 vols., Madrid, 1885/1926). See Schaeffer, *Indice*, below for index of same.

COLD. Colección de libros y documentos referentes á la historia de América, ed. Manuel Serrano Sanz (Madrid, 1908). **8**, *Relaciones historicas y geográficas de América Central.*

CONTRERAS, JUAN DE (MARQUES DE LOZOYA). 1934. *Historia del arte hispánico* (**2**, Barcelona).

*CÓRDOBA, FR. MATIAS DE. 1798. *Utilidad de que todos los indios y ladinos se vistan y calcen a la española y medios de conseguirlo sin violencia, coacción ni mandato* (Guatemala).

*CORTÉS Y LARRAZ, PEDRO DE (*fl.* late 18th century). "Descripción Geográfico-Moral de la Diócesis de Goathemala, hecha por su Arzobispo el Illmo. Sr. dn. Pedro Cortez y Larras del Consejo de S. M. en el tiempo que la visitó y fué desde al día 3 de noviembre de 1768 hasta el día 1° de Julio de 1769 / desde el día 22 de noviembre de 1769 hasta el día 9 de Febrero de 1770 / desde el día 6 de Junio de 1770 hasta el día 22 de Agosto del año 1770." A manuscript in Archivo de Indias, Seville, *Aud. de Guat.*, *leg.* 948, 3 tomes in pergamine, with 112 maps.
This manuscript was published by the Sociedad de Geografía e Historia de Guatemala, the title of which is *Descripción geográfico-moral* (2 vols., Guatemala, 1958).

DGE. Dirección General de Estadística. República de Guatemala. *Departamentos, municipios, ciudades, villas, pueblos, aldeas y caserios de la república de Guatemala* (Guatemala, 1953). A gazetteer with 22 loose sheets being maps of the departments of Guatemala.

DÍAZ, VICTOR MANUEL. 1927. *La romántica ciudad colonial* (Guatemala).

——1934. *Las bellas artes en Guatemala* (Guatemala).

*DÍAZ DEL CASTILLO, BERNAL (*fl.* 16th century). 1933, 1934. *Verdadera y notable relación del descubrimiento y conquista de la Nueva España y Guatemala* (2 vols., Guatemala).

DÍAZ DURÁN, J. C. 1942/43. "Historia de la Casa de Moneda del Reino de Guatemala, desde 1731 hasta 1773." *ASGH* **18**: pp. 191–224.

*DÍEZ, RAMÓN PASCUAL (*fl.* 18th century). 1932. "Arte de hacer el estuco jaspeado, o imitar jaspes a poca costa, y con la mayor propiedad." Published by José Gabriel Navarro, *Archivo Español de Arte y de Arqueología*, **8**: pp. 237–257.

*DÍEZ DE LA CALLE, JUAN (*fl.* mid-17th century). 1646. *Memorial, y* *noticias sacras, y reales del imperio de las Indias Occidentales . . .* (Madrid).

—— 1648. *Memorial y compendio breve del libro intitulado noticias sacras y reales de los imperios de la Nueva España, el Perú y sus islas de las Indias Occidentales* (Madrid). Also in 1654 with slightly different title.

*Diez de Navarro, Luis (*fl.* 18th century). 1744. "Información rendida por el Ing. Luis Diez de Navarro de su viaje por las provincias de Guatemala. Año 1744." A document, *AGG*, A 1.17.3 (1744) 17.508–2335.

—— 1850. *Relación sobre el antiguo Reino de Guatemala, hecha por el ingeniero Luis Diez de Navarro en 1745* (Guatemala, Imprenta Nueva de L. Luna).

Dunlop, R. G. 1847. *Travels in Central America* (London).

Dunn, Henry. 1828. *Guatimala, or the United Provinces of Central America in 1827–8* (New York).

Efem. See Pardo, *Efemérides*, below.

Elliot, L. E. 1924. *Central America: New Paths in Ancient Lands* (London).

*Fuentes y Guzmán, Francisco Antonio de (*fl.* 17th century). 1932/33. *Recordación Florida: Discurso historial y demonstración natural, material, militar y política del Reyno de Guatemala* (3 vols., Guatemala).

"Fundación de Pueblos en el Siglo XVI," *Boletín del Archivo General de la Nación* **6**, 3 (Mexico, 1935): pp. 331–360.

Gaceta de Guatemala. First volume published in 1797, continued until 1854. Sometimes also given as *Gazeta de Goathemala.*

*Gage, Thomas (*fl.* 17th century). 1677. *A New Survey of the West Indies* (3rd. ed., London).

*García de la Concepción, P. Fr. José. 1723. *Historia bethlemítica. Vida ejemplar y admirable del venerable siervo de Dios y padre Pedro de San Joseph Bethancur, fundador del Regular Instituto de Bethlén en las Indias Occidentales* (Seville).

García Peláez, Francisco de Paula (*fl.* 1785/1867). 1943/44. *Memorias para la historia del antiguo Reino de Guatemala* (3 vols., 2nd ed., Guatemala).

Gavarrete, Francisco. 1868. *Geografía de Guatemala* (Guatemala).

*Goicoechea, José Antonio (*fl.* 1735/1814). 1936/37. "Relación del R. P. Dr. Fr. José Antonio Goicochea, sobre los indios gentiles de Pacura, en el obispado de Comayagua." *ASGH* **13**: pp. 303–315.

*González Bustillo, Juan (*fl.* late 18th century). 1774. *Extracto ó relación methódica y puntual de los autos de reconocimiento, practicados en virtud de comisión del señor Presidente de la real audiencia de este reino de Guatemala* (Antonio Sánchez Cubillas, Mixco, Guatemala).

—— 1774. *Razón particular de los templos, casas de comunidades, y edificios públicos y por mayor del número de los vecinos de la capital Guatemala; y del deplorable estado a que se hallan reducidos por los terremotos de la tarde del veinte y nueve de julio, trece y catorce de diciembre del año próximo pasado de setenta y tres* (Antonio Sánchez Cubillas, La Hermita [Guatemala City]). See also *AGG*, A 1.18.6 (1774) 38.306–4502, and for a modern copy, *AGG*, A 1.18.6 (1904) 14.001–2021.

—— 1774. *Razón puntual de los sucesos más memorables de los estragos, y daños que ha padecido la ciudad de Guatemala, y su vecindario, desde que se fundó en el parage llamado Ciudad Vieja o Almolonga, y de donde se trasladó a el en que actualmente se halla* (Mixco, Guatemala).

*González Dávila, Gil (*fl.* 1578/1658). 1649/55. *Teatro ecclesiástico de la primitiva iglesia de las Indias occidentales* (2 vols., Madrid).

González Mateos, María Victoria. 1946. "Marcos Ibáñez, arquitecto español en Guatemala." *An. Estudios Amer.* **3**, reprinted in *ASGH* **24** (1949): pp. 49–75.

Guerra Trigeros, Alberto. 1938. "The Colonial Churches of El Salvador." *Bulletin of the Pan-American Union* **72**: pp. 271–279.

Guido, Angel. 1944. *Redescubrimiento de América en el arte* (3rd. ed., Buenos Aires).

Hale, J. 1826. *Six Months Residence and Travels in Central America through the Free States of Nicaragua, and particularly Costa Rica, giving an interesting account of that beautiful country* (New York).

Hanke, Lewis. 1952. *Bartolomé de las Casas, Historian* (Gainesville, Florida).

Harth-Terré, Emilio. 1945. *Artífices en el virreinato del Perú* (Lima).

Hernández Díaz, José, and Antonio Sancho Corbacho, and Francisco Collantes Terán. 1939, 1943, 1951, 1955. *Catálogo arqueológico y artístico de la provincia de Sevilla* (4 vols., Seville).

*Herrera y Tordesillas, Antonio de (*fl.* 1559/1625). 1934/35. *Historia general de los hechos de los castellanos en las islas y terrafirme del Mar Oceano* (5 vols., Madrid).

*Hidalgo, Joseph Domingo (*fl.* late 18th century). 1797, 1798. "Memoria para hacer una descripción puntual del Reino de Guatemala." Reprinted in *ASGH* **26** (1952): pp. 383–413 from *Gaceta de Guatemala*, **1** and **2**.

*Hincapie Meléndez, Cristóbal. 1717. *Relación de la ruina de la ciudad de Santiago de los Caballeros en Goatemala por el terremoto y quatro volcanes el dia 17 de Agosto de 1717* (Guatemala [Antigua]).

Irisarri, Antonio José de. 1847. *El cristiano errante* (Bogotá, Colombia). Reprinted in *ASGH* **9** (1932/33): pp. 101 ff., 249 ff., 354 ff., 492 ff.; **10** (1933/34): pp. 245 ff., 381 ff., 517 ff.; **11** (1934/35): pp. 224 ff., 367 ff., 497 ff.

**Isagoge histórica apologética de las Indias Occidentales y especial de la provincia de San Vicente de Chiapa y Guatemala, de la orden de Predicadores* (Guatemala, 1935). Written soon after 1700 by an unknown author.

Jeffreys, W. 1953. "Precolumbian Negroes in America." *Scientia* **7–8**, reprinted in *El Imparcial* (Guatemala), October 14, 1953.

JSAH. Journal of the Society of Architectural Historians **1**– (1941/).

Juarez Muñoz, J. Fernando. 1942/43. "Peregrinación por las ruinas de la Antigua Guatemala." *ASGH* **18**: pp. 148 ff.

*Juarros, Domingo (*fl.* late 18th and early 19th centuries). 1936/37. *Compendio de la historia de la ciudad de Guatemala* (2 vols., Guatemala).

Kelemen, Pal. 1941. "Colonial Architecture in Guatemala." *Bulletin of the Pan-American Union* **65**: pp. 437–448.

—— 1942. "Guatemala Baroque." *Magazine of Art* **35**: pp. 22–25.

—— 1944. "Some church façades of colonial Guatemala." *Gazette des Beaux Arts*, ser. 6, **25**, 924: pp. 113–126.

—— 1951. *Baroque and Rococo in Latin America* (New York).

Kroeber, A. L. 1934. "Native American Populations." *American Anthropologist* **36** (New York).

Kubler, George. 1942. "Population Movements in Mexico, 1520–1600." *Hispanic American Historical Review*, **32**: pp. 606 ff.

—— 1948. *Mexican Architecture of the Sixteenth Century* (New Haven).

—— 1957. *Arquitectura de los siglos XVII y XVIII* (Madrid).

—— 1961. "On the Colonial Extinction of the Motifs of Pre-Columbian Art." In Lothrop, Samuel K., and others, *Essays in Pre-Columbian Art and Archaeology* (Cambridge, Mass.). pp. 14–34.

Kubler, George, and Martin Soria. 1959. *Art and Architecture in Spain and Portugal and their American Dominions, 1500 to 1800* (Baltimore).

Lamadrid, Fr. Lázaro. 1942/43. "Los estudios franciscanos en Antigua Guatemala." *ASGH* **18**: pp. 279–305.

Lanning, John Tate. 1955. *The University in the Kingdom of Guatemala* (Ithaca).

—— 1956. *The Eighteenth Century Enlightenment in the University of San Carlos de Guatemala* (Ithaca).

*Larrazábal, Antonio (*fl.* early 19th century). 1953/54. "Apuntamientos sobre agricultura y comercio del Reyno de Guatemala." Reprinted in *ASGH* **27**: pp. 87–109, being a document dated October 20, 1810.

—— 1953/54. "Bosquejo estadístico del arzobispado de Goatemala y obispados sufragéneos." *ASGH* **27**: pp. 113–125.

—— 1953/54. "Discurso que el Sr. Diputado en Cortes (de Cádiz) por la provincia de Guatemala don Antonio Larrazábal dijo en la sesión del 29 de marzo de 1813, abogando por la libertad de comercio en las colonias de España." *ASGH* **27**: pp. 79–86.

LARREINAGA, MIGUEL. 1857. *Prontuario de todas las reales cédulas, cartas acordadas, y órdenes comunicadas a la audiencia del antigua Reino de Guatemala, desde el año 1600 hasta 1818* (Guatemala).

*LEMOINE VILLACAÑA, ERNESTO. 1961. "Historia sucinta de la construcción de la catedral de Guatemala. Escrita en 1677 por Don Gerónimo de Betanzos y Quiñones." *Boletín del Archivo General de la Nación*, no. 3 (México): pp. 405–430, figs. 1–8.

Libro de Actas, see Rafael de Arévalo, ed.

LLAGUNO Y AMÍROLA, EUGENIO. 1829. *Noticias de los arquitectos y arquitectura de España desde su restauración* (4 vols., Madrid).

*LÓPEZ DE VELASCO, JUAN (*fl.* 16th century). 1894. *Geografía y descripción universal de las Indias, recopilada por el cosmógrafo-cronista Juan López de Velasco desde el año 1571 al de 1574* (Madrid).

*FR. LORENZO DE SAN NICOLÁS, see San Nicolás, Fr. Lorenzo de.

LUJÁN, MUÑOZ, LUIS. 1961. "Noticia breve sobre la segunda catedral de Guatemala." *ASGH* **34**: pp. 61–82.

MARCO DORTA, ENRIQUE. 1945. "Iglesias renacentistas en las riberas del lago Titicaca." *An. Estudios Amer.* **2**: pp. 701–717.

MARKMAN, SIDNEY D. 1956. "Santa Cruz, Antigua, Guatemala and the Spanish Colonial Architecture of Central America." *JSAH* **15**: pp. 12–19.

—— 1961. "Las Capuchinas: an Eighteenth-Century Convent in Antigua." *JSAH* **20**: pp. 27–33.

—— 1963. *San Cristóbal de las Casas* (Seville).

MCANDREW, JOHN. 1949. "The Relationship of Mexican Architecture to Europe: Problems in the Field of Colonial Studies." *Museum of Modern Art, Proceedings of a Conference on Studies in Latin American Art* (American Council of Learned Societies, Washington).

—— 1965. *The Open Air Churches of Sixteenth-Century Mexico* (Cambridge, Mass.).

MCBRYDE, FELIX WEBSTER. 1945. *Cultural and Historical Geography of Southwest Guatemala* (Washington).

MEDINA, J. TORIBIO. 1897/1907. *Biblioteca hispano-americana, 1493–1810* (7 vols., Santiago de Chile).

—— 1910. *La imprenta en Guatemala, 1660–1821* (Santiago de Chile).

MENCOS GUAJARDO-FAJARDO, FRANCISCO XAVIER. 1950. "Arquitectos de la epoca colonial en Guatemala." *An. Estudios Amer.* **7**: pp. 163–209.

—— *La arquitectura hispano-americana en la Capitanía General de Guatemala* (Tesis doctoral, Facultad de Filosofía y Letras, Universidad de Madrid). Unpublished manuscript.

MESA, JOSÉ and TERESA GISBERT. 1963. "El edificio circular de Capuchinas en Antigua, Guatemala." *Anales del Instituto de Arte Americano e Investigaciones Estéticas: Universidad de Buenos Aires, Facultad de Arquitectura y Urbanismo* **6**: pp. 13–27.

*MOLINA, FR. ANTONIO DE (d. 1683). 1943. *Memorias del m. r. p. maestro Antonio de Molina continuadas y marginadas por fray Agustín Cano y fray Francisco Ximénez, de la orden de Santo Domingo*, ed. by Jorge del Valle Matheu (Guatemala).

*MONTERO DE MIRANDA, FRANCISCO. 1575. *Relación dirigida al Illmo. Señor Palacio . . . sobre la provincia de la Verapaz ó Tierra de Guerra*, see Bandelier, *op. cit.*, p. 104. Reprinted in *ASGH* **22** (1953/54): pp. 342–358, being a transcription of a MS in the library of the University of Texas: Latin American Section, xx, Central America, no. 3, "Descripción de la provincia de la Verapaz por Fray Francisco Montero de Miranda."

MONTGOMERY, G. W. 1839. *Narrative of a Journey to Guatemala, in Central America in 1838* (New York).

MUYBRIDGE, EADWARD. 1876. *The Pacific Coast of Central America and Mexico; the Isthmus of Panama; and the cultivation and shipment of coffee* (San Francisco). An album of 144 mounted photographs. Five sets of the photographs were made by Muybridge of which two no longer exist. One set is in the Library of Congress, and another in the Art Gallery and Museum, Stanford University, Stanford, California. A third loose set is in the possession of Prof. Walter R. Miles, Yale University, who kindly permitted me to make photographic copies and to publish those which appear here as figs. 16, 17, 21, 106, 117, and 201.

NEUMEYER, ALFRED. 1948. "The Indian Contribution to Architectural Decoration in Spanish Colonial America." *Art Bulletin* **30**: pp. 104–121.

NOEL, MARTÍN S. 1923. *Contribución a la historia de la arquitectura hispano-americana* (Buenos Aires).

OLVERA, JORGE. 1950. "Joyas de arquitectura colonial en Chiapas." *Chiapas* **2**, 13 (Tuxtla Gutiérrez, Chiapas): pp. 14–17 and 29 ff.

—— 1956. "Copanaguastla, Joya del plateresco en Chiapas." *Ateneo* **1**, 2 (Tuxtla Gutiérrez, Chiapas): pp. 115–136.

O'RYAN, JUAN ENRIQUE. 1897. *Bibliografía de la imprenta en Guatemala en los siglos XVII y XVIII* (Santiago de Chile).

OSBORNE, LILY DE JONGH. 1945. "Arterías comerciales." *ASGH* **20**: pp. 320–325.

*OVIEDO Y VALDÉS, GONZALO FERNÁNDEZ (*fl.* 1478/1557). 1950. *Sumario de la natural historia de las Indias* (Mexico).

*PALACIO, LIC. (*fl. ca.* 1576). 1927/28. "Relación hecha por el licenciado Palacio al Rey D. Felipe II, en la que describe la provincia de Guatemala, las costumbres de los indios y otras cosas notables." Reprinted in *ASGH* **4**: pp. 71 ff., taken from *CODI* **6**: pp. 5–40.

PARDO, J. JOAQUÍN. 1939/40. "Indice de documentos existentes en el Archivo de Indias de Sevilla, que tienen interés para Guatemala." *ASGH* **16**: pp. 401–424.

—— 1941. *Prontuario de reales cédulas, 1529/99* (Guatemala).

—— 1944. *Efemérides para escribir la historia de la Muy Noble y Muy Leal Ciudad de Santiago de los Caballeros del Reino de Guatemala* (Guatemala). See *Efem.*

PARDO, J. JOAQUÍN, and PEDRO ZAMORA CASTELLANOS. 1943. *Guía turística de las ruinas de la Antigua Guatemala* (Guatemala).

*PEDRAZA, CRISTÓBAL DE (*fl.* 1544). 1916. "Relación de varios sucesos ocurridos en Honduras, y del estado en que se hallaba esta provincia. Gracias a Dios, 18 de mayo de 1539." *Relaciones históricas de América. Primera mitad del Siglo XVI*, published by Sociedad de Bibliófilos Españoles (Madrid).

PÉREZ VALENZUELA, PEDRO. 1934. *La Nueva Guatemala de la Asunción* (Guatemala).

*PINEDA, JUAN DE (*fl.* late 16th century). 1924. "Descripción de la provincia de Guatemala, 1594." *COLD* **8**: pp. 415–471, reprinted in *ASGH* **1**: pp. 327–363.

*PONCE, FR. ALONSO (*fl.* late 16th century). 1873. *Relación breve y verdadera de algunas cosas de las muchas que sucedieron al padre Fray Alonso Ponce en las provincias de Nueva España, siendo comisario general de aquellas partes* (2 vols., Madrid). All citations are from **1** only.

POOLE, D. M. 1951. "The Spanish Conquest of Mexico: Some Geographical Aspects." *Geographical Journal* **117**, 1: pp. 27 ff.

RECINOS, ADRIAN. 1949. "La ciudad de Guatemala, 1524–1773." *AntHistGuat* **1**, 1: pp. 48–56.

**Reglamento General de artesanos de la Nueva Guatemala que la junta comisionada para su formación propone a la general de la Real Sociedad* (Nueva Guatemala, 1798). See Medina, *Imprenta*, p. 6799.

*"Relación de los caciques principales del pueblo de Atitlán, 1° de Febrero de 1571." *ASGH* **26** (1952): pp. 437 ff.

*REMESAL, FR. ANTONIO DE (*fl.* early 17th century). 1932. *Historia general de las Indias Occidentales, y particular de la gobernación de Chiapa y Guatemala* (2nd ed., 2 vols., Guatemala).

ROBERTS, ORLANDO W. 1827. *Narrative of Voyages and Excursions on the East Coast and in the Interior of Central America* (Edinburgh).

*RODRÍGUEZ, JUAN (?) (*fl.* 16th century). 1543. *Relación del espantable terremoto que agora nuevamente ha acontecido en las Indias en una ciudad llamada Guatimala* (Toledo).

RODRÍGUEZ BETETA, VIRGILIO. 1925/26. "Nuestra bibliografía colonial." *ASGH* **2**: pp. 83–98 and 227–238.

ROSENBLATT, ANGEL. 1938/39, 1939/40. "El desarrollo de la población indígena de América." Reprinted in *ASGH* **15**: pp. 367 ff. and 486 ff., **16**: pp. 114 ff., from *Tierra Firme* **1**, 1: pp. 115–133, no. 2: pp. 117–148, and no. 3: pp. 109–141.

ROYS, RALPH L. 1952. "Conquest Sites and the Subsequent Destruction of Maya Architecture in the Interior of Northern Yucatan." *Carnegie Institution: Contributions to American Anthropology and History* **11**, 54 (Washington): pp. 129–182.

SAENZ DE LA CALZADA, CONSUELO. 1956."El retablo barroco español y su terminología artística, Sevilla." *Archivo Español de Arte* **29**: pp. 211–242.

*SAGREDO, DIEGO LÓPEZ DE (*fl.* 16th century). 1526.*Medidas del romano necesarias alos oficiales seguir las formaciones delas basas, columnas, capitales y otras pieças delos edificios antiguos* (Toledo).

*SALAZAR, FR. JUAN JOSÉ (*fl.* mid-18th century). 1754. *Piedra fundamental del templo del Sacrosanto Cuerpo de Christo, Señor S. Joseph. En cuyo dia celebró . . . los sumptuosos reparos, a que se restituyó su templo de Guathemala de las ruinas, que causó el temblor del año 1751* (Imprenta de Sebastián Arévalo, Guatemala [Antigua]).

SALAZAR, RAMÓN A. 1897. *Historia del desenvolvimiento intelectual de Guatemala* (Guatemala).

SAMAYOA GUEVARA, HECTOR HUMBERTO. 1960. "La reorganización gremial guatemalense en la segunda mitad del siglo XVIII." *AntHistGuat* **12**, 1: pp. 63–106.

—— 1962. *Los gremios de artesanos en la ciudad de Guatemala* (Guatemala).

SÁNCHEZ G., DANIEL. 1920. *Católogo de los escritores franciscanos de la Provincia Seráfica del Santísimo Nombre de Jesús de Guatemala* (Guatemala).

SANCHO CORBACHO, ANTONIO. 1949. "Leonardo de Figueroa y el patio de San Acasio, de Sevilla." *Archivo Español de Arte* **22**: pp. 341–352.

—— 1952. *Arquitectura barroca sevillana del siglo XVIII* (Madrid).

SANFORD, TRENT ELWOOD. 1947. *The Story of Architecture in Mexico* (New York).

*SAN NICOLÁS, FR. LORENZO DE (*fl.* 17th century). 1796. *Arte y uso de la arquitectura, con el primer libro de Euclides traducido en castellano* (4th ed., 2 vols., Madrid).

SAPPER, KARL. 1924. "Die Zahl und die Volksdichte der indianischen Bevölkerung in Amerika." *Proceedings of the Twenty-first International Congress of Americanists* (The Hague).

SCHÄFER, ERNESTO. 1946/47. *Indice a la colección de documentos inéditos de Indias* (2 vols., Madrid).

SCHERZER, CARL. 1857. *Travels in the Free States of Central America: Nicaragua, Honduras, and Salvador* (2 vols., London).

SCHUBERT, OTTO. 1924. *Historia del barroco en España* (Madrid).

SECZY, JANOS. 1953. *Santiago de los Caballeros en Almolonga* (Guatemala).

Seminario Centroamericano de Crédito Agricola: Dirección General de Estadística, *Síntesis estadística de Guatemala* (Guatemala, 1952).

Servicios Geográfico e Histórico del Ejercito. See *Cartografía de ultramar* etc.

SOLÁ, MIGUEL. 1935. *Historia del arte hispanoamericano* (Barcelona).

SQUIER, EPHRAIM GEORGE. 1855. *Notes on Central America* (New York).

STANISLAWSKI, DAN. 1946. "The Origin and Spread of the Grid-Pattern Town." *Geographical Review* **36**: pp. 105–120.

—— 1947. "Early Spanish Town Planning in the New World." *Geographical Review* **37**: pp. 94–105.

STEPHENS, JOHN L. 1841. *Incidents of Travel in Central America, Chiapas and Yucatan* (2 vols., New York).

STOLL, OTTO. 1886. *Guatemala, Reisen und Schilderungen aus den Jahren 1878–1883* (Leipzig).

TERMER, FRANZ. 1934/35. "La habitación rural en la América del centro, a través de los tiempos." *ASGH* **11**: pp. 391–409.

THOMPSON, G. A. 1829. *Narrative of an Official Visit to Guatemala from Mexico* (London).

TOLEDO PALOMO, RICARDO. 1959. "Capilla abierta o capilla de indios." *AntHistGuat* **11**, 1: pp. 40–43.

*TORQUEMADA, FR. JUAN DE (*fl.* 17th century). 1943. *Monarquía indiana* (3rd ed., Mexico).

TORRE REVELLO, JOSÉ. 1956. "Tratados de arquitectura utilizados en Hispano-América (siglos XVI a XVIII). *Revista Interamericana de Bibliografía* (Washington) **6**, 1: pp. 3–24.

TORRES BALBÁS, LEOPOLDO. 1949. *Arte almohade—Arte nazarí—Arte mudéjar* (Madrid).

TORRES LANZAS, PEDRO. 1903. *Relación descriptiva de los mapas, planos, etc. de la audiencia y Capitanía General de Guatemala, existentes en el Archivo General de Indias* (Madrid).

TOUSSAINT, MANUEL. 1946. *Arte mudéjar en América* (Mexico).

——1948. *Arte colonial en México* (Mexico).

*TOVILLA, MARTÍN ALFONSO (*fl.* 1635). 1960. *Relación histórica descriptiva de las provincias de la Verapaz y de la del Manché, escrita por el capitán D. Martín Alfonso Tovilla* (Guatemala).

*VÁSQUEZ, FR. FRANCISCO (*fl.* late 17th and early 18th centuries). 1937, 1938, 1940, 1944. *Crónica de la provincia del Santísimo Nombre de Jesús de Guatemala de la orden de N. Seráfico Padre San Francisco en el reino de la Nueva España* (2nd ed., 4 vols., Guatemala).

*VÁSQUEZ DE ESPINOSA, ANTONIO (d. 1630). 1943. *La audiencia de Guatemala. Primera parte. Libro quinto del compendio y descripción de las Indias Occidentales, por Antonio Vásquez de Espinosa, año de 1629* (Guatemala).

*VIANA, FR. FRANCISCO DE (*fl.* 16th century). 1955. "Relación de la provincia de la Verapaz hecha por los religiosos de Santo Domingo de Cobán, 7 de diciembre de 1574." *ASGH* **28**: pp. 18–31. Transcribed from a document in the Library, University of Texas.

VILLACORTA C., J. ANTONIO. 1942. *Historia de la Capitanía General de Guatemala* (Guatemala).

—— 1944. *Bibliografía guatemalteca* (Guatemala).

VILLEGAS, VICTOR MANUEL. 1956. *El gran signo formal del barroco: ensayo histórico del apoyo estípite* (Mexico).

VIÑAS Y MEY, CARMELO. 1929. *El estatuto del obrero indígena en la colonización española* (Madrid).

WETHEY, HAROLD E. 1949. *Colonial Architecture and Sculpture in Peru* (Cambridge, Mass.).

*XIMÉNEZ, FR. FRANCISCO (*fl.* late 17th and early 18th centuries). 1929/31. *Historia de la provincia de San Vicente de Chiapa y Guatemala de la Orden de Predicadores* (3 vols., Guatemala).

YPSILANTI DE MOLDAVIA, GEORGE. 1937. *Monografía de Comayagua, 1537–1937* (Tegucigalpa).

*ZARAGOZA Y EBRI, AGUSTÍN BRUNO. See Brizguz y Bru, above.

ZAVALA, SILVIO. 1947. "Contribución a la historia de las instituciones coloniales en Guatemala." *ASGH* **22**: pp. 206–257, reprinted from *Jornadas*, no. 36 (Mexico, 1945).

ZUCKER, PAUL. 1959. *Town and Square* (New York).

CATALOGUE OF DOCUMENTS CITED IN THE TEXT

From the Archivo General del Gobierno de Guatemala

A1.1 (1607) 1–1. Providencia ordenando que ciertos pueblos de la región occidental, proporcionen indios para trabajar en las obras de la ciudad, con motivo de la ruina del 9 de octubre de 1607.

A 1.1 (1758) 3–156. Cédula del 3 de julio de 1753, acerca de que el ingeniero Luis Diez de Navarro le sean restituidos 700 pesos que gastó en la provincia de Costa Rica en el tiempo que fué gobernador de élla.

A 1.1 (1760) 24871–2817. Don Pedro de Sala y Uruena, informa de las medidas que el ingeniero Luis Diez de Navarro, usó en el mapa del valle de Guatemala.

A 1.1 (1780) 17990–2374. José Ramírez, maestro de albañilería, rinde su informe acerca del valor de un predio de Nicolás Cervantes. Agregado un plano.

A 1.1 (1796) 24902–2817. Reglamento general para el establecimiento de la Escuela de Dibujo adscrita a la Sociedad Económica.

A 1.1 (1797) 24904–2817. Bando dando a conocer las nuevas rutas de los correos.

A 1.1 (1798) 514–18. El ingeniero José Sierra, pide que le sean cancelados las emoluciones que devengó en la reconstrucción del edificio de la fábrica de pólvora de la Antigua Guatemala.

A 1–1 (1802) 660–22. Providencia ordenando que ningún edificio público se inicie en su construcción sin haber sido aprobado el plano respectivo.

A 1–1 (1806) 702–23. Don Santiago Marqui, arquitecto, siguió autos para probar que no se retardó más de lo necesario de España a Guatemala.

A 1–1 (1810) 5212–221. El arquitecto Santiago Marqui, solicitando cierta suma del Fondo de Comunidades, para cubrir parte del edificio del Educatorio de Indias.

A 1.1 (1810) 18009–2377. El maestro de carpintería, Diego de Nájera, da un informe acerca del valor de la fábrica (obras) de carpintería, de la casa que fué de don Juan Ramírez.

A 1–1 (1820) 922–30. El arquitecto Santiago Marqui, solicita la devolución de la tercera parte de su sueldo que está embargado, por tener que cancelar cierta suma a doña Francisca Ferrer.

A 1–2 (1663) 951–39. El síndico del ayuntamiento, don Luis Abarca Paniagua, solicita autorización para el gasto de la construcción de dos puentes: uno de San Lázaro y otro de Santa Lucía.

A 1.2 (1697) 15793–2211. El Fiel Ejecutor, Capitán don Francisco Antonio Fuentes y Guzmán, informa del estado de los cajones de la plaza y pulperías de la ciudad de Guatemala.

A 1.2 (1723) 15806–2212. Instancia del síndico de la ciudad de Guatemala, acerca de que no sean arrastradas por las calles las maderas.

A 1.2 (1732) 15808–2212. Autos relativos a ceder varias pajas de agua por parte del ayuntamiento, al beaterio de Santa Rosa.

A 1.2 (1790) 15833–2213. Renuncia presentada por el maestro mayor de obras públicas, don Bernardo Ramírez.

A 1.2 (1803) 15857–2214. El maestro Bernardo Ramírez cobra la suma de 40 pesos del ayuntamiento.

A 1.2 (1813) 11.815–1805. Acerca de ocupar la catedral para asiento de la Parroquia de San José. [Also 11.818–1805, 11.819–1805, 11.820–1805, 11.821–1805, all of which deal with the same matter.]

A 1.2–1 (1684) 25064–2824. Instancia del capitán Cristóbal Fernández de Rivera, sobre que se le paguen los gastos hechos en reparaciones de los puentes, edificio del matadero y calles.

A 1.2–2 (1611) 11.766–1772. El alcalde don García de Castellanos, opina que no es conveniente la fundación del monasterio de San Agustín.

A 1.2–2 (1668) 11.775–1781. El ayuntamiento acuerda celebrar fiestas de plaza, conmemorando el estreno de la iglesia de Belén y hospital de convalecientes.

A 1.2–2 (1669) 11.774–1780. El ayuntamiento acuerda contribuir con 200 pesos anuales para los gastos de la obra de la catedral.

A 1.2–2 (1674) 11.774–1780. El ayuntamiento informa que está de acuerdo con la fundación del convento de Santa Teresa ya que el sitio y las casas han sido donadas por el presbítero maestro don Bernardino de Obando.

A 1.2–2 (1677) 11.775–1781. El ayuntamiento acuerda asistir al acto inaugural del convento de Santa Teresa de Jesús.

A 1.2–2 (1679) 11.776–1782. Introducción del agua de Santa Ana y construcción de la pila de la Alameda del Calvario.

A 1.2–2 (1687) 11.777–1783. Acuerda el ayuntamiento hacer festividades con motivo del estreno de la iglesia y convento de Santa Teresa, agradeciendo a don José de Aguilar y Rebolledo, haber costeado dicha obra.

A 1.2–2 (1689) 11.778–1784. Providencias dictadas por el ayuntamiento, con motivo de la ruina ocasionada por los temblores del 12 de febrero.

A 1.2–2 (1689) 11.778–1784–14. El ayuntamiento nombra comisionados para que señalen los predios en que deben ser construidas las capillas de los pasos.

A 1.2–2 (1691) 11.778–1784. Acuerdos y determinaciones del ayuntamiento acerca de la fundación del colegio de Misioneros de Propaganda Fide de Cristo Crucificado.

A 1.2–2 (1693) 11.778–1784. El convento de Santa Catalina Mártir solicita del ayuntamiento licencia para hacer construir un arco, sobre la calle que conduce de la plaza mayor a la iglesia de Nuestra Señora de las Mercedes.

A 1.2–2 (1695) 11.779–1785. El síndico del ayuntamiento presenta la licencia otorgada por la audiencia, para la reconstrucción de las casas consistoriales.

A 1.2–2 (1696) 952–39. El síndico del ayuntamiento don Francisco Xavier Folgar pide autorización para gastar 400 pesos, con el fin de concluir la construcción de las casas consistoriales.

A 1.2–2 (1698) 11.779–1785. Cabildo del 21 de octubre de 1698.—El ayuntamiento acuerda asistir a las festividades de la inauguración del nuevo templo "... tan costoso como oseado y curioso ..." de la Compañía de Jesús, y por lo tanto quedará suspendido el cabildo del 5 de diciembre.

A 1.2–2 (1699) 11.776–1782. El ayuntamiento acuerda nombrar comisionados para que atiendan y reciban a las monjas que han de fundar el convento de Santa Clara.

A 1.2–2 (1701) 11780–1786. In a *cabildo* dated February 4, 1701, the *ayuntamiento* agrees to help in the cost of bringing an oil painting of Christ from Esquipulas to the church of El Carmen.

A 1.2–2 (1701) 11.780–1786. Recolección ". . . licencia para fundarse en el barrio de San Jerónimo . . ." 30 de mayo de 1701.

A 1.2–2 (1702) 11.780–1786. El ayuntamiento cede tres mil maravedis para la reedificación de la iglesia de las Benditas Animas.

A 1.2–2 (1703) 11.780–1786. Licencia concedida para ensanchar la "hermita de la Cruz del Milagro." Cabildo con fecha 8 de Junio de 1703.

A 1.2–2 (1703) 11.780–1786. Es presentada al ayuntamiento una solicitud acerca de la reconstrucción de la ermita de Nuestra Señora de Los Dolores.

A 1.2–2 (1704) 11780–1786. In a *cabildo* dated January 29, 1704, the oil painting of Christ in the church of El Carmen is mentioned.

A 1.2–2 (1704) 11780–1786. The accounts for paving the *plaza mayor* are presented to the *ayuntamiento*. The amount is 600 *pesos*.

A 1.2–4 (1538) 2196–138. 20 de diciembre de 1538.—Dispónese que en la construcción de las casas, solamente sea empleada la piedra y el ladrillo y que los techos sean de teja; las salas amplias y los patios con sol.

A 1.2–4 (1538) 15.752–44, vuelto. 20 de diciembre de 1538.—Ordena su majestad que a los seis meses de promulgarse en la ciudad de Santiago, esta cédula, todos los vecinos hagan construir sus viviendas de piedra, ladrillo y teja.

A 1.2–4 (1541) 15752–52. 20 de enero de 1541.—Ordénase a los ayuntamientos que tengan bajo su cuidado la construcción y conservación de los caminos.

A 1.2–4 (1547) 2196–153. 20 de Junio de 1547.—Que la administración de los fondos provenientes de las encomiendas que fueron de doña Beatriz de la Cueva y de su esposo don Pedro de Alvarado, y destinados para la obra de la catedral de Guatemala, le sea a cargo de Juan Pérez Dardón.

A 1.2–4 (1548) 2196–122. 25 de Julio de 1548.— Ordenando al ayuntamiento de la ciudad de Santiago de la provincia de Guatemala, que formule un reglamento para el pago de los jornales que devenguen los oficiales mecánicos.

A 1.2–4 (1556) 2196–127. 18 de julio de 1556.— Que el ayuntamiento de la ciudad de Santiago de la provincia de Guatemala, pueda establecer dos barreros (obrajes) para labrar teja y ladrillo.

A 1.2–4 (1587) 2195–247. 20 de Julio de 1587.—Su majestad ordena a la real audiencia, que dé toda su ayuda económica a los que se están encargado de la construcción y dotación de la ermita de Nuestra Señora de los Remedios, en la capital.

A 1.2–4 (1723) 16.192–2245. Información acerca del patronato que goza el ayuntamiento en el convento de Nuestra Señora de la Limpia Concepción de María.

A 1.2–5 (1718) 15.766–2207–16. El cabildo, habiendo informado al Rey del general estrago de esta ciudad con los terremotos del año diez y siete, lo hace en particular de la iglesia y convento de Nuestra Señora de la Merced.

A 1.2–5 (1719) 15776–2207–71. El cabildo informa haber reparado la ciudad, de las ruinas que padeció con los terremotos de 1717.

A 1.2–6 (1604) 11.810–1804. Becerro del asiento general y particular de las cuadras, casas y vecinos que hay en ellas de la ciudad de Santiago de los Caballeros de la provincia de Guatemala, en que se ha de repartir la alacabala que está obligada a pagar su majestad.

A 1.2–6 (1703) 25573–2848. The convent of Santa Clara petitions the *ayuntamiento* for a larger water allotment.

A 1.2–6 (1716) 29993–4000. El hermano Tomás García, en nombre de la Tercera Orden de San Francisco pide tributo de agua de goza para la casa y templo del Calvario.

A 1.2–9 (1698) 25348–2840. El ayuntamiento de Guatemala, da 150 pesos para celebrar el estreno de la iglesia de la Compañía de Jesús.

A 1.3 (1680) 12.245–1885. Autos sobre la posesión de las cátedras de la real universidad de San Carlos, asignación de materias e inauguración de los cursos.

A 1.3 (1763) 1157–45. Autos acerca de que la universidad sea trasladada a la casa que había ocupado don José de Alcantará, ubicada al sur de la catedral.

A 1.3 (1782) 1163–45. Instancia del señor Dr. don Felipe Romana y Herrera, fiscal de esta real audiencia en razón de haberse puesto en las paredes públicas de la real universidad el blasón de la silla apostólica alternando con las armas reales de su majestad.

A 1.3–1 (1676) 12.235–1882. Autos de la merced y fundación de la real universidad de San Carlos de esta ciudad de Santiago de Guatemala. Cédula de la erección de 31 de enero de 1676.

A 1.3–3 (1683) 12388–1896. Agustín Núñez, *maestro ensamblador*, enters a bid of 1,600 *pesos* to construct a retable for the University. He also submits some drawings. [*Were not found with the document.*]

A 1.3–4 (1782) 4410–49. Autos acerca de la petición del fiscal de la audiencia y contra dichos por el rector Dr. Juan Antonio Dighero relativos a los blasones en la universidad.

A 1.3–8 (1681) 12445–1899. One professor protests regarding the poor class attendance of students, especially of those studying theology.

A 1.3–20 (1681) 13145–1956. Contra el doctor Juan Bautista de Urquiola, por ciertos gastos hechos en la construcción de la capilla y otras dependencias del edificio de la universidad.

A 1.3–21 (1751) 13160–1957. Autos del reconocimiento del estado del edificio de la universidad, debido a los temblores del día 4 de marzo.

A 1.3–21 (1751) 13161–1957. Copia del informe rendido a S. M. del estado ruinoso en que quedó el edificio de la universidad, debido a los temblores del 4 de marzo.

A 1.3–21 (1764) 13162–1957. Avaluo y remate del sitio y edificio que ocupó la universidad en las inmediaciones de Santo Domingo.

A 1.3–21 (1773) 13163–1957. Reconocimiento del estado en que quedó el edificio de la universidad debido al terremoto de Santa Marta (29 de julio).

A 1.3–21 (1788) 13164–1957. Nicolás Monzón informa de que el techo de la biblioteca de la universidad está ruinoso.

A 1.3–21 (1790) 13165–1957. Borrador del costo que tuvo el edificio de la universidad en la Antigua Guatemala.

A 1.3–21 (1790) 13166–1957. Instancia del claustro y rector de la universidad, para que el superior gobierno le ayude en la construcción del nuevo edificio.

A 1.3–21 (1809) 13168–1957. Cuenta del producto de la suscripción para terminar la construcción del edificio de la universidad. Fueron recogidos 28 pesos.

A 1.3–21 (1815) 13170–1957. Relación de los gastos y arbitrios invertidos en la construcción del edificio de la universidad.

A 1.3–25 (1775) 13253–1961. Acerca de la traslación de la universidad al llano de la Virgen.

A 1.7 (1752) 1296–52. El maestro Juan de Dios Estrada, quien había rematado la obra de albañilería del hospital, se queja por no habérsele admitido la fianza.

A 1.9 (1731) 1380–54. Permuta de la casa e iglesia de las Capuchinas por la casa del Colegio de Niñas en tanto que es concluida la obra del monasterio de las Capuchinas.

A 1.10 (1773) 1535–55. Padrón que determina, por parroquias, la población existente en la Antigua Guatemala.

A 1.10 (1773) 18.773–2444. Carta del 31 de agosto de 1773, dirigida por el ayuntamiento a su majestad, informándole de la ruina acaecida el 29 de julio y solicitando algunas providencias en favor del vecindario.

A 1.10 (1775) 1543–56. El maestro Fr. Simón Reina, presenta su informe acerca del estado de algunos materiales que existen aprovechables de los edificios de la Antigua Guatemala.

A 1.10 (1776) 1548–56. Providencias reglamentando la elevación, terraplenos y materiales de construcción que debían ser empleados en la Nueva Guatemala.

A 1.10 (1776) 1567–58. Sumaria instruida a las justicias de Santa Catarina Pinula, por no haber acatado la orden de enviar un repartimiento de indios que debían trabajar en las obras de la ciudad.

A 1.10 (1776) 1568–58. Providencia acerca del repartimiento de indios entre los vecinos de la Nueva Guatemala.

A 1.10 (1776) 1569–59. Providencia a fin de que el alcalde mayor de Verapaz remita cierta cantidad de indios con destino a los trabajos de la Nueva Guatemala.

A 1.10 (1776) 4470–62. El justicia mayor de Sacatepéquez expone que varios indios han solicitado que se les dispense del trabajo en la obras de la nueva ciudad por tener que dedicarse a sus siembras.

A 1.10 (1776) 31361–4049. Método regular en formación de los cimientos. Sobre colocación de horcones. Instrucciones para construir.

A 1.10 (1777) 1571–59. Los indios de Chimaltenango solicitan no venir a los trabajos de Guatemala por estar ocupados en sus siembras.

A 1.10 (1777) 1572–59. The Indians of Rabinal ask to be excused from coming to work in Guatemala because they are busy sowing their fields.

A 1.10 (1777) 1575–59. Autos ordenando la cancelación de los gastos que hicieron don Marcos Ibáñez (arquitecto) y don Antonio Bernasconi (dibujante) para su venida.

A 1.10 (1777) 4473–62. Los indios de varios pueblos de la alcaldía mayor de Sololá solicitan la dispensa en las obras de la Nueva Guatemala, por tener que dedicarse a sus siembras.

NOTE: The following documents from A 1.10 (1777) 4474–62 to A 1.10 (1778) 4490–63, all deal with the same problem; namely, the Indians resident in the towns indicated all ask to be excused from forced labor in the construction of Nueva Guatemala.

A 1.10 (1777) 4474–62: Quetzaltenango
A 1.10 (1777) 4475–62: Patzún
A 1.10 (1777) 4476–63: San Pedro Las Huertas
A 1.10 (1777) 4477–63: San Andrés Izapa
A 1.10 (1777) 4478–63: Ciudad Vieja (Antigua)
A 1.10 (1777) 4479–63: Izapa
A 1.10 (1777) 4480–63: Santiago de Mataesquintla
A 1.10 (1777) 4481–63: San Pedro Pinula
A 1.10 (1777) 4482–63: Santa Apolonia (Chimaltenango)
A 1.10 (1777) 4483–63: San Raymundo de las Castillas (Sac.)
A 1.10 (1777) 4485–63: Petapa
A 1.10 (1777) 4486–63: Chimaltenango
A 1.10 (1777) 4487–63: Chimaltenango, order to arrest Indians who refuse.
A 1.10 (1778) 4488–63: Sololá
A 1.10 (1778) 4489–63: Sololá
A 1.10 (1778) 4490–63: Cubulco

A 1.10 (1777) 4524–69. Autos llevados a cabo por el oidor don Manuel Antonio de Arredondo acerca de que sean trasladados, de la Antigua Guatemala, los escombros de los edificios reales.

A 1.10.1 (1711) 14902–2101. Cuenta de lo gastado en la reconstrucción del palacio.

A 1.10.1 (1736) 14903–2101. Cuenta de lo gastado en el Palacio de los Capitanes.

A 1.10.1 (1736) 14904–2101. Cuenta de lo gastado en el Palacio de los Capitanes.

A 1.10.1 (1740) 14905–2101. Cuenta de lo gastado en la construcción de algunos cuartos del Palacio de los Capitanes.

A 1.10.1 (1746) 14906–2101. Planillas de gastos. Reconstrucción del palacio real.

A 1.10–1 (1760) 1421–64. Providencia ordenando que se reconozca el estado que tiene el palacio.

A 1.10–1 (1761) 1422–64. Providencia autorizando el gasto de la reparación de la portada del palacio. Este trabajo lo dirigió Diez de Navarro.

A 1.10.1 (1768) 31.220–4044. Acerca de que sea desocupada la pieza destinada a la oficina de almonedas, en vista de la reedificación del palacio real.

A 1.10.1 (1768) 31.222–4044. Por haber quedado terminada la reedificación del palacio real, dispónese el traslado del real tribunal, de las habitaciones de la Casa de Moneda a sus dependencias permanentes.

A 1.10.1 (1768) 31.223–4044. Comprobante de haber recibido el ayuntamiento de Guatemala los comprobantes a las cuentas de la construcción de real palacio.

A 1.10.1 (1769) 184–8. Testimonio de la cédula de 8 de enero de 1763 acerca de otorgar licencia para el gasto de 65,183 pesos en la construcción del Palacio de los Capitanes Generales.

A 1.10–1 (1769) 1423–64. Autos hechos en razón de lo ordenado en cédula de 8 de enero de 1763, autorizando el gasto para la reparación del Palacio de los Capitanes Generales (65,183 pesos).

A 1.10–1 (1779) 6458–307. A ledger with accounts of the materials collected from the ruined buildings in Antigua Guatemala to be used for construction in the new capital.

A 1.10–1 (1804) 1484–65. El arquitecto Pedro Garci-Aguirre, solicita que le sean cancelados los sueldos que devengó en carácter de dirigente de los trabajos de reparación de la casa que habita el oidor Francisco Camacho.

A 1.10.2 (1751) 18769–2447. Juan de Dios Estrada, maestro mayor de obras del ayuntamiento de Guatemala, pide se controle la edificación de casas de particulares.

A 1.10.2 (1766) 18771–2447. El maestro de obras del ayuntamiento de Guatemala, Francisco de Estrada, presenta el proyecto sobre construir de arcadas el portal del cabildo en su extremo poniente.

A 1.10.2 (1774) 1642–66. Providencia acerca de que los escombros del palacio y real aduana y hospital, sean conservados para ser empleados en la construcción de edificios en la Nueva Guatemala.

A 1.10–2 (1777) 1650–67. Juan Medina solicita que se le dé alguna gratificación por los trabajos que efectuó con el objeto de acopiar materiales.

A 1.10.2 (1777) 4523–69. Legajo de la correspondencia cruzada entre don Manuel de Arredondo y don José Manuel de Barroetea, acerca de la conducción de escombros, maderas, puertas, etc. de la Antigua Guatemala.

A 1.10.2 (1777) 4524–69. Autos llevados a cabo por el oidor don Manuel Antonio de Arredondo acerca de que sean trasladados de la Antigua Guatemala los escombros de los edificios reales.

A 1.10–2 (1783) 1660–68. El maestro Bernardo Ramírez, propone que el portal de las casas consistoriales, sea fabricado siguiendo el mismo estilo que el que tiene el de la Real Aduana.

A 1.10–2 (1786) 1669–68. Don Manuel de la Bodega, es nombrado superintendente de la obra de la catedral.

A 1.10–2 (1788) 1670–68. Por fallecimiento de don Sebastián Gamundi, director interino de la obra de la catedral, es nombrado el ingeniero José Sierra. En este cuaderno está el proyecto general de la obra debido al maestro José Arroyo.

A 1.10–2 (1797) 1672–68. Don Manuel del Campo y Rivas, solicita que él es el llamado a ser el superintendente de la obra de la catedral.

A 1.10–2 (1798) 1673–68. Providencia nombrando el maestro Bernardo Ramírez, para que dirija la obra de la catedral.

A 1.10–2 (1800) 1674–68. Habiéndose enfermado el ingeniero Antonio Porta, quien había sido comisionado para ir a Granada y revisar la catedral de aquella ciudad, fué nombrado el ingeniero José Sierra.

A 1.10–2 (1802) 1677–68. El ingeniero José Sierra, solicita que le sean cancelados los sueldos que devengó como director de la obra de la catedral. Sierra pensaba marcharse a España.

A 1.10–2 (1802) 1678–68. El ingeniero Pedro Garci-Aguirre, uno de los encargados de la obra de la catedral, solicita que ya no se explote la cantera de Arrivillaga sino la de el Naranjo.

A 1.10.3 (1667) 31.253–4046. Los alcaldes y regidores del pueblo de San Agustín de la Real Corona, del corregimiento de Acasaguastlán, piden parte de sus tributos para edificar su iglesia de artezón y no de paja como es.

A 1.10.3 (1672) 31.255–4046. Informe acerca del estado de la edificación de la nueva catedral de Guatemala.

A 1.10.3 (1672) 31.256–4046. Providencia asignando fondos del ramo de encomiendas vacantes, con destino a la obra de la catedral de Guatemala.

A 1.10.3 (1672) 31.258–4046. El maestro de albañilería José de Porras, director de la obra de la catedral de Guatemala, pide aumento de sueldo.

A 1.10.3 (1679) 31.267–4046. El capitán José de Aguilar de Revolledo, alcalde de la ciudad de Guatemala, expone que se le notificó un auto para que en unión del capitán Francisco Antonio de Fuentes y Guzmán, colecten limosnas para sufragar los gastos de la edificación del templo de Nuestra Señora de los Remedios.

A 1.10.3 (Siglo XVII) 31.386–4051. Cuenta de lo invertido en la obra de la catedral de Guatemala.

A 1.10.3 (1691) 31.271–4046. El presbítero Nicolás Díaz, solicita al ayuntamiento de Guatemala, cierta cantidad de madera (limosna) para el techo de Santa Lucía.

A 1.10–3 (1693) 31.272–4046. El procurador del convento de San Francisco de Guatemala, pide dos canteros del pueblo de Santa María de Jesús, para la obra del templo.

A 1.10–3 (1695) 31.275–4046. El capitán José Domínguez, mayordomo de la ermita de Santa Lucía, a cuyo cargo corre la obra, pide licencia para pedir limosna y finalizar la construcción.

A 1.10.3 (1702) 31.278–4047. Los indígenas del barrio del Espíritu Santo, de la ciudad de Guatemala, sobre que se les conceda la cuarta parte de sus tributos para edificar su templo.

A 1.10.3 (1703) 31.279–4047. Juan Antonio Barahona, Felipe de Herrera, Benito de Santa María y Juan Ventura, vecinos de Guatemala, piden al ayuntamiento la cantidad de 10 varas más de tierra, para ampliar la ermita de la Santa Cruz a espaldas del convento de Concepción.[Santa Cruz del Milagro.]

A 1.10.3 (1705) 31280–4047. José de Santa María y María de Loaiza, ambos pardos libres, piden licencia para recaudar limosnas con destino a la construcción del templo de Mazagua.

A 1.10.3 (1712) 16543–2280. El ayuntamiento cede a favor del templo de Santa Lucía en la ciudad de Guatemala un predio para lonja.

A 1.10.3 (1720) 18803–2448. Sobre la construcción del templo de San Lázaro (Antigua).

A 1.10.3 (1720) 31.291–4047. Bernardo Manuel y José de Larios, vecinos de la ciudad de Guatemala, piden ayuda para reedificar la ermita de Santa Lucía, arruinada en Septiembre de 1717.

A 1.10.3 (1720) 31.297–4047. Acerca de la reedificación de la ermita del Calvario de la ciudad de Guatemala.

A 1.10.3 (1725) 16.544–2280. El presbítero Manuel de Morga, solicita al ayuntamiento la limosna de agua con destino al templo de Nuestra Señora del Carmen.

A 1.10.3 (1727) 18804–2448. Sobre la reconstrucción del templo de la Cruz del Milagro.

A 1.10.3 (1728) 18805–2448, fol. 2. 23 de Julio de 1728.—Cédese a favor de la ermita de Nuestra Señora de los Dolores (Cruz del Milagro) dos varas de terreno y una más sobre la banda donde corre el río Pensativo, obligándose el prioste y cofrades de esta ermita, según disposición del ayuntamiento a hacer construir una calzada en la margen derecha del citado río.

A 1.10.3 (1738) 31.313–4047. El P. Manuel de Herrera, rector del colegio de la Compañía de Jesús, pide al ayuntamiento ayuda económica para reconstruir este centro de estudios.

A 1.10.3 (1739) 18806–2448. El ayuntamiento recibe la cédula por la cual se le ordena ayude a la reconstrucción del convento y monasterio de San Agustín.

A 1.10.3 (1739) 31.318–4047. Acerca de ceder la cuarta parte de tributos para reconstruir el templo de Santa Ana.

A 1.10.3 (1743) 31.322–4048. Visita de ojos para determinar el estado del templo de Santa Ana, extramuros de la ciudad de Guatemala.

A 1.10.3 (1745) 16.545–2280. Instancia de los vecinos de la calle Ancha ante el ayuntamiento de Guatemala sobre que se permita cubrir la Cruz de Piedra. Hay un plano. [See A 1.11 (1753) 2091–98, below.]

A 1.10.3 (1745) 31.334 and 31.335–4048. Fr. Juan de San Mateo, prior del convento y hospital de Nuestra Señora de Belén, indica que para perpetuar la memoria del Hermano Pedro, han dispuesto la reedificación de la ermita de las Animas.

A 1.10.3 (1751) 127–8, fol. 72. An order to estimate the cost of the repairs required on the Cathedral ruined in the earthquake of March 4, 1751.

A 1.10.3 (1751) 4215–33. Sobre informar a su magestad sobre el estado en que quedó la catedral metropolitana con los terremotos de 4 de marzo de 1751.

A 1.10.3 (1751) 18807–2448. Instancia del P. Miguel de los Ríos, prepósito de la Congregación de San Felipe Neri, sobre que el ayuntamiento ayude para la reconstrucción del templo.

A 1.10.3 (1751) 31.346–4049. Sor Josefa María Sta. Gertrudis, priora del convento de las Carmelitas Descalzas (Santa Teresa) pide ayuda para reconstruir el templo arruinado con los terremotos del 4 de febrero de 1751.

A 1.10.3 (1751) 31.347–4049. Se solicita ayuda para reconstruir la ermita de Santa Lucía.

A 1.10.3 (1751) 31349–4049. Informes acerca del estado en que quedó la catedral con motivo del terremoto del 4 de marzo de 1751.

A 1.10.3 (1753) 18808–2448. Se solicita ayuda para reconstruir la iglesia de San Felipe Neri.

A 1.10.3 (1753) 31351–4049. Real provisión otorgando licencia a Fr. Diego de Iruve, O.P., cura del templo de la Candelaria, de la ciudad de Guatemala, para que pueda pedir limosna y reconstruir el templo.

A 1.10.3 (1761) 4679–236, fol. 1. The Indians of Jocotenango ask to be exempted from payment of extraordinary contributions because they are engaged in the construction of their church.

A 1.10.3 (1773) 31.358–4049. Sobre finalizar la construcción del templo de Jocotenango.

A 1.10.3 (1775) 4536–74. Carta del ingeniero Diez de Navarro, solicitando los implementos necesarios para el delineamiento de la nueva ciudad.

A 1.10–3 (1776) 4544–75. Los fabricantes de teja y ladrillo, solicitan mejor pago por el millar de estos productos.

A 1.10.3 (1776) 31.361–4049. Método regular en formación de los cimientos. Sobre colocación de horcones. Instrucciones para construir. [Same as A 1.10 (1776) 31.361–4049, above.]

A 1.10–3 (1777) 4571–76. Don Benito Matute escribe desde la antigua Guatemala, que una parte del ex-palacio se había hundido por el excesivo peso de los materiales guardados en él.

A 1.10–3 (1781) 1722–73. El cura de la iglesia de Nuestra Señora de la Asunción de Jocotenango, solicita ayuda para amueblar la casa conventual.

A 1.10.3 (1784) 15091–2123. Acerca de la zona que corresponde al sagrario. Hay un ante-proyecto de la fachada de la catedral de Antigua Guatemala.

A 1.10–3 (1790) 1729–73. El cura de Jocotenango, Juan Gaya, solicita providencia acerca de la formal traslación del pueblo, éste estaba en las inmediaciones de la antigua Guatemala.

A 1.10.3 (1790) 18823–2448. El presbítero Antonio García Redondo, pide se le permita trasladar de la ermita de Santa Lucía (de Antigua) las imágenes con destino al templo de San Sebastián.

A 1.10.3 (1790) 18824–2448. Permission is asked to transfer some ornaments from the church of Santa Lucía in Antigua to the church of San Sebastián in Nueva Guatemala.

A 1.10–3 (1797) 18827–2448. Nicolás Monzón pide se le paguen sus honorarios devengados por dirigir la obra del templo de Santo Domingo Xenacoj.

A 1.10.3 (1803) 2356–109. Autos para llevar a cabo la creación (construcción) del Beaterio de Indias. Hay un plano.

A 1.10.3 (1820) 18831–2448. El cabildo eclesiástico de Nueva Guatemala, concede al ayuntamiento de Antigua permiso para reconstruir la ex-catedral.

A 1.11 (1710) 2043–94. Demanda entablada por el decano y cabildo eclesiástico contra los religiosos mercedarios, dominicos y franciscanos para que estos enteren el 3% del fruto de sus curatos, para substancia del Seminario Tridentino.

A 1.11 (1710) 2044–94. Autos hechos a pedimento de Andrés Ruiz de la Cota, mayordomo de la obra de la iglesia de los Remedios, contra el fundidor José de Arria. Este no había cumplido con la entrega de una campana.

A 1.11 (1720) 16.779–2292. Informe del ayuntamiento acerca de la fundación del convento de las Capuchinas. [This document also contains similar information concerning the other religious communities in Antigua—Escuela de Cristo, San Agustín, etc.]

A 1.11 (1741) 5025–211. Relación histórica de la provincia de Nuestra Señora de la Merced, Redención de Cautivos de la Presentación de Guatemala.

A 1.11 (1752) 2082–97. El rector del colegio de la Compañía de Jesús (Colegio de San Lucas), solicita cierta ayuda para reconstruir el edificio que ocupa dicho centro, arruinado con motivo de los temblores de San Casimiro. (4 Marzo 1751.)

A 1.11 (1752) 2087–97. El cabildo eclesiástico solicita que los oficiales reales sean los que descuenten el 3% de los sínodos dados a los curas, para que así se cumpla con la dádiva destinada al Seminario Tridentino.

A 1.11 (1753) 2091–98. Varios vecinos de la calle Ancha de Herreros solicitan construir de bóveda la ermita de Santa Cruz (Cruz de Piedra). [See A 1.10.3 (1745) 16.545–2280, above.]

A 1.11 (1763) 2113–98. Instancia seguida en el superior gobierno por el real fisco con la provincia de Nuestra Señora de la Merced sobre la exhibición de la real licencia de su Mag. para la fundación del colegio de San Jerónimo que por este defecto se declaró extento.

A 1.11 (1772) 2117–98–7. Construcción y bendición del beaterio de Nuestra Señora del Rosario de la ciudad de Guatemala de la tercera orden de Santo Domingo.

A 1.11 (1774) 2117–98–13 vuelto. Destrucción del edificio del beaterio de Nuestra Señora del Rosario, de la tercera orden de Santo Domingo.

A 1.11 (1779) 2117–98–9 vuelto. Traslación de las Beatas Indias a la Nueva Guatemala de la Asunción.

A 1.12 (1695) 16.795–2294. Cabildo. Año de 1695. Sobre la fundación de los Rvdos. PP. Misioneros en esta ziudad.

A 1.16 (1591) 1751, fol. 26 v. Auto de 14 de octubre de 1591, por el cual queda prohibida que indígenas contraten a oficiales plateros, etc., para hacer obras en sus iglesias y casas de comunidad las cuales "son impertinentes y no necesarias: que ningún oficial platero, bordador, albañil, carpintero, . . . ni otro ningún oficial . . . . puedan hacer ni hagan ninguna obra a pueblo de indios sin previa licencia de la Audiencia."

A 1.16 (1773) 2830–148. El maestro de arquitectura, don Bernardo Ramírez, expone la necesidad de arreglar la oficina de albañilería.

A 1.16 (1798) 2905–149. Fr. Antonio de San José Muñoz solicita la impresión de una memoria acerca de las ventajas que resultan a los indios y ladinos de que se calcen.

A 1.16.1 (1761) 17133–2312. Solicitan los ermitaños de la orden de San Agustín, sobre que los albañiles presten su cooperación en la reconstrucción del templo.

A 1.16.4 (1747) 2811–148. El hermano de la tercera orden Franciscana, Tomás de Morales, maestro fundidor de campanas y cañones, pide que nadie en Guatemala pueda ejercer su oficio

A 1.16.22 (1641) 38298–4500. Salvador García, oficial carpintero, pide examen para el grado de maestro.

A 1.17 (1740) 112–6. Real provisión circulada a las autoridades del reino y RR. PP. de las órdenes religiosas, para que formulen descripciones geográficas e históricas.

A 1.17.1 (1740) 5002–210. Relación geográfica del valle de Goathemala por Guillermo Martínez de Pereda corregidor y alcalde mayor de dicho valle.

A 1.17.3 (1744) 17.508–2335. Información rendida por el ingeniero Luis Diez de Navarro de su viaje por las provincias de Guatemala. Año de 1744.

A 1.17.3 (1756) 38.302–4501. Relación e informe sobre el camino que de la capital al puerto de San Fernando Omoa, puede servir para el comercio y defensa del reino. Por el ingeniero Luis Diez de Navarro.

A 1.17.3 (1793) A 1.25–21389–2603, fol. 4. Estado ó Razón de las distancias que hay desde la capital a las ciudades de este reino y cabezas de partidos: días en que entran y salen los correos en ellas y leguas que cruzan a las poblaciones por donde transitan los correos de a caballo y a pie.

A 1.18 (1718) 1400–2021. Breve y verdadera historia del incendio del volcán . . . y terremotos de la ciudad . . . . 27 de septeiembre de 1717.

A 1.18 (1740) 5022–211. Testimonio de las diligencias hechas en la curia eclesiástica de Guatemala en orden a las relaciones de los 4 conventos de religiosas sugetas al ordinario, sus erecciones, órdenes a que pertenecen, número de religiosas de que se componen y sus rentas con la comprobación de dichas relaciones.

A 1.18 (1740) 07/5022–211. Relación del administrador de Santa Catalina.

A 1.18 (1740) 5022–211. Relación del convento de religiosas Capuchinas.

A 1.18 (1740) 5022–211. Relación del administrador de Nuestra Señora de la Concepción.

A 1.18 (1740) 5027–211. Relación histórica del Colegio de Misioneros de Cristo Crucificado de la Ciudad de Guatemala por Fr. Antonio de Andrade.

A 1.18 (1740) 5031–211. Relación histórica del colegio de la Compañía de Jesús por el P. Manuel Herrera.

A 1.18 (1741) 5029–211. Relación histórica del convento de San Felipe Neri de la ciudad de Santiago de Guatemala. Por el presbítero Pedro Martínez de Molina.

A 1.18.6 (1774) 38306–4502. "Razón particular de los Templos, Casas de Comunidades, y Edificios Públicos y por mayor del número de los vecinos de la capital de Guatemala; y del deplorable estado a que se hallan reducidos por los terremotos de la tarde del veinte y nueve de julio, trece y catorce de diciembre del año próximo pasado de setenta y tres. Por Juan González Bustillo, año 1774." [See bibliography above.]

A 1.20 (1602) 10.171–16 vuelto. Escritura de la toma de posesión del pueblo de Jocotenango por parte de los frailes de Santo Domingo, para la fundación de un convento y vicaria otorgada el 21 de enero de 1602.

A 1.20 (1626) 757. Escritura de concierto para la edificación del templo de Santa Catalina Mártir.

A 1.20 (1636) 690–53. El maestro pintor Pedro de Liendo se obliga a hacer el retablo de la capilla de Nuestra Señora La Antigua, en la iglesia de Santo Domingo. 30 de mayo de 1636.

A 1.20 (1636) 690–69. Escritura otorgada el 18 de junio de 1636 por los canteros Juan Bautista de Vallejo y Martín de Autillo, obligándose a hacer la obra de cantería del altar mayor de la iglesia de Santo Domingo.

A 1.20 (1648) 694–668. Escritura de concierto otorgada por Martín de Ugalde el 14 de octubre de 1648, a favor del convento de Santo Domingo, comprometiéndose a terminar la obra de la capilla mayor de la iglesia de dicho convento.

A 1.20 (1668) 1480–14 vuelto. Poder general de la tercera orden de San Francisco, otorgado a favor de Antonio de la Cruz, para que gestione en el Real Consejo de Indias, la aprobación de la posesión sobre el templo del Calvario.

A 1.20 (1673) 476–10. Los carpinteros Nicolás y Juan López se obligan . . . de hacer de nuevo el techo de la iglesia del convento de San Francisco de la ciudad de Santiago de Guatemala.

A 1.20 (1675) 477–32. Escritura de concierto de la obra de la portada de la iglesia del convento de San Francisco de la ciudad de Santiago de Guatemala.

A 1.20 (1675) 1322–803. Fernando de Cuellas dona cierta suma para la fundación del convento de Santa Teresa.

A 1.20 (1675) 9093–600. Poder otorgado por el maestro don Bernardino de Obando, clérigo presbítero para la fundación del convento de Santa Teresa de Jesús.

A 1.20 (1678) 14.480–28. Historical data relative to the founding of the convent of Santa Teresa. Dated April 23, 1678.

A 1.20 (1678) 1480–35. Presbítero Antonio de Salazar dona 1,000 pesos para la fundación del convento de Santa Teresa. 30 de abril de 1678.

A 1.20 (1678) 1480–40 vuelto. Testimonio en favor de la fundación del convento de Santa Teresa con fecha 1 de mayo de 1678.

A 1.20 (1679) 683–191. Escritura de dotación otorgada por don Sebastián Alvarez, Alfonso Rosica de Caldas, presidente de la real audiencia y otros vecinos, a cargo de la fundación del convento de las Carmelitas Descalzas, bajo la advocación de Santa Teresa de Jesús. [An additional number of testimonials are included, all dated in September of 1679.]

A 1.20 (1690) 695–119. Escritura otorgada el 13 de octubre de 1690 por el maestro ensamblador Vicente de la Parra . . . obligándose a hacer el retablo de Nuestra Señora de la Natividad de la iglesia de Santa Catalina Mártir.

A 1.20 (1691) 1189–120 vuelto. El alférez Ramón de Molina, se obliga a hacer el retablo de la capilla del Santo Cristo, de la iglesia de Nuestra Señora de las Mercedes.

A 1.20 (1692) 696–23. Poder otorgado por el difuntorio de la provincia del Santísimo Nombre de Jesús de la regular observancia del Señor San Francisco, el 29 de enero de 1692, para el cobro de las asignaciones destinadas a la fundación del convento de Santa Clara.

A 1.20 (1703) 738–24 vuelto. Agustina Ramona, india, cede un sitio al colegio de Cristo Crucificado.

A 1.20 (1704) 739–183 vuelto. Pedro Lorenzo y José Vélez, se obligan a dorar el altar mayor de la iglesia de Nuestra Señora de las Mercedes.

A 1.23 (1546) 1511–33. 26 de marzo de 1546.— Dispone su magestad que los caminos de las provincias de Guatemala y Honduras "se aderezen dos vezes al año."

A 1.23 (1549) 1511–142. 13 de junio de 1549.— El Licenciado Cerrato, presidente de la audiencia, informa a su majestad que había iniciado la construcción de un camino que conduciría de la capital a la costa del mar del norte.

A 1.23 (1550) 1511–148. 4 de agosto de 1550.— Su majestad dispone que para finalizar la obra de la catedral de Guatemala, su costo sea distribuido así: un tercio sea tomado de los fondos reales; un tercio de las encomiendas y tributos de la jurisdicción del obispado y el resto por los vecinos e indios, éstos estando encomendados bajo la corona real.

A 1.23 (1553) 1511–193. 27 de abril de 1553.— Que todo español avecindado de la provincia de Guatemala, sea obligado a ejercer su oficio o a cultivar la tierra, so pena de ser enviado a España bajo partida de registro.

A 1.23 (1558) 1511–238. 1 de agosto de 1558.—Ordenando que se dé ayuda económica, y no indios en repartimiento, para la construcción de los monasterios.

A 1.23 (1559) 1512–270. 18 de diciembre de 1559.— La real audiencia con el objeto de fomentar la agricultura y evitar la holgazanería de los españoles, propone repartir tierras en el valle de Guatemala.

A 1.23 (1574) 1513–523. 27 de abril de 1574.— Por carta de esta fecha su majestad recomienda lo siguiente: a) que los ayuntamientos tenga a su cargo la construcción y conservación de puentes y caminos; . . .

A 1.23 (1575) 1512–475. 31 de enero de 1575.— Su majestad pide informes acerca del estado en que se encuentra la obra de la iglesia y conventos franciscanos.

A 1.23 (1574) 1512–447. 27 de abril de 1574.— Ordénase que los negros, bozales y esclavos tambien paguen tributos.

A 1.23 (1575) 1512–474. 31 de enero de 1575.— En vista que los franciscanos solicitaron a su majestad ayuda económica para reconstruir su iglesia y convento, se pide informes a la real audiencia.

A 1.23 (1586) 1513–660. 30 de mayo de 1586.— Baltazar Estévez de Santa María, ermitaño de la ermita de Nuestra Señora de los Remedios, de la ciudad de Santiago de los Caballeros, informó a su majestad que a su llegada en el año 1574, inició la reconstrucción de dicha ermita, que estaba cubierta de paja, y que ahora estaba cubierta de artesón, y que para establecer la festividad de Nuestra Señora de la O, necesitaba cierta ayuda de costa. Su majestad se la otorgó.

A 1.23 (1594) 5113–750. 29 de mayo de 1594.— Ordénase al presidente de la real audiencia y al obispado de la provincia de Guatemala, que informen sobre si es conveniente erigir en parroquia la ermita de Nuestra Señora de los Remedios.

A 1.23 (1675) 10975–1520–220. Real cédula de 13 de febrero de 1676, autorizando la fundación del convento de las Carmelitas Descalzas, en vista de lo acordado el 22 de junio de 1675.

A 1.23 (1680) 10.076–1521–221. Real cédula mediante la cual su majestad aprueba lo determinado por la junta de la universidad, en lo relativo a la fábrica del edificio y creación de dos cátedras más y el escudo de la universidad.

A 1.23 (1727) 1526–210. Su majestad aprueba la fundación del convento de religiosas de Nuestra Señora del Pilar de Zaragoza (Monjas Capuchinas).

A 1.23 (1733) 1526–310. Su majestad pide informes acerca de la instancia de Fr. Francisco Seco (O.F.M.) acerca de asignar fondos al convento e iglesia de Santa Clara.

A 1.37 (1818) 17517–2335. El gobernador intendente de la provincia de Honduras, don Ramón Anguiano, informa acerca de dicha provincia.

A 1.62.2 (1706) 48.139–5556. Agustín Núñez entrega dos retablos al templo de la Compañía de Jesús.

A 1.69.3 (1723) 48141–5556. It is proposed that those who desire the title "maestro de arquitectura" be given examinations.

A 1.69.3 (1752) 48142–5556. Juan de Dios Estrada, maestro mayor de obras, pide al ayuntamiento que se prohiba dirigir obras de construcción a los oficiales que no han sido examinados.

A 3.2 (1673) 15207–825. Memoria y padrón de las doctrinas y conventos que administra la religión de Nuestro Padre San Francisco, en el obispado de Guatemala. 24 de octubre de 1673.

A 3.17 (1734) 26924–1655, fol. 32. Plano de la fábrica de la Casa de Moneda. Autor, Diego de Porras.

A 3.17 (1738) 26924–1655, fol. 266. Plano de la Casa de Moneda de la ciudad de Guatemala. Autor, Diego de Porras.

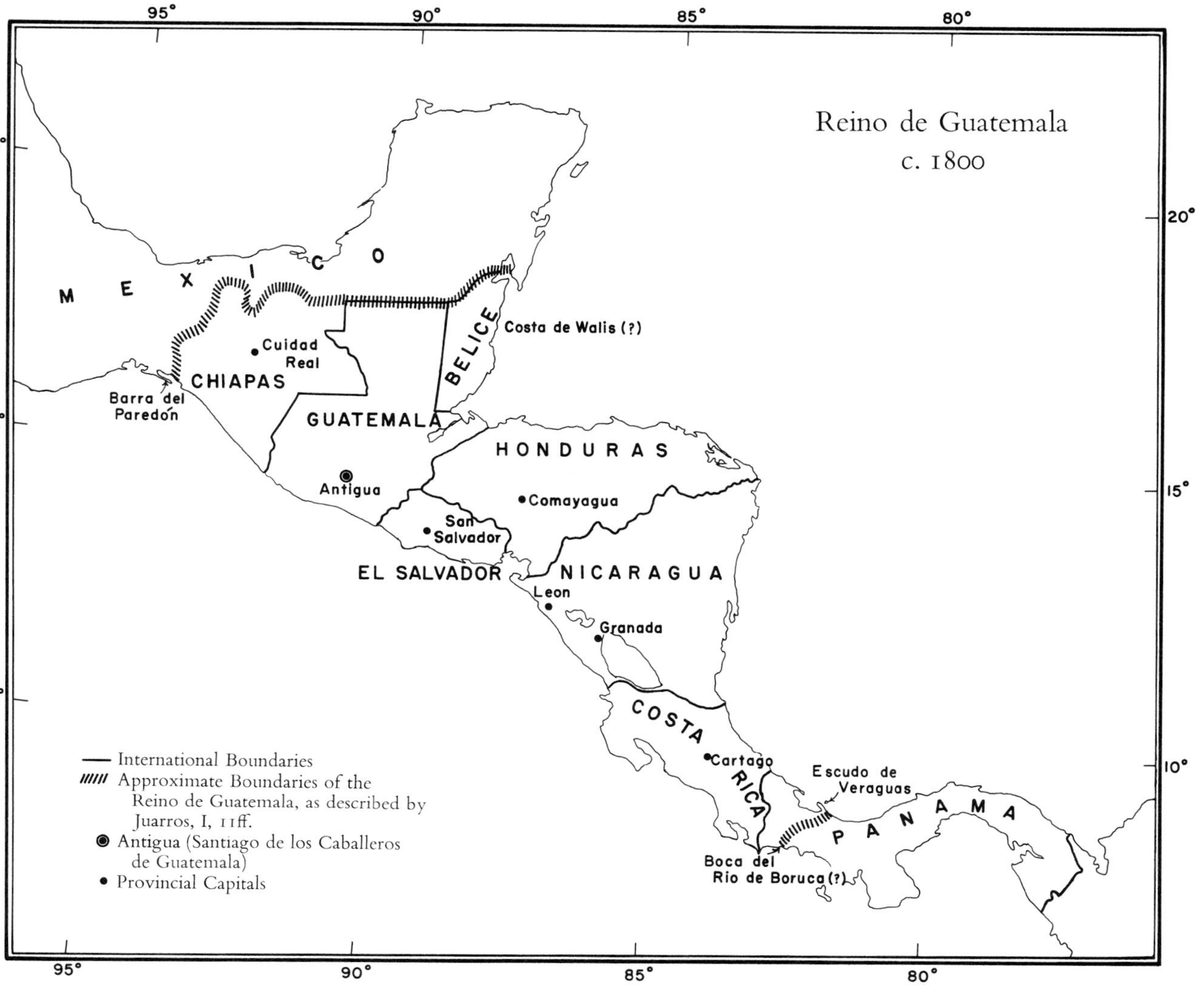

Fig. 12. Map of Central America. (Reino de Guatemala.)

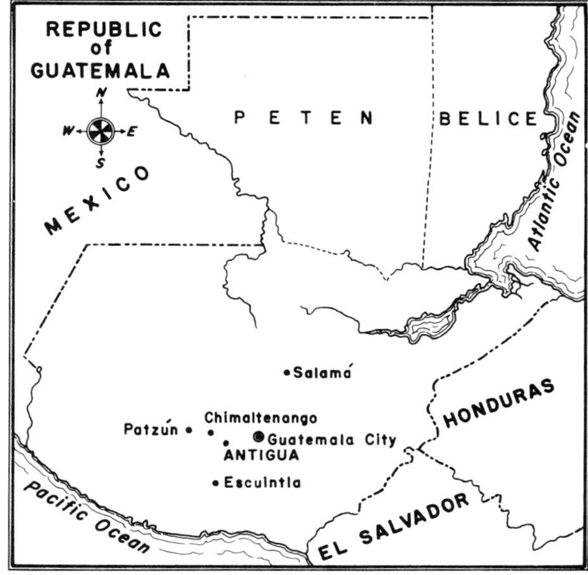

Fig. 13. Map of Guatemala.

Fig. 14. Air view of the Pacific Coast of Guatemala extending from volcanoes Agua, Acatenango, and Fuego to the Mexican border. Antigua, far right. (Photo Alvarez, Guatemala City.)

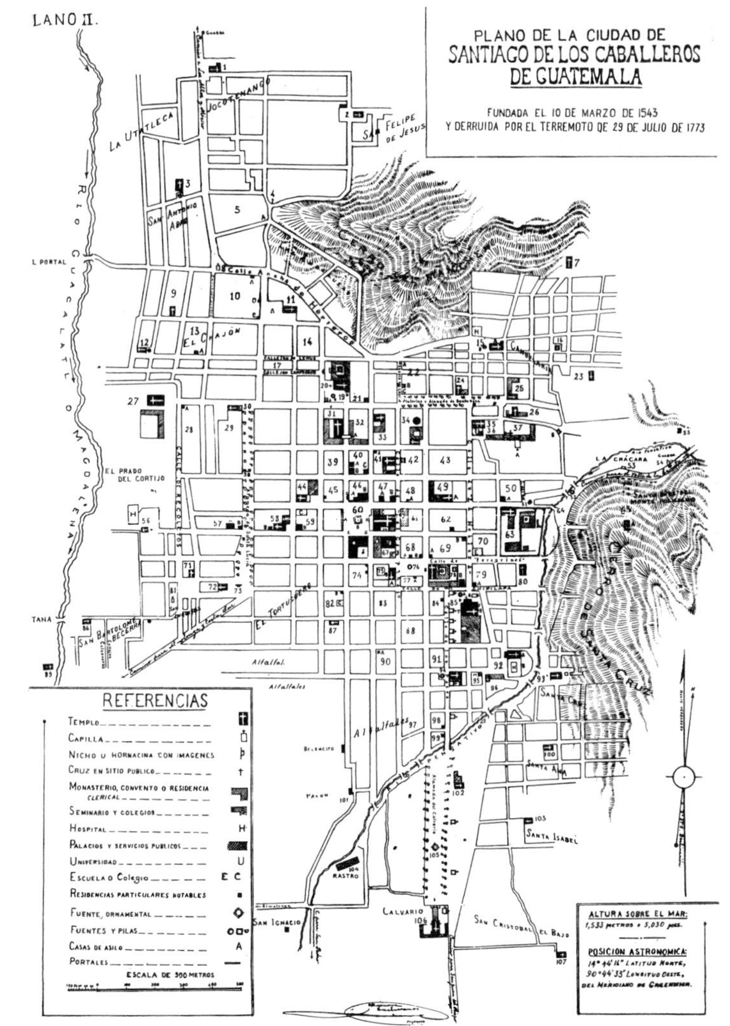

Fig. 15. Plan of Antigua. (After Pardo, *Guía*, frontispiece.)

FIG. 16. Antigua, *ca.* 1876, with Volcán de Agua to south. (After Muybridge, *Pacific Coast*, negative no. 4361.)

FIG. 17. Antigua, *ca.* 1876, with volcanoes of Fuego and Acatenango to southwest. (After Muybridge, *Pacific Coast*, negative no. 4362.)

Fig. 18. Antigua. View of Volcán de Agua through arch of Santa Catalina.

Fig. 19. Antigua. Plaza Mayor, *ca.* 1840. Drawn by Catherwood. (After Stephens, *Central America*, facing p. 286.)

FIG. 20. Antigua. Plaza Mayor, 1963. Same view as that drawn by Catherwood, who places the volcanoes too far to the east.

FIG. 21. Antigua. Plaza Mayor with the Ayuntamiento on the north side. (After Muybridge, *Pacific Coast*, negative no. 4372.)

Fig. 22. Santa Catalina, 1625 and 1647. Church. East side showing doors which gave access to nave.

Fig. 23. Santa Catalina. Church. South door.

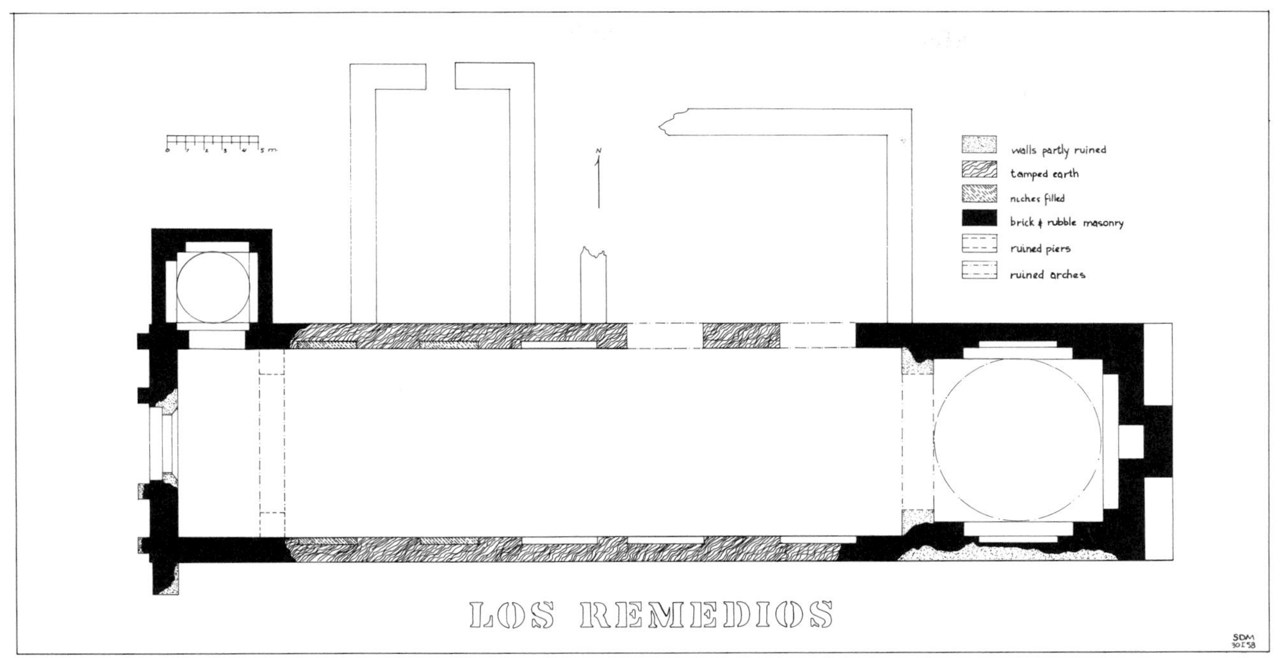

Fig. 24. Los Remedios, *ca.* 1650 and after 1679. Plan.

Fig. 25. Los Remedios. West façade.

Fig. 27. Los Remedios. West façade. Detail, niche, north side bay, lower story.

Fig. 26. Los Remedios. West façade. Detail, central retable.

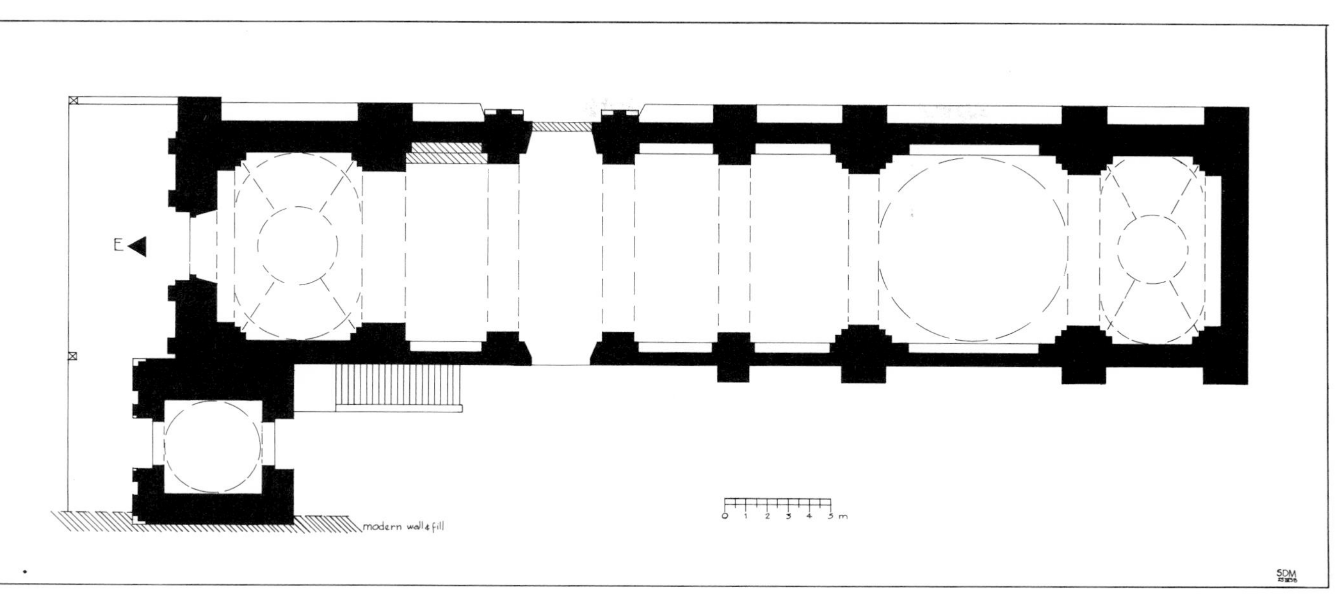

Fig. 28. San Agustín, 1657 and *ca.* 1761. Plan.

Fig. 29. San Agustín. West façade. Detail, lower story.

Fig. 30. San Agustín. West façade.

Fig. 31. San Agustín. West façade. Detail, niche, south bay, lower story.

FIG. 32. San Pedro, 1662, 1675, and 1869. Church façade and hospital entrance.

Fig. 33. San Pedro. Church façade.

Fig. 34. San Pedro. Hospital entrance.

Fig. 35. Belén, 1670. West façade.

Fig. 36. Belén. West façade. South tower. Detail, window, lower story.

FIG. 37. Santuario de Guadalupe, 1874. North façade.

Fig. 38. San Cristóbal el Bajo, seventeenth century. West façade and north side.

Fig. 39. San Cristóbal el Bajo. West façade. Detail, door.

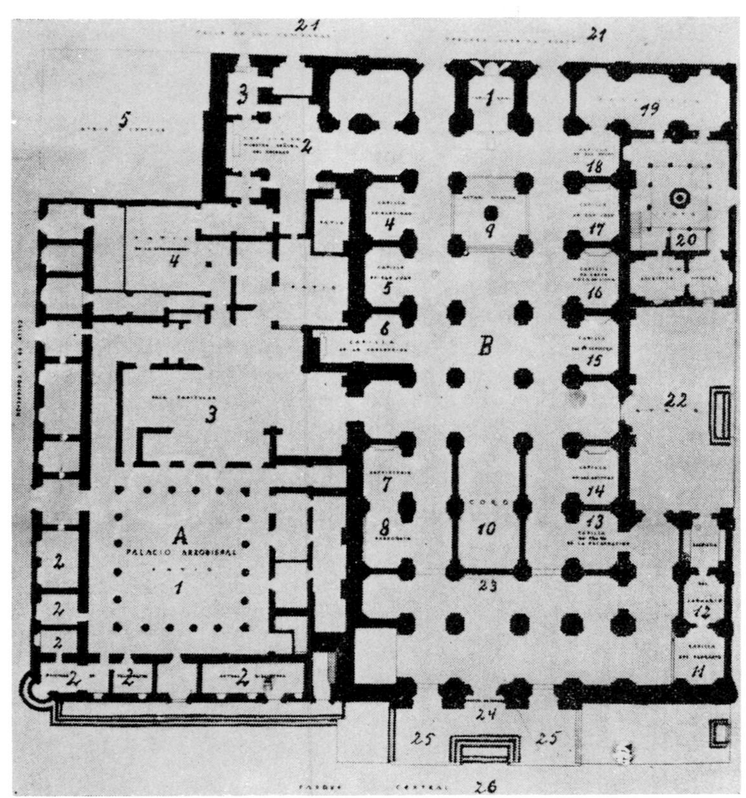

Fig. 40. Cathedral, 1680. Plan, including Palacio Arzobispal. (After Villacorta, *Historia*, p. 318.)

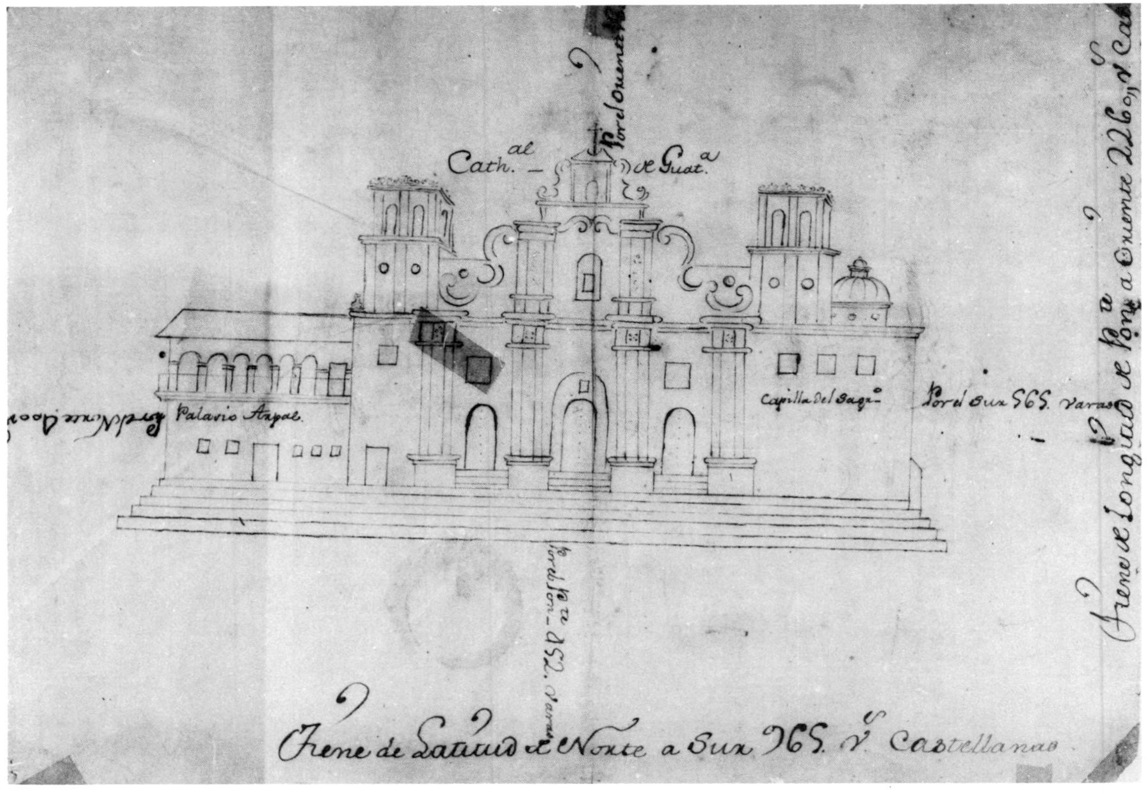

Fig. 41. Cathedral. Drawing of façade. (From a document, *AGG*, A 1.10.3 (1784) 15.091–2123.)

FIG. 42. Cathedral. West façade.

Fig. 44. Cathedral. South side. Detail, door to side aisle and nave.

Fig. 43. Cathedral. West façade. Detail, door, central bay.

Fig. 45. Cathedral. Nave, looking toward east.

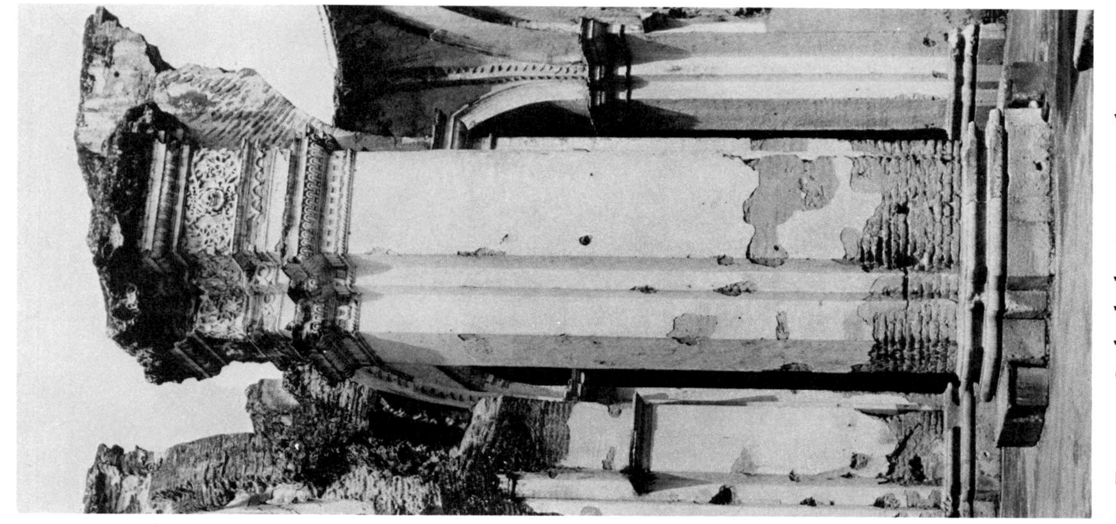

Fig. 46. Cathedral. Nave. Detail, pier.

Fig. 47. Cathedral. Nave. Pier. Detail, pilaster capital and entablature.

Fig. 48. Cathedral. Crossing. Detail, southeast pendentive.

FIG. 49. Cathedral. Crossing. Detail, pendentives with sculptures and reliefs in plaster.

FIG. 50. Cathedral. Choir. Detail, window piercing west façade.

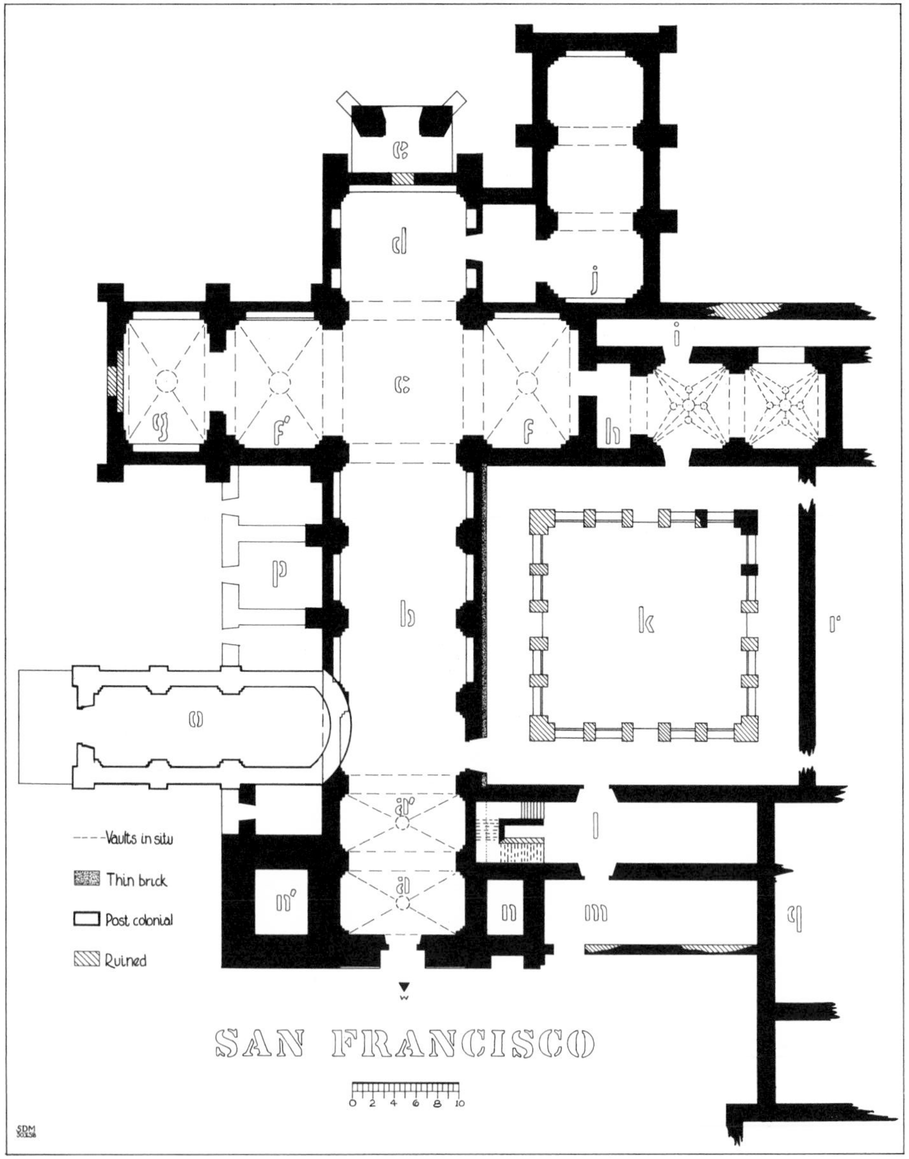

FIG. 51. San Francisco, 1675, 1690, and 1702. Plan.

FIG. 52. San Francisco. West entrance to convent grounds and church atrium.

FIG. 53. San Francisco. West façade.

FIG. 54. San Francisco. West façade. Detail, upper story of central retable.

FIG. 55. San Francisco. Nave. Detail, pilaster capitals and entablature.

Fig. 56. San Francisco. Nave, looking toward east.

Fig. 57. San Francisco. Crossing. Detail, pilasters, southeast corner.

Fig. 58. San Francisco. Lower choir. Detail, pilasters.

Fig. 59. San Francisco. Entrance room to convent. Detail, mural paintings.

Fig. 60. San Francisco. Nave. Detail, clerestory window.

Fig. 61. San Francisco. Choir. Detail, plaster decoration, spandrel of arch facing nave.

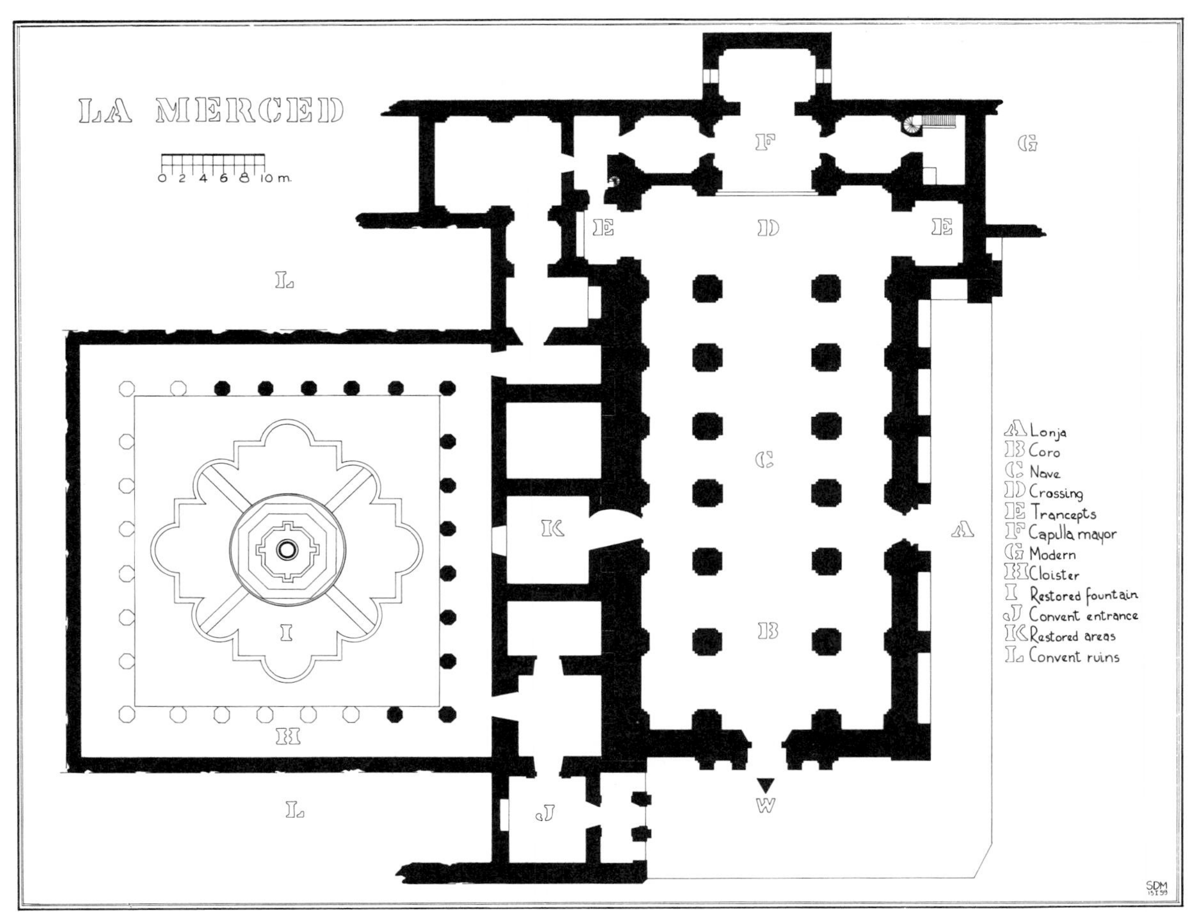

Fig. 62. La Merced, *ca.* 1650/70 and 1767. Plan.

Fig. 63. La Merced. Cloister. Detail, south arcade.

Fig. 64. La Merced. West façade. Roofs of tower belfries were rebuilt about 1960.

FIG. 65. La Merced. West façade. Detail, central retable.

Fig. 66. La Merced. West façade. Detail, south tower.

Fig. 67. La Merced. West façade. Central retable. Detail, plaster decoration, upper side bay. Statue in niche-window with original head.

Fig. 68. La Merced. West façade. Central retable. Detail, niche-window, central bay, second story. Head of the figure of the Virgin was replaced shortly before 1957.

Fig. 69. La Merced. West façade. Central retable. Detail, niche, south bay, lower story.

Fig. 70. La Merced. North side. Detail, north side aisle roof with small domes and lanterns.

Fig. 71. La Merced. Cupola over crossing. Note glazed terra-cotta lions at either side of window pediments. In foreground, exterior of nave vaults.

FIG. 72. La Merced. West façade. North tower. Detail, nonarchitectonic pilasters on south side above nave roof. The mixtilinear element of the bottom member was added about 1960.

FIG. 73. La Merced. Convent entrance, north church tower, and atrium with stone cross.

Fig. 74. La Merced. Convent entrance. Detail, door.

Fig. 75. La Merced. Cloister. South arcade, interior showing semihexagonal vault. See fig. 6, p. 137, for cross section of vault.

Fig. 76. Santa Teresa, 1687 and after 1738. West façade.

Fig. 77. Santa Teresa. Nave. Detail, pilaster with springing of arches.

Fig. 78. Santa Teresa. Cloister. Detail, southwest angle.

FIG. 79. San Sebastián, 1692. West façade. Plan and elevation.

FIG. 80. San Sebastián. West façade. Detail, upper story.

Fig. 81. San Sebastián. West façade.

Fig. 82. San Sebastián. West façade. Detail, niche, south bay, lower story.

FIG. 83. Compañía de Jesús, 1698. East façade, reconstruction. (After Villacorta, *Historia*, frontispiece.)

FIG. 84. Compañía de Jesús. East façade. Detail, upper story, and south side.

Fig. 85. Compañía de Jesús. East façade. Detail, archivolt and lunette of main door.

Fig. 86. Compañía de Jesús. Nave, looking west.

Fig. 87. Compañía de Jesús. Crossing. Detail, southwest pilaster.

Fig. 88. Pila de la Alameda del Calvario, 1679. Gatehouse of the church of El Calvario in the distance.

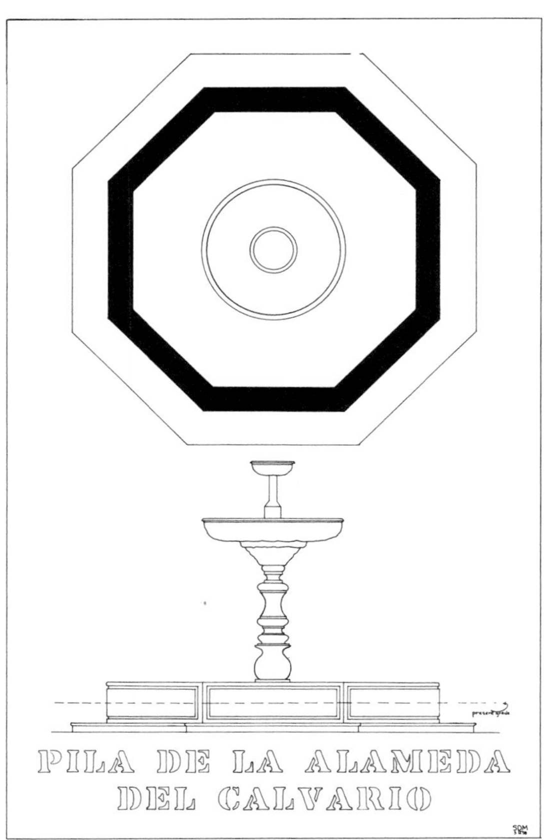

FIG. 89. Pila de la Alameda del Calvario. Plan and elevation. Broken line represents present grade level. Scale, 1 m.

FIG. 90. Pila de la Alameda del Calvario. Detail, bowl.

Fig. 91. Fuente de las Delicias, seventeenth century. Entrance to convent of La Concepción to right.

Fig. 92. Fuente de los Dominicos, 1618(?).

FIG. 93. Fuente de los Dominicos. Detail, basin, showing Dominican blazon.

FIG. 94. Fuente de los Dominicos. Detail, bowl.

FIG. 95. Fountain near La Merced, seventeenth century.

FIG. 96. Fountain near La Merced. Detail, basin showing mermaids.

Fig. 97. Fountain near La Merced. Detail, stem and bowl.

Fig. 98. Fountain near La Merced. Detail, basin, showing nude male figure.

Fig. 99. Espíritu Santo, after 1702. West façade. Detail, central and south bays.

Fig. 100. Nuestra Señora de los Dolores del Cerro, 1710. West façade.

FIG. 101. Nuestra Señora de los Dolores del Cerro. *Capilla mayor*. Detail, vaulting.

FIG. 102. Palacio Arzobispal, 1711. Cloister. Detail, east corridor. See fig. 40 above for plan.

Fig. 103. Palacio Arzobispal. Cloister. Detail, column with painted decoration.

Fig. 104. Palacio Arzobispal. West side. Detail, doors.

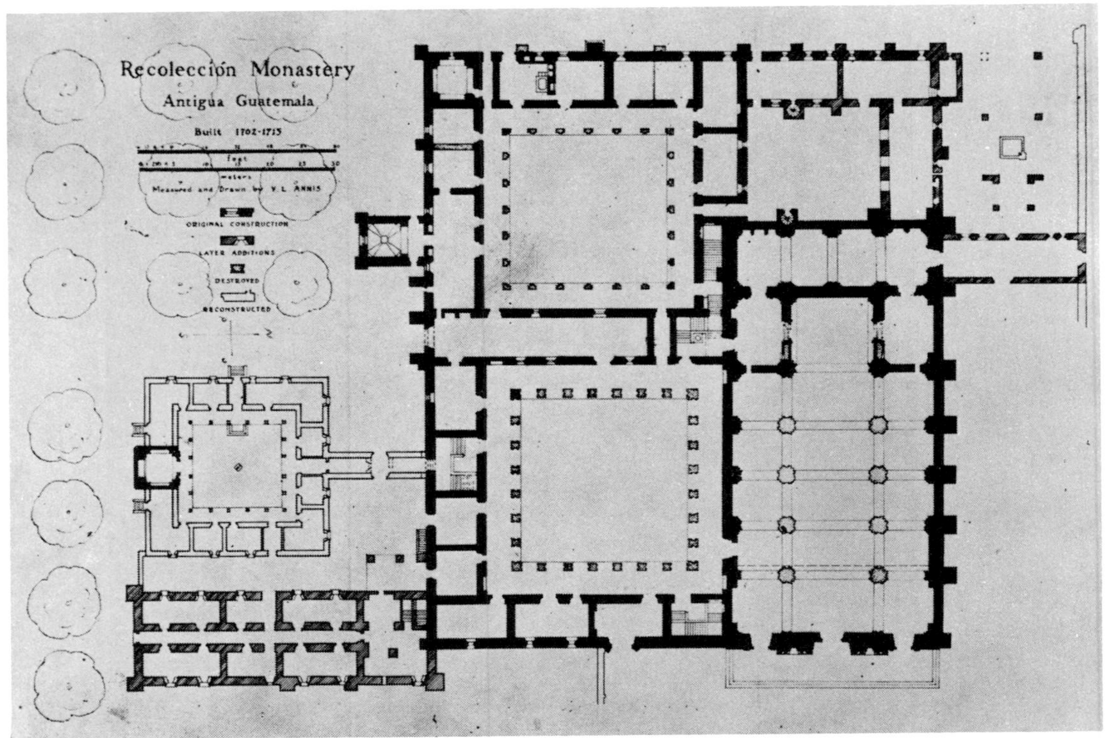

FIG. 105. La Recolección, 1717. Plan. (After Villacorta, *Historia*, p. 323.)

FIG. 106. La Recolección. Ruins of church, *ca.* 1876. (After Muybridge, *Pacific Coast*, negative no. 4367.)

FIG. 107. La Recolección. East façade. Detail, center door.

FIG. 109. La Recolección. Nave. Detail, entablature and clerestory.

FIG. 108. La Recolección. Nave. Detail, pilasters and arch of crossing.

FIG. 110. El Calvario, 1720. Gatehouse. North façade.

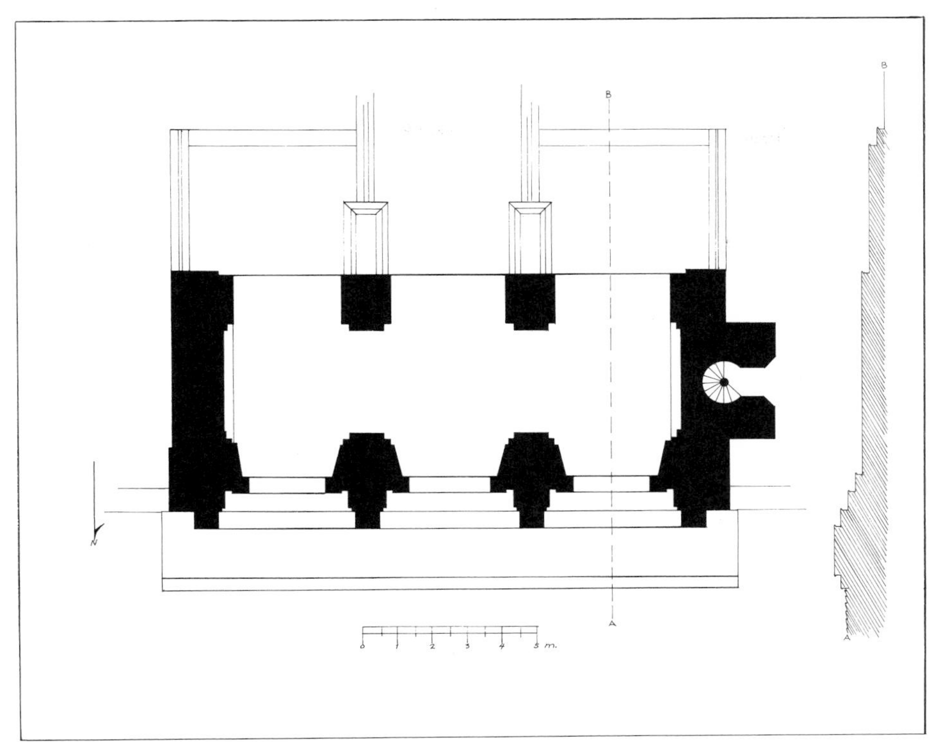

FIG. 111. El Calvario. Gatehouse. Plan.

FIG. 112. El Calvario. Gatehouse. South side.

Fig. 113. El Calvario. Gatehouse. North façade. Detail, west bay.

Fig. 114. El Calvario. Gatehouse. Detail, interior, pilasters.

Fig. 115. El Calvario. Church. North façade.

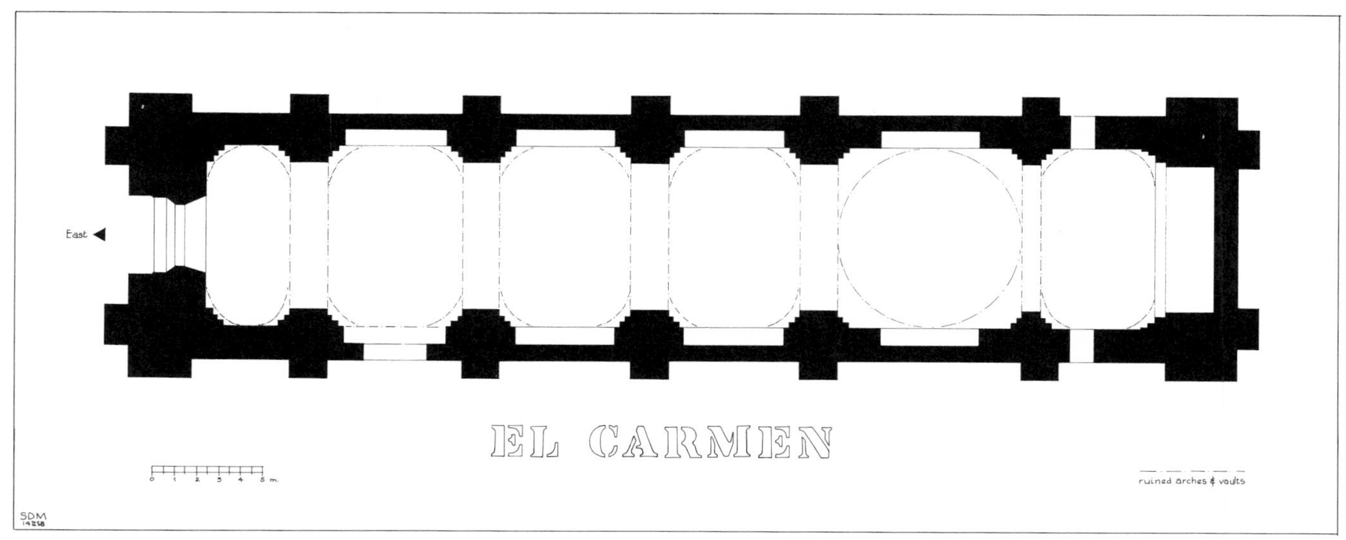

FIG. 116. El Carmen, 1728. Plan.

FIG. 117. El Carmen. East façade and south side, *ca.* 1876. (After Muybridge, *Pacific Coast*, negative no. 4365.)

Fig. 118. El Carmen. East façade.

Fig. 119. El Carmen. Nave. Detail, pilasters, entablature, and clerestory.

FIG. 120. El Carmen. East façade. Detail, upper story.

FIG. 121. El Carmen. East façade. Detail, archivolt of door.

Fig. 122. La Candelaria, late seventeenth and early eighteenth centuries. West façade, church and chapel(?).

Fig. 123. La Candelaria. Chapel(?). West façade. Detail, door header.

Fɪɢ. 124. La Candelaria. Chapel(?). West façade. Detail, door.

Fɪɢ. 125. La Candelaria. Church. West façade. Detail, engaged columns and niche, south side bay.

FIG. 126. La Concepción, 1694. Convent entrance.

FIG. 127. La Concepción. Church. Detail, north side showing buttresses.

Fig. 128. La Concepción. Convent entrance. Detail, upper part.

FIG. 129. La Concepción. Church. Nave, looking west.

FIG. 130. La Concepción. Church. Nave. Detail, entablature.

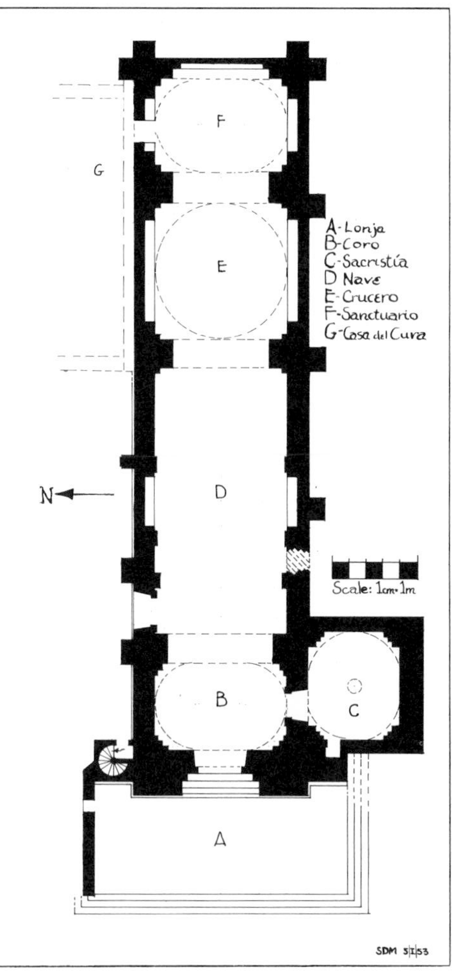

Fig. 131. Santa Cruz, 1662 and 1731. Plan.

Fig. 132. Santa Cruz. West façade.

Fig. 133. Santa Cruz. West façade. Detail, south side bay, lower story.

Fig. 134. Santa Cruz. Nave and choir, looking west.

FIG. 135. Santa Cruz. West façade. Detail, showing plaster decoration.

FIG. 136. Santa Cruz. West façade. Detail, door jamb and base of engaged column.

FIG. 137. Santa Cruz. West façade. Detail, archivolt of door and engaged column of south side bay.

FIG. 138. Santa Cruz. Choir. Detail, pilaster and door to sacristy.

FIG. 139. Santa Cruz. Crossing. Detail, pendentives.

FIG. 140. Santa Clara, 1734. South façade. Detail, east side bay, lower story.

FIG. 141. Santa Clara. West side. Detail, door.

FIG. 142. Santa Clara. Nave. East wall. Detail, pilasters and clerestory.

FIG. 143. Santa Clara. Cloister. Detail, arcade, southeast angle.

FIG. 144. Santa Clara. Cloister, northeast angle.

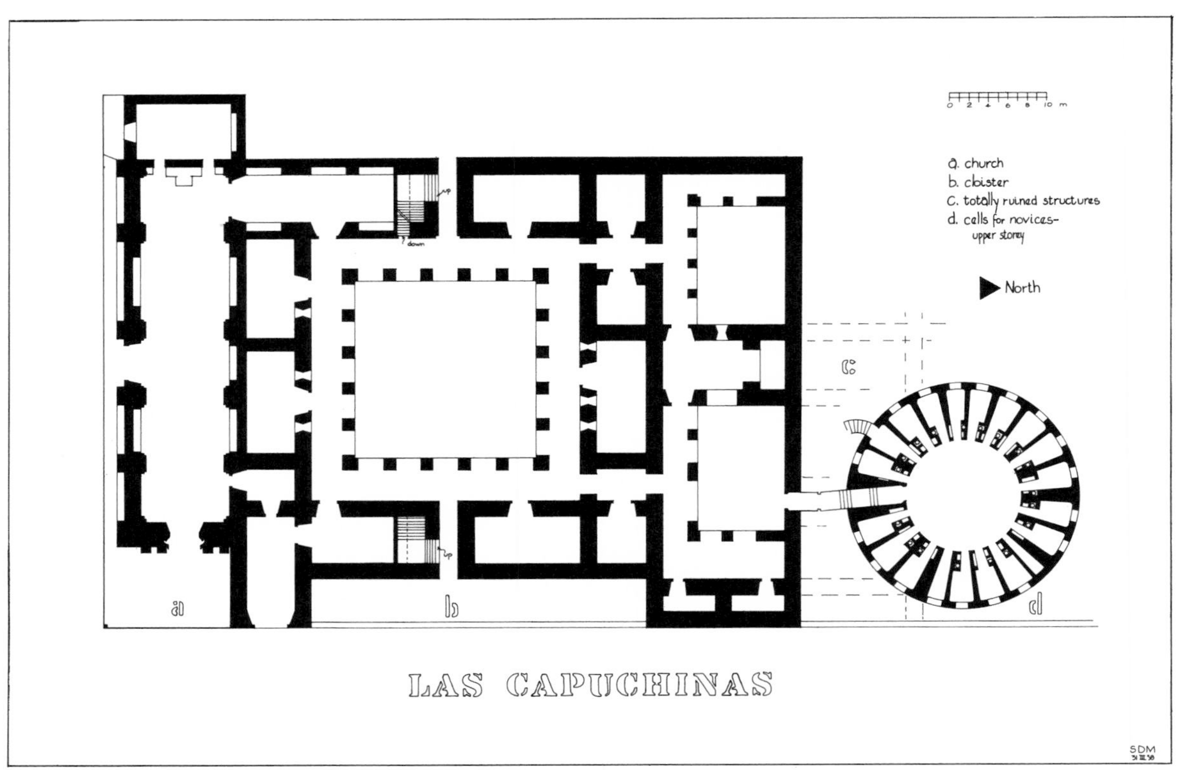

Fig. 145. Las Capuchinas, 1736. Plan.

Fig. 146. Las Capuchinas. East façade and south side.

FIG. 147. Las Capuchinas. Nave. Detail, vaulting and clerestory.

FIG. 148. Las Capuchinas. Nave and crossing. Detail, vaulting.

Fig. 149. Las Capuchinas. Nave. Detail, pilasters.

Fig. 150. Las Capuchinas. Cloister. Detail, corner pier with responds.

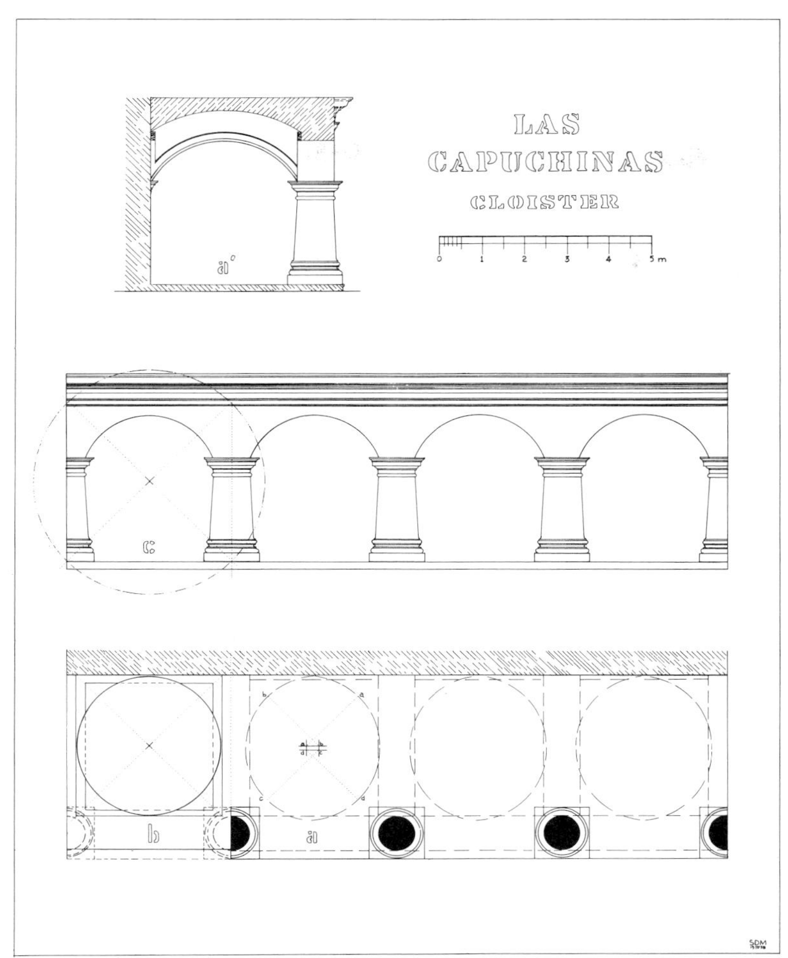

FIG. 151. Las Capuchinas. Plan. Detail of cloister.

FIG. 152. Las Capuchinas. Cloister. Detail, southwest angle.

Fig. 153. Las Capuchinas. Cloister. Detail, north corridor.

Fig. 154. Las Capuchinas. Circular building for novices. Detail, patio and cells of upper story.

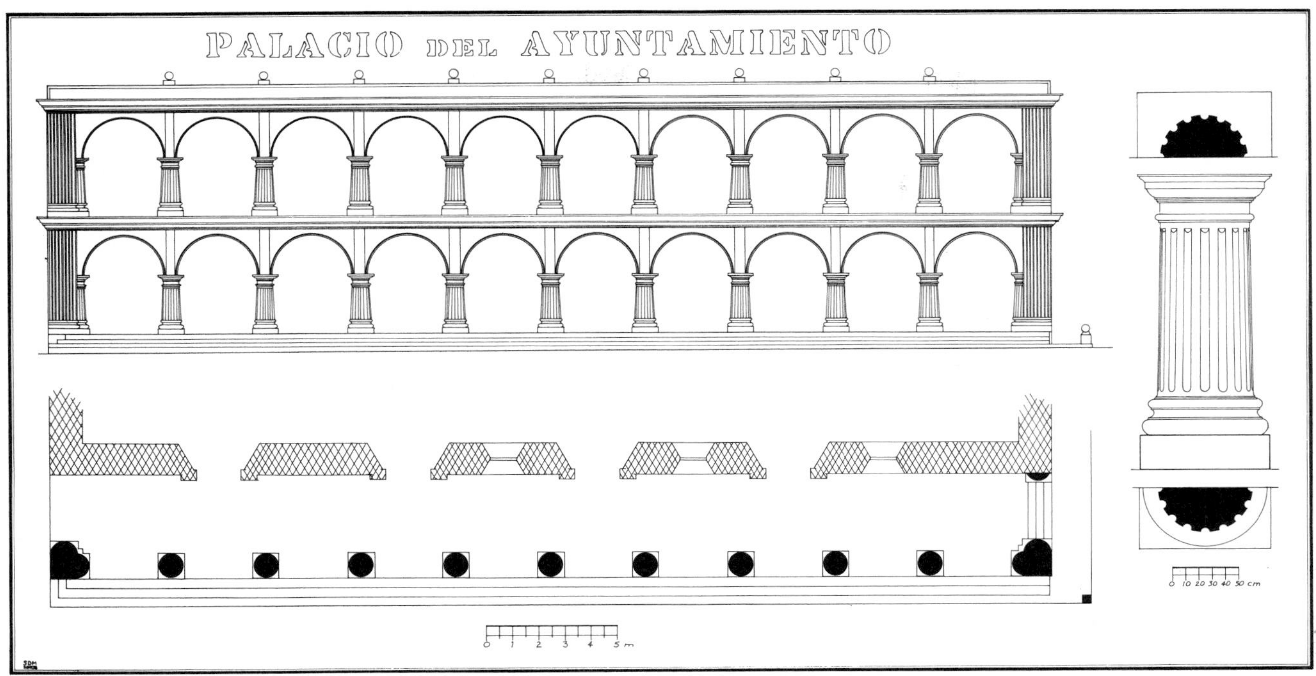

FIG. 155. Ayuntamiento, 1743. Plan and elevation of arcaded south façade.

FIG. 156. Ayuntamiento. South façade.

FIG. 157. Ayuntamiento. South façade. Detail of arcades.

FIG. 158. Ayuntamiento. South façade. Detail, east corner pier with respond.

Fig. 159. Ayuntamiento. South façade. Lower corridor.

Fig. 160. Ayuntamiento. South façade. Detail, column, lower corridor.

Fig. 161. Escuela de Cristo. After 1720 and after 1740. West façade and convent entrance.

Fig. 162. Escuela de Cristo. Cloister. Detail, south corridor, lower story. Restored in 1957.

Fig. 163. Escuela de Cristo. West façade. Detail, north side bay, lower story.

Fig. 164. Escuela de Cristo. Convent. Entrance. Detail, door and pilasters.

Fig. 166. Escuela de Cristo. Cloister. Detail, columns, north corridor, upper story, before restoration.

Fig. 165. Escuela de Cristo. Cloister and church.

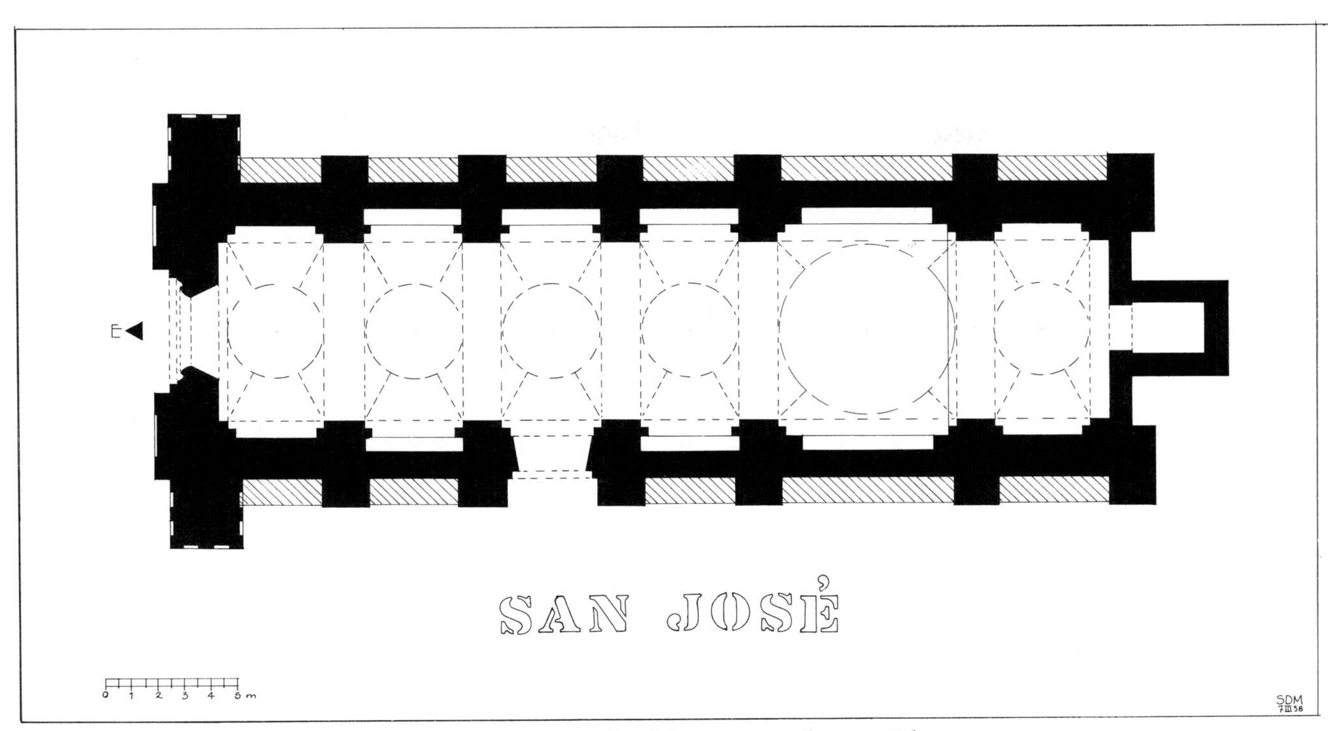

Fig. 167. San José el Viejo, 1740 and 1761. Plan.

Fig. 168. San José el Viejo. East façade.

Fig. 169. San José el Viejo. East façade. Detail, engaged columns and niche, north side bay, lower story.

Fig. 170. San José el Viejo. East façade. North tower. Detail, entablature and belfry.

FIG. 171. Santa Rosa de Lima, *ca.* 1750. West façade.

Fig. 172. Santa Rosa de Lima. West façade. Detail, niche-window in upper story central bay and third story *remate*.

Fig. 173. Santa Rosa de Lima. West façade. Detail, north side bay, lower story.

Fig. 174. Santa Rosa de Lima. Nave. Detail, pendentive and vaulting.

Fig. 175. Santa Rosa de Lima. Nave. Detail, soffit of transverse arch, pendentive, and pilaster entablature.

Fig. 176. Santa Ana, mid-eighteenth century. West façade.

Fig. 177. Santa Ana. West façade. Detail, south side bay, lower story.

Fig. 178. Santa Isabel, mid-eighteenth century. West façade.

Fig. 179. Santa Isabel. West façade. Detail, south side bay, lower story.

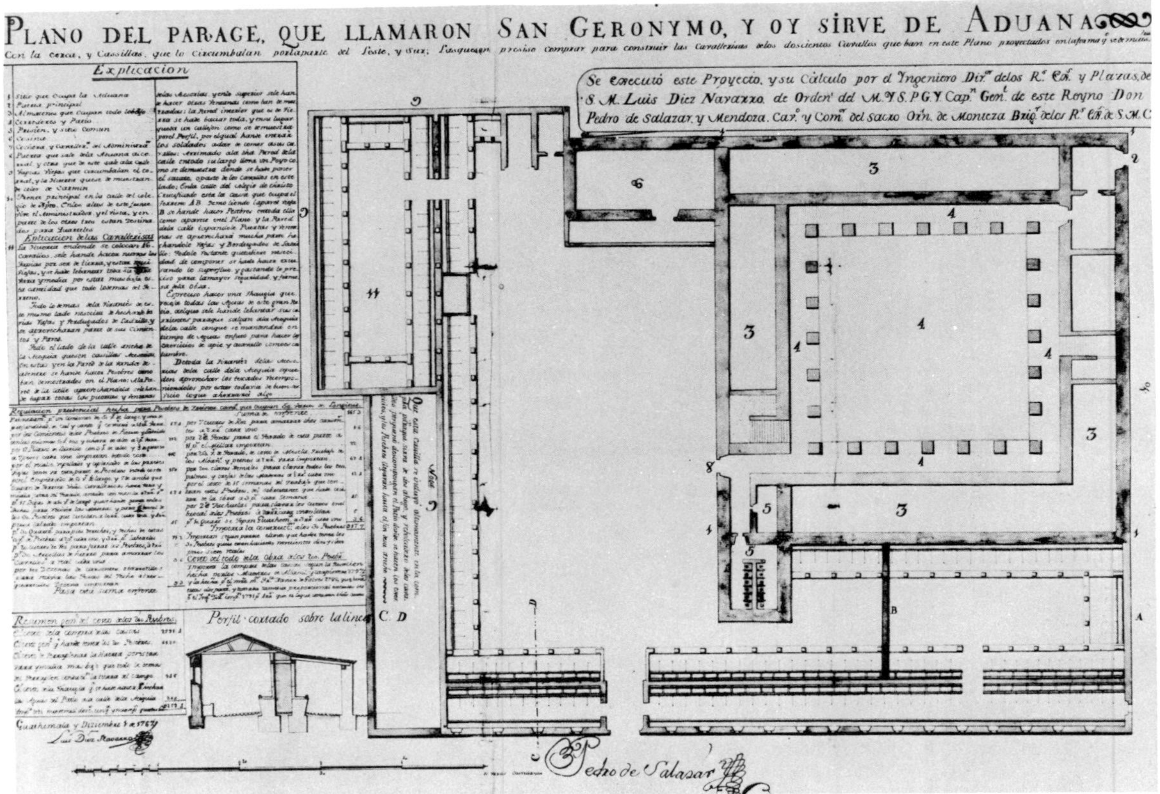

FIG. 180. San Jerónimo. Plan of 1767. (After Angulo, *Planos*, pl. 163.)

FIG. 181. San Jerónimo. Church, late seventeenth century. West façade.

Fig. 183. Seminario Tridentino, *ca.* 1758. North side. Detail, main door.

Fig. 182. San Jerónimo. Door on north side of present ruins.

FIG. 184. Seminario Tridentino. Cloister. Detail, north arcade.

FIG. 185. Seminario Tridentino. Chapel. Decoration of window reveal.

FIG. 186. Seminario Tridentino. Chapel. Detail, pilaster and entablature.

FIG. 187. Seminario Tridentino. Chapel. Detail, ceiling decoration.

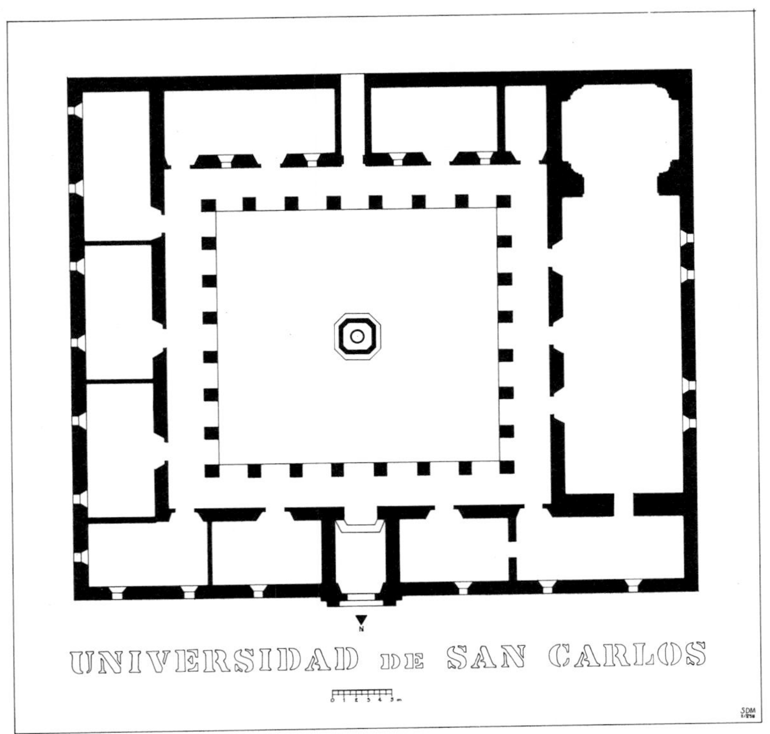

FIG. 188. University of San Carlos, *ca.* 1763 and later. Plan.

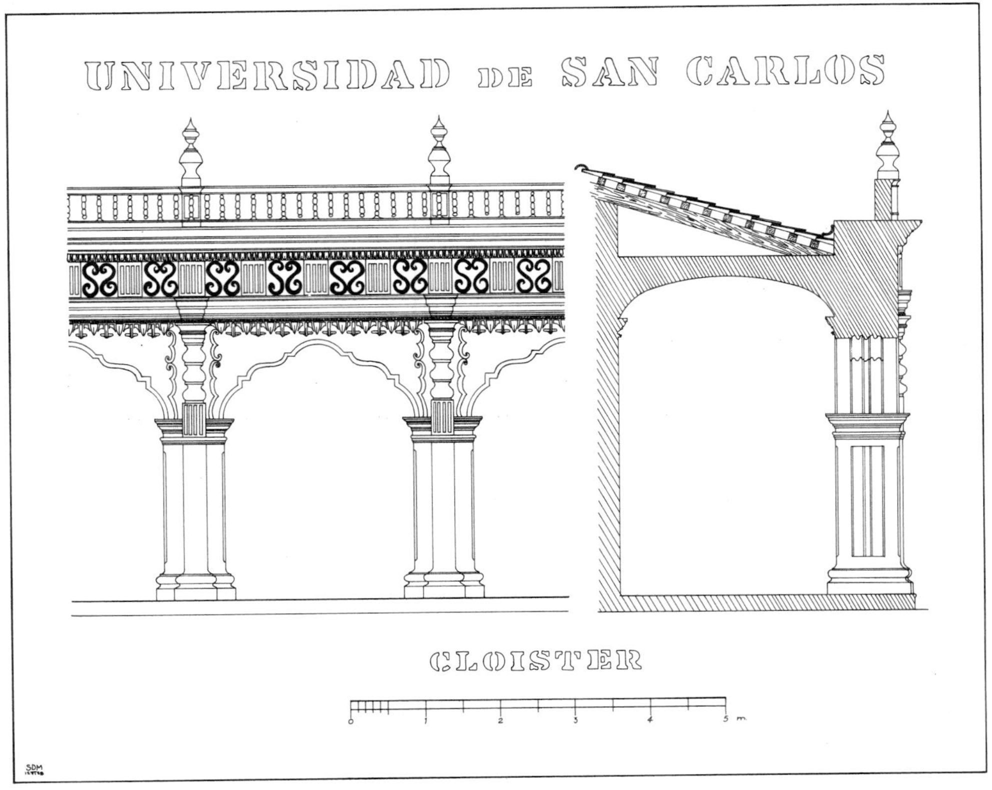

FIG. 189. University of San Carlos. Plan. Detail, cloister.

Fig. 190. University of San Carlos. Exterior, from northeast.

Fig. 191. University of San Carlos. Exterior. Detail, northeast corner.

Fig. 192. University of San Carlos. Cloister. Detail, east corridor.

Fig. 193. University of San Carlos. Cloister. Volcán de Agua in background to south.

Fig. 194. University of San Carlos. Cloister. Detail, southeast angle.

Fig. 195. University of San Carlos. Cloister. West side. Detail, papal coat-of-arms.

Fig. 196. University of San Carlos. Cloister. South side. Detail, entablature and royal coat-of-arms.

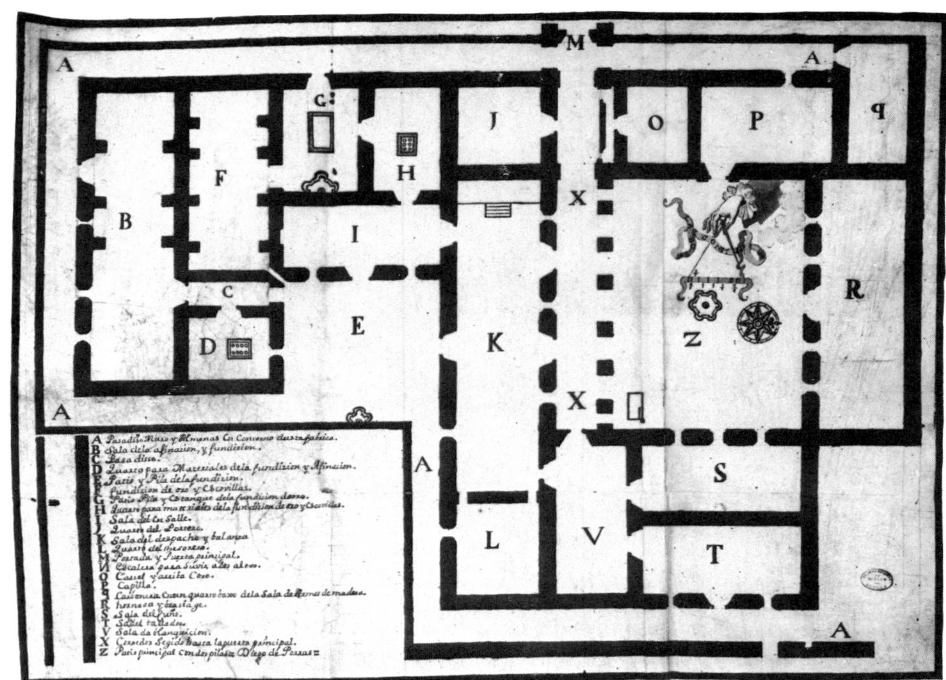

Fig. 197. Capitanía. Detail, Casa de Moneda, 1733. Plan drawn by Diego de Porras. (From a document, *AGI*, Guatemala, 314 tomo de 1740, fol. 94. Two exact copies are in Guatemala—*AGG*, A 3.17 (1734) 26924–1655, fol. 32 and A 3.17 (1738) 26924–1655, fol. 266.)

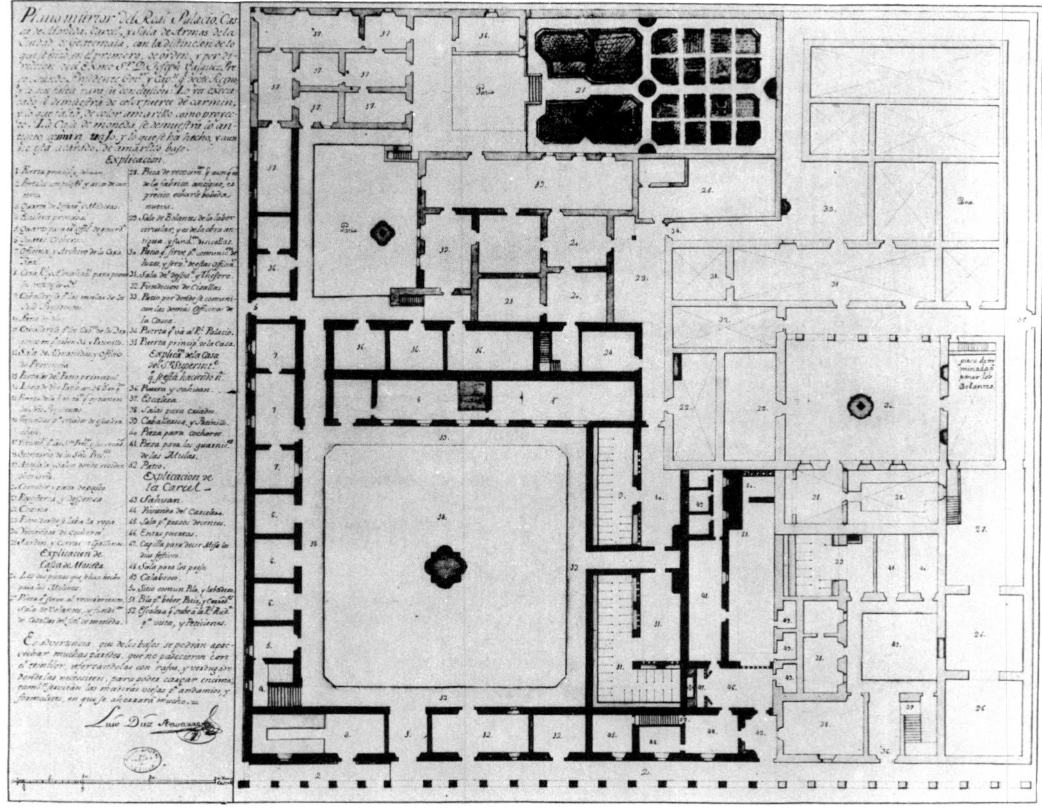

Fig. 198. Capitanía, 1769. Plan, ground floor. Drawn by Diez de Navarro in 1755. (After Angulo, *Planos*, pl. 157.)

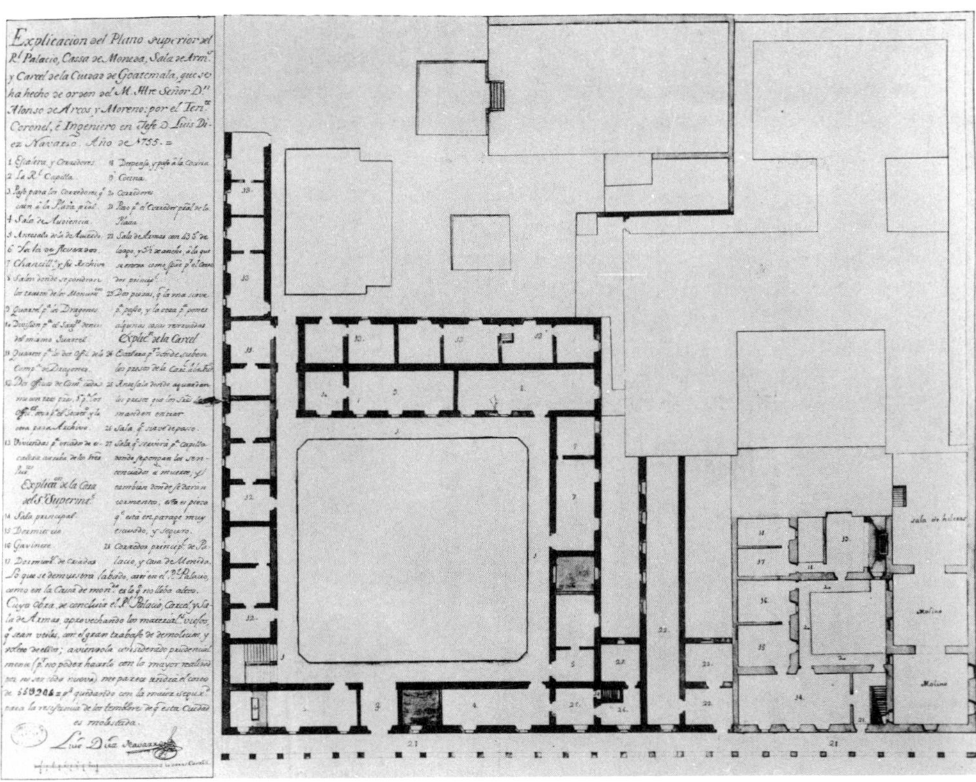

FIG. 199. Capitanía. Plan, second floor of plan in fig. 198. (After Angulo, *Planos*, pl. 158.)

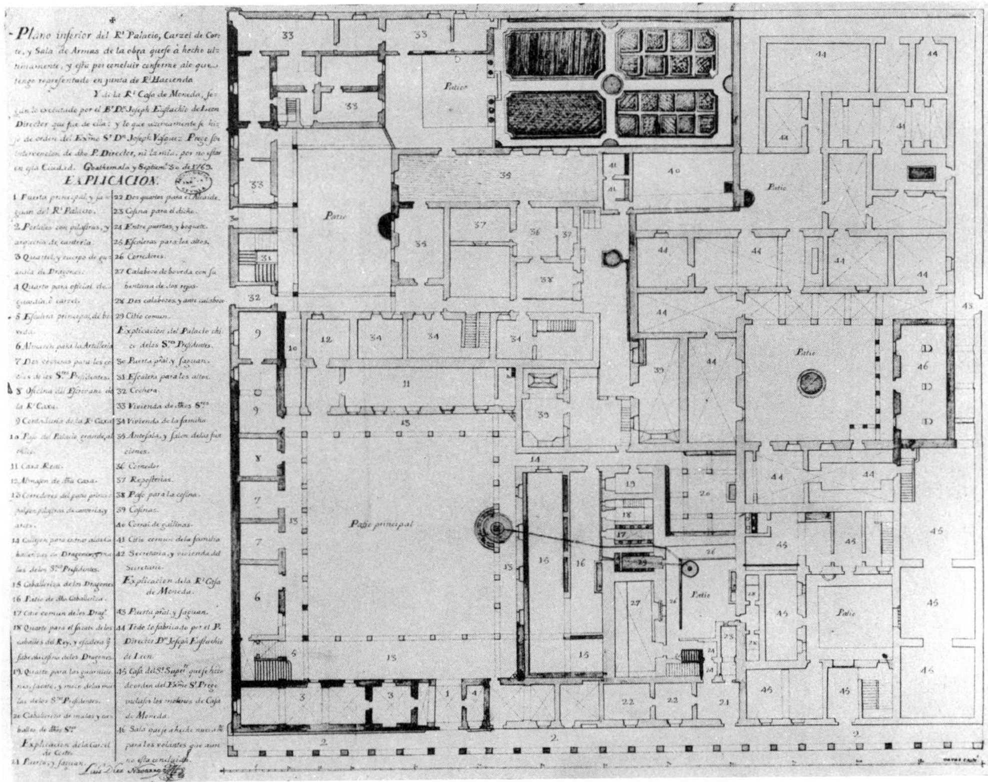

FIG. 200. Capitanía. Plan, ground floor. Drawn by Diez de Navarro in 1769. (After Angulo, *Planos*, pl. 161.)

Fig. 201. Capitanía. North façade, *ca.* 1876. (After Muybridge, *Pacific Coast,* negative no. 4264.)

Fig. 202. Capitanía. North façade. From Cathedral roof. Volcanoes Fuego and Acatenango covered with clouds in background.

Fig. 203. Capitanía. North façade. Detail, central bays.

Fig. 204. Capitanía. North façade. Central bay. Detail, column, lower story.

Fig. 205. Capitanía. North façade. Lower corridor looking west. Reconstructed portion of arcade in foreground.

Fig. 206. Capitanía. North façade. Lower corridor looking east. Original construction in foreground.

Fig. 207. Santísima Trinidad, late eighteenth century. Nave. Detail, pilaster, clerestory, and arch over crossing.

Fig. 208. Jocotenango, seventeenth century and eighteenth century. Church. West façade.

Fig. 209. Jocotenango. West façade and south side.

Fig. 210. Jocotenango. West façade. Detail, north side bay, lower story.

Fig. 211. Jocotenango. Capilla de la Virgen. West façade.

Fig. 212. Jocotenango. Capilla de la Virgen. West façade. Detail, central bay, upper part.

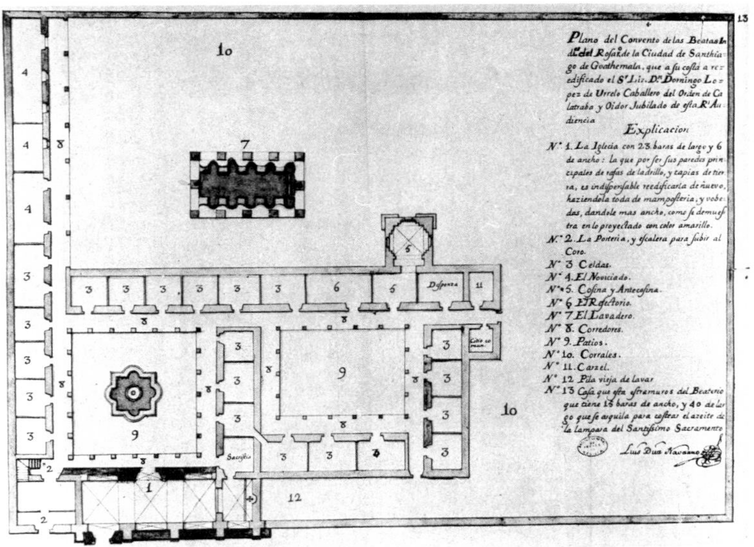

Above: FIG. 213. Beaterio de Indias del Rosario, 1762 and after 1771. Plan. Drawn by Diez de Navarro in 1762. (After Angulo, *Planos*, pl. 154.)

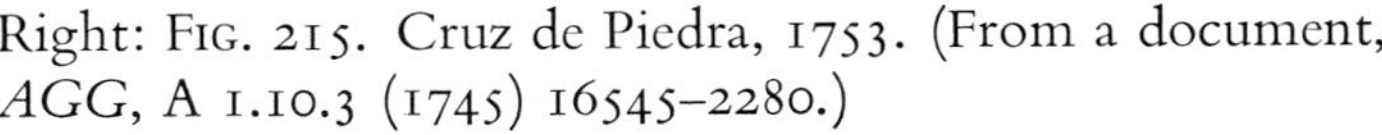

Right: FIG. 215. Cruz de Piedra, 1753. (From a document, *AGG*, A 1.10.3 (1745) 16545–2280.)

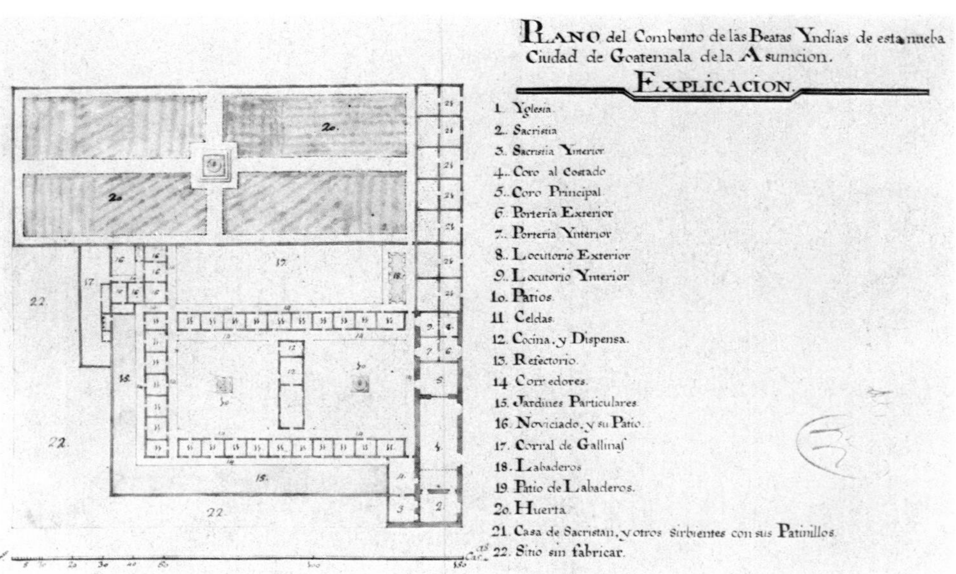

FIG. 214. Beaterio de Indias. Plan. Drawn in 1788. (From a document, *AGG*, A 1. 10. 3 (1803) 2356–109. The same plan exists in Seville, *AGI*, Guatemala 362 and 228. See Angulo, *Planos*, pl. 155.)

INDEX

2000
copies printed at
The Stinehour Press,
Lunenburg, Vermont
with illustrations by
The Meriden Gravure Company
and binding by
J. F. Tapley Co.